AF823096

Biological and Environmental Chemistry of DMSP and Related Sulfonium Compounds

Biological and Environmental Chemistry of DMSP and Related Sulfonium Compounds

Edited by

Ronald P. Kiene
University of South Alabama
Mobile, Alabama

Pieter T. Visscher
University of Connecticut
Groton, Connecticut

Maureen D. Keller
Bigelow Laboratory for Ocean Sciences
West Boothbay Harbor, Maine

and

Gunter O. Kirst
University of Bremen
Bremen, Germany

Plenum Press • New York and London

Library of Congress Cataloging-in-Publication Data

Biological and environmental chemistry of DMSP and related sulfonium compounds / edited by Ronald P. Kiene ... [et al.].
p. cm.
"Proceedings of the First International Symposium on DMSP and Related Sulfonium Compounds, held June 5-8, 1995, in Mobile, Alabama"--T.p. verso.
Includes bibliographical references and index.
ISBN 0-306-45306-1
1. Dimethylpropiothetin--Congresses. 2. Sulphonium compounds--Congresses. I. Kiene, Ronald P. II. International Symposium on DMSP and Related Sulfonium Compounds (1st : 1995 : Mobile, Ala.)
QK898.D54B56 1996
581.5'222--dc20 96-21632
CIP

Proceedings of the First International Symposium on DMSP and Related Sulfonium Compounds, held June 5–8, 1995, in Mobile, Alabama

ISBN 0-306-45306-1

© 1996 Plenum Press, New York
A Division of Plenum Publishing Corporation
233 Spring Street, New York, N. Y. 10013

All rights reserved

10 9 8 7 6 5 4 3 2 1

No part of this book may be reproduced, stored in a retrieval system, or transmitted in any form or by any means, electronic, mechanical, photocopying, microfilming, recording, or otherwise, without written permission from the Publisher

Printed in the United States of America

PREFACE

It is with great pleasure that I introduce this volume, which represents the proceedings of the First International Symposium on DMSP and Related Sulfonium Compounds, which was held in Mobile, Alabama, in June 1995. The thirty-six chapters in this book cover a broad range of topics related to dimethylsulfoniopropionate (DMSP), and provide extensive background as well as the latest research and ideas on the subject.

Scientific interest in DMSP has accelerated in recent years, in part because this compound has been identified as a major precursor of dimethylsulfide (DMS), a volatile form of sulfur which, by its degradation products, affects atmospheric chemistry and global climate. In addition, DMSP and related sulfonium compounds are of great interest to biological chemists because: 1) they are used by organisms to combat osmotic (water) stress; 2) they are potential methyl donors; and 3) they form a large pool of organic sulfur in some environments. Despite this interest, much of the basic biology and environmental chemistry of these compounds remains poorly understood.

Research on DMSP and related compounds is currently being carried out in many different disciplines, and the number of researchers working on these compounds has increased rapidly in recent years. For these reasons, among others, there was a clear need for a focused symposium on the subject. The idea of holding a symposium on DMSP coalesced out of casual conversations held between myself and many colleagues over the years, particularly those with my close associates Pieter Visscher and Maureen Keller. In the Fall of 1993, I had a key phone conversation with Andrew Hanson, then of the University of Montreal, which made it clear that there were others outside my main discipline who were interested in getting together to share ideas about DMSP. I rushed to my Department Chairman, Dr. Bob Shipp, the idea of a symposium and received from him encouragement and a promise of some financial support to get things started. The organizing committee (Kiene, Visscher, Keller and Kirst) was formed shortly thereafter, and we set about to assess the interest level in a symposium dealing mainly with DMSP. The response to a preliminary mailing was very strong, and we therefore set about to raise more funds and to iron out details of the meeting. All the while I, as organizer, became newly acquainted with many colleagues by fax and e-mail.

After nearly a year of planning, the First International Symposium on DMSP and Related Sulfonium Compounds was scheduled for early June, 1995. Sixty-four participants representing 12 nations were on the final list of attendees for the symposium. Because this was the first symposium of its kind, and because a diversity of scientists (ranging from oceanographers to sulfur biochemists and a high school student to emeritus professors) were planning to attend, I was filled with some trepidation as to whether the interdisciplinary format would work. My fears were allayed during the pre-symposium mixer when all the participants gathered for the first time. Though most were newly acquainted, it was as if everyone had known each other for years. Admixtures of oceanographers, biochemists, microbiologists, and plant physiologists gathered in small groups and all were engaged in ardent conversation related to what else – DMSP! As I mingled, I noticed that the composition of the small groups changed over time, but the conversations didn't let up. The energy level was intense right from the start, and it was clear then that the symposium would be a success. By all accounts it was, and this volume and the fact that future conferences of this type are currently being planned are testament to the first symposium's success.

ACKNOWLEDGMENTS

The success of the symposium and the preparation of this book owe a great deal to the efforts of folks at the University of South Alabama and also at the Dauphin Island Sea Lab, including Amada Gonzales, Carolyn Wood, Georgia Mallon, Jim Cowan, Bob Shipp, George Crozier, Alan Foster, Glen Miley, Lee Stanton, Jennifer Bachmann, and Alma Waggoner.

All of the papers in this book were submitted by participants in the symposium and, for the most part, the material in the papers is that which was presented in Mobile. All of the papers were subjected to peer review by one or more reviewers after the symposium. I would like to express deep appreciation to my coeditors of this volume and to all the reviewers who provided timely and helpful comments on the manuscripts.

Financial support for this "first of its kind" symposium was crucial and was provided by: U.S. Office of Naval Research, U.S. Department of Energy, U. S. Department of Agriculture, University of South Alabama, The Dauphin Island Sea Lab, and the National Science Foundation EPSCoR program.

In particular, the dissemination of this book to participants in the symposium was made possible by funding provided by the National Institute for Global Environmental Change – Southwest Regional Center.

Ronald P. Kiene
Mobile, Alabama

CONTENTS

1

30 YEARS OF RESEARCH ON DIMETHYLSULFONIOPRIOPIONATE

A Personal Retrospective

Yuzaburo Ishida

Department of Fisheries, Faculty of Agriculture
Kyoto University
Kyoto, Japan

SUMMARY

In 1961, with an interest in the "sea smell" as food flavor component, we commenced to do research on dimethylsulfoniopropionate (DMSP) in marine unicellular algae. After we identified the DMSP crystal from *Crypthecodinium cohnii* by NMR and IR and clarified the characteristics of a DMSP cleaving enzyme, we examined the biological significance of DMSP in this alga. It was difficult, however, to find sufficient experimental evidence. After we had conducted an ecological study on harmful red tide algae for some time, we resumed our studies on DMSP, with the emphasis on algal DMS evolution in the sulfur cycle and bioactivities of DMSP. We have since been doing research on the biosynthesis of DMSP and its regulation at the biochemical and molecular levels, in order to understand its biological role in the microalgae.

INTRODUCTION

It is truely a great honor and an enormous pleasure to be asked to present a plenary talk and paper under this title. The research in this field is making rapid progress nowadays, beyond comparison with its pace in the past.

First, I will mention the circumstances under which I became interested in the subject of DMSP. In 1961 I had a position as a research associate at the Laboratory of Applied Microbiology, Food Research Institute, Kyoto University, and studied the utilization of microalgae as food of fish and humans (17). At first I isolated and collected unicellular algae as axenic cultures from natural sea waters, and examined culture conditions to accomplish mass cultivation of microalgae as food for young fish in the larval stages after hatching out. Simultaneously, I bred the larvae hatched out from fertilized eggs of black sea bream. Just at that time Prof. Kadota brought back a strain of heterotrophic dinoflagellate *Gyrodinium cohnii (Crypthecodinium cohnii)* with three other dinoflagellates, which were kindly provided by Prof. Luiji Provasoli, Haskins Laboratory, USA.

Biological and Environmental Chemistry of DMSP and Related Sulfonium Compounds
edited by Ronald P. Kiene et al., Plenum Press, New York, 1996

Shortly after I began the study, I noticed the sea smell "Iso-no-kaori", which evolved from the packed cells of *C. cohnii*. This smell is very familiar to the Japanese as the smell of Nori (*Porphyra terena*) and sea urchin, which feed on *Ulva* and *Enteromorpha*. In addition, I took an interest in volatile sulfur compounds, as I had studied the ecology of sulfate-reducing bacteria in estuaries. H_2S was the main sulfur compound of interest at that time since sulfate reduction rates in Maizuru Bay were very high. Organic forms of sulfur, such as DMS were not recognized as being important in the global sulfur cycle.

Several early studies on volatile organic sulfur compounds were published by Haas (10), Challenger and Simpson (4) and Obata et al. (36). They reported dimethyl sulfide (DMS) production from marine macroalgae, *Polysiphonia* spp, *Enteromorpha* sp. and *Ulva pertusa*. The first sulfonium compound was isolated in pure form by Challenger and Simpson (4) from a member of the Rhodophyta, *Polysiphonia fastigiata*, and it proved to be DMSP. Although there had been no reports on DMS evolution from unicellular algae, I predicted that they produced DMS, based on existing literature (4, 36). Immediately after I started my studies, two reports suggesting that the petroleum odour in salmon and the Blackberry problem in cod derived from DMS from marine algae were published by Motohiro (27) and Sipos and Ackman (40). It was interesting but very strange that DMS smelled differently in the macroalgae, the sea urchin and fish. Later I obtained the preliminary result that the "good" flavor of sea urchin, but not the petroleum odour in fish, was attributed to association of DMS with neutral esters extracted from *C. cohnii*.

Identification of DMSP

In 1965 I sent a letter to Prof. Challenger that I had found that DMS was produced from marine unicellular algae such as *C. cohnii* and had isolated its precursor, DMSP as a crystal. In return I received an enthusiastic answer from him, in January, 1966. I still remember having been encouraged by him. A few months later Ackman et al. (1) published an article on DMS production from unicellular marine algae one year earlier than we did. They and we confirmed that some marine unicellular algae belonging to dinoflagellates, haptophyta and diatoms produced DMS (1,11). Furthermore, we proved that the precursor of DMS in *C. cohnii* was DMSP (12). I can never forget that excitement when I found a small transparent crystal in the desiccator (Fig. 1). The crystal was analyzed by use of D_2-NMR. The NMR spectrum in D_2O showed $(CH_3)_2$-protons at 7.01 ppm, the S^+-CH_2-protons at 6.40

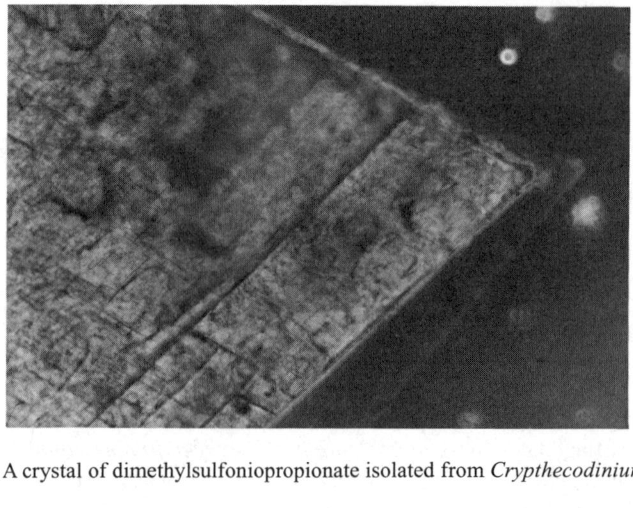

Figure 1. A crystal of dimethylsulfoniopropionate isolated from *Crypthecodinium cohnii.*

ppm, and CH_2COO-protons at 6.98 ppm. The infrared(IR) spectrum of this crystal corresponded with that of synthetic DMSP. In addition, the stoichiometry of DMSP degradation with cold alkali also supported the findings obtained by NMR and IR spectra.

BIOLOGICAL SIGNIFICANCE OF DMSP

At that time, we had made the assumption that there were three possible functions of DMSP: 1) DMS evolution, 2) transmethylation, and 3) osmoregulation.

DMS Evolution

DMS evolution by unicellular algae was limited to marine types such as dinoflagellates, diatoms and haptophyta, and was not seen in the freshwater types. The effect of salts on DMS production from DMSP was examined, and the enzymatic activity of DMS evolution was markedly activated by increases in various inorganic and organic salts at concentrations from 0.1 M to 0.4 M. Table 1 shows the order of increasing effectiveness.

Table 1. Characteristics of DMSP cleaving enzymes in cell free extract of *C. cohnii*

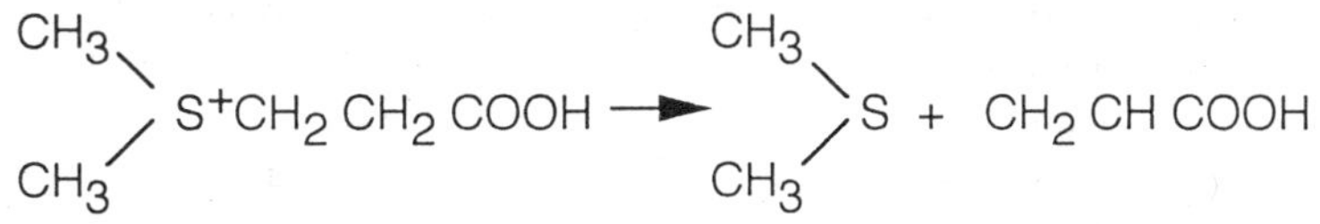

Optimum pH of the activity : 6-7
pH value on maximum stability of the enzyme: 5.1
Optimum temperature : 27℃
Optimum of NaCl conc : $>$0.4 M
The evolution of DMS was activated by addition of inorganic and organic salts at high concentrations in an order of increasing effectiveness;

◎$NaCl > KCl > MgCl_2 = CaCl_2 > LiCl > NaBr > LiBr > NaNO_3 >> NaI > KI > Na_2SO_4 > Li_2SO_4 > CH_3COONa > CH_2CHCOONa$

◎$Cl^- > Br^- > NO_3^- >> I^- > SO_4^= > CH_3COO^- > CH_2CHCOO^-$

Inhibition of the enzymatic activity:
strong inhibition—pCMB($5x10^{-4}$M), IAA($5x10^{-4}$M), KCN($2x10^{-3}$M)
week inhibition—EDTA($1x10^{-2}$M)

The inhibition was released by homocysteine($2x10^{-3}$M) or 2-mercaptoethanol($2x10^{-3}$M) — SH-enzyme

Km : 1.5×10^{-3}M

pCMB: p-chloromercuribenzoic acid
IAA: iodoacetamide
EDTA: ethylendiaminetetraacetate-Na_2

This table shows the characteristics of the DMSP-cleaving enzyme in *C. cohnii* (17). DMS production may contribute to the regulation of the over-production of DMSP, or DMS and DMSP might act as an attractant or a repellent for the other organisms.

Transmethylation

In general, the sulfonium compounds such as DMSP in macroorganisms are known to act as methyl donors in the biological transmethylation reactions and are often involved in synthesis of methionine and other compounds (6, 7, 22 26). In the 1960's many researchers concerned with sulfonium compounds took a great interest in the potential transmethylation reactions, but there was no evidence for transmethylation in microalgae.

In *C. cohnii* the amount of DMSP in the cells decreased considerably with an increase of glycine betaine in the medium from 0 to 3.3 mM. At higher betaine concentrations (>3.3 mM), DMSP levels did not change further. The final cell yield decreased by about 40% in the medium without betaine. Choline also acted as well as betaine at lowering DMSP levels, but methylmethionine (MMSC), methionine and others did not (Fig. 2). These findings suggested that DMSP plays an important role in the transmethylation reaction (13). In fact, when $^{14}CH_3$-DMSP was reacted with homocysteine in the cell-free extract of *C. cohnii,* ^{14}C of $^{14}CH_3$-DMSP was incorporated into the amphoteric fraction (methionine and its sulfoxide) and anionic fraction (methylthiopropionic acid and its sulfone). This suggested that one methyl group of DMSP was transferred to homocysteine and that methylthiopropionic acid was produced as a metabolic product. Unexpectedly the activity was very low, and moreover, the incorporation into methionine was only 0.3% of the methyl group of DMSP in the absence of betaine, in comparison with 6.8% in the presence of betaine.

We assumed that DMSP did not substitute for betaine as a methyl donor in this alga. Then we concluded that DMSP in the algae did not act as a methyl donor in the algae themselves, even if DMSP might act as a methyl donor in organs of animals which ingested these algae.

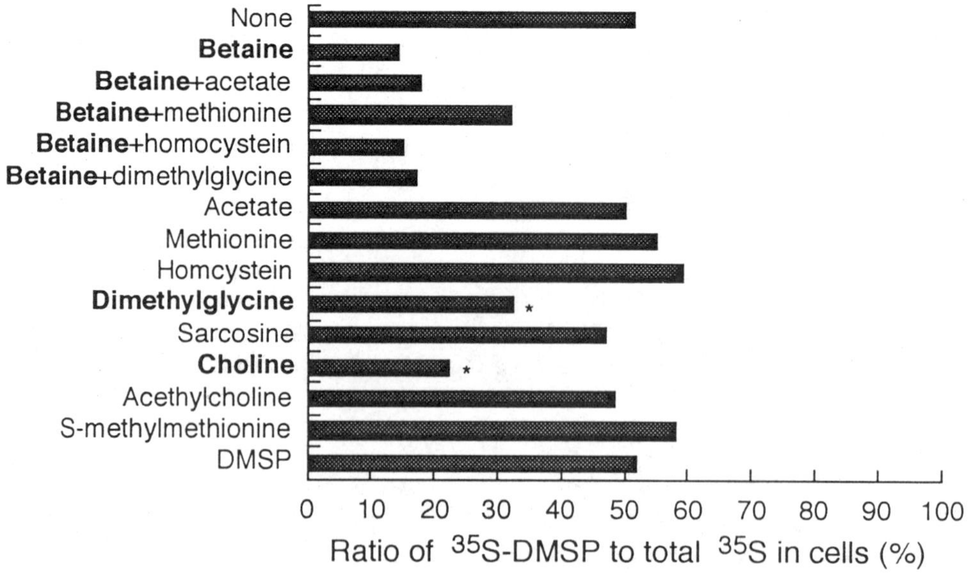

Figure 2. Ratio of ^{35}S-DMSP to total ^{35}S in the *C. cohnii* Cells incubated with $^{35}SO_4$ and a substrate in place of betaine.

Osmoregulation

We made a preliminarily examination of the osmoregulatory function of DMSP. Recently, excellent works on osmoregulation have been published by Kinne (21), Gröne and Kirst (9), etc., but around 1968 there were no reports on the osmoregulation or compatible solute functions of onium compounds such as betaine. In 1953 and 1960 Nicolai and Preston (34, 35) suggested that a portion of DMSP is firmly associated with some macromolecular compounds.

As described above (Fig.2), DMSP content in the *C. cohnii* cells decreased about 20% from about 60% after the addition of betaine. To clarify whether DMSP was decomposed or not, and to which fraction the decreased DMSP was transferred if it was not decomposed, we determined the behavior of ^{35}S-DMSP in the cells of *C. Cohnii* (13). As Fig. 3 shows, by addition of betaine the sulfur of the acid-insoluble fraction (macromolecular compounds) in the cells clearly increased, while on the contrary DMSP proportionally decreased. Then, a preliminary analysis of the ^{35}S-compounds in the acid-insoluble fraction was made, as shown in Fig. 4 (15). The cationic fraction which was released by hydrolysis of the acid-insoluble fraction was applied to paper chromatography in parallel with ^{35}S-DMSP as an authentic sample. We confirmed that one of the ^{35}S-compounds in the acid-insoluble fraction coincided with ^{35}S-DMSP. We assumed from this finding that a portion of DMSP in this alga was bound with some macromolecular compounds.

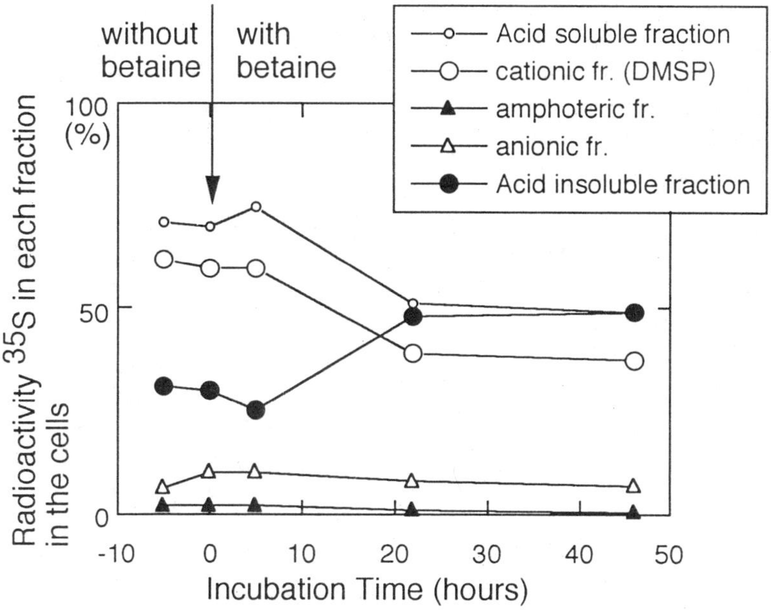

Figure 3. Behavior of ^{35}S-DMSP and other fraction in the *C. cohnii* Cells transferred to betaine-medium from betaine-free medium. The ^{35}S-labeled cells cultivated in betaine free-medium were reincubated in the medium with betaine. The ^{35}S-labeled cells were harvested by centrifugation and extracted with 5% cold perchloric acid (PCA). The residue is the PCA insoluble fraction. The extract(acid soluble fraction) was passed through an Amberlite IR 120 (H^+) column(through fraction is anionic fraction), and then eluted with 2N HCl. After removing HCl by evaporation, the solution was passed through an Amberlite IRA 400 (OH^-) column. The through fraction (cationic fraction) was further treated with charcoal and evaporated to a syrup (^{35}S-DMSP). The fraction eluted with HCl is amphoteric fraction.

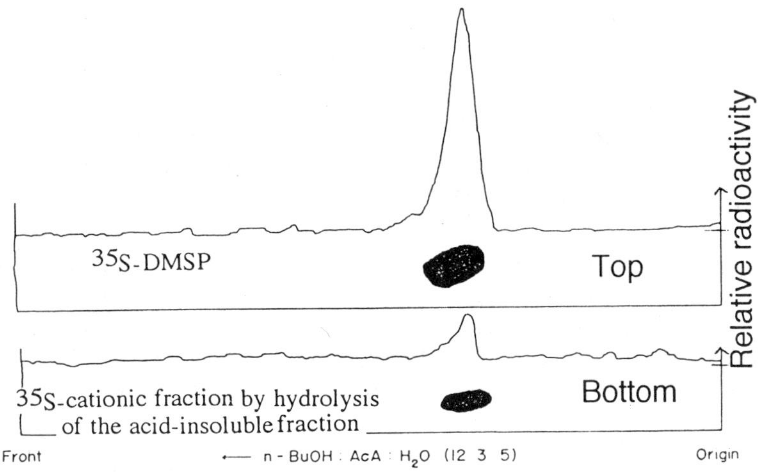

Figure 4. Preliminary analysis of ^{35}S-compounds in the acid-Insoluble fraction by means of paper chromatography. Bottom; ^{35}S-cationic fraction released by hydrolysis of the acid-insoluble fraction. Top; authentic ^{35}S-DMSP. After hydrolysis with 6N HCl at 105°C for 18 h, the hydrolyzate was fractionated by anionic and cationic resins. The ^{35}S-cationic fraction was applied to paper chromatography. Radioactivities of the spots which were examined by the iodide-platinium method was scanned using Aloka PCS-2 type Scanner.

Next, we examined the effect of salts on the content of DMSP in the cells incubated in the medium with or without betaine. DMSP was accumulated in the cells in response to the NaCl concentration in the medium without betaine, but not so much in the presence of betaine. If we had determined the intracellular concentration of betaine, we might have obtained a finding similar to that obtained in *Escherichia coli* (23). In addition, I had tried to prepare a DMSP-deficient mutant of *C. cohnii* by ultraviolet irradiation, but could not.

Table 2. List of microalgae containing DMSP

Microalgae	DMSP	Ref
Dinoflagellates		
Crypthecodinium cohnii (M); axenic culture	19 mg /g wet cells	44
Amphidinium carteri (M); axenic culture	5.8 mg /g wet cells	1
Gyrodinium aureolum (M); field sample	ca 4.5 μg S/l (11 ng S/10^4 cells)	41
Peridinium bipes (F); field sample	0.3 mg /g wet cells	44
Haptophyta		
Emiliania huxleyi (M); field sample	6 ng S/10^4 cells	41
Hymonomonas carterae (M); unialgal culture(?)	380 ng S/10^4 cells	46
Syracosphaera carterae (M); axenic culture	12 mg /g wet cells	1
Diatom		
Skeletonma costatum (B); axenic culture	6.8 mg /g wet cells	1

M , Marine ; F, Freshwater

On the other hand, DMSP was accumulated not only in marine algae but also in freshwater dinoflagellates such as *Peridinium bipes* (Table 2).

According to Schlenk (38), some yeasts, *Candida utilis* and *Saccharomyces cerevisiae* accumulated S-adenosylmethionine, a sulfonium compound, at the high concentrations of 20 to 40 mM, and 35 to 50 mM, respectively, in a medium containing 10 mM methionine, independent of the salt concentration. This was similar to the findings for DMSP in *C. cohnii.* Judging from these findings it was somewhat doubtful that the accumulation of DMSP was attributable to osmoregulation at least in unicellular algae. At that time I thought that an approach different from the existing one should be taken to solve this problem.

DMS in the Sulfur Cycle

Just at that time in 1968 I moved to the present department from the Food Research Institute. This department located in Maizuru campus is ca. 100 km away from Kyoto campus, and both campuses were in the height of the famous student riots. Furthermore, harmful algal blooms sometimes occurred in Maizuru Bay. I was compelled to discontinue studies on DMSP at that time.

Fifteen years later in 1984, Uchida and I were invited to join the project of "Man's Activity and Sulfur Cycle" in the Research Programme on Environmental Science, supported by the Grant in Aid of Scientific Research from the Ministry of Education, Culture and Science, Japan. Through this project we intended to study an ecological role of DMSP and DMS by unicellular algae. Since Lovelock (25) in 1972, the transfer of sulfur from the sea through the air to the land surfaces has been assigned to DMS in place of H_2S. In 1987 Shaw (39) proposed that the aerosol produced by the oxidation of sulfur gases from the algae may affect the climate. With this information, we again began to study DMS evolution from DMSP in unicellular algae.

In the open ocean in which DMS-producing dinoflagellates, diatoms and haptophyta are generally dominant, we detected 100-1,500 ng $DMS{\cdot}l^{-1}$ in the surface water (43), as observed by Andreae et al. (2) and Cline and Bates (5). The correlation between DMS and chlorophyll *a* for all water samples was not significant (r=0.37, n=154). The maximum of DMS sometimes shifted with the chlorophyll *a* maximum at the depth of around 50 to 100 m. However, in the coastal waters of Maizuru Bay, the DMS concentration exhibited a well defined seasonal variation. There was a marked increase in concentration of 10 to 20 g $DMS{\cdot}l^{-1}$ from June to August when *Prorocentrum micans* was dominant, followed by a decrease to the minimal value of 0.5 g $DMS{\cdot}l^{-1}$ in December (42). This trend coincided with the trend of chlorophyll *a* in the region. Above all, the correlation between DMS and chlorophyll *a* is very clear in the sea water in which red tide dinoflagellates were abundant.

In the case of Gokasho Bay where the red tide of *Gymnodinium mikimotoi* occurred and of Maizuru Bay where the red tide of *P.micans* occurred, high correlations were found for both dinoflagellates; r=0.96 for *G. mokimotoi* and r=0.91 for *P. micans* (42). The same result was obtained in the case of freshwater bloom of *P. bipes* occurring in reservoirs.

From the knowledge of the oceanic content of DMS, we could determine the evolution of the regional flux of DMS from the offshore, tropical and subtropical Pacific Ocean to the atmosphere. Using the equation of Liss and Slater (24), the mean flux of DMS was calculated as ca 6.6 ng $DMS{\cdot}m^{-2}{\cdot}s^{-1}$ (296 g $S{\cdot}m^{-2}{\cdot}d^{-1}$) referring to a mean surface concentration of 265 ng $DMS.l^{-1}$ in the open Pacific Ocean. The calculated fluxes of DMS across the sea surface of Maizuru Bay in summer and winter were ca 2,700 and 40 g $S^{-1}.m^{-2}.d^{-1}$, respectively (42). We have estimated from several reports (4, 22, 43, 44) that the most abundant DMS producers are the haptophyta and dinoflagellates.

Bioactivities of DMSP

Finally, I will introduce our studies on some biological activities of DMSP. In 1984, Uchida and I joined the group "Regulation Mechanisms on the Ecological Interaction system" supported by a Grant in Aid for a Special Project Research for Ministry of Education, Culture and Science, Japan and for Scientific Research (A). We began a study on feeding attraction and growth promotion of fish by microalgae and their intracellular low molecular components. This study was carried out in association with Dr. Nakajima who had an interest in this project. He has studied the striking behavior of fish by a kimograph and the electrical response from fish olfactory tract to various amino acids (28). Dr. Nakajima will present his data in detail, elsewhere in this volume.

We attempted to detect among various sulfur compounds some substances attractive to freshwater and marine fish. First, we found that the feeding activity in crucian carp and goldfish was most efficiently stimulated by DMSP and, to a lesser extent, by dimethylthetin (dimethylsulfonioacetate), dipropyl sulfide, dimethylsulfoxide and dimethylsulfone (29, 30, 31). This result was confirmed in the experiments conducted on the olfactory tracts of anaesthetized carp.

Next, we examined the effect of dietary supplemented DMSP on the growth of marine fish. Growth of red sea bream, yellowtail and flounder as significantly promoted by the addition of 5 mM DMSP to the diet. This suggested that DMSP is a growth-promotive compound for fish, acting like a vitamin or hormonal agent (32).

Another bioactivity of DMSP that I will illustrate here is that that DMSP acts as an antiulcer agent (16). Methylmethionine sulfonium chloride (MMSC), or vitamin U, is a commonly used antiulcer drug, which was originally discovered in terrestrial vegetables such as cabbage, and onion. Since DMSP is a chemical analogue of MMSC, DMSP and its derivatives will also be expected to be antiulcer drugs. A crude extract of *C. cohnii* and synthetic DMSP-Br significantly and dose-dependently protected the gastric mucosa against HCl-ethanol-induced lesions in rats, when these materials were administered 1 hr before HCl-ethanol administration. At 0.7 $ml{\cdot}kg^{-1}$ of a crude extract containing 300 $mg{\cdot}kg^{-1}$ DMSP, the development of gastric lesions was markedly prevented, the inhibition being 90.5%. Synthesized DMSP-Br, significantly and dose-dependently protected the gastric mucosa against the lesions, as well as the crude extracts of *C. cohnii.* DMSP at 300 and 600 $mg{\cdot}kg^{-1}$ had 66.6% and 99.2% of inhibitory effect on the lesions, respectively. The photograph (Fig.

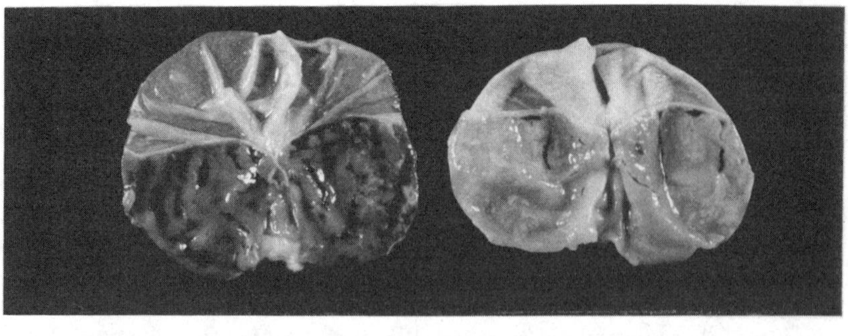

Figure 5. Effect of DMSP on HCl-Ethanol-induced gastric lesions in rats. Left; Gastric mucosal lesions were produced by giving 1.0 ml/200g of body wt. of 60% ethanol(v/v) in 150 mM Hcl (HCl-ethanol). The animals were killed 1 h later, and their stomachs were removed. After removal of the gastric contents, the stomachs were inflated by injecting 8 ml of 2% formalin for 10 min. The gastric mucosa was severely damaged 1 h after administration of HCl ethanol. Right; The DMSP (300 mg/kg rat) was given to the rat 0.5 h before HCl-ethanol administration. It significantly protected the gastric mucosa against HCl-ethanol-induced lesions.

5) shows that DMSP acted locally to protect the gastric mucosa of rats. Since indomethacin pretreatment reduced the protective effect to some extent, endogenous prostaglandins might be partly involved in the mechanisms of action. Both the crude extract and DMSP significantly increased gastric secretion.

Biosynthesis of DMSP

Previously Green (8) showed that methionine was an efficient precursor for the biosynthesis of DMSP in *Ulva lactuca*, and that the methyl group and sulfur of DMSP were derived from those of methionine. Kahn (19) reported that the carbon at the carbon-2 of glycine was incorporated into the methyl group of DMSP. We also demonstrated that methyl carbon and carboxyl carbons of DMSP were derived from the methyl and carboxyl groups, respectively, of acetate when *C. cohnii* was allowed to take up acetate (14).

Starting in 1987, Uchida and I began a study which is trying to elucidate the mechanism of biosynthesis of DMSP in *C. cohnii*, at the level of biochemistry and molecular biology. We showed that the methyl-, C3- and C2-carbon and sulfur of methionine except C1-carbon were converted to DMSP, and methylthiopropionic acid was produced as an intermediate (45). Then L-methionine decarboxylase as a key enzyme was isolated and purified. A single band of this purified enzyme was confirmed by polyacrylamide gel electrophoresis. The enzyme required pyridoxal phosphate and 1 mM Mg^{++}, and the activity was not dependent on NaCl or KCl concentration. The molecular weight was 230 kDa and the enzyme consisted of a homodimer. The N-terminal amino acid sequence was Ala-Leu-Cys-Try-Ser-Asp-Ile-Ser-Pro—, but this 9 amino acid sequence was too short to make the DNA probe for detecting the gene of methionine decarboxylase. Now we are attempting to repeat this experiment. Further details on methionine decarboxylase are presented in Uchida et al, this volume.

In order to clarify the role of DMSP in such processes as osmoregulation, it is very important to prepare DMSP-deficient or methionine decarboxylase deficient mutants, and to analyze the factors which regulate the level of transcription of the methionine decarboxylase gene.

In relation to DMSP, I must point also to gonyauline (cis-2-dimethylsulfoniocyclopropanecarboxylate) in a luminous dinoflagellate, *Gonyaulax polyedra* (33). This gonyauline is a unique cyclopropane derivative of DMSP, and has been identified as an active substance which shortens the period of bioluminescent circadian rhythm in the cultured marine dinoflagellate *G. polyedra* (37). The chemical structure of gonyauline is close to that of DMSP, and gonyauline specifically occurred only in *G. polyedra,* but not in the other dinoflagellates such as *P. lima* etc, which contain DMSP. Biosynthesis of gonyauline was examined by use of ^{13}C-NMR, and it could be proved that methionine was converted to gonyauline via DMSP, and gonyol was converted from DMSP with acetate (Nakamura, unpublished data). He suggested that C1-carbon on gonyauline derives from CO_2.

It is interesting to note that gonyauline expressed the period-shortening function in *G. polyhedra* only when it was added externally, even though this alga itself contained intracellular gonyauline equivalent to the amount which ellicited expression. Whether or not gonyauline in *G. polyedra* plays the same role as DMSP in other dinoflagellates, in addition to the circadian rhythm remains to be determined.

CONCLUSION

I will point out the following questions about DMSP:

1. Why do aquatic plants, including algae, accumulate DMSP but not MMSC, while terrestrial plants (asparagus, cabbage, onion etc), accummulate MMSC (except for sugarcane which accumulates DMSP but not MMSC (37))?
2. Why is DMSP accumulated in all marine and freshwater dinoflagellates and haptophyta, and a part of the diatoms, but not in unicellular chlorophyta except for some macroalgae cholorophyta such as *Enteromorpha, Ulva* etc?

In order to address these questions, I am hoping to continue our efforts towards clarifying DMSP function; first, by means of analysis of the DNA sequence of methionine decarboxylase and a regulatory factor at the level of the transcription, and second, by means of preparation of DMSP deficient mutants and methionine decarboxylase deficient mutants.

ACKNOWLEDGMENTS

This work was supported by the Grants in Aid of Scientific Research from the Ministry of Education, Culture and Science, Japan (Nos. 5903001, 6003001, 6103001, 60129034, 61134044, 62124040, 63440015) and the Grants from the Ministry of Agriculture, Forestry and Fisheries, Japan. I especially wish to express cordial thanks to Emeritus Prof. H. Kadota and Dr. A. Uchida. Thanks are also due to Dr. T. Kawai, Dr. K. Nakajima, Dr. S. Okabe, Mr. T. Ooguri, Mr. T. Ishida and Mr. H. Kitaguchi for their kind help.

REFERENCES

1. Ackman, R.G., C.S. Tocher and J. McLachelan. 1966. Occurrence of dimethyl—propiothetin in marine phytoplankton. J. Fish. Res. Bd. Canada. 23: 357-364.
2. Andreae, M.O., W.R. Barnard and J.M. Ammons. 1983. The biological production of dimethylsulfide in the ocean and its role in the global atmospheric sulfur budget. Environ. Biogeochem. Ecol. Bull. 35: 167-177.
3. Barnard, W.R., M.O. Andreae and R.L. Iverson. 1984. Dimethylsulfide and *Phaeocystis poucheti* in the southeastern bering sea. Cont. Shelf Res. 3: 103-113.
4. Challenger, F. and M.I. Simpson. 1948. Studies on biological methylation. Part XII. A precursor of the dimethyl sulfide evolved by *Polysiphonia fastigiata*. Dimethyl-2-carboxyethyl sulphonium hydroxide and its salts. J. Chem. Soc. 1948: 1591-1597.
5. Cline, J.D., T.S. Bates. 1983. Dimethyl sulfide in the equatorial Pacific Ocean: A natural source of sulfur to the atmosphere. Geophys. Res. Lett. 10: 949-952.
6. Dubnoff, J.W. and H. Borsook. 1948. Dimethylthetin and dimethyl—propiothetin in methionine synthesis. J. Biol. Chem. 176: 789-796.
7. Durell, J., D.G. Anderson and G.L. Cantoni. 1957. The synthesis of methionine by enzymatic transmethylation 1. Purification and properties of thetin homocysteine methylpherase. Biochim. Biophys. Acta. 26: 270-281.
8. Greene, R.C. 1962. Biosynthesis of dimethyl—propiothetin. J. Biol. Chem. 237: 2251-2254.
9. Grone T., G.O. Kirst. 1991. Aspects of dimethylsulphoniopropionate effects on enzymes isolated from the marine phytoplankter *Tetraselmis subcordiformis* (Stein). J. Plant. Physiol. 138: 85-91.
10. Haas, P. 1935. The liberation of methyl sulphide by sea-weed. Biochem. J. 29: 1297-1299.
11. Ishida, Y. and H. Kadota. 1967. Production of dimethyl sulfide from unicellular algae. Bull. Japan. Soc. Sci. Fish. 33: 782-787.
12. Ishida, Y. and H. Kadota. 1967. Isolation and identification of dimethyl—propiothetin from *Gyrodinium cohnii*. Agr. Biol. Chem. 31: 756-757.
13. Ishida, Y. and H. Kadota. 1968. Participation of dimethyl—propiothetin in transmethylation reaction in *Gyrodinium cohnii*. Bull. Japan. Soc. Sci. Fish. 34: 699-705.
14. Ishida, Y. and H. Kadota. 1968. A note on biogenesis of dimethyl—propiothetin in *Gyrodinium cohnii*. Mem. Res. Inst. Food Sci. 29: 67-68.

15. Ishida, Y. 1968. Physiological studies on evolution of dimethyl sulfide from unicellular marine algae. Mem. Coll. Agr., Kyoto Univ. No. 94: 47-82.
16. Ishida, Y., Y. Ogihara and S. Okabe. 1990. Effects of a crude extract of a marine dinoflagellate, containing dimethyl—propiothetin, on HCl-ethanol-induced gastric lesions and gastric secretion in rats. Japan. J. Pharmacol. 54: 333-338.
17. Kadota, H. 1963. Planktons as Foods. Food Technol. 18: 25.
18. Kadota, H. and Y. Ishida. 1968. Effect of salts on enzymatical production of dimethyl sulfide from *Gyrodinium cohnii*. Bull. Japan. Soc. Sci. Fish. 34: 512-518.
19. Kahn, V. 1964. Glycine as a methyl donor in dimethyl—propiothetin synthesis. J. Expl. Bot. 15: 225-231.
20. Keller, M.D., W.K. Bellows and R.R.L. Guillard. 1989. Dimethyl sulfide production in marine phytoplankton, p.167-182. In E.S. Sattzmahn and W.J. Cooper (ed.), Biogenic sulfur in the environment. American Chemical Society, (Symposium Series 393), Washington D.C.
21. Kinne, R. K. H. 1993. The note of organic osmolytes in some regulation from bacteria to mammals. J. Exp. zoology. 265: 346-355.
22. Klee, W.A. 1965. Thetin-homocysteine methylpherase: A study in molecular organization, p.220-229. In S. K. Shapiro and F. Schlenk (ed.), Transmethylation and methionine biosynthesis. Univ. of Chicago Press.
23. Le Rudulier, D., A.R. Strom, A.M. Dandekar, L.T. Smith and R.C. Valentine. 1984. Molecular biology of osmoregulation. Science. 224: 1064-1068.
24. Liss, P.S. and P.G. Slater. 1974. Flux of gases across the air-sea interface. Nature. 247: 181-184.
25. Lovelock, J.E., R.J. Maggs and R.A. Rasmussen. 1972. Atmospheric dimethylsulfide and the natural sulfur cycle. Nature. 237: 452-453.
26. Maw, G.A. 1956. Thetin-homocysteine transmethylase. A preliminary manometric study of the enzyme from rat liver. Biochem. J. 63: 116-124.
27. Motohiro, T. 1962. Studies on the petroleum odour in canned chum salmon. Mem. Fac. Fish. Hokkaido Univ. 10: 1-65.
28. Nakajima, K. 1987. New determination method for striking behavior of fish. Mem. Koshien Univ. 15: 13-17.
29. Nakajima, K., A. Uchida and Y. Ishida. 1989. Effects of heterocyclic sulfur-containing compounds on the striking response of goldfish. Mem. Koshien Univ. 16: 7-10.
30. Nakajima, K., A. Uchida and Y. Ishida. 1989. A new feeding attractant, dimethyl—propiothetin, for freshwater fish. Nippon Suisan Gakkaishi. 55: 689-695.
31. Nakajima, K., A. Uchida and Y. Ishida. 1989. Effect of supplemental dietary feeding attractant, dimethyl—propiothetin, on growth of goldfish. Nippon Suisan Gakkaishi. 55: 1291.
32. Nakajima, K., A. Uchida and Y. Ishida. 1990. Effect of a feeding attractant dimethyl—propiothetin, on growth of marine fish. Nippon Suisan Gakkaishi. 56: 1151-1154.
33. Nakamura, H. 1993. Bioactive substances from marine organisms. Nippon Nogeikagaku Kaishi. 67: 1-6.
34. Nicolai, E. and R.D. Preston. 1953. Variability in the X-ray diagram of the cell walls of the marine alga *Spongomorpha*. Nature. 171: 752-753.
35. Nicolai, E. and R.D. Preston. 1960. Cell-wall studies in the Chlorophyceae. III. Differences in structure and development in the Cladophoraceae. Proc. Roy. Soc. B. 151: 244-255.
36. Obata, Y., H. Igarashi and K. Matano. 1951. Studies on the flavor of seaweeds 1. On the component of the flavor of some green algae. Bull. Japan. Soc. Sci. Fish. 17: 60-62.
37. Paquet, L. et al. 1994. Accumulation of the compatible solute 3-dimethylsulfoniopropionate in sugarcane and its relatives, but not in other gramineous crops. Aust. J. Plant Physiol. 21: 37-48.
38. Roenneberg, T., H. Nakamura, L.D. Cranmer III, K. Ryan, Y. Kishi and J.W. Hastings. 1990. Gonyauline: A novel endogenous substance shortening the period of the circadian clock of a unicellular alga. Experientia. 47: 103-105.
39. Schlenk, F. 1965. Biochemical and cytological studies with sulfonium compounds. p.48- . In S.K. Shapiro & F. Schlenk (eds.), Transmethylation and methionine biosynthesis. The University of Chicago Press., Chicago & London.
40. Shaw, G. E. 1987. Aerosols as climate regulators: a climate biosphere linkage? Atmos. Environ. 21:985-986.
41. Sipos, J.C. and R.G. Ackman. 1964. Association of dimethyl sulfide with the "Blackberry" problem in cod from the Labrador area. J. Fish. Res. Bd. Canada. 21: 423-425.
42. Turner, S.M., G. Malin, P.S. Liss, D.S. Harbour and P.M. Holligan. 1988. The seasonal variation of dimethyl sulfide and dimethyl sulfoniopropionate concentration in nearshore waters. Limnol. Oceanogr. 33: 364-375.
43. Uchida, A., T. Ooguri, T. Ishida and Y. Ishida. 1992. Seasonal variations in dimethylsulfide in the water of Maizuru Bay. Nippon Suisan Gakkaishi. 58: 255-259.

44. Uchida, A., T. Ooguri and Y. Ishida. 1992. The distribution of dimethylsulphide in the waters off Japan and in the subtropical and tropical Pacific Ocean. Nippon Suisan Gakkaishi. 58: 261-265.
45. Uchida, A. 1992. Relation to environmental problems. p.102-115. In K. Yamaguchi (ed.), Utilization of microalgae (Japanese). Koseisha-Kouseikaku Publ., Tokyo.
46. Uchida, A., T. Ooguri, T. Ishida and Y. Ishida. 1993. Incorporation of methionine into dimethylthio-propanoic acid in the dinoflagellate *Crypthecodinium cohnii*. Nippon Suisan Gakkaishi. 59: 851-855.
47. Variavamurthy, A., M.O. Andreae and R.L. Iverson. 1985. Biosynthesis of dimethylsulfide and dimethyl propiothetin by *Hymenomonas carterae* in relation to sulfur source and salinity variation. Limnol. Oceanogr. 30: 59-70.

2

CHEMICAL AND BIOCHEMICAL PROPERTIES OF SULFONIUM COMPOUNDS

Arthur J. L. Cooper

Department of Biochemistry
Cornell University Medical College
New York, New York 10021

SUMMARY

Several sulfonium compounds are found in Nature. The most biologically versatile is *S*-adenosyl-L-methionine (AdoMet). Much of this chapter is concerned with AdoMet, but includes a discussion of other sulfonium compounds. Some of the more interesting chemical properties of sulfonium compounds are discussed as well as how these properties are exploited in biochemical reactions. Finally, the chemistry of sulfonium compounds is discussed in relation to the biosynthesis of 3-dimethylsulfoniopropionate (DMSP). In this brief review references and topics are necessarily selective.

DISCOVERY OF ADOMET

For many years methionine was known to be a major source of methyl groups in biological methylation reactions. However, it was also recognized that methionine had to be activated first. Cantoni (9, 10) was the first to propose that "active methionine" is *S*-adenosylmethionine - a "high energy" sulfonium compound. The proposal was later confirmed by degradation of the natural product (5) and by chemical synthesis (4).

SYNTHESIS OF ADOMET

Two biological reactions are known in which the adenosyl moiety of ATP is transferred to an acceptor: 1) Synthesis of AdoMet catalyzed by AdoMet synthetase (ATP: L-methionine *S*-adenosyltransferase). 2) Synthesis of adenosylcobalamin - the cofactor in the methylmalonyl CoA mutase reaction. Triphosphate is formed at the active site during the reaction catalyzed by AdoMet synthetase but is not released. Rather it is hydrolyzed to phosphate and pyrophosphate (eq. 1). The latter is then hydrolyzed to free phosphate by the

Biological and Environmental Chemistry of DMSP and Related Sulfonium Compounds
edited by Ronald P. Kiene et al., Plenum Press, New York, 1996

action of pyrophosphatases. As a result, two "high-energy" phosphate bonds are hydrolyzed, providing the free energy for the synthesis of "high-energy" AdoMet.

$$\text{L–Methionine} + \text{ATP} \rightarrow \text{AdoMet} + \text{PPP}_i \; (+ \text{H}_2\text{O} \rightarrow \text{P}_i + \text{PP}_i) \qquad (1)$$

Mammalian AdoMet synthetase has been well characterized. The enzyme of mammalian kidney (II) is distinct from that of the liver (I and III) and the two forms are coded by separate genes. Isoform I and isoform III are products of the same gene and are homotetramers and homodimers, respectively. [For a review see Hoffman (41)]. cDNA or genomic clones of the enzyme from *Escherichia coli*, *Saccharomyces cerevisiae*, *Arabidopsis thaliana*, and parsley (*Petroselinum crispum*) have also been isolated (45 and references cited therein).

CHEMISTRY OF SULFONIUM COMPOUNDS

This subject is covered in a two volume treatise edited by Stirling (74). Only selected topics are discussed here.

Stereochemical Considerations. Sulfonium compounds exhibit pyramidal bonding and therefore can exist as enantiomeric forms. [For reviews of chiral organosulfur compounds see Andersen (2) and Mickolajczyk and Drabowicz (57)]. de la Haba et al. (21) showed that AdoMet synthesized by liver or yeast AdoMet synthetase was completely consumed in the guanidinoacetate methyltransferase reaction. However, only about a half of synthetically-prepared AdoMet was consumed in the reaction. Thus, only one of the two possible diastereoisomers (i.e., *(S)*,L- or *(R)*,L-) of AdoMet is a substrate of guanidinoacetate methyltransferase. Moreover, the same diastereoisomer was also shown to participate in the catechol *O*-methyltransferase (COMT) reaction (21). These experiments, however, did not establish the absolute configuration of the sulfonium center of biologically active AdoMet. The problem was solved by Cornforth et al. (19), who showed that the sulfonium center is of the (*S*)-configuration.

Theoretically, enzyme-catalyzed transfer of a methyl group from AdoMet to a suitable acceptor can occur with retention or inversion of configuration. Woodard and coworkers synthesized AdoMet in which the methyl group is asymmetric and contained one H, one D and one T (29). When epinephrine was converted to metanephrine in a reaction catalyzed by COMT in the presence of [(*R*)-D,T-*methyl*]AdoMet, the product contained labeled methyl in the (*S*)-configuration. Inversion of configuration occurred in the five enzyme-catalyzed methyl transfer reactions investigated. Thus, each enzyme operated by a mechanism involving direct transfer of a methyl group. Evidently, the reaction does not proceed via a ping-pong mechanism nor does it involve a methylated enzyme intermediate (29).

Although the most important biological methyl donor is AdoMet, other methylsulfonium compounds can sometimes be used in transmethylation reactions, especially when the methyl is donated to sulfur (see below). For example, some yeasts and plants possess an enzyme that catalyzes the transfer of a methyl group of *S*-methyl-L-methionine (SMM) to L-homocysteine (eq. 2) (38, 60). The use of SMM in biological methylations raises an interesting issue. Although the two methyls of SMM are chemically equivalent, they should be enzymatically distinguishable. By introducing label into one or other of the methyls, diastereoisomers of SMM may be synthesized. Grue-Sørensen et al. (38) synthesized the two diastereoisomers of [2-^{2}H, *methyl*-^{13}C]SMM and showed that the reaction catalyzed by jack bean (*Canavalia ensiformis*) *S*-methylmethionine: homocysteine *S*-methyltransferase proceeds with a preference for the removal of the *pro*-(*R*)-methyl group to the extent of 94% or more.

$$S-\text{Methyl–L–methionine} + \text{L–homocysteine} \rightarrow 2\text{L–methionine} + H^+ \quad (2)$$

Acidity of Protons Attached at the Carbon A *to the Sulfur.* Loss of a proton α to the sulfur will yield an ylide. Preparation of the ylide of $PheS^+(CH_3)_2$ requires the presence of a strong base and rigorously anhydrous conditions (31). Coward and Sweet (20) estimated that the pK_a of sulfonium compounds of the type $ArS^+(CH_3)_2$ is typically quite high (>14 to ~17), but still considerably less than that of a typical RCH_3 group. Despite negligible ylide formation, α protons in many sulfonium compounds (e.g. trimethylsulfonium iodide) are labile in aqueous solution (22).

The acidity of the α protons is increased by appropriate substitution at the β carbon. β-(Dimethylsulfonio)pyruvic acid behaves as a diprotic acid (pK_{a1} ~ 9; pK_{a2} ~ 5), whereas the ester behaves as a monoprotic acid with a pK_a of ~9 (8). The unusually high acidity of the α protons in β-(dimethylsulfonio)pyruvic acid is unlikely to be due solely to the inductive effect of the electron-withdrawing dimethylsulfonio group. Resonance stabilization of the conjugate base must also contribute. The dimethylsulfonio group is conjugated with the keto group by expansion of the sulfur valence shell (eqs. 3 and 4).

$$(CH_3)_2S^+CH_2C(=O)CO_2^- + OH^- \rightarrow H_2O + (CH_3)_2S^+CH = C(O^-)CO_2^- \quad (3)$$

$$(CH_3)_2S^+CH = C(O^-)CO_2^- \leftrightarrow (CH_3)_2S^+CH^-C(=O)CO_2^- \leftrightarrow (CH_3)_2S = CHC(=O)CO_2^- \quad (4)$$

Vinylsulfonium ions are reactive and undergo nucleophilic addition at the C2 position of the vinyl group (41) (eq. 5). This reactivity is due to stabilization of the carbanion at C-1 in the transition state (41, 73).

$$CH_3CH_2\underset{\underset{CH_3}{|}}{S^+}CH{=}CH_2 + Nuc^- + H^+ \rightarrow CH_3CH_2\underset{\underset{CH_3}{|}}{S^+}CH_2CH_2\text{-}Nuc \quad (5)$$

As a result of the lability of the α protons and exchange with solvent protons, the metabolic fate of deuterium or tritium labeled in the position α to the sulfur of a typical sulfonium compound might be difficult to follow in some cases. By avoiding harsh conditions, however, Hanson et al. (39) were able to use [^{2}H-*methyl*]SMM in studies of DMSP synthesis in plants.

Elimination Reactions. The electron-withdrawing properties of the alkyl sulfonium group and the acidity of the protons α to the sulfur contribute to the ease with which elimination reactions occur. These reactions are analogous to those exhibited by quaternary ammonium salts, which may be cleaved to the corresponding olefin in the presence of a strong base (e.g. PheLi, KNH_2 in liquid NH_3). The mechanism is an α′,β-elimination and the reaction may proceed via a nitrogen ylide (see references quoted in March (51)). Sulfonium compounds may undergo a similar α′,β elimination reaction to yield olefin and dialkyl sulfide. The reaction with sulfonium hydroxides is readily accomplished by heating (48) (eq. 6). Alternatively, the reaction may proceed via a sulfur ylide in the presence of a strong base (eq. 7) (30).

(6)

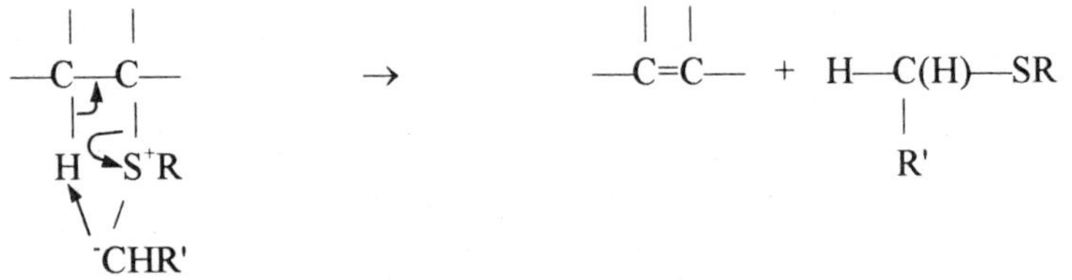

(7)

Conversion of Sulfonium Compounds to Thioethers by Reaction with Nucleophiles. The free energy change ($\Delta G^{\circ\prime}$) for the conversion of a sulfonium compound to a thioether is ~ -7 to -10 kcal/mol. [See the discussion by Coward and Sweet (20).] This value is similar to that for the hydrolysis of the "high energy" bonds of ATP. Despite a favorable free energy change, however, conversion of a sulfonium compound to a thioether is usually very slow in water. Large activation energies prevent the reaction from occurring except under forcing conditions. Coward and Sweet (20) investigated the attack of nucleophiles (nuc$^-$) on $ArS^+(CH_3)_2$ to yield $ArSCH_3$ and CH_3nuc. The reaction of the sulfonium compound with oxygen-containing nucleophiles was found to be very slow. The relative reactivity of the added nucleophiles was in the order: O <N <S (20).

Zappia et al. (82) investigated the methyl donor specificity of representative *O*-, *N*- and *S*-methyltransferases. AdoMet: acetylserotonin *O*-methyltransferase (an *O*-methylase) exhibits a strict requirement for AdoMet. On the other hand, AdoMet: histamine *N*-methyltransferase (an *N*-methylase) utilizes *S*-inosyl-L-methionine to a small extent in place of AdoMet. Finally, AdoMet: homocysteine *S*-methyltransferase (an *S*-methylase), utilizes *S*-inosyl-L-methionine, *S*-adenosyl-(5′)-3-methylthiopropylamine and SMM almost as effectively as AdoMet (82). The data suggest that in the case of the *O*-methylase, the methyl donor must meet strict requirements. These requirements are satisfied only by AdoMet, which must interact at several points on the active site of the enzyme (82). Evidently, for effective catalysis of transmethylation reactions involving oxygen nucleophiles, sufficient lowering of the activation energy can only be brought about with a methyl donor capable of interaction on the enzyme at multiple points. On the other hand, where methylation is kinetically favored, as in the case of the *S*-methylase, catalysis requires less stringent binding of methyl donor to the active site, and SMM can replace AdoMet.

NATURALLY-OCCURRING SULFONIUM SALTS (OTHER THAN ADOMET)

For earlier reviews see Schlenk (68) and Maw (52). A brief summary, especially of more recent references, is presented here.

S-Methyl-L-Methionine (SMM). This compound was first identified as a constituent of cabbage leaves and asparagus in the early 1950s and is now known to occur in algae and many higher plants. SMM is formed from AdoMet by the enzyme AdoMet: methionine *S*-methyltransferase in many higher plants (eq. 8):

AdoMet + L-methionine → *S*-methyl-L-methionine + *S*-adenosyl-L-homocysteine (8)

SMM is catabolized by two routes. It can be converted back to L-methionine by the action of SMM: L-homocysteine *S*-methyltransferase (eq. 2) or it can be converted to dimethylsulfide and L-homoserine by the action of SMM hydrolase (eq. 9). For original references and discussion see Mudd and Datko (60).

$$S\text{-Methyl-L-methionine} + H_2O \rightarrow (CH_3)_2S + \text{L-homoserine} + H^+ \quad (9)$$

A combination of AdoMet:L-methionine *S*-methyltransferase (eq. 8) and SMM: L-homocysteine *S*-methyltransferase (eq. 2) together with *S*-adenosyl-L-homocysteine hydrolase (eq. 10) yields a net reaction in which AdoMet is converted to L-methionine and adenosine (eq 11).

$$S\text{-Methyl-L-methionine} + \text{L-homocysteine} \rightarrow 2\ \text{L-methionine} + H^+ \quad (2)$$

$$\text{AdoMet} + \text{L-methionine} \rightarrow S\text{-methyl-L-methionine} + S\text{-adenosyl-L-homocysteine} \quad (8)$$

$$\underline{S\text{-Adenosyl-L-homocysteine} + H_2O \leftrightarrow \text{Adenosine} + \text{L-homocysteine} \quad (10)}$$

$$\text{Net: AdoMet} + H_2O \rightarrow \text{Adenosine} + \text{L-methionine} + H^+ \quad (11)$$

According to Mudd and Datko (60) this "*S*-methylmethionine cycle" provides the plant with a means of sustaining a pool of soluble methionine even when overshoot occurs in the conversion of soluble methionine to AdoMet. These reactions ensure that the levels of methionine and AdoMet are tightly regulated, but at the cost of consumption of ATP. The cycle appears to operate at a low level, however, in the leaves of *Wollastonia biflora* (L.) even though the plants contain SMM (39).

Farooqui et al. (27) showed that *Euglena gracilis* contains two methyltransferases capable of methylating specific residues of cytochrome *c*. Enzyme I methylates a methionine residue at position 65 and enzyme II methylates an arginine at position 38. At the time of the report by Farooqui et al. (27), no examples of the natural occurrence of SMM in protein molecules had been recorded and the authors were unable to suggest a biological role for enzyme I. Of interest is the recent finding of Katz and Gerhardt (44) that the abnormal storage protein which accumulates in hereditary ceriod-lipofuscinosis (Batten's disease) is rich in *S*-methylmethionine.

Finally, strong evidence implicates SMM as the precursor of dimethylsulfoniopropionate in the angiosperm *W. biflora* (39).

Dimethylsulfoniopropionate (DMSP). Many marine algae and certain salt-tolerant angiosperms contain DMSP (= dimethyl-β-propiothetin) as an osmoprotectant (e.g. 42). As discussed by Hanson and colleagues (42), DMSP is known to be a major precursor of dimethylsulfide, an important component of the global sulfur cycle (1), a contributor to acid precipitation (65), and the main source of cloud condensation nuclei over oceans (12).

Two routes are known for the catabolism of DMSP in bacteria and algae (46). DMSP may be converted to acrylate and dimethylsulfide; the latter is oxidized to dimethylsulfoxide. Alternatively, DMSP may be demethylated (or transmethylated) to give 3-methylthiopropionate (MTP). DMSP may function as a methyl donor to nitrogen-containing compounds in algae. Thus, Chillemi et al. (13) noted that feeding of the red alga (*Chondria coerulescens*) with [$^{14}CH_3$]DMSP resulted in the formation of methyl-labeled 4-hydroxy-*N*-methylproline.

DMSP can provide essential methyl groups in experimental animals (53). Transmethylation of DMSP is catalyzed by kidney and liver extracts (23). DMSP is likely, however, to be limited in the diet of non-cetacian mammals. Therefore, this pathway is probably of little importance in L-methionine synthesis from L-homocysteine in most mammals.

Phosphatidylsulfocholine. "Sulfocholine" the sulfur analog of choline was synthesized almost 50 years ago (54). The compound could not support the growth of rats on a methionine and choline-deficient diet. Phosphatidylsulfocholine (PSC) along with phosphatidylcholine (PC) has been detected in several photosynthetic diatom species. In one non-photosynthetic species *Nitzschia alba*, PSC was found to completely replace PC. [See references quoted by Bisseret et al. (7)].

Recently Described Algal Sulfonium Compounds. 4-Dimethylsulfonio-2-methoxybutyrate and 5-dimethylsulfonio-4-hydroxy-2-aminovalerate have been reported to be present in some Mediterranean red algae (70). Gonyauline (*cis*-2-(dimethylsulfonio)cyclopropane carboxylate) is a novel zwitterion that has recently been identified as an endogenous period-shortening substance in the bioluminescent circadian rhythm of the dinoflagellate *Gonyaulax polyedra* (62). Another sulfonium compound named gonyol (3*S*-5-dimethylsulfonio-3-hydroxypentanoate) has been isolated from *Gonyaulax polyedra* grown on a source rich in methionine and acetate (63).

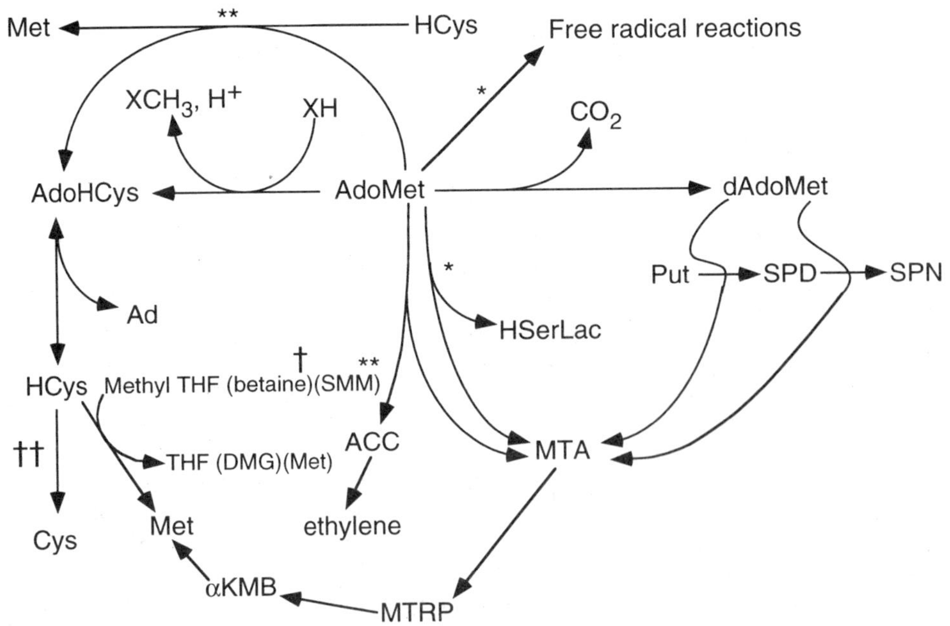

Figure 1. Composite scheme of the major metabolic reactions involving *S*-adenosyl methionine. Abbreviations: ACC, 1-aminocyclopropane-1-carboxylate; Ad, adenosine; AdoMet, *S*-adenosyl-L-methionine; αKMB, α-keto-γ-methiolbutyrate; dAdoMet, decarboxylated AdoMet; DMG, dimethylglycine; AdoHCys, *S*-adenosyl-L-homocysteine; HCys, L-homocysteine; HSerLac, L-homoserine lactone; MTA, 5′-methylthioadenosine; MTRP, methylthioribose phosphate; Put, putrescine; SMM, *S*-methyl-L-methionine; SPD, spermidine; SPN, spermine; THF, tetrahydrofolate; X, methyl acceptor. Although many of the reactions shown in the scheme are common to both prokaryotes and eukaryotes, several are more restricted. Reactions denoted by a superscript are thought to be largely limited in occurrence as follows: *bacteria; **plants; †mammals; ††mammals and bacteria.

TRANSFER REACTIONS INVOLVING ADOMET/ADOHCYS

A schematic diagram of AdoMet metabolism is shown in Fig. 1. Most of the biologically important pathways are initiated by breakage of one or other of the three S bonds at the sulfonium center, either directly on AdoMet (methyl-, 3-amino-3-carboxypropyl-, adenosyl transfers) or on dAdoMet - the decarboxylated form of AdoMet (3-aminopropyl transfer) (e.g. 36).

Transfer of the Methyl Group of AdoMet to Low-Molecular-Weight Compounds. Transmethylation reactions result in the transfer of a methyl group of AdoMet to a suitable acceptor (at S, N, O, C or halide) with the formation of *S*-adenosyl-L-homocysteine (AdoHCys) (Fig. 1). These reactions are quantitatively the most important route for the metabolism of AdoMet. The methyl may be transferred to relatively small molecules, as, for example, in the synthesis of creatine, phosphatidyl choline, epinephrine (*N*-methylations), 3-*O*-methoxyepinephrine (*O*-methylation), and L-methionine (*S*-methylation of L-homocysteine). In addition, AdoMet is the methyl donor in the synthesis of a number of biologically-important *C*-methylated small molecules (e.g., methylcobalamin, ubiquinone, plastoquinone, methyl fatty acids, cyclopropyl fatty acids, C_{28} and C_{29} sterols, C_{31} triterpenes, vitamin K, and tocopherol) (49). Finally, AdoMet is the methyl donor in a reaction catalyzed by *Brassica oleracea* AdoMet:halide/bisulfide methyltransferase (3). In this reaction, the methyl group is transferred to halide (X^-) or HS^- to generate CH_3X or CH_3SH, respectively. Iodide is an especially effective methyl acceptor.

Methylations involving AdoMet may have clinical relevance. Parkinson's disease, which is characterized by tremors and rigidity, results from excessive loss of dopaminergic neurons in the substantia nigral region of the brain. The disease usually occurs in elderly patients. Recently, however, after self-administration of a contaminated illicit drug related to meperidine several young drug abusers developed a classic Parkinson syndrome. The contaminant was identified as 1-methyl-4-phenyl-1,2,3,6-tetrahydropyridine (MPTP) which is oxidized to toxic 1-methyl-4-phenylpyridinium ion (MPP^+). MPP^+ is toxic to the substantia nigra and causes Parkinsonium symptoms in primates. MPP^+ interferes with the mitochondrial electron-transport chain. It is possible that adult-onset Parkinson's disease is due to life-time exposure to low levels of environmental toxins or protoxins. Ingestion of 4-phenylpyridine or of 4-phenyl-1,2,3,6-tetrahydropyridine (MTP) is expected to generate MPP^+ from AdoMet-dependent *N*-methylation/oxidation and *N*-methylation reactions, respectively. 4-Phenylpyridine and MTP may be regarded, therefore, as protoxins which are not toxic of themselves but which are metabolized to neurotoxic compounds. Other protoxins of the substantia nigra include 1,2,3,4-tetrahydro-β-carboline (THBC) and 1,2,3,4-tetrahydroisoquinoline (TIQ), both of which occur naturally. AdoMet-dependent *N*-methylations and oxidation reactions on tetrahydrocarbolines and tetrahydroisoquinolines will give rise to potentially neurotoxic compounds that will interfere with the electron-transport chain (41).

Mammalian tissues contain two AdoMet-dependent enzymes that methylate sulfhydryls to form thioethers and an AdoMet-dependent enzyme that methylates thioethers to yield methyl sulfonium compounds (41). Thiopurine methyltransferase is cytoplasmic and methylates the sulfur of aromatic and heterocyclic thiols (e.g., 6-mercaptopurine). Thiol methyltransferase is microsomal and catalyzes the methylation of the sulfur of aliphatic thiols (e.g., mercaptoethanol, captopril, D-penicillamine). Both enzymes are considered to be detoxifying. Thioether methyltransferase catalyzes the methylation of compounds of the type RXR′, where X is S, Se or Te and R (or R′) does not contain a carboxyl or a direct attachment to S at an aromatic ring (41). The enzyme may be involved in bioactivation of xenobiotics. Thus, vinyl- and allyl sulfides are potentially toxic when converted to the corresponding sulfonium ions; these ions may form adducts with macromolecules (41).

Transfer of the Methyl Group of AdoMet to Macromolecules. In proteins, methylation may be at lysine, arginine, histidine, or glutamate (27). ε-Trimethyllysine is a precursor of carnitine - the carrier molecule for acyl groups between cytosol and mitochondria. Polysaccharides may be modified by *O*-methylation (6). AdoMet is also involved in the methylation of nucleic acids. Methylations of tRNA may be at N, C (base or ring) or O (ribose) (67).

Transfer of the 3-(3-Amino-3-Carboxypropyl) Group of AdoMet. This process releases 5′-methylthioadenosine (MTA). [For a review see Schlenk (69)]. An example is the synthesis of one of the base modifications in *E. coli* tRNA (64) (eq. 12).

$$\text{tRNA}^{\text{phe}}\text{(uridine)} + \text{AdoMet} \rightarrow \text{tRNA[3-(3-amino-3-carboxypropyl)uridine]} + \text{MTA} \quad (12)$$

Formation of MTA can also occur by intramolecular nucleophilic attack and cyclization (59). For example, many microorganisms contain an enzyme that converts AdoMet to MTA and α-amino-γ-butyrolactone (L-homoserine lactone; Hserlac) (59, 69, 72). The carbon of HSerlac can be recycled back to AdoMet (71) (eq. 13):

$$\text{HSerlac} \rightarrow O\text{-succinylhomoserine} \rightarrow \text{cystathionine} \rightarrow \text{HCys} \rightarrow \text{Met} \rightarrow \text{AdoMet} \quad (13)$$

An alternative internal nucleophilic attack gives rise to 1-aminocyclopropane-1-carboxylate (ACC) and MTA in a reaction catalyzed by ACC synthase. ACC is then oxidized to ethylene, CO_2 and cyanide by ACC oxidase. Two ACC oxidases occur in avocados. One has been characterized and shown to require O_2, Fe^{II} and ascorbate (56). Ethylene is a hormone in higher plants that has many functions including stimulation of fruit ripening (24).

Transfer of the 3-Aminopropyl Group of AdohCys. In this process, AdoMet is decarboxylated to AdoHCys; the latter serves as a 3-aminopropyl donor in the synthesis of polyamines spermidine and spermine from putrescine (1,4-diaminobutane). Transfer of the propylamine moiety results in the formation of MTA (Fig. 1). The biological importance of polyamines is still debated, but as discussed in an earlier review by Tabor and Tabor (76), the concentration of polyamines rises when growth rate increases. Moreover, polyamines have a high affinity for nucleic acids, stimulate protein synthesis, and stabilize membranes (76). More recently, spermidine has been recognized as a precursor of hypusine (N^{ε}-(4-amino-2-hydroxybutyl)lysine), a post-translationally modified lysine residue of eukaryotic translation initiation factor 5A (eIF-5A) (80).

Transfer of the Deoxyadenosyl Group. This reaction is rare (only 3 examples are presently known) but nonetheless fascinating. The overall process may be regarded formally as a transfer of the deoxyadenosyl moiety to a hydride ion (47) (eq. 14).

$$\text{AdoMet} + H^- \rightarrow \text{L-Methionine} + \text{5'-deoxyadenosine} \quad (14)$$

Pyruvate-formate lyase catalyzes a key step in glucose fermentation in *Escherichia coli* (i.e., pyruvate + CoA → acetyl CoA + formate). The enzyme is inactive in the non-radical (EH) form. Under anaerobic conditions, however, it is converted to the active radical form ($E^\bullet$) by the action of pyruvate-formate-lyase-activating enzyme. The activating enzyme contains iron and utilizes reduced flavodoxin to reductively cleave AdoMet to 5′-deoxyadenosine-5′-yl radical and L-methionine. The 5′-deoxyadenosine-5′-yl radical is then converted to 5′-deoxyadenosine by abstraction of the *pro* (*S*) hydrogen of a specific glycine residue of pyruvate-formate lyase (33). The net reaction is shown in eq. 15.

$$\text{E(gly)H} + \text{AdoMet} + e^- \rightarrow \text{E(gly)}^{\bullet} + \text{5'-deoxyadenosine} + \text{L-methionine} \qquad (15)$$

Aerobic ribonucleotide reductase of *E. coli* is a heterodimer (α_2,β_2) in which the β subunit contains a tyrosine free radical. In contrast, anaerobic (class 3) ribonucleotide reductase (*nrdD*) of *E. coli* is a homodimer which has an iron-sulfur center and a free radical (almost certainly glycine) at the active site of each subunit. The anaerobic enzyme is activated by AdoMet, NADPH and a reducing agent (61). As in pyruvate formate lyase, AdoMet plays a role in the formation of the active site radical.

Clostridium SB4 lysine 2,3-aminomutase catalyzes the interconversion of lysine and β-lysine. The enzyme utilizes pyridoxal 5′-phosphate (PLP), AdoMet, Co(II) and an Fe-S cluster (32, 66) as cofactors. According to Frey and colleagues, activation of lysine-2,3-aminomutase proceeds in two steps. Lysine binds to the enzyme through Schiff's base linkage to PLP; the metal cofactors (Co, FeS) are reduced by reaction with a suitable reductant (glutathione, dihydrolipoate) and react with incoming AdoMet. In the second step, AdoMet is cleaved to 5′-deoxyadenosyl-metal cofactor and enzyme-bound methionine. The 5′-deoxyadenosyl-metal cofactor undergoes reversible homolytic cleavage to yield 5′-deoxyadenosine-5′-yl radical. The 5′-deoxyadenosine -5′-yl radical then initiates a rearrangement reaction by abstracting a hydrogen atom from carbon 3 of the lysine-PLP imine to yield 5′-deoxyadenosine. The substrate-PLP radical then undergoes a 1,2-imino migration to yield β-lysine-PLP imine radical. Abstraction of hydrogen from 5′-deoxyadenosine results in the formation of β-lysine-PLP imine and reformation of the 5′-deoxyadenosine-5′-yl radical. Hydrolysis of the imine releases β-lysine to complete the reaction sequence.

The lysine-2,3-aminomutase reaction has several unusual features. First, the 5′-deoxyadenosine-5′-yl radical is proposed to play a direct role in the aminomutase reaction mechanism, whereas in the pyruvate-formate lyase- and anaerobic ribonucleotide reductase reactions the 5′-deoxyadenosine-5′-yl radical plays an indirect role by maintaining the enzyme in an active (free radical) form. Secondly, although PLP was previously recognized to be the most versatile biological cofactor it was not known to participate in free-radical reactions. With the the recent realization that it can indeed participate in free radical reactions it seems that the versatility of PLP is even greater than previously suspected. Thirdly, AdoMet must now be considered to share catalytic features with adenosylcobalamin in that both are converted to 5′-deoxyadenosine-5′-yl radical during mutase-type reactions. Frey (32) has suggested that during the course of evolution the relatively simple AdoMet may have arisen first as a sole source of 5′-deoxyadenosine-5′-yl radicals. This role was then generally superseded later in evolution by the more elegant, but more complex, adenosyl cobalamin. If this is true, then lysine-2,3-aminomutase may be an evolutionary holdout.

METHIONINE SALVAGE PATHWAY

As mentioned above, several reactions involving scission at the sulfonium center of AdoMet or dAdoMet result in the formation of MTA (Fig. 1). Giulidori et al. (37) estimated from tracer studies that 25% of administered AdoMet in rats is metabolized via the aminopropylation pathway and formation of MTA. If MTA were not metabolized then its synthesis might lead to a drain on scarce sulfur and methyl moieties of methionine in mammals. In mammalian liver, however, MTA is converted back to methionine through a salvage pathway. MTA is also produced in microorganisms and plants in substantial amounts. Indeed, it has been estimated that the methionine concentration in plants is too low to support sustained ethylene production and methionine sulfur and methyl are also salvaged from MTA in plants (28, 58). The process whereby MTA is recycled back to methionine is interesting. In mammals, MTA is reacted with phosphate to yield methylthioribose-1-phosphate (MTRP) plus adenine. MTRP is then converted to α-keto-γ-methiolbutyrate (αKMB), which, in turn, is converted to L-methionine via transamination with glutamine. Cooper and Meister were the first to point out that rat liver and kidney contain aminotransferases capable of catalyzing rapid transamination between glutamine and α-keto-γ-methiolbutyrate (14, 15). It now seems that glutamine (and to a lesser extent asparagine) is also the preferred amino donor in the methionine salvage pathway of bacteria (78) and plants (11, 58). In the mammalian salvage cycle (i.e., Met → AdoMet → dAdoMet → MTA → MTRP → X → αKMB → Met), the original methyl and sulfur of methionine are preserved but the 4-carbon unit of αKMB (and Met) is formed anew from four of the five ribose carbons of MTRP. [See Cooper and Meister (16) for a discussion]. In bacteria the pathway is almost identical. The only difference is that MTA is hydrolyzed to methylthioribose which is then reacted with ATP to yield MTRP (78). The nature of the intermediates (X) in the conversion of MTRP to αKMB in mammals and bacteria has been elucidated by Abeles and colleagues (34, 81). The pathway involves some interesting chemistry. MTRP (α-anomer) is isomerized by aldose-ketose isomerase to methylthioribulose-1-phosphate. This compound is dehydrated and ketonized to 2,3-diketo-5-methylthio-1-phosphopentane, enolized to a phosphoene-diol and hydrolyzed to an aci-reductone ($CH_3SCH_2CH_2C(O)C(OH)=C(O^-)H$). In rat liver, the aci-reductone is cleaved by the action of a dioxygenase to yield αKMB ($CH_3SCH_2CH_2C(O)CO_2^-$) and formate. In *Klebsiella* this reaction probably is non-enzymatic. *Klebsiella* (but not rat liver) possesses a different dioxygenase that converts the aci-reductone to 3-methylthiopropionate, CO and formate.

ELIMINATION REACTIONS INVOLVING BIOLOGICALLY-OCCURRING SULFONIUM COMPOUNDS AND FORMATION OF OLEFINS

As noted above, one pathway for the metabolism of AdoMet in microorganisms begins with its conversion to homoserine lactone by internal nucleophilic displacement by carboxylate ion. The enzyme that catalyzes this reaction in yeast appears to be largely specific for AdoMet (59). Mazelis et al. (55) showed that a soil bacterium possesses an enzyme (methionine sulfonium lyase) that can catalyze the conversion of both AdoMet and SMM to L-homoserine. No evidence was found for homoserine lactone formation in the reaction with SMM and it was suggested that in this case, nucleophilic attack on the γ-carbon is by hydroxide ion and not by carboxylate ion (55). The hydrolase activity with SMM has also been shown to be present in cabbage leaves (50) and onion seedlings (40). Conceivably, the enzyme contains PLP and the reaction proceeds through a β,γ-elimination of the SMM-PLP imine followed by addition of water to the resulting β,γ double bond. However, Mazelis et al. (55) found no evidence that the reaction proceeds through a PLP-imine intermediate. On the other hand, Gessler and Bezzubov (35) reported that SMM lyase partially purifed from cabbage leaves loses all its activity after Sephadex chromatography, but that 20% of the activity may be restored by addition of PLP. Gessler and Bezzubov (35) measured dimethylsulfide production from SMM but did not measure homoserine or its lactone. Therefore, it is not clear whether dimethylsulfide was truly formed via a PLP-dependent lyase. It is possible that the reaction is actually due to PLP-dependent transamination followed by non-enzymatic decomposition of the α-keto acid analogue of SMM (see below). Loss of activity after Sephadex filtration could then have been due to removal of α-keto acid substrate.

Gessler and Bezzubov (35) also noted that rat tissue homogenates were able to convert SMM to dimethylsulfide. Interestingly, activity was stimulated 3-fold and 1.5-fold by addition of pyruvate and α-ketoglutarate, respectively. Therefore, it seems almost certain that the dimethylsulfide arose in large part, if not totally, from transamination followed by non-enzymatic β,γ-elimination (see below). [However, this possibility was not recognized by the authors.] Evidently, more work needs to be carried out on the mechanism of dimethylsulfide formation from SMM in plants and animals.

Several amino acids with good leaving groups in the γ position undergo β,γ-elimination when oxidized at the α-carbon. The C-H bond adjacent to the resulting α-imino (or α-keto acid) is activated, facilitating a β,γ-elimination reaction (43). For example, oxidation of L-methionine-S,R-sulfoximine (18), L-homocysteine (17), and L-canavanine (43) with L-amino acid oxidase results in the formation of methane sulfinamide, H_2S, and hydroxyguanidine, respectively.

This propensity of γ-substituted amino acids to fragment after oxidation at the α carbon has been exploited in the design of prodrugs (i.e. compounds that are initially unreactive but become metabolized to active drug in the target organ). Thus, Elfarra and Hwang (25) have designed several prodrugs of the anticancer drug 6-mercaptopurine that are directed to the kidneys. One of these is the *S*-homocysteine conjugate which is transaminated in the kidney. The resulting α-keto acid spontaneously undergoes β,γ-elimination to yield 6-mercaptopurine. In a similar reaction, *S*-(1,2-dichlorovinyl)-L-homocysteine is transaminated to the corresponding α-keto acid by a rat kidney homogenate. The α-keto acid spontaneously undergoes β,γ-elimination (26).

Finally, Stoner and Eisenberg (75) showed that AdoMet is the amino donor in the transamination of 7-oxo-8-aminopelargonic acid (an intermediate in the biotin biosynthetic pathway) to 7,8-diaminoperlargonic acid in *E. coli*. The expected α-keto acid (*S*-adenosyl-α-keto-γ-methylthiobutyric acid) could not be detected but the elimination fragment (MTA) was identified.

In each of the above mentioned elimination reactions involving "oxidized" γ-substituted amino acids, the expected olefinic product is vinylglyoxylate (2-oxo-3-butenoic acid). This compound has apparently not been synthesized (although vinylglycolate and vinylglycine are well known compounds) and appears to be very unstable, generating a product with a masked carbonyl. However, vinylglyoxylate can be readily trapped with a suitable nucleophile such as 2-mercaptoethanol (18).

Although DMSP has a potentially good leaving group (dimethylsulfonio moiety $[(CH_3)_2S^+\text{-}]$), it is relatively stable at neutral and acid pH and only decomposes to acrylic acid and dimethylsulfide in the presence of base (79). As mentioned above, however, some bacteria and algae possess an enzyme capable of catalyzing the conversion of DMSP to acrylate and dimethylsulfide. Wagner and Daniels (79) suggested that the electrophilic nature of the sulfonium group in DMSP is diminished by its tendency to form an internal salt with the carboxyl group. On the other hand, esters of DMSP do not form this internal salt link, resulting in facile elimination of dimethyl sulfide (79). Thus, the pantotheine ester of DMSP decomposes to acrylylpantotheine and dimethylsulfide in solution at pH values of 5.0 or above.

The above discussion raises some interesting questions regarding the synthesis of DMSP from SMM. Several possible routes to DMSP starting from SMM are theoretically possible (39, 42). 3-Dimethylsulfoniopropylamine, 3-dimethylsulfoniopropionamide and 3-dimethylsulfoniohydroxybutyrate are possible intermediates. However, Hanson and colleagues (42) have presented evidence that these compounds are not on the main pathway from SMM to DMSP in *W. biflora*. On the other hand, evidence was presented that 3-dimethylsulfoniopropionaldehyde (DMSP-ald) fits the bill as a possible intermediate. This compound is unstable, however, and spontaneously decomposes to dimethylsulfide. DMSP-ald is probably present in the plant at naturally very low levels. Interestingly, DMSP-ald and SMM are converted to dimethylsulfide much more rapidly than is DMSP in *W. biflora* leaf disks. This raises the possibilty that biogenic dimethylsulfide production may be a by-product of DMSP biosynthesis (42).

Hanson and colleagues (42) suggest that DMSP-ald arises from SMM via α-carbon oxidation/decarboxylation. This process could occur in one step via decarboxylation-transamination. This reaction would be analogous to that catalyzed by dialkylglycine decarboxylase (77). Alternatively, SMM could be converted to 4-dimethylsulfonio-2-ketobutyrate (DMSKB) by the action of an L-amino acid oxidase, transaminase or a dehydrogenase, followed by decarboxylation. However, because the dimethylsulfonio moiety is an excellent leaving group and the β-CHs in the DMSKB are activated one would predict that this α-keto acid will spontaneously decompose to vinylglyoxylate and dimethylsulfide. Indeed, incubation of SMM with snake venom L-amino acid oxidase and catalase at neutral pH results in the formation of considerably less α-keto acid than ammonia with the concomitant formation of dimethylsulfide (A. J. L. Cooper and A. D. Hanson, unpublished data). Additionally, SMM is a substrate (albeit a poor one) of rat kidney glutamine transaminase K. Vinylglyoxylate was identified in the reaction mixture by trapping with mercaptoethanol (A. J. L. Cooper, unpublished data). This instability of DMSKB will make it difficult to determine whether it is a free intermediate on the pathway of SMM to DMSP.

ACKNOWLEDGMENTS

Some of the work quoted from the author's laboratory was supported by NIH grant DK-16739. I thank Dr. Andrew Hanson for providing me with many helpful suggestions.

REFERENCES

1. Aneja, V. P. and W. J. Cooper. 1989. Biogenic sulfur emissions, pp. 2-13. *In* E. S. Saltzman and W. J. Cooper (eds.) Biogenic sulfur in the environment. American Chemical Society, Washington, D. C.
2. Andersen, K. K. 1981. Stereochemistry and chiroptic properties of organosulfur compounds, pp. 229-312. *In* C. J. M. Stirling (ed.), The Chemistry of the Sulphonium Group. Part 1. Wiley, New York.
3. Attieh, J. M., A. D. Hanson and H. S. Saini. 1995. Purification of a novel methyltransferase responsible for biosynthesis of halomethanes and methanethiol in *Brassica oleracea*. J. Biol. Chem. *270*: 9250-9257.
4. Baddiley, J. and G. A. Jamieson. 1954. Synthesis of "active methionine". J. Chem. Soc. 4280-4284.
5. Baddiley, J., G. L. Cantoni and G. A. Jamieson. 1953. Structural observations on "active methionine". J. Chem. Soc. 2662-2664.
6. Ballou, C. E. 1977. Carbohydrate methyltransferase reactions, pp. 435-450. *In* F. Salvatore, E. Borek, V. Zappia, H. G. Williams-Ashman and F. Schlenk (eds), The Biochemistry of Adenosylmethionine. Columbia University Press, New York.
7. Bisseret, P., S. Ito, P.-A. Tremblay, B. E. Volcani, D. Dessort and M. Kates. 1984. Occurrence of phosphatidylsulfocholine, the sulfonium analog of phosphatidylcholine in some diatoms and algae. Biochim. Biophys. Acta *796*: 320-327.
8. Blau, N. F. and G. G. Stuckwisch. 1957. The conjugative effect of the dimethylsulfonio group in an aliphatic system. J. Org. Chem. *22:* 82-83.
9. Cantoni, G. L. 1952. The nature of the active methyl donor formed enzymatically from L-methionine and adenosinetriphosphate. J. Am. Chem. Soc. *74:* 2942-2943.
10. Cantoni, G. L. 1953. *S*-Adenosylmethionine; a new intermediate formed enzymatically from L-methionine and adenosinetriphosphate. J. Biol. Chem. *204*: 403-406.

11. Chapple, C. C. S., J. R. Glover and B. E. Ellis. 1990. Purification and characterization of methionine:glyoxylate aminotransferase from *Brassica carinata* and *Brassica napus*. Plant Physiol. *94*: 1887-1896.
12. Charlson, R. J., J. E. Lovelock, M. O. Andreae and S. G. Warren. 1987. Oceanic phytoplankton, atmospheric sulphur, cloud albedo and climate. Nature *326*: 655-661.
13. Chillemi, R., A. Patti, R. Morrone, M. Piattelli and S. Sciuto. 1990. The role of methylsulfonium compounds in the biosynthesis of *N*-methylated metabolites in *Chondria coerulescens*. J. Nat. Prod. *53*: 87-93.
14. Cooper, A. J. L. and A. Meister. 1972. Isolation and properties of highly purified glutamine transaminase. Biochemistry *11*: 661-671.
15. Cooper, A. J. L. and A. Meister. 1974. Isolation and properties of a new glutamine transaminase from rat kidney. J. Biol. Chem. *249*: 2554-2561.
16. Cooper, A. J. L. and A. Meister. 1984. Glutamine transaminases, pp. 3-15. *In* A. E. Evangopoulos (ed.), Chemical and Biological Aspects of Vitamin B_6 Catalysis. Part B: Metabolism, Structure, and Function of Transaminases. Alan R. Liss, New York.
17. Cooper, A. J. L. and A. Meister. 1985. Enzymatic oxidation of L-homocysteine. Arch. Biochem. Biophys. 239: 556-569.
18. Cooper, A. J. L., R. A. Stephani and A. Meister. 1976. Enzymatic reactions of methionine sulfoximine. Conversion to the corresponding α-imino and α-keto acids, and to α-ketobutyrate and methane sulfinamide. J. Biol. Chem. *251*: 6674-6682.
19. Cornforth, J. W., S. A. Reichard, P. Talalay, H. L. Carrell and J. P. Glusker. 1977. Determination of the absolute configuration at the sulfonium center of *S*-adenosylmethionine. Correlation with the absolute configuration of the diastereomeric *S*-carboxymethyl-(*S*)-methionine salts. J. Am. Chem. Soc. *99*: 7292-7300.
20. Coward, J. K. and W. D. Sweet. 1971. Kinetics and mechanism of methyl transfer from sulfonium compounds to various nucleophiles. J. Org. Chem. *36*: 2337-2346.
21. de la Haba, G., G. A. Jamieson, S. H. Mudd, and H. H. Richards. 1959. *S*-Adenosylmethionine: The relation of configuration at the sulfonium center to enzymatic reactivity. J. Am. Chem. Soc. *81*: 3975-3980.
22. Doering, W. von E. and A. K. Hoggman. 1955. d-Orbital resonance. III. Deuterium exchange in methyl "onium" salts and in bicyclo[2.2.1]heptane-1-sulfonium iodide. J. Am. Chem. Soc. *77*: 521-526.
23. Dubnoff, J. W. and H. Borsook. 1948. Dimethylthetin and dimethyl β-propiothetin in methionine synthesis. J. Biol. Chem. *176*: 789-796.
24. Ecker, J. R. 1995. The ethylene signal transduction pathway in plants. Science 268: 667-675.
25. Elfarra, A. A. and Y. Hwang. 1993. Targeting of 6-mercaptopurine to the kidneys. Metabolism and kidney-selectivity of *S*-(6-purinyl)-L-cysteine analogs in rats. Drug Metab. Dispos. *21*: 841-845.
26. Elfarra, A. A., L. H. Lash and M. W. Anders. 1986. Metabolic activation and detoxication of nephrotoxic cysteine and homocysteine *S*-conjugates. Proc. Natl. Acad. Sci. USA *83*: 2667-2671.
27. Farooqui, J. Z., M. Tuck and W. K. Paik. 1985. Purification and characterization of enzymes from *Euglena gracilis* that methylate methionine and arginine residues of cytochrome *c*. J. Biol. Chem. *260*: 537-545.
28. Flores, H. E., C. M. Protacio and M. W. Signs. 1989. Primary and secondary metabolism of polyamines in plants, pp. 329-393. *In* J. E. Poulton, J. T. Romeo and E. E. Conn (eds.), Plenum Press, New York.
29. Floss, H. G., L. Mascaro, M-D. Tsai and R. W. Woodard. 1979. Stereochemistry of enzymatic transmethylation, pp. 135-141. *In* E. Usdin, R. T. Borchardt and C. R. Creveling (Eds.), Transmethylation. Elsevier/North-Holland, New York.
30. Franzen, V. and C. Mertz. 1960. Zum Mechanismus der Hofmann-Eliminierung bei Sulfoniumsalzen. Chem. Ber. *93*: 2819-2824.
31. Franzen, V. and H. E. Driessen. 1963. Umsetzung von Sulfonium-Yliden mit polaren Doppelbindungen. Chem. Ber. *96*: 1881-1890.
32. Frey, P. A. 1993. Lysine 2,3-aminomutase: Is adenosylmethionine a poor man's adenosylcobalamin? FASEB J. *7*: 662-670.
33. Frey, M., M. M. Rothe, A. F. Volker Wagner and J. Knappe. 1994. Adenosylmethionine-dependent synthesis of the glycyl radical in pyruvate formate-lyase by abstraction of the glycine C-2 pro *S*-hydrogen atom. Studies of [^{2}H]glycine-substituted enzyme and peptides homologous to the glycine 734 site. J. Biol. Chem. *269*: 12432-12437.
34. Furfine, E. S. and R. H. Abeles. 1988. Intermediates in the conversion of 5′-methylthioadenosine to methionine in *Klebsiella pneumoniae*. J. Biol. Chem. *263*: 9598-9606.

35. Gessler, N. N. and A. A. Bezzubov. 1988. Study of the activity of *S*-methylmethionine sulfonium hydrolase in plant and animal tissues. Prikl. Biokhim. Mikrobiol. *24*: 240-260. (English translation by Plenum, pp. 200-205.)
36. Giovanelli, J., S. H. Mudd and A. H. Datko. 1980. Sulfur amino acids in plants, pp. 453-505. *In* B. J. Miflin (ed.), The Biochemistry of Plants: A Comprehensive Treatise. Vol. 5. Academic Press, New York.
37. Giulidori, P., M. Galle-Kile, E. Catto and G. Stramentinoli. 1984. Transmethylation, transsulfuration, and aminopropylation reactions of *S*-adenosyl-L-methionine *in vivo*. J. Biol. Chem. *259*: 4205-4211.
38. Grue-Sørensen, G., E. Kelstrup, A. Kjær and J. Ø. Madsen. 1984. Diastereospecific, enzymatically catalyzed transmethylation from *S*-methyl-L-methionine to L-homocysteine, a naturally occurring process. J. Chem. Soc. Perkin Trans. *1*: 1091-1097.
39. Hanson, A. D., J. Rivoal, L. Paquet, and D. A. Gage. 1994. Biosynthesis of 3-dimethylsulfoniopropionate in *Wollastonia biflora* (L). Evidence that *S*-methylmethionine is an intermediate. Plant Physiol. *105*: 103-110.
40. Hattula, T. and B. Granroth. 1974. Formation of dimethylsulfide from *S*-methylmethionine in onion seedlings (*Allium cepa*). Sci. Fd. Agric. *25*: 1517-1521.
41. Hoffman, J. L. 1994. Bioactivation by *S*-adenosylation, *S*-methylation, or *N*-methylation. Adv. Pharmacol. *27*: 449-477.
42. James, F., L. Paquet, S. A. Sparace, D. A. Gage and A. D. Hanson. 1995. Evidence implicating dimethylsulfoniopropionaldehyde as an intermediate in dimethylsulfoniopropionate biosynthesis. Plant Physiol. *108*: 1- .
43. Hollander, M. M., A. J. Reiter, W. H. Horner and A. J. L. Cooper. 1989. Conversion of canavanine to α-keto-γ-guanidinooxybutyrate and to vinylglyoxylate and 2-hydroxyguanidine. Arch. Biochem. Biophys. *270*: 698-713.
44. Katz, M. L. and K. O. Gerhardt. 1990. Storage protein in hereditary ceriod-lipofuscinosis contains *S*-methylated methionine. Mech. Ageing Develop. *53*: 277-290.
45. Kawalleck, P., G. Plesch, K. Hahlbrock and I. E. Somssich. 1992. Induction by fungal elicitor of *S*-adenosyl-L-methionine synthase and *S*-adenosyl-L-homocysteine hydrolase mRNAs in cultured cells and leaves of *Petroselinum crispum*. Proc. Natl. Acad. Sci. USA *89*: 4713-4717.
46. Kiene, R. P. 1993. Microbial sources and sinks for methylated sulfur compounds in the marine environment, pp. 15-33. *In* D. P. Kelly and J. C. Murrell (eds.), Microbial Growth on C_1 Compounds. Intercept, Andover, U. K.
47. Knappe, J., and T. Schmitt. 1976. A novel reaction of *S*-adenosyl-L-methionine correlated with the activation of pyruvate formate-lyase. Biochem. Biophys. Res. Commun. *71*: 1110-1117.
48. Knipe A. C. 1981. Reactivity of sulfonium salts, pp. 313-385. *In* C. J. M. Stirling (ed.), The Chemistry of the Sulphonium Group. Part 1. Wiley, New York.
49. Lederer, E. 1977. Biological *C*-alkylation reactions, pp. 89-126. *In* F. Salvatore, E. Borek, V. Zappia, H. G. Williams-Ashman and F. Schlenk (eds), The Biochemistry of Adenosylmethionine. Columbia University Press, New York.
50. Lewis, B.G., C. M. Johnson and T. C. Broyer. Cleavage of *Se*-methylselenomethionine selenonium salt by a cabbage leaf enzyme fraction. Biochim. Biophys. Acta *237*: 603-605.
51. March, J. 1985. Advanced Organic Chemistry. Reactions, Mechanisms and Structure. 3rd Edition, pp. 908-909. Wiley, New York.
52. Maw, G. A. 1981. The biochemistry of sulphonium salts, pp. 703-770. *In* C. J. M. Stirling (ed.), The Chemistry of the Sulphonium Group. Part 2. Wiley, New York.
53. Maw, G. A. and V. du Vigneaud. 1948. Compounds related to dimethylthetin as sources of methyl groups. J. Biol. Chem. *176*: 1037-1045.
54. Maw, G. A. and V. du Vigneaud. 1948. An investigation of the biological behavior of the sulfur analogue of choline. J. Biol. Chem. *176*: 1029-1036.
55. Mazelis, M., B. Levin and N. Mallison. 1965. Decomposition of methyl methionine sulfonium salts by a bacterial enzyme. Biochim. Biophys. Acta *105*: 106-114.
56. McGargey, D. J. and R. E. Christofferson. 1992. Characterization and kinetic parameters of ethylene-forming enzyme from avocado fruit. J. Biol. Chem. *267*: 5964-5967.
57. Mikolajczyk, M. and J. Drabowicz. 1982. Chiral organosulfur compounds. Top. Stereochem. *13*: 333-468.
58. Miyazaki, H. and S. F. Yang. 1987. Metabolism of 5-methylthioribose to methionine. Plant Physiol. *84*: 277-281.
59. Mudd, S. H. 1959. The mechanism of the enzymatic cleavage of *S*-adenosylmethionine to α-amino-γ-butyrolactone. J. Biol. Chem. *264*: 1784-1786.
60. Mudd, S. H. and A. H. Datko. 1990. The *S*-methylmethionine cycle in *Lemna paucicostata*. Plant Physiol. *93*: 623-630.

61. Mullier, E., M. Fontecave, J. Gaillard and P. Reichard. 1993. An iron-sulfur center and a free radical in the active anaerobic ribonucleotide reductase of *Escherichia coli*. J. Biol. Chem. *268*: 2296-2299.
62. Nakamura, H., M. Ohtoshi, O. Sampei, Y. Akashi and A. Murai. 1992. Synthesis and absolute configuration of (+)-gonyauline: A modulating substance of bioluminescent circadian rhythm in the unicellular alga *Gonyaulax polyedra*. Tet. Lett. *33*: 2821-2822.
63. Nakamura, H., K. Fujimaka, O. Sampai and A. Murai. 1993. Gonyol: Methionine-induced sulfonium accumulation in a dinoflagellate *Gonyaulax polyedra*. Tet. Lett. *34*: 8481-8484.
64. Nishimura, S. 1977. Characterization & enzymatic synthesis of 3-(3-amino-3-carboxypropyl)-uridine in transfer tRNA: Transfer of 3-amino-3-carboxypropyl group of adenosylmethionine, pp. 510-520. *In* F. Salvatore, E. Borek, V. Zappia, H. G. Williams-Ashman and F. Schlenk (eds), The Biochemistry of Adenosylmethionine. Columbia University Press, New York.
65. Nriagu, J. O., D. A. Holdway and R. D. Coker. 1987. Biogenic sulfur and the acidity of rainfall in remote areas of Canada. Science *237*: 1189-1192.
66. Petrovich, R. M., F. J. Ruzicka, G. H. Reed and P. A. Frey. 1991. Metal cofactor of lysine-2,3-aminomutase. J. Biol. Chem. *266*: 7656-7660.
67. Salvatore F. And F. Cimino. 1977. Bacterial tRNA methyltransferases: Properties and biological roles, pp. 187-215. *In* F. Salvatore, E. Borek, V. Zappia, H. G. Williams-Ashman and F. Schlenk (eds), The Biochemistry of Adenosylmethionine. Columbia University Press, New York.
68. Schlenk, F. 1965. The chemistry of biological sulfonium compounds. Fortschr. Chem. Org. Naturst. *23*: 61-112.
69. Schlenk, F. 1983. Methylthioadenosine. Adv. Enzymol. *54*: 195-265.
70. Sciuto, S., R. Chillemi, R. Morrone, A. Patti and M. Piattelli. 1989. Dragendorff-positive compounds in some Mediterranean red algae. Biochem. System. Ecol. *17*: 5-10.
71. Shapiro S. K. and A. J. Ferro. 1977. Metabolism of adenosylmethionine during the life cycles of *Enterobacter aerogenes* and *Saccharomyces cerevisae*, pp. 58-82. *In* F. Salvatore, E. Borek, V. Zappia, H. G. Williams-Ashman and F. Schlenk (eds), The Biochemistry of Adenosylmethionine. Columbia University Press, New York.
72. Shapiro S. K. and A. A. Mather. 1958. The enzymatic decomposition of *S*-adenosyl-L-methionine. J. Biol. Chem. *133*: 631-633.
73. Stirling, C. J. M. 1977. Sulfonium salts, pp. 473-525. *In* S. Oae (ed.), Organic Chemistry of Sulfur. Plenum Press, New York.
74. Stirling, C. J. M. 1981. The Chemistry of The Sulphonium Group. Parts 1 and 2. Wiley, New York.
75. Stoner, G. L. and M. A. Eisenberg. 1975. Purification and properties of 7,8-diaminopelargonic acid aminotransferase, an enzyme in the biotin biosynthetic pathway. J. Biol. Chem. *250*: 4029-4036.
76. Tabor, C. W. and H. Tabor.1976. 1,4-Diaminobutane (putrescine), spermidine, and spermine. Annu. Rev. Biochem. *45*: 285-305.
77. Toney, M. D., E. Hohenester, S. W. Cowan, and J. N. Jansonius. 1993. Dialkylglycine decarboxylase structure: Bifunctional active site and alkali metal sites. Science *261*: 756-759.
78. Tower, P. A., D. B. Alexander, L. L. Johnson and M. K. Riscoe. 1993. Regulation of methylthioribose kinase in *Klebsiella pneumoniae*. J. Gen. Microbiol. *139*: 1027-1031.
79. Wagner C. and C. A. Daniels. 1965. Synthesis and properties of dimethyl-β-propiothetin thiolesters. Biochemistry *4*: 2485-2490.
80. Wolff, E. C., M. H. Park and J. E. Folk. 1990. Cleavage of spermidine as the first step in deoxyhypusine synthesis. The role of NAD^+. J. Biol. Chem. *265*: 4793-4799.
81. Wray J. W. and R. H. Abeles. 1995. The methionine salvage pathway in *Klebsiella pneumoniae* and rat liver. Identification and characterization of two novel dioxygenases. J. Biol. Chem. *270*: 3147-3153.
82. Zappia, V., C. R. Zydeck-Cwick and F. Schlenk. 1969. The specificity of *S*-adenosylmethionine derivatives in methyl transfer reactions. J. Biol. Chem. *244*: 4499-4509.

3

CHARACTERIZATION OF 3-DIMETHYLSULFONIOPROPIONATE (DMSP) AND ITS ANALOGS WITH MASS SPECTROMETRY

D. A. Gage[1] and A. D. Hanson[2]

[1] Department of Biochemistry
Michigan State University
East Lansing, Michigan 48824
[2] Horticultural Sciences Department
University of Florida
Gainesville, Florida 32611

SUMMARY

Identification of 3-dimethylsulfoniopropionate (DMSP) most often relies upon indirect methods, such as gas chromatographic analysis of dimethyl sulfide (DMS) released after treatment with base. Conventional electron ionization mass spectrometry (EI-MS) has traditionally not played a major role in the direct characterization of tertiary sulfonium compounds, such as DMSP, because of the low inherent volatility of these molecules. The development of desorption/ionization MS techniques, which do not require the sample to be introduced in the gas phase, has permitted DMSP and its analogs to be analyzed directly by mass spectrometry. Fast atom bombardment mass spectrometry, introduced in the early 1980's, is a simple technique that provides ions representative of the intact molecule with little fragmentation. With stable isotope labeled analogs as internal standards, we have used FAB-MS to routinely identify and quantify DMSP levels in a variety of plant tissues. FAB-MS has also been employed recently to characterize synthetic potential precursors of DMSP and to follow isotope incorporation patterns in biosynthetic studies. Other alternative ionization techniques, including plasma desorption and electrospray ionization mass spectrometry have also shown some promise for the characterization of DMSP. Recently, a new approach based on gas chromatography/mass spectrometry (GC-MS) has been developed to analyze DMSP and its analogs. The carboxyl group is first protected by *t*-butyldi-methysilylation, and after co-injection of the derivatized sample with a catalyst, a nucleophile-assisted on-column demethylation step converts nonvolatile tertiary sul-fonium compounds to their volatile S-methyl analogs. The GC-MS method has an

Biological and Environmental Chemistry of DMSP and Related Sulfonium Compounds
edited by Ronald P. Kiene et al., Plenum Press, New York, 1996

advantage over direct MS analysis, since retention times and mass spectral data provide two dimensional characterization. Further, this method is particularly suitable for mixture analysis and the required instrumentation is widely available.

INTRODUCTION

Direct analysis of 3-dimethylsulfoniopropionate (DMSP) and related sulfonium compounds from biological samples is difficult because of their zwitterionic character and consequent nonvolatility. As a result, DMSP has been traditionally detected and quantified by chromatographic (1,2), electrophoretic (4) or indirect methods. Usually the latter involves analyzing dimethylsulfide (DMS) released from DMSP upon treatment with base (5,18,25). With this approach, biological samples or purified DMSP preparations are incubated in a septum-sealed vial with cold concentrated NaOH for 1-4 hours at room temperature. A headspace sample is then injected into a gas chromatograph and the evolved dimethyl sulfide is then measured by a flame ionization detector (FID) or flame photometric detector (FPD). The indirect assay is sensitive, with linear detection between 0.005-2 μmol of DMSP per vial (17). Although this indirect procedure has low detection limits and is easy to carry out, detection of DMS does not unequivocally demonstrate the presence of DMSP (25). Other sulfonium compounds, such as S-methylmethionine (SMM) will also yield DMS when subjected to treatment with base, albeit in some cases with different efficiency. Thus, direct methods to identify and quantitate DMSP and other sulfonium compounds are preferable.

In this chapter, we focus on the use of mass spectrometry for the structural characterization of sulfonium compounds. Although direct analysis of DMSP and its analogs by conventional mass spectrometry using electron impact ionization is not possible because of the nonvolatile character of tertiary sulfonium compounds, alternative MS-based approaches are capable of characterizing their intact structures. We will discuss the use of three different MS methodologies: fast atom bombardment desorption ionization, electrospray ionization (ESI) and a novel gas chromatography-mass spectrometry (GC-MS) technique involving an "on-column" demethylation to produce volatile S-methyl analogs of dimethylsulfonium compounds. Another useful technique, plasma desorption mass spectrometry (PD-MS), will be not be covered here, as it is the subject of another chapter of this volume (22, see also reference 3). With their structural specificity, these analytical methods are becoming increasingly important for studies of the biosynthesis of DMSP (9,10,13). The ability to follow stable isotope incorporation patterns from labeled precursors to intermediates is important in establishing the pathway to DMSP (7, 10, 13). Accurate and sensitive analytical methods are also important in surveying organisms for the presence of DMSP and novel sulfonium compounds (17). We will also describe how novel mass spectrometric techniques were used to investigate the origin of a DMSP analog, dimethylsulfoniopentanoate in *Diplotaxis tenuifolia* (L.)DC (8).

SAMPLE PREPARATION

For our studies of DMSP, we have employed a simple purification scheme (6,17,20) that provides preparations that are suitable for direct mass spectrometric analysis. Briefly, this procedure involves extraction with methanol:chloroform:water. The aqueous phase is then passed through a mixed bed ion exchange resin (Dowex-1-OH^- and Biorex-70-H^+) in tandem with a second Dowex 50 column, H^+ form. Zwitterions, such as DMSP, pass through the mixed bed resin, and are retained on the Dowex 50 column. DMSP and other zwitterions can then be eluted from the Dowex 50 column with 2.5 M HCl. The eluant is then lyophilized and taken up in water prior to analysis. If excessive salt is present in the sample, an additional mixed bed resin step (eluting with water) can be added.

FAST ATOM BOMBARDMENT MASS SPECTROMETRY (FAB-MS)

Unlike conventional electron ionization mass spectrometry (EI-MS), the fast atom bombardment experiment does not depend upon introducing the sample in the gas phase. Therefore, it is particularly suitable for characterizing nonvolatile compounds, such as DMSP. For FAB-MS analysis, 1-2 μl of a solution of the sample to be analyzed are mixed on the FAB probe tip with 1-2 microliters of a viscous, nonvolatile matrix in which the analyte is soluble, such as glycerol. The sample probe is then introduced through a vacuum lock into the mass spectrometer's ion source where the target is bombarded with a beam of high energy atoms or ions (typically 6 keV ^{0}Xe or 10-25 keV Cs^{+}). This process results in desorption of matrix-solvated analyte molecules, which are then ionized by proton transfer from the matrix during desolvation. It should be noted that the FAB desorption/ionization process as described here oversimplifies the actual events involved in analyte ionization. Whether the protonation occurs in the gas phase or in solution is still not clear. In addition, for some classes of molecules, odd electron ($M^{+\cdot}$) species can be produced with some matrices. This description also does not address the formation of negative ions, $[M-H]^{-}$ or $M^{-\cdot}$, which can be detected by FAB-MS.

Once gas phase ions are produced, conventional mass spectral analysis is performed. In the case of DMSP, the protonated zwitterion, MH^{+}, with a net positive charge, is formed and appears in the spectrum at m/z 135. If the sample contains a sufficient quantity of salts, metal cation adducts ($[M+Na]^{+}$ or $[M+K]^{+}$) can also be observed, and these can dominate the spectrum if salt concentrations are high. Fragments formed from the MH^{+} ion are not

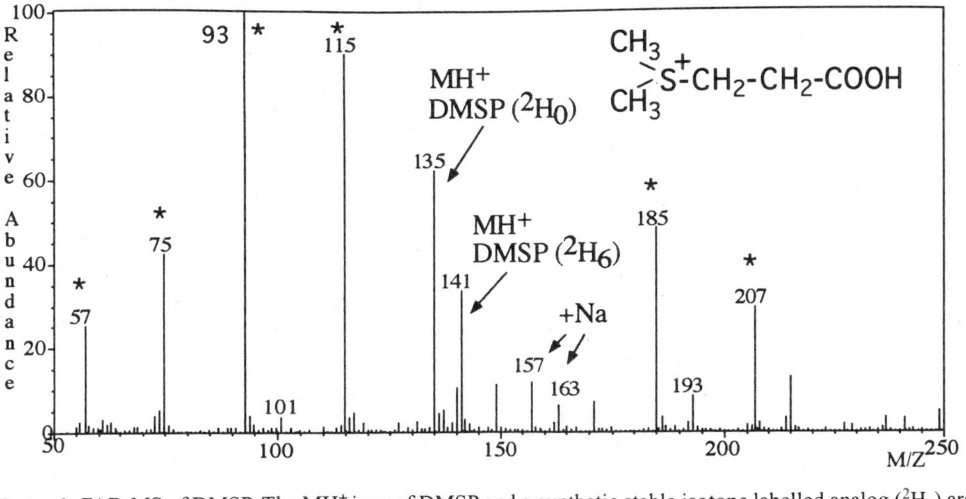

Figure 1. FAB-MS of DMSP. The MH^{+} ions of DMSP and a synthetic stable isotope labelled analog ($^{2}H_{6}$) are detected at m/z 135 and 141, respectively. Sodium adducts, $[M+Na]^{+}$, for both compounds are also present in the spectrum at m/z 157 and 163. These data were obtained in a glycerol matrix and matrix-related peaks are marked with an asterisk.

prominent in the spectrum of DMSP, although this is not necessarily true of other sulfonium analogs (see below).

Two factors enhance the detectability of analyte molecules in FAB-MS analysis. First, in positive ion mode analysis, molecules with fixed positive charges are preferentially detected, even in mixtures. The dimethyl sulfonium moiety of DMSP and its analogs provides this advantage, although the net positive charge is acquired only after protonation of the zwitterion. The second factor that enhances detectability is a high concentration of the analyte on the surface of the matrix solution droplet. This is because desorption occurs primarily from the surface of the matrix. In the hydrophilic matrices required to solubilize polar analytes, enhancing the hydrophobic character of the analyte can increase surface concentration. Thus, the preparation of butyl ester derivatives of zwitterionic betaines prior to their analysis by FAB-MS significantly improves detectability (19,20). For DMSP, however, the reaction conditions for esterification promote elimination of DMS, so that this approach cannot be used. Nevertheless, there is sufficient sensitivity to readily detect DMSP at the μmol level without derivatization (9,17,24). Detection limits are well below 100 nmol applied to the probe.

Quantification of DMSP and related compounds by FAB-MS is possible with the use of internal standards. We have typically used stable isotope-labeled analogs of DMSP, such as the d_6 ($S\text{-}(C^2H_3)_2$) or d_3 ($S\text{-}C^2H_3,CH_3$) compounds (9,17,24) (figure 1). These internal standards can be readily synthesized from commercially available precursors (10,17). Peaks are observed in the FAB spectrum at m/z 141 (MH^+ of $^2H_6$-DMSP) and m/z 138 (MH^+ of $^2H_3$-DMSP) for these two deuterated standards. The use of other labeled analogs for quantification is possible, but for maximum accuracy at low levels, compounds differing by only two mass units (*e.g.*, $^{13}C_2$- or $^2H_2$-labeled DMSP) should be avoided due to interference from the natural abundance ^{34}S (+2) isotope of DMSP (giving a peak at m/z 137 that is approximately 4% of the unlabeled MH^+ peak at m/z 135). Quantification of DMSP is performed by adding a known quantity of the internal standard to the material to be analyzed before extraction or sample workup. Following FAB-MS analysis, the peak areas of the MH^+ ions of unlabeled DMSP and the labeled internal standard are measured. In conjunction with a standard curve, the ratio of these areas can be used to determine the level of DMSP in the sample. Using 1 μmol of a stable isotope labeled analog, the standard curve is linear between 100 nmol and 2 μmol of DMSP, and is usable up to 5 μmol.

FAB-MS IN BIOSYNTHETIC STUDIES OF DMSP

Isotope tracer studies are critical to the elucidation of biosynthetic pathways. We have investigated the biosynthetic route to DMSP with both radiolabeled and stable isotope-labeled precursors and intermediates. For these studies, FAB-MS has played several roles. First, this mass spectrometric technique has been used for confirmation of the isotopic enrichments and structures of putative precursors and intermediates prepared by synthesis. For synthetic radiolabeled precursors, isotope enrichment is insufficient (less than 0.001% enrichment on an atom molar basis) to detect by conventional mass spectrometric means, but FAB-MS has been useful in optimizing synthetic protocols with cold reagents prior to the radiosynthesis. A second application has been to follow the pattern of isotope incorporation by analysis of products and intermediates following stable isotope-labeled precursor feeding studies.

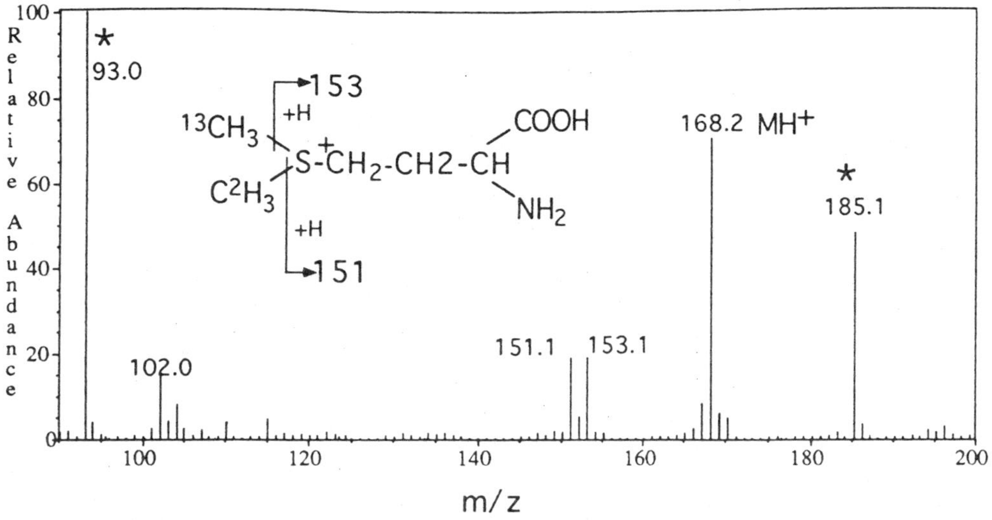

Figure 2. AB-MS of synthetic [C^2H_3,$^{13}CH_3$]-S-methylmethionine. The protonated molecule is detected at m/z 168. Unlike DMSP, fragments from the loss of the S-methyl groups of SMM (m/z 151.1 and 153.1) are observed in the spectrum, since the amino group in remaining portion of the molecule provides a site for protonation. These fragments help to confirm the labelling pattern of the synthetic product.

STRUCTURAL CONFIRMATION OF SYNTHETIC PRECURSORS

The validity of isotope labeling results in biosynthetic studies is dependent on feeding precursors or intermediates of known structure. Administering a mixture of compounds, rather than a single desired precursor can easily confound interpretations of precursor-product relationships. Thus, we have routinely used FAB-MS to characterize synthetic intermediates before feeding studies. The spectrum of the labeled analog of S-methylmethionine (SMM) containing one trideuterio S-methyl group (C^2H_3) and one ^{13}C -labeled S-methyl group ($^{13}CH_3$) is shown in Figure 2. The spectrum displays an MH^+ ion at m/z 168, confirming that no partially labeled intermediates were formed during the synthesis. In addition, two fragments in the spectrum at m/z 151 and 153 confirm the expected labeling of the two S-methyl groups. Other putative synthetic intermediates (Figure 3) have also been synthesized for these biosynthetic studies and structurally characterized by FAB-MS (10,13). However, not all possible intermediates are sufficiently stable to synthesize for feeding studies or FAB-MS analysis. Dimethylsulfonioα-ketobutyrate (DMSKB) is one such compound. Another intermediate, DMSP aldehyde (DMSP-ald), has a very short half life in solution, but was stable enough to analyze by FAB-MS (Figure 4). The FAB spectrum shows an M^+ ion for DMSP-ald at m/z 119 (desorption of DMSP-ald from the FAB matrix yields a positively-charged species without protonation) as well as a glycerol matrix adduct at m/z 211 ($[M+92]^+$) and the diethylacetal (formed by exposure to trace levels of ethanol during purification) at m/z 193. Although the quantity of the diethylacetal in the sample is small, its response (*i.e.*, peak intensity) in FAB-MS is significantly greater than that of the parent aldehyde because of its increased hydrophobicity. A fragment peak in the spectrum at m/z 103 is formed by loss of a methyl group and

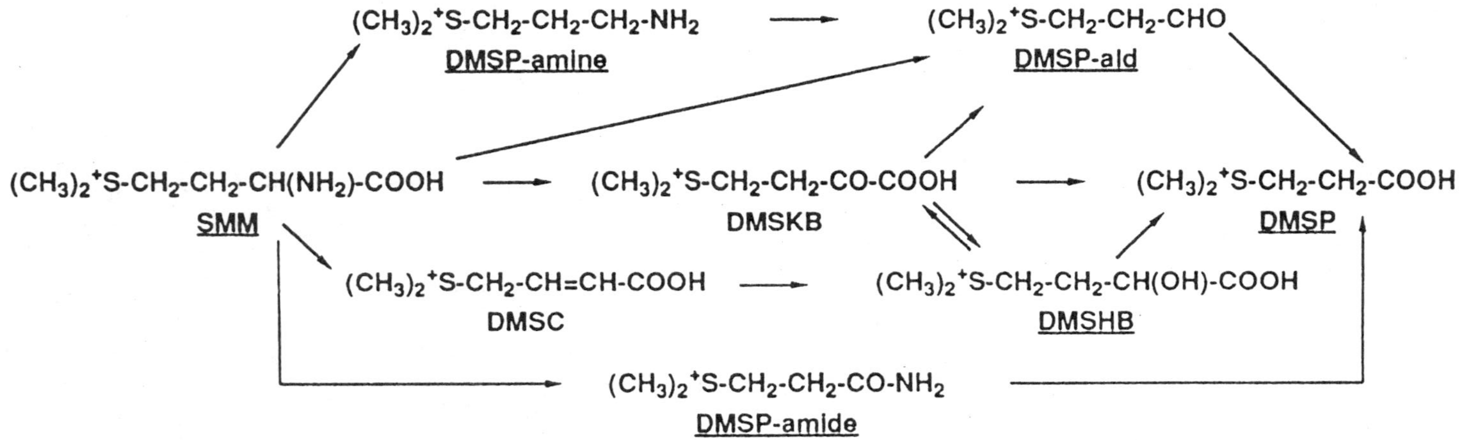

Figure 3. Possible biosynthetic routes to DMSP from methionine. Underlined compounds have been synthesized and characterized by FAB-MS for use in feeding studies (7).

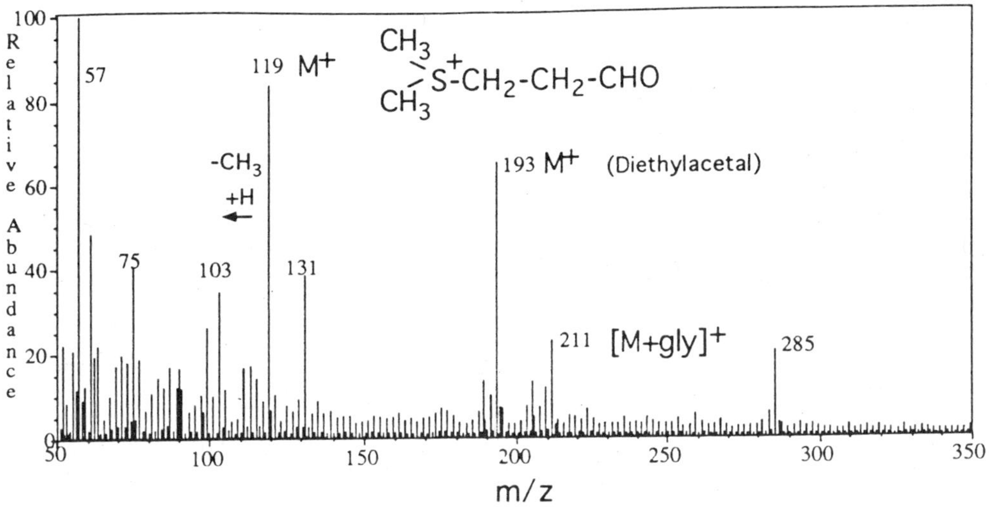

Figure 4. FAB-MS of the unstable biosynthetic intermediate DMSP aldehyde (DMSP-ald) in glycerol. The molecular ion is observed at m/z 119 and a second peak representing the M^+ ion of the diethylacetal, formed during sample workup, is found at m/z 193. Glycerol adducts to these two compounds are seen at m/z 211 and 285.

hydrogen to leave a methylenesulfoniopropionaldehyde fragment. By monitoring DMSP-ald with FAB-MS under different solvent conditions, we were able to optimize the stability of this intermediate prior to leaf disk feeding studies (13).

FOLLOWING STABLE ISOTOPE INCORPORATION WITH FAB-MS

Following isotope incorporation provides the advantage of structural specificity in the analysis of labeled products. The ability to determine labeling patterns, as well as isotope enrichment, has been helpful in distinguishing direct *vs* indirect precursor-product relationships. A good example of the power of this analytical approach was the demonstration that SMM was the first committed precursor in DMSP biosynthesis (7,10). In experiments with *Wollastonia biflora* leaf disks, feeding a stable isotope analog of SMM, containing a ^{13}C label in one S-methyl group and a trideuterio label in the other, produced a time dependent increase in labeled DMSP (figure 5). Initially at T_0, only the constitutive unlabeled DMSP (m/z 135) was observed in the spectrum and no labeled DMSP derived from this putative precursor (m/z 139) could be detected. However, a peak corresponding to DMSP with the SMM precursor's labeling pattern was detected after one day (T_1) and increased as a relative proportion of total DMSP on the second day (T_2). No evidence for partially labeled products ([^{13}C]-DMSP or [$^2H_3$]-DMSP was observed. That the methyl labeling pattern was maintained in the product, without any evidence of methyl group scrambling, strongly supports the notion that SMM is a direct precursor of DMSP. If the C^2H_3, $^{13}CH_3$ methyl group pairs were not retained in the DMSP produced, SMM might have contributed to DMSP synthesis only indirectly, via the SMM cycle (15). These data clearly indicate that this was not the case in *Wollastonia biflora*.

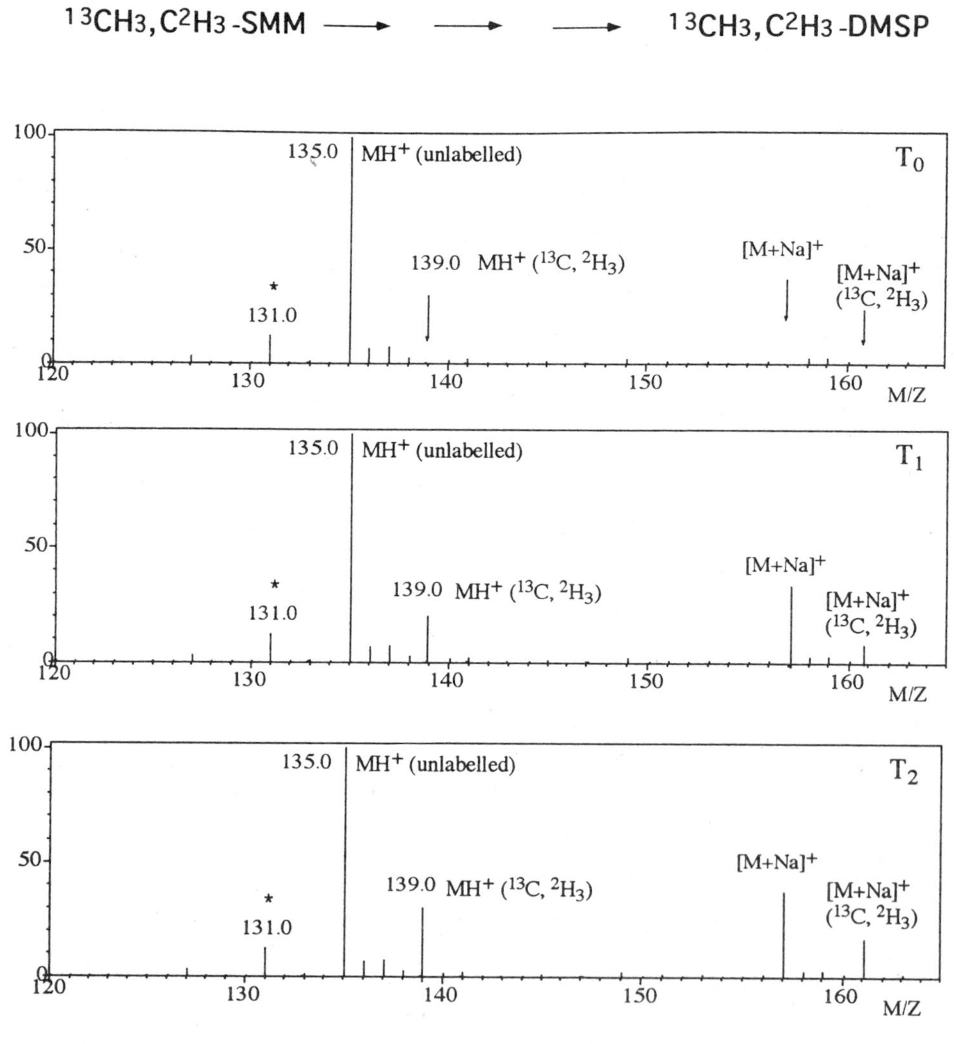

Figure 5. FAB-MS data from time course feeding experiments with C^2H_3,$^{13}CH_3$-labelled SMM. These data confirm that the double label is incorporated in DMSP without scrambling (see text for discussion).

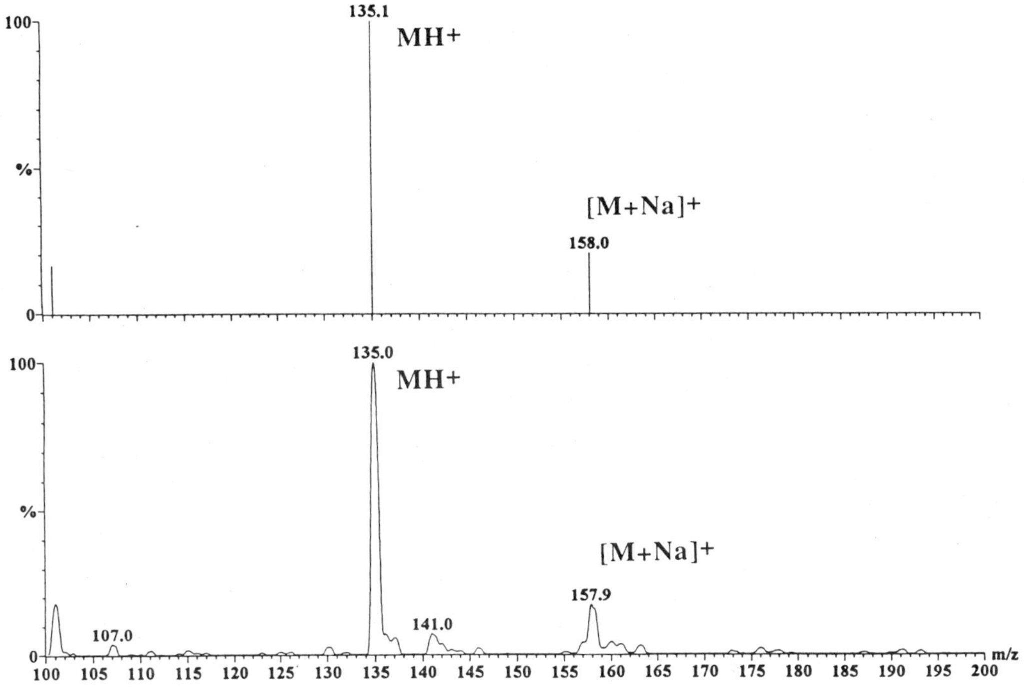

Figure 6. ESI-MS of DMSP. The spectrum was obtained by averaging spectra acquired during infusion of a 10 μl sample containing 13.6 ng (10 pmol) of DMSP. Analog spectral data are shown below.

ELECTROSPRAY IONIZATION MASS SPECTROMETRY (ESI-MS)

This extremely sensitive mass spectrometric ionization technique (23) can potentially lower the limits of detection for DMSP and sulfonium analogs by several orders of magnitude relative to FAB-MS. The sensitivity of this technique results from highly efficient ionization of the analyte. In ESI-MS the sample is introduced in acidic aqueous solution, and sprayed via a fine needle, held at high electrical potential, into a region of the ion source at relatively high pressure. In this process charged droplets are produced containing solvated analyte molecules. As the droplets are moved further into the lower pressure regions of the ion source under the influence of an electric field, evaporation increases charge density in the droplets to a critical point where the droplets explode into smaller droplets. This process continues until only desolvated, charged analyte molecules remain. Thus, nearly every molecule introduced is ionized and available for mass analysis. For DMSP 10 μl of a 1 pmol/μl solution (1 pmol=1.3 ng) produces a strong MH^+ ion (Figure 6). A sodium adduct, $[M+Na]^+$, is also detected at m/z 157. Because carrier solvents containing high ionic strength buffers or salts can interfere with the electrospray ionization process, it is likely that some sample cleanup will always be necessary prior to analysis of DMSP samples. Detection limits for ESI-MS analysis of DMSP are remarkable; reasonable signals are obtained even from low fmol/μl solutions (data not shown). Thus, the sensitivity of ESI-MS is sufficient to detect DMSP directly in seawater, but it is likely that some microscale desalting procedure will have to developed before this mass spectrometric technique can be used for this purpose.

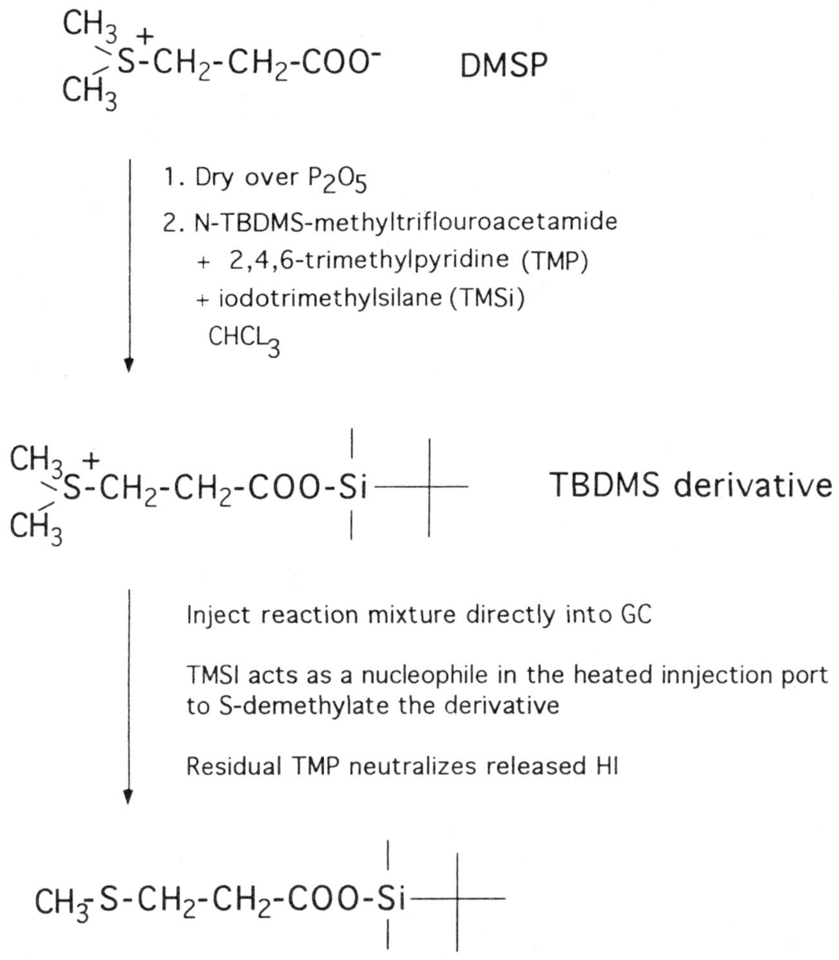

Volatile S-demethylated product

Figure 7. Reaction scheme for the derivatization and catalytically-assisted "on-column" demethylation of DMSP for GC-MS analysis.

GAS CHROMATOGRAPHY MASS SPECTROMETRY OF S-METHYL DERIVATIVES OF DIMETHYLSULFONIUM COMPOUNDS

One of the benefits of combined gas chromatography-mass spectrometric analysis, is that a compound's identification can be confirmed by both its chromatographic retention time and mass spectrometric characteristics. This can be particularly useful in the analysis of mixtures. However, molecules containing a sulfonium or quaternary ammonium group are generally insufficiently volatile to be directly amenable to GC analysis. To overcome this limitation, we recently developed a procedure to convert DMSP to a volatile derivative suitable for GC-MS (8,17). The procedure involves two steps: the conversion of the carboxyl group of DMSP to a *t*-butyldimethylsilyl (TBDMS) ester, and following injection into the

GC-inlet, an "on-column," catalytically-assisted S-demethylation (figure 7). The product, S-methyl, *t*-butyldimethylsilylpropionate, has good chromatographic properties and can be readily analyzed by conventional electron ionization mass spectrometry. The approach used here is based on a method developed to analyze acylcarnitines (3-acyloxy, 4-trimethylaminobutyrates) (11,12).

The experimental procedure for GC-MS analysis of DMSP is relatively simple. The sample is first taken to dryness and then further dried over P_2O_5 *in vacuo*. For µg level samples, the TBDMS derivative is prepared by adding 20 µl of a freshly made solution containing N-(*t*-butyldimethylsilyl), N-methyltriflouro-acetamide/ 10% 2,4,6-trimethylpyridine in $CHCl_3$/ 10% 2,2-dimethoxypropane in $CHCl_3$ 10% iodotrimethylsilane in $CHCl_3$ (10:2:1:1, v/v/v/v) (17). The silylation reaction is complete within 5 minutes at room temperature and the derivative is stable for several hours. The TBDMS derivative is preferred over the commonly used trimethylsilyl (TMS) derivative, since the former is less subject to hydrolysis in the presence of residual water or high humidity.

For the GC-MS analysis, a 1 µl aliquot of the reaction mixture is injected directly into the GC inlet (splitless, 260°C). The iodotrimethysilane in the reagent mixture acts as the demethylation catalyst upon injection in the heated inlet, forming the volatile S-monomethyl analog of the TBDMS ester of DMSP. A stronger nucleophile, I^-(in the form of KI), was previously used to promote demethylation in the GC-MS analysis of acylcarnitines (11), but this was found to cause degradation of DMSP. Typically in our laboratory, a methyl silicone (0.25 µm film thickness) capillary column (15 m X 0.32 mm id DB-1, J&W Scientific, Folsom CA) is used for the separation; this is connected *via* a heated transfer line directly to the ion source of a JEOL AX-505 mass spectrometer. The initial column temperature is held at 100°C for 2 minutes and then ramped to 300°C at 10°/minute. The S-monomethyl derivative of DMSP elutes in approximately 8 minutes (figure 8a).

The mass spectrum of the volatile DMSP derivative is obtained by conventional electron ionization mass spectrometry (EI-MS), using either 70 eV, or preferably 30 eV, ionization energy. The mass spectrum of the peak corresponding to the DMSP derivative is shown on figure 8b. The molecular ion ($M^{+\cdot}$) for the S-methyl, *t*-butyldimethylsilylpropionate (expected at m/z 234) is not detected, but high mass ions corresponding to $[M\text{-}15]^+$ at m/z 219 (loss of methyl group) and $[M\text{-}57]^+$ at m/z 177 (loss of the *t*-butyl group, $(CH_3)_3C.$) are observed. Additional diagnostic fragments are seen in the spectrum at m/z 129, $[M\text{-}(CH_3)_3C\text{-}CH_3SH]^+$, 103, $[CH_3SCH_2CH_2CO]^+$, 75, $[CH_3S{=}CHCH_3]^+$, and 61, $[CH_3S{=}CH_2]^+$. Note that the low abundance peak at m/z 219 may not be detected when higher ionization energies (*e.g.*, 70 eV) are used.

Betaines similarly form their corresponding TBDMS esters and with the derivatizing reagents and undergo N-demethylation in the injection port ("on-column") to yield volatile N,N-dimethyl analogs (17). We have used this procedure to analyze sugarcane samples containing both glycine betaine and DMSP (17). Two separate peaks are detected in the chromatogram for the glycine betaine and DMSP derivatives. A diagnostic mass spectrum is obtained for glycine betaine (data not shown, see 17). With appropriate internal standards, the methodology can be readily used to quantitate onium compounds in mixtures. The common availability of GC-MS instrumentation and simplicity of the derivatization protocol should permit this analytical approach to be widely used for the direct analysis of DMSP and its analogs.

SULFONIUM ANALOGS IN *DIPLOTAXIS TENUIFOLIA*

Flowers of the plant, *Diplotaxis tenuifolia* (L.) DC. (Cruciferae) have been reported to produce 5-dimethylsulfoniopropionate, a homolog of DMSP (14). We used both FAB-MS

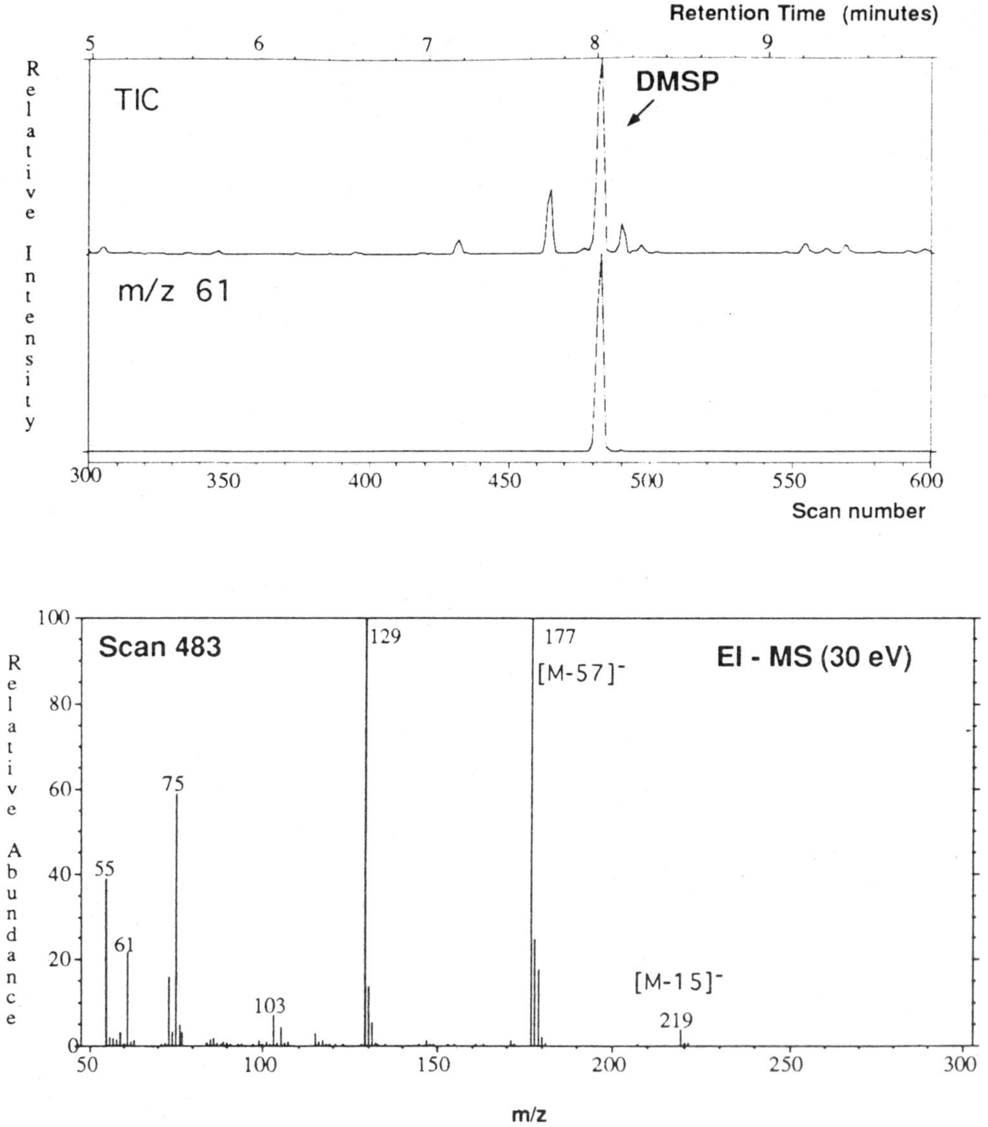

Figure 8. GC-MS analysis of derivatized DMSP (17). Figure 8a (top panel) shows the total ion and the mass chromatogram for the diagnostic ion at m/z 61. Figure 8b (lower panel) is the mass spectrum obtained from the derivatized DMSP (S-methyl, *t*-butyldimethylsilylpropionate) peak. See text for discussion.

and the newly developed GC-MS methodology to investigate the production of sulfonium compounds in this species and related salt tolerant crucifers (8). When our standard extraction and purification procedure was employed, no zwitterionic compounds could be detected by either MS method. In line with these results, TLC analysis of the initial extract indicated that choline was the only Dragendorff-positive compound present. The extraction procedure used in the original study differed from ours in that the aqueous fraction from the initial methanol:chloroform:water extract was heated for 16-20 hours in 6N HCl (14). Using this procedure, a new Dragendorff-positive spot was detected by TLC, and FAB-MS analysis of the zwitterionic fraction obtained after ion exchange chromatography indicated the presence of two DMSP analogs. The spectrum displayed peaks at m/z 163 and 177 (data not shown), which represented the MH^+ ions of dimethylsulfoniopentanoate and its higher homolog, dimethylsulfoniohexanoate. The structures of these two compounds were further characterized by the GC-MS method (8). Two chromatographic peaks (P′ and H′ in figure 9a) corresponding to the sulfonium compounds were observed and their EI-MS spectra confirmed the expected derivative structures (figures 9b and 9c). The spectrum of the first component displayed peaks at m/z 247 and 205 for the $[M\text{-}15]^+$ and $[M\text{-}57]^+$ ions, respectively, of the S-methyl, *t*-butyldimethylsilylpentanoate (figure 9b). Other fragments in the

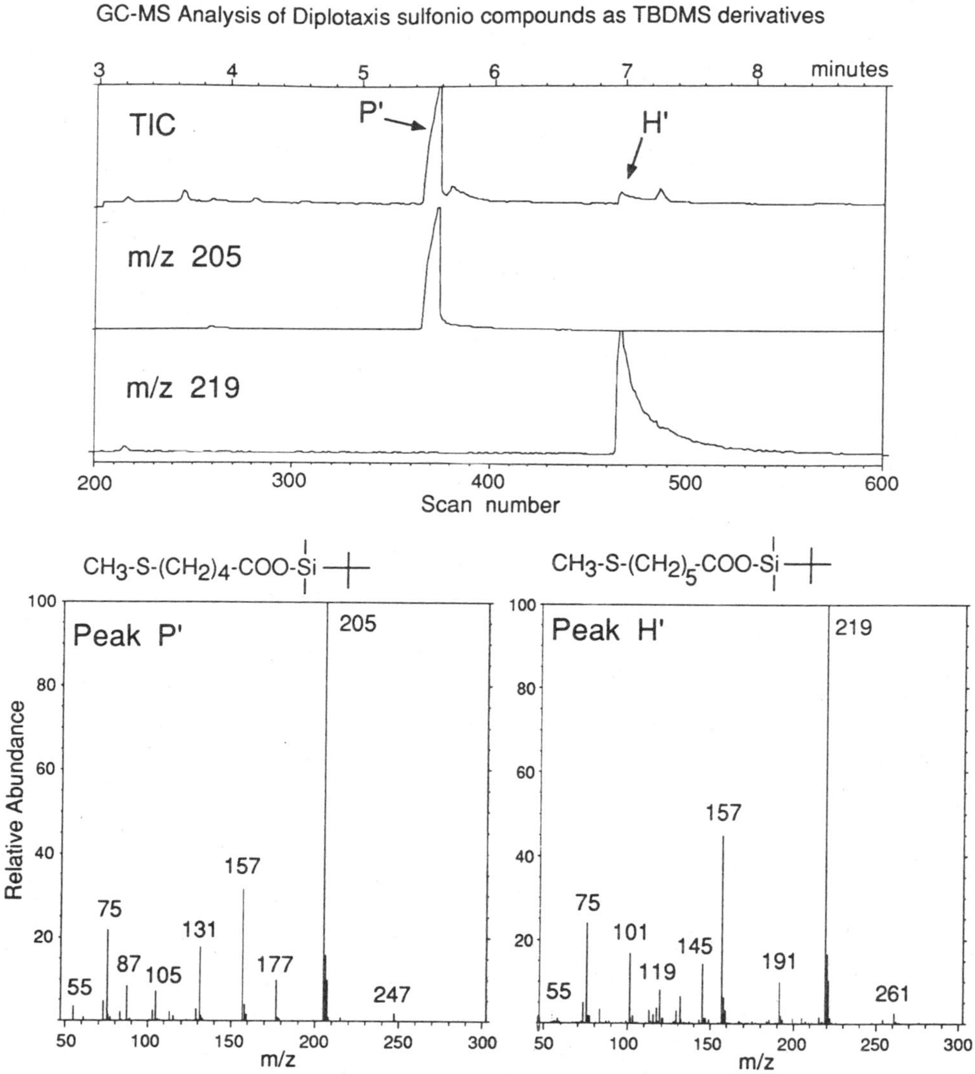

Figure 9. GC-MS analysis of derivatized sulfonium compounds isolated from *Diplotaxus tenuifolia* (8). In figure 9a (top panel) peaks P′ and H′, represent the S-methyl, *t*-butyldimethylsilyl derivatives of dimethylsulfoniopentanoate and dimethylsulfoniohexanoate, respectively. The spectra of these two components are shown below (figures 9b and 9c).

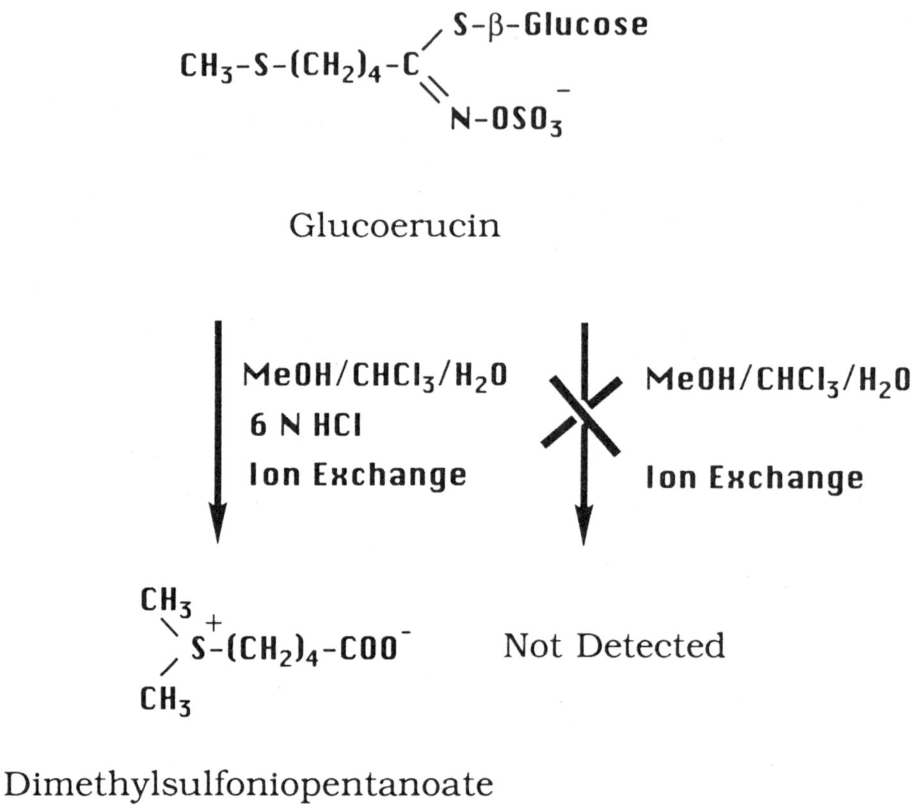

Figure 10. The formation of dimethylsulfoniopentanoate from the glucosinolate, glucoerucin, upon heating with 6N HCl. The sulfonium compound is not formed if 6N HCl is omitted from the extraction procedure.

spectrum at m/z 157, $[M-(CH_3)_3C-CH_3SH]^+$, 131, $[CH_3S(CH_2)_4CO]^+$, and 75, $[CH_3S{=}CHCH_3]^+$, supported this structural assignment. The spectrum of the second sulfonium derivative contained high mass peaks at m/z 261 and 219, representing $[M-15]^+$ and $[M-57]^+$, respectively, of the S-methyl, *t*-butyldimethylsilyl derivative of 6-dimethylsulfoniohexanoate (figure 9c). As expected, the fragments at m/z 157 and 75 were also present in the spectrum of this second sulfonium compound's S-methyl derivative. The appearance of a new fragment at m/z 145, $[CH_3S(CH_2)_5CO]^+$, provided additional evidence that this peak was the C_6 homolog of DMSP.

Because the two sulfonium compounds were detected in *Diplotaxus tenuifolia* and not in related crucifers, and further, were observed only after heating with 6N HCl, it seemed likely that these DMSP analogs were produced during the extraction procedure from precursors restricted to this species. The glucosinolates, glucoerucin and glucoberteroin, were likely candidate precursors (17). Treatment of an authentic standard of glucoerucin with 6N HCl yielded dimethylsulfoniopentanoate (figure 10), confirming the artifactual origin of this DMSP homolog in *Diplotaxus tenuifolia*. The detailed structural information provided by the GC-MS procedure makes it particularly suited for investigation of these and other novel sulfonium compounds.

CONCLUSIONS

Mass spectrometric techniques are now available for the direct characterization of DMSP and its analogs. Newer ionization methods, including fast atom bombardment (FAB) and electrospray ionization (ESI), provide a means to identify these nonvolatile zwitterions by producing MH^+ ions, representing the intact molecules. FAB-MS also has been useful in stable isotope labeling studies to investigate the biosynthesis of DMSP, both to characterize stable isotope labeled precursors for feeding studies and to analyze labeling patterns in products. To determine the structure of DMSP homologs, particularly in mixtures, a novel GC-MS procedure has been developed. Each of these techniques has a role to play in future investigations of DMSP in living organisms and in the environment.

ACKNOWLEDGMENTS

The authors would like to thank Beverly Chamberlin for the FAB-MS analyses and Jim Bradford for the ESI-MS spectra. This research was supported by grants from the Natural Sciences and Engineering Council of Canada and from the National Institutes of Health. A grant (P41-RR00480) from the latter agency supports, in part, the MSU Mass Spectrometry Facility through the National Center for Research Resources, Biotechnology Research Technology Program.

REFERENCES

1. Blunden, G., M.M. ElBarouni, S.M. Gordon, W.F.H. McLean and D.J. Rogers. 1981. Extraction, purification and characterization of Dragendorf-positive compounds from some British marine algae. Botanica Mar. *24:* 451-456.
2. Blunden, G. and S.M. Gordon. 1986. Betaines and their sulfonio analogues in marine algae. Prog. Phycol. Res. *4:* 39-80.
3. Bonham, C.C., K.V. Wood, W.-J. Yang, A. Nadolska-Orczyck, Y. Samaras, D.A. Gage, J. Poupart, M. Burnet, A.D. Hanson and D. Rhodes. 1995. Identification of quaternary ammonium and tertiary sulfonium compounds by plasma desorption mass spectrometry. J. Mass Spectrom. (in press)
4. Gorham, J., S.J. Coughlan, R. Storey and R.G. Wyn-Jones. 1981. Estimation of quaternary ammonium and tertiary sulfonium compounds by thin layer electrophoresis and scanning reflectance densitometry. J. Chromatogr. *210:* 550-554.
5. Grone, T. and G.O. Kirst. 1992. The effect of nitrogen deficiency, methionine and inhibitors of methionine metabolism on the DMSP contents of *Tetraselmis subcordiformis* (Stein) Mar. Biol. *112*, 497-503.
6. Hanson, A. D. and D. A. Gage. 1991. Identification and determination by fast atom bombardment mass spectrometry of the compatible solute choline-*O*-sulphate in *Limonium* species and other halophytes. Aust. J. Plant Physiol. *18:* 317-327.
7. Hanson, A. D., and D. A. Gage. 1996. 3-Dimethylsulfoniopropionate biosynthesis and use by flowering plants. *In* R. P. Kiene, P. T. Visscher, M. D. Keller, and G. O. Kirst (ed.), Biological and environmental chemistry of DMSP and related sulfonium compounds. Plenum, New York.
8. Hanson, A. D., Z. H. Huang and D. A. Gage. 1993. Evidence that the putative compatible solute 5-dimethylsulfoniopentanoate is an extraction artifact. Plant Physiol. *101*: 1391-1393.
9. Hanson, A. D., J. Rivoal and D. A. Gage. 1993b. Synthesis and accumulation of 3-dimethylsulfonio-propionate in the halophyte *Wedelia biflora*. Plant Physiol. *102:*(S) 905.
10. Hanson, A. D., J. Rivoal, L. Paquet and D. A. Gage. 1994. Biosynthesis of 3-dimethylsulfoniopropionate in *Wollastonia biflora* (L.) DC: Evidence that S-methylmethionine is an intermediate. Plant Physiol. *105:* 103-110.
11. Huang, Z. H., D. A. Gage, L. L. Bieber and C. C. Sweeley. 1991. Analysis of acylcarnitines as their N-demethylated ester derivatives by gas chromatography-chemical ionization mass spectrometry. Anal. Biochem. *199:* 98-105.

12. Huang, Z. H., C. C. Sweeley, L. L. Bieber, D. A. Gage, C. Tallarico, G. Bruno and A. Marzo. 1992. A general method for the simultaneous analysis of free carnitine and acylcarnitines by GC-MS of their N-demethylated silylated derivatives. *In* Proceedings of the 40th ASMS Conference on Mass Spectrometry and Allied Topics, Washington, DC, p. 1091-1092.
13. James, F., L. Paquet, S. A. Sparace, D. A. Gage and A. D. Hanson. 1995. Evidence implicating dimethylsulfonio-propionaldehyde as an intermediate in dimethylsulfoniopropionate biosynthesis. Plant Physiol., in press.
14. Larher, F. and J. Hamelin. 1979. L'acide diméthylsulfonium-5 pentanoïque de *Diplotaxis tenuifolia*. Phytochemistry *18:* 1396-1397.
15. Mudd, S.H. and A.H. Datko. 1990. The S-methylmethionine cycle in *Lemna paucicostata*. Plant Physiol. *93:* 623-630.
16. Paquet, L., B. Rathinasabapathi, H. Saini, D. A. Gage and A. D. Hanson. 1993. Accumulation of the compatible solute 3-dimethylsulfoniopropionate in sugarcane. Plant Physiol. *102:*(S) 936.
17. Paquet, L., B. Rathinasabapathi, H. Saini, L. Zamir, D. A. Gage, Z.-H. Huang and A. D. Hanson. 1994. Accumulation of the compatible solute 3-dimethylsulfonio-propionate in sugarcane and its relatives, but not other gramineous crops. Aust. J. Plant Physiol. *21:* 37-48.
18. Reed, R. H. 1983. Measurement and osmotic significance of β-dimethyl-sulphionio-propionate in marine macroalgae. Marine Biology Letters *4:* 173-181.
19. Rhodes, D. 1990. Fast atom bombardment mass spectrometry, p. 95-123. *In* H. F. Linskens and J. F. Jackson, (eds.), Modern Methods of Plant Analysis, New Ser. Vol. 11: Physical Methods in Plant Science. Berlin/Heidelberg: Springer-Verlag.
20. Rhodes, D. and A. D. Hanson. 1993. Quaternary ammonium and tertiary sulfonium compounds in higher plants. Ann. Rev. Plant Physiol. Plant Mol. Biol. *44:* 357-384.
21. Rhodes, D., P. Rich, A. C. Myers, C. C. Reuter and G. C. Jamieson. 1987. Determination of betaines by fast atom bombardment mass spectrometry: identification of glycine betaine deficient genotypes of *Zea mays*. Plant Physiol. *84:* 781-788.
22. Sheets, E. B., and D. Rhodes. 1996. Determination of DMSP and other onium compounds in *Tetraselmis subcordiformis* by plasma desorption mass spectrometry. *In* R. P. Kiene, P. T. Visscher, M. D. Keller, and G. O. Kirst (ed.), Biological and environmental chemistry of DMSP and related sulfonium compounds. Plenum, New York.
23. Smith, R. D., J. A. Loo, Ch. G. Edmonds, Ch. J. Barinaga and H. R. Udseth. 1990. New developments in biochemical mass spectrometry: electrospray ionization. Anal. Chem. *62:* 882-899.
24. Storey, R., J. Gorham, M. G. Pitman, A. D. Hanson and D. A. Gage. 1993. Response of *Melanthera biflora* to salinity and water stress. J. Exp. Botany *44:* 1551-1560.
25. White, R. H. 1982. Analysis of dimethyl sulfonium compounds in marine algae. J. Mar. Res. *40:* 529-536.

4

NMR SPECTROSCOPY AS A PROBE FOR DMSP AND GLYCINE BETAINE IN PHYTOPLANKTON CELLS

C. J. Macdonald, R. Little, G. R. Moore, and G. Malin

School of Chemical Sciences and School of Environmental Sciences
University of East Anglia
NR4 7TJ United Kingdom

SUMMARY

The aim of this work was to evaluate nuclear magnetic resonance as a probe for studying osmoregulatory compounds in marine phytoplankton cells. To date we have largely focused on methods for harvesting and preparing cell extracts and the development of techniques applicable to the use of ^{13}C NMR. DMSP was detected and quantified in intact cells using natural abundance ^{13}C NMR. Subsequently labelled cell extracts were prepared using $NaH^{13}CO_3$ which increased sensitivity and enabled characterisation of unknown organic solute signals. The effects of varying the nitrogen level in cultures was also examined. Finally the ^{13}C NMR spectra of selected phytoplankton species are compared.

INTRODUCTION

Natural abundance ^{13}C and ^{1}H nuclear magnetic resonance spectroscopy offer the possibility of simultaneously detecting a range of compatible solutes, including DMSP and glycine betaine (GBT) within marine phytoplankton cells. Although these methods have been used to study e.g. osmoregulation in unicellular cyanobacteria (2,10,11,12), there are few reports of NMR being used for studies on marine phytoplankton. This is perhaps due to the relative insensitivity of the technique which requires cell densities in the range of 10^6 to 10^9 cm^3.

MATERIALS AND METHODS

Cell Cultures

The following phytoplankton species were used in this study. Haptophytes: *Pleurochrysis carterae* (Braarud et Fagerl.) Braarud, CCMP645. *Emiliania huxleyi* (Lohm.) Hay et Mohler, CCMP370. (CCMP=Provasoli-Guillard Center for culture of Marine Phytoplankton, Maine, USA.). Dinoflagellate: *Amphidinium carterae* Hulbert, Plymouth 450. (Plymouth Marine Laboratory Culture Collection, Plymouth, UK.) Prasinophyte: *Tetraselmis (Platymonas) subcordiformis* (Stein), Goetingen 161-1a. (Goetingen Culture Collection, Germany). Diatom: *Phaeodactylum tricornutum* Bohlin 1987, CCAP1052/1A (Culture Collection of Algae and Protozoa, Dunstaffnage Marine Research Laboratory, Scotland.)

The phytoplankton were grown in stationary batch culture on F/2 enriched seawater medium (8) at 15 °C using a 14 hour light 10 hour dark cycle. Light was supplied by cool white fluorescent tubes with a photon flux density of approximately 40 mmol $m^{-2}s^{-1}$. Cultures were swirled daily. For initial NMR experiments on intact cells of *Pleurochysis carterae* , 2 dm^3 of culture was required to produce a cell pellet of sufficient volume (3 cm^3) and density. This was subsequently reduced to a minimum of 250 cm^3 as the methodology was refined.

For ^{13}C labelled growths 0.5 - 5.0 mM $NaH^{13}CO_3$ was added to F/2 medium which was then filter sterilised. Similar cultures of *P.carterae* were also grown with ^{13}C-methyl labelled methionine (L-methionine-methyl-^{13}C Sigma-Aldrich) which was added to a concentration of 100 mM (7). The effects of high and low levels of available nitrate nutrient on cultures of *P.carterae* and *T.subcordiformis* were also examined using F/2 with nitrate added to a concentration of 820 mM and filtered aged seawater as a base for F/2, (nitrate concentration <5 mM).

Harvesting

For whole cell experiments cultures were harvested in the late log phase (usually after about two weeks growth) by gravity sedimentation over approximately 36 hours. In subsequent experiments centrifugation was used. For small volumes (max. 50 cm^3 per tube) 600g was applied for 5 min, and for larger volumes (max. 260 cm^3 per bottle) 600g for 15 min.(7).

Preparation of Extracts

Harvested material was immediately snap frozen by immersion in liquid nitrogen for 2 minutes. Samples were then stored at -20°C for no more than a few days before freeze drying in a Hetosicc Freezedryer (Heto Lab Equipment supplied by MDH.). Culture volumes of 250 cm^3 are now routinely used, and these produce a minimum of 200 mg of freeze dried material, which provides sufficient labelled cell extract for several NMR samples. Subsequent storage was in the dark at 4°C. Due to the limited information available in the literature an evaluation of extraction methods relevant to preparing phytoplankton cells for NMR analysis was carried out (detailed in results section). This resulted in the following procedure. Dried material was thoroughly ground and homogenised with a mortar and pestle and 50-150 mg weighed into a micro centrifuge tube, 1 cm^3 of D_2O was then added and the tubes briefly vortexed and then shaken vigorously for 30 minutes. This was immediately followed by centrifugation at 12000 rpm for 10 minutes (Eppendorf microcentrifuge) and the supernatant removed for NMR analysis.

NMR

A JEOL GX-400 NMR spectrometer was used throughout this study. Intact cell spectra were acquired in 10 mm non-spinning NMR tubes which required a minimum cell pellet volume of 3 cm^3. Sample temperature was controlled at 15±0.1°C. ^{13}C spectra were obtained in the broadband proton decoupled mode using Waugh (15) decoupling to minimise sample heating. Typically observation pulses of 60° and recycle times of 1s were used. Under these conditions considerable saturation occurred, particularly of the carbonyl resonances.

Field/frequency lock was provided by a sample of deuterated benzene and tetramethylsilane contained within an external capillary tube. This also allowed chemical shift referencing. Exponential line broadening of 30-50 Hz was generally applied prior to fourier transformation.

Cell extract spectra were acquired in 5mm tubes and the D_2O used in the extraction provided the lock signal. Line broadening in these spectra was reduced to 1.5-3 Hz. Although recycle times of 3s were used, saturation effects were still present.

^1H NMR spectra were obtained only on D_2O extract samples. Solvent suppression was used to reduce the residual solvent signal using homonuclear gated decoupling (presaturation).

2-D carbon-proton shift correlation (COSY) spectra were acquired using a spectral width of 10500 Hz in f2 over 1K data points, and 3300 Hz in f1 over 128 or 256 data points. If necessary data matrices were zero-filled to give a maximum 2 Mword matrix and 64-400 scans per increment were collected. 2-D proton-proton shift correlation (COSY) were acquired using similar size matrices with spectral widths of 2000 Hz in both dimensions and zero-filled to give <4 Hz resolution in f1 and f2. Between 16 and 64 scans per increment were collected. In these 2-D shift correlation experiments the pulse delay and number of f1 increments were adjusted to suit the available experiment time (2.5 - 14 hr).

2-D double quantum carbon-carbon correlation spectra (INADEQUATE) were also obtained for *P.carterae* extracts. These experiments suffer from inherently poor sensitivity and resolution. The sensitivity problem was somewhat alleviated by the high degree of ^{13}C labelling, and the spectral width in f1 was reduced (allowing signals to foldback) to improve digital resolution in this dimension. Data matrices were 4K x 128 zero-filled to 2 Mwords. Accumulation times of 64 hours were required using 1200 scans per t1 increment. Spectral widths of f1 and f2 were 12626 Hz and 25252 Hz respectively.

RESULTS

The ^{13}C proton decoupled spectrum of *P.carterae* is shown in Fig.1a. Signals from DMSP can be seen at 25 ppm 31 ppm and 41 ppm (CH_3, a CH_2 and b CH_2 respectively). Specific signals at 183, 69, 20, 70 and 35 ppm and those in the region 60 to 110 ppm are discussed below. Other signals were only observed in early spectra. These may have been due to the harvesting and sample handling techniques used at that time, since they disappeared as methods were further developed. Glycine betaine was not apparent in these spectra. The stability of the whole cell samples was examined by monitoring hourly for changes in the spectrum over 15 hours. Towards the end of this experiment signals at 183, 69 and 20 ppm had slightly increased intensities. After the experiment the cells were removed and examined by microscopy and were found to be highly motile suggesting that they had therefore not been adversely affected by the NMR experiment.

DMSP levels were then quantified; the NMR spectrum of a 3 cm^3 pellet of *P.carterae* cells was obtained and then the sample was sonicated, the cell debris filtered off and the filtrate spectrum acquired. Standard additions of small volumes of a 1M DMSP solution were then made and the integrals of the DMSP peaks calculated. The calibration line produced is shown in fig.2b. As a further check on linearity and limits of detection a large volume cell pellet was divided into 6 samples of between 0.5 and 3.0 cm^3 and each was diluted to 3.0 cm^3 prior to data acquisition. The methyl signal intensities in each of these samples were measured and plotted in fig.2a.

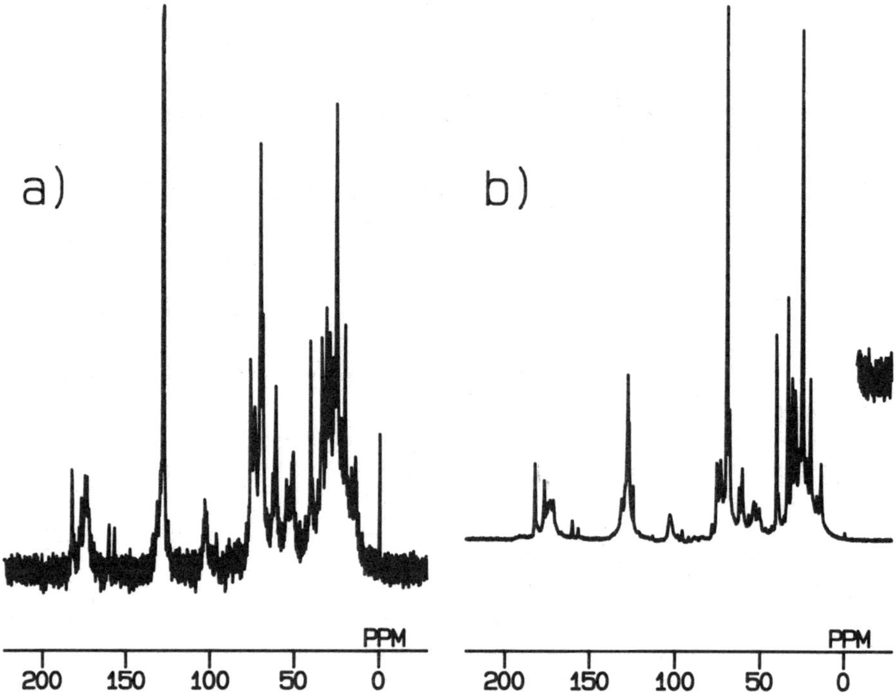

Figure 1. ^{13}C-Spectra of *P.carterae* whole cells, a) unlabelled b) grown on 5.0 mM $NaH^{13}CO_3$. Inset is with Y-gain x30.

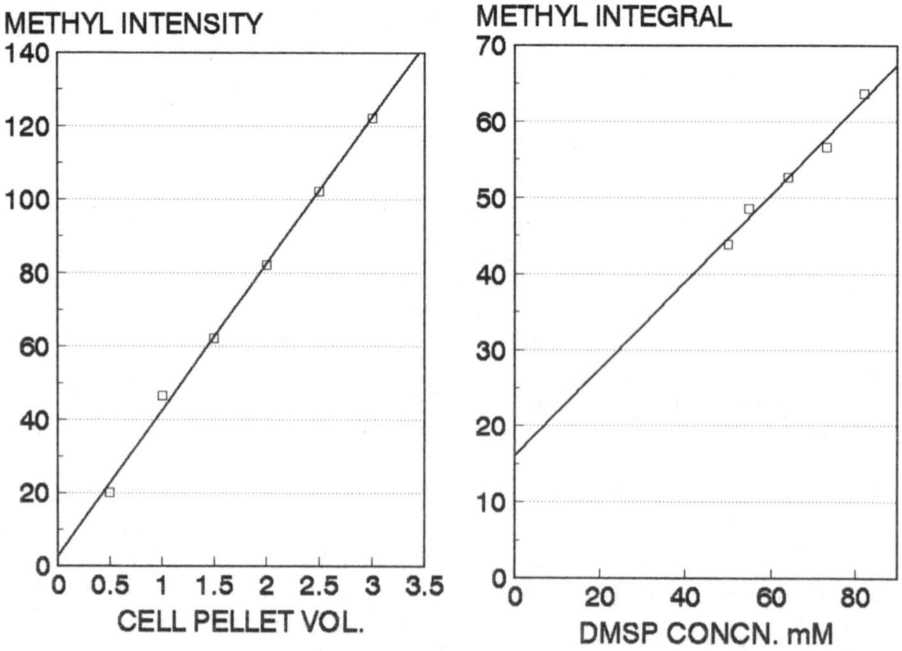

Figure 2. Plots of quantification experiments. a) DMSP methyl signal intensity in intact cells against cell pellet volume. b) DMSP methyl signal integral in supernatant after sonication and spinning against concentration of DMSP (mM).

Table 1. Summary of carbon-proton shift correlations and other NMR spectraldata found for unknown signals in ^{13}C-observed 1 and 2D spectra of *P. carterae* cell extracts

^{13}C Chemical Shifs	183 PPM	70 PPM	69 PPM	35 PPM	20 PPM
Experiment					
Dept	C	CH	CH	CH_2	CH_3
CH-COSY	None	3.8 PPM	4.2 PPM	1.9 PPM	1.35 PPM
CC-INAD	69 PPM	35 PPM	183 20 PPM	70 PPM	69 PPM

The natural abundance of ^{13}C is only 1.01% and the remainder (^{12}C) is NMR silent, therefore only this amount of any molecular species is detectable by ^{13}C NMR. If the ^{13}C level can be increased the effective concentration of the sample is increased making detection easier. Cells of *Synechococcus* N100 grown on 90% $NaH^{13}CO_3$ have been shown to incorporate the ^{13}C label (10) and therefore in order to improve signal-to-noise ratio in the spectra, and reduce experiment time, ^{13}C labelled cells were produced. A comparison of labelled and unlabelled cells is shown in fig.1. The substantial improvement in spectral quality facilitated the identification and characterisation of the unknown signals.

Producing cell extracts increases further the power of NMR spectroscopy for probing cellular systems, resolution is dramatically increased, samples are more conveniently handled and may be repeatedly subjected to longer accumulations of 2 D NMR data. We evaluated typical extraction methods (2,5,9,10,12) using *P.carterae* ; 80% Ethanol, D_2O and 5%, 10%, 20% and 50% trichloroacetic acid in D_2O were compared as extraction solvents. Only very minor differences in the relative intensities of some signals could be detected in the ^{13}C spectra of the various extraction systems. These were deemed to be trivial so other aspects of these solvents were considered. Ethanol produces large intensity signals close to regions of interest and (unless fully deuterated) obscured signals in the ^{1}H spectrum. Although TCA extracted samples seemed to contain less finely suspended solid material and may have been more stable over longer time periods, this too produces ^{13}C resonances. Therefore, in this work D_2O was preferred.

Harvesting by snap freezing and normal (slow) freezing were also compared. Spectra of snap frozen samples lacked signals at 183, 69 and 20 ppm which were seen in the slowly frozen samples, and also increased in intensity with time during the 15 hour experiment discussed previously. It was therefore concluded that these signals arise when the cells are perturbed for extended periods of time. For these reasons the methodology described in the methods section was adopted.

2-D hetero and homonuclear shift correlation (COSY),2-D carbon-carbon double quantum correlation (INADEQUATE) and polarisation transfer experiments (DEPT) were performed on labelled cell extract samples in order to characterise the unknown signals in the ^{13}C spectrum. Correlations and further information obtained is summarised in Tables 1 and 2.

Figs.3 and 4 show a comparison of *P.carterae* and *T.subcordiformis* grown under conditions of high and low nitrogen ($NaNO_3$) levels. DMSP is present in all spectra, but marked differences can be seen and are discussed below.

Fig.5 shows a comparison of the spectra of *P.carterae* grown with the addition of 100 mM ^{13}C-methyl labelled methionine or 0.5mM $NaH^{13}CO_3$. These spectra were not acquired

Table 2. Summary of proton-proton shift correlations and other NMR spectral data found for unknown signals in ^{1}H-observed 1 and 2D spectra of *P. carterae* cell extracts

^{1}H Chemical shifts	4.2PPM	3.8PPM	1.9PPM	1.35PPM
EXperiment				
HH-COSY	1.35PPM	1.9PPM	3.8PPM	4.2PPM
Integrals	1	1	1	3
Multiplicty	Quartet	Br.Singlet	Br.Singlet	Doublet

over the same experiment time or on a similar level of ^{13}C label and therefore are not directly comparable, nevertheless relative intensities of signals within spectra are revealing.

DISCUSSION

The quantification data shown in Fig.2a generates a least squares fit line with a regression coefficient of 0.998 and an intercept of 2.76. Obviously this line should pass through the origin. If the data point corresponding to 1.0 cm^3 is omitted (this sample was inadvertently left at room temperature for over 1 hour before data acquisition) an even better fit is obtained ($r = 0.9999$ and intercept of 0.16). This indicates the linearity of the method. A DMSP concentration of 16.8 mM was determined using gas chromatography on the cell pellet used for the data point at 3.0 cm^3. Therefore DMSP methyl signal intensities obtained under the conditions of this experiment could be related to absolute concentrations of the osmolyte. This plot also shows the limit of detection to be at least as low as 2.8 mM DMSP.

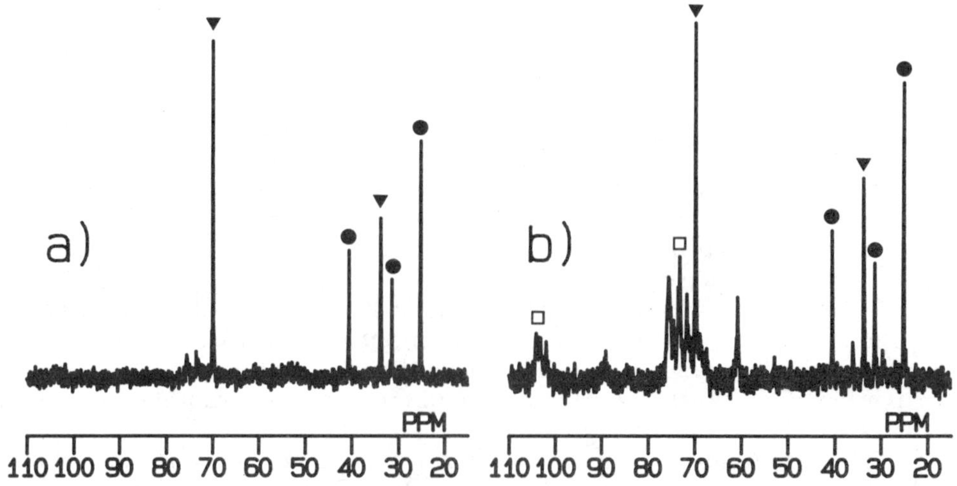

Figure 3. ^{13}C-Spectra of *P.carterae* cell extracts grown on a) high and b) low levels of nitrate. DMSP signals are indicated by filled circles, GBT by open circles, cyclohexanetetrol by filled triangles and carbohydrate signals by open squares.

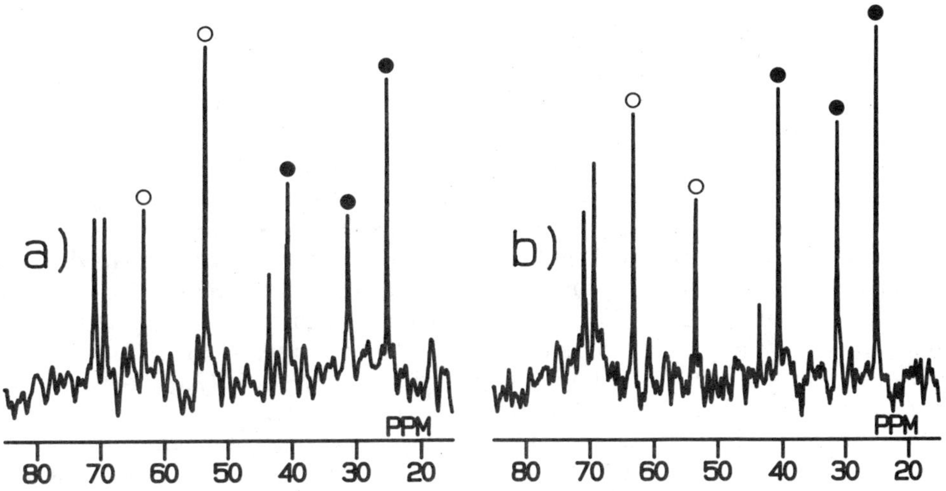

Figure 4. ^{13}C-Spectra of *T.subcordiformis* cell extracts grown on a) high and b) low levels of nitrate. DMSP signals are indicated by filled circles, GBT by open circles.

From the experimental results shown in Fig.2b a DMSP concentration in this cell pellet of 29.8 mM can be directly calculated; this compares well with the value obtained by gas chromatography of 28.2 mM. This value can be related to an approximate intracellular concentration of DMSP of 8 pg per cell by taking into consideration cell counts, (these were done by light microscopy using a haemocytometer). Following this experiment the cell pellet was sonicated and filtered and the cell debris resuspended in D_2O and re-examined. Even after extended accumulation no DMSP was detected and it therefore seemed likely that the majority of the DMSP is free within the cytoplasm. NMR can therefore quantitatively determine DMSP in intact phytoplankton cells, and providing standards and/or complimentary analytical methods are available, other osmoregulatory compounds could be treated in a similar way.

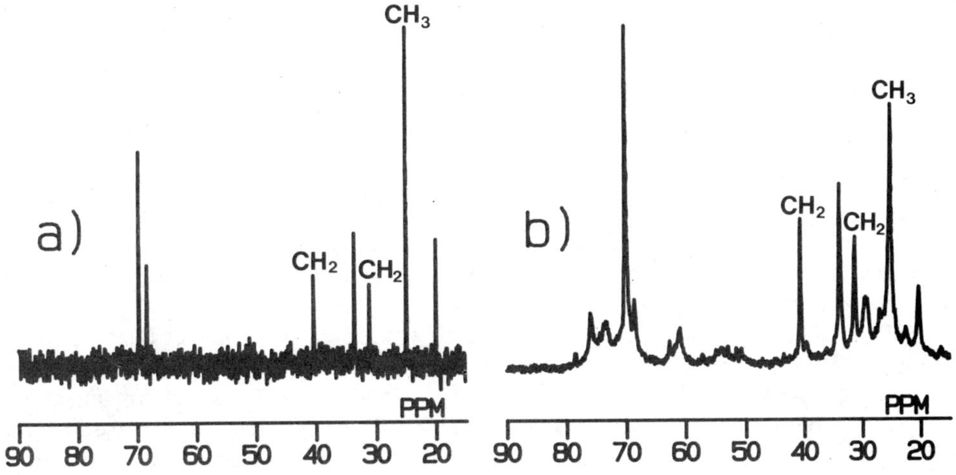

Figure 5. ^{13}C-Spectra of *P.carterae* cell extracts grown on a) 100 mM ^{13}C-methyl-labelled methionine and b) 0.5 mM $NaH^{13}CO_3$. The methyl and methylene signals of DMSP are indicated.

Qualitative spectra of unlabelled intact cells are generally available within 20 minutes, quantification obviously requires longer to obtain accurate integrations. Fig.1 shows a ~30 fold increase in signal to noise obtained with the spectrum of intact cells grown on 5.0 mM $NaH^{13}CO_3$. The reduction in accumulation time that this degree of labelling provides means that spectra suitable for quantitative determinations may be obtained in 2 minutes. It is therefore feasible to monitor responses of cells within the NMR spectrometer in real time with a resolution of 2 minutes. The construction of an NMR probe to allow these types of experiments is presently under consideration.

Quantifying osmolyte concentrations in samples produced using $NaH^{13}CO_3$ is more complicated and requires knowledge of, amongst other factors, the degree of ^{13}C enrichment at the site used in the determination. In the NMR spectra of cell extracts this may be possible by studying ^{13}C satellite signals. Close inspection of the DMSP signals in spectra of samples grown with 5.0 mM $NaH^{13}CO_3$ show a "triplet" for the carbonyl resonance, the central singlet (relative intensity 2) arises from the singly labelled $^{12}\beta C$-$^{12}\alpha C$-$^{13}COOH$ isotopomer and the outer doublet lines (relative intensity 1) from the doubly labelled $^{12}\beta C$-$^{13}\alpha C$-$^{13}COOH$ isotopomer ($^1J_{CC}$ = 50 Hz). A similar pattern is seen for the β-carbon ($^1J_{CC}$ = 35 Hz). The α-carbon shows a nine line multiplet with relative intensities 1:2:2:1:4:1:2:2:1 arising from the $^{12}\beta C$-$^{13}\alpha C$-$^{12}COOH$ (singlet), $^{13}\beta C$-$^{13}\alpha C$-$^{12}COOH$ (doublet $^1J_{CC}$ = 35 Hz), $^{12}\beta C$-$^{13}\alpha C$-$^{13}COOH$ (doublet $^1J_{CC}$ = 50 Hz), and a doublet of doublets from the fully labelled 13βC-13αC-$^{13}COOH$ isotopomers. Analysis of the intensities of these lines indicates that the DMSP molecule is equally labelled at these 3 sites with an enrichment factor of 50%. Further information might be revealed from ^{13}C satellite signals regarding biosynthetic pathways and cell metabolism from the selective incorporation of ^{13}C into particular sites within a molecule. For example, our work suggests that in the molecule cyclohexanetetrol, (discussed below), the ^{13}C label is initially incorporated into the CH carbons rather than equally distributed between all carbons.

In our work, ^{13}C labelling has proved invaluable although other workers (12) have discussed the detrimental effects of high levels of ^{13}C labelling (increased spectral complexity and overlap of resonances). Indeed without the 50% enrichments achieved here the double quantum carbon correlation experiments would have been impossible.

From the COSY and INADEQUATE connectivity data it is possible (also considering chemical shift and molecular symmetry constraints) to propose two structures to account for the 5 major unknown signals in the ^{13}C spectrum of *P.carterae*. The signals at 70 and 35 ppm, which are present at high levels in all spectra would appear to originate from the compound 1,2,4,5-cyclohexanetetrol. This has been observed in the related species *Monochrysis lutheri* (3) and an osmoregulatory role for it was proposed. The signals at 183, 69 and 20 ppm are likely to arise from 3-hydroxy propanoic acid. The remaining unknown signals between 60 and 110 ppm also show connectivities in the 2D spectra although the lower intensities of these signals makes identification more difficult. However the pattern of signals and ^{13}C chemical shifts are characteristic of a carbohydrate. Mass spectral analysis will be used to confirm these preliminary identifications.

The effect of nitrogen deficiency on DMSP in *T.subcordiformis* has been discussed previously (7). We therefore examined the spectra of *T.subcordiformis* and also *P.carterae* grown on high and low levels of nitrate nutrient (Figs. 3 and 4). Cyclohexanetetrol and DMSP are present in both spectra of *P.carterae* , the major difference being the increase in signal intensity between 60 and 110 ppm at low nitrate levels from the as yet unidentified carbohydrate. These signals are almost completely absent when high levels of nitrate are available. In the spectra of *T.subcordiformis* both DMSP and glycine betaine (at 63 and 54 ppm) are detected at both levels but under high nitrate conditions the GBT is seen to increase

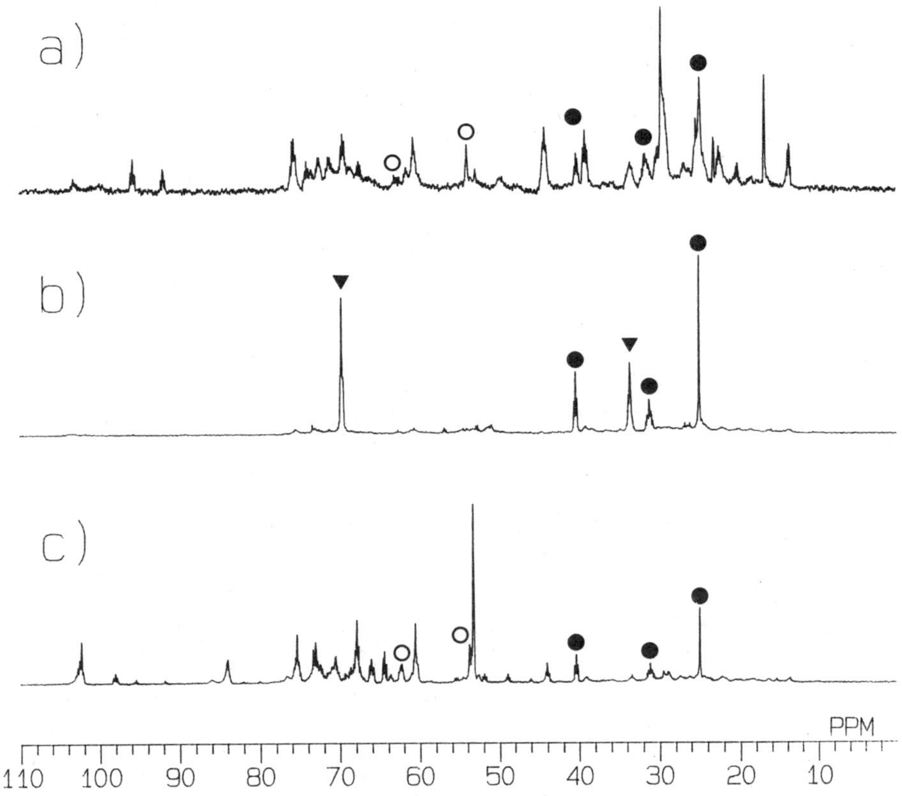

Figure 6. High field region of ^{13}C-spectra of cell extracts of a) *A.carterae* b) *E.huxleyi* and c) *P.tricornutum*. DMSP signals are indicated by filled circles, GBT by open circles, cyclohexanetetrol by filled triangles.

relative to DMSP. This is in general agreement with the hypothesis put forward by Andrea (1), and the observations of Turner et al (13) on *E.huxleyi* . Dickson and Kirst (4) also report the presence of homarine in *T.subcordiformis* but this compound was not detected in our work.

Methionine has been shown to be a precursor for DMSP in e.g. the macroalga *Ulva lactuca* (6) and microalgae such as *T.subcordiformis* (7), and a biosynthetic pathway has been proposed (14). We have also shown this to be the case for *P.carterae.* Fig.5 shows the increase of the methyl signal of DMSP relative to the methylene signals in the spectrum of the cells grown on ^{13}C labelled methionine. The methionine methyl group is therefore used by the organism as a source for the methyl of DMSP. However, no intermediates in this biosynthetic pathway were detected.

The high field region of some representative spectra of other phytoplankton are shown in Fig.6. *E.huxleyi* produces a very similar spectrum to *P.carterae* , containing signals from cylclohexanetetrol and DMSP. No GBT was found in this organism. In contrast the spectra from *A.carterae* and *P.tricornutum* are much more complex, DMSP was detected in both species along with a considerable number of other signals including an indication of GBT. We believe these data demonstrate the usefulness of the NMR technique in simultaneously detecting a range of organic solutes. A full analysis of these spectra is now being carried out.

ACKNOWLEDGMENTS

The authors wish to thank Helen Reynolds and Gary Creissen of the John Innes Centre, Norwich, U.K. for freeze drying the cultured cells. We would like to thank Prof. Gunter Kirst, University of Bremen, Germany for the kind gift of the *Tetraselmis subcordiformis* culture.

REFERENCES

1. Andreae, M.O. 1986. The ocean as a source of atmospheric sulphur compounds, p. 331-362. *In* P.Buat-Menard (ed.), The role of air-sea exchange in geochemical cycling. Reidel.
2. Borowitzka, L.J., S. Demmerle, M.A. Mackay and R.S. Norton. 1980. Carbon-13 nuclear magnetic resonance study of osmoregulation in a blue-green alga. Science *210*: 650-651.
3. Craigie, J.S. 1969. Some salinity-induced changes in growth, pigments, and cyclohexanetetrol content of *Monochrysis lutheri* . J. Fish. Res. Bd. Canada *26*: 2959-2967.
4. Dickson, D.M. and G.O. Kirst. 1986. The role of β-dimethylsulphoniopropionate, glycine betaine and homarine in the osmoacclimation of *Platymonas subcordiformis*. Planta *167*:536-543.
5. Ghoul, M., T.Bernard and M. Cormier. 1990. Evidence that *Escherichia coli* accumulates glycine betaine from marine sediments. Appl. Environ. Microbiol. *56*: 551-554.
6. Green, R.C. 1962. Biosynthesis of dimethyl-β-propiothetin. J. Biol. Chem. *237:* 2251-2254.
7. Grone, T. and G.O. Kirst. 1992. The effect of nitrogen deficiency, methionine and inhibitors of methionine metabolism on the DMSP contents of *Tetraselmis subcordiformis* (Stein). Mar. Biol. *112*. 497-503.
8. Guillard, R.R.L. (1975) Culture of phytoplankton for feeding marine invertebrates. p29-60. *In* W.L.Smith and M.H.Chanley (ed.), Culture of marine invertebrate animals. Plenum, New York.
9. Landfald, B. and A.R. Strom. 1986. Choline-glycine betaine pathway confers a high level of osmotic tolerance in *E.coli*. J. Bacteriol. *165*: 849-855.
10. Mackay, M.A. and R.S. Norton. 1987. ^{13}C Nuclear magnetic resonance study of biosythesis of glucosylglycerol by a cyanobacterium under osmotic stress. J. Gen. Microbiol. *133:* 1535-1542.
11. Norton, R.S., M.A. Mackay and L.J. Borowitzka. 1982. The physical state of osmoregulatory solutes in unicellular algae. Biochem. J. *202*: 699-706.
12. Tel-Or, E., S. Spath, L. Packer and R.J. Mehlhorn. 1986. Carbon-13 NMR studies of salt shock-induced carbohydrate turnover in the marine cyanobacterium *Agmenellum quadriplicatum*. Plant Phys. *82*: 646-652.
13. Turner, S.M., et. al. 1988 The seasonal variation of dimethyl sulfide and dimethylsulfoniopropionate concentrations in nearshore waters. Limnol. Oceanogr. *33* : 364-375.
14. Uchida, A., T. Ooguri, T. Ishida and Y. Ishida. 1992. Incorporation of methionine into dimethylthiopropanoic acid in the dinoflagellate *Crypthecodinium cohnii*. Nippon Suisan Gakkaishi *58*: 255-259.
15. Waugh, J.S. 1982. Broadband spin decoupling. J.Magn. Resn. *50*: 30.

5

DETERMINATION OF DMSP AND OTHER ONIUM COMPOUNDS IN *TETRASELMIS SUBCORDIFORMIS* BY PLASMA DESORPTION MASS SPECTROMETRY

Elisa B. Sheets[1] and David Rhodes[2]

[1] West Lafayette Senior High School
West Lafayette, Indiana 47906
[2] Department of Horticulture
Purdue University
1165 Horticulture Building
West Lafayette, Indiana 47907-1165

SUMMARY

Tetraselmis subcordiformis (formerly *Platymonas subcordiformis*) is a β-dimethylsulfoniopropionate (DMSP) accumulating unicellular marine green alga, and represents a useful model system for investigating factors controlling DMSP production. This study has focused on: 1) developing a method for quantifying DMSP (and other onium compounds) in *Tetraselmis* extracts using plasma desorption mass spectrometry (PD-MS), and 2) applying this method to analysis of DMSP (and other onium compound) levels of *Tetraselmis* cells grown at different temperatures. Onium compounds were purified from methanol extracts of *Tetraselmis* by methanol: chloroform: water phase separation and ion exchange chromatography, and were analyzed by PD-MS. Samples included two internal standards; $^2H_9$-glycinebetaine ($M + H^+ = m/z$ 127), and $^2H_6$-DMSP ($M + H^+ = m/z$ 141). Purified onium fractions from *Tetraselmis* cell extracts were found to contain DMSP ($M + H^+ = m/z$ 135), glycinebetaine ($M + H^+ = m/z$ 118), and trigonelline (nicotinic acid betaine) and/or homarine (picolinic acid betaine) ($M + H^+ = m/z$ 138). Glycinebetaine was quantified from the ratio of ions at m/z 118 : 127, while DMSP was quantified from the m/z 135 : 141 ion ratio. As growth temperature was increased from 5 to 19°C, glycinebetaine levels increased from 30.0 to 44.2 fmol/cell, whereas DMSP levels decreased from 121.2 to 14.8 fmol/cell. The high levels of DMSP at low growth temperatures appear consistent with the proposed role for DMSP as a cryoprotectant.

Biological and Environmental Chemistry of DMSP and Related Sulfonium Compounds
edited by Ronald P. Kiene et al., Plenum Press, New York, 1996

INTRODUCTION

The tertiary sulfonium compound β-dimethylsulfonio-propionate (DMSP) is a precursor of dimethylsulfide (DMS), recognized as the most important volatile sulfur compound produced by marine algae and certain higher plant halophytes (3,9-13,17,22,23,25). DMS released from oceans and salt marshes to the atmosphere is strongly implicated in the photochemical reactions which lead to formation of cloud-condensation nuclei which may influence global albedo and hence climate (2). Factors controlling DMSP synthesis and degradation by marine algae and higher plants are therefore of great interest in terms of climate modeling (2,14).

Salinity, irradiance and temperature are major determinants of DMSP contents of marine macroalgae (9-13). Because DMSP is accumulated to osmotically significant levels in marine macroalgae it is strongly implicated as a compatible osmotic solute (11,18). It may also represent an important cryoprotectant (13). Thus, DMSP tends to be accumulated to higher levels in cold-water marine macroalgae, and light-dependent accumulation of DMSP is stimulated in these organisms up to 5-fold when their growth temperature is decreased from 10°C to 0°C (13).

The unicellular, flagellated marine alga, *Tetraselmis subcordiformis,* is known to accumulate the onium compounds DMSP, glycinebetaine and homarine as organic solutes in response to hypersaline stress (4). DMSP, glycinebetaine and homarine are implicated as compatible osmolytes in this organism, collectively contributing significantly to osmotic adjustment (4). DMSP has been shown to be more compatible to certain enzymes of *Tetraselmis* than equivalent concentrations of NaCl (6). However, DMSP may be a compatible solute only under conditions of moderate temperature (15). Whereas DMSP stabilizes certain enzymes against cold-induced denaturation, it is not an effective stabilizer of protein structure under conditions of heat denaturation (15). This contrasts with glycinebetaine which not only offers partial protection of enzymes against NaCl or KCl inhibition, but also stabilizes membranes and proteins against heat inactivation (19).

DMSP accumulation is induced in *Tetraselmis* in response to nitrogen deficiency, suggesting that the nitrogen-free DMSP may substitute for the nitrogen-containing osmolyte glycinebetaine under conditions of nitrogen deficiency (7). This may also apply to higher plants (8,24). Moreover, it has been suggested that DMSP may play a role in storage of excess sulfur in certain halophytic higher plants (26).

Dickson and Kirst employed a semi-quantitative thin layer chromatography method for estimating glycinebetaine, homarine and DMSP contents of *Tetraselmis* (4). Although DMSP can be readily quantified by gas chromatography following the liberation of DMS under alkaline conditions (7,17,18), this method is not applicable to simultaneous quantification of the quaternary ammonium compounds, glycinebetaine and homarine. Recently, desorption mass spectrometry methods including fast atom bombardment (FAB-MS) (8) and plasma desorption mass spectrometry (PD-MS)(1,27) have been applied to the analysis of quaternary ammonium and tertiary sulfonium compounds in higher plants. Here we tested the applicability of the PD-MS method to analysis of onium compounds of *Tetraselmis subcordiformis*. Because the effect of growth temperature on DMSP and glycinebetaine contents of *Tetraselmis* does not appear to have been previously investigated, we specifically applied this method to the analysis of the levels of these compounds as a function of growth temperature.

MATERIALS AND METHODS

Chemicals

$^{2}H_9$-Glycinebetaine was synthesized as described (20). $^{2}H_6$-DMSP was synthesized as follows. 3-Mercaptopropionate (0.1273 g) was dissolved in 10 mL methanol containing 0.5g $NaHCO_3$, and chilled on ice for 5 min. Deuterium methyl iodide ($C^{2}H_3I$) (2 mL) was added and the sample stirred for 23 h at room temperature in a closed vial. After reaction, the sample was filtered through Whatman No. 4 filter paper, and the filtrate was then dried under a stream of air using gentle heat (< 25°C). The sample was extracted 3 times with 5 mL acetonitrile : methanol (10:1 v/v), and filtered through Whatman No. 4 filter paper, discarding insoluble material. The filtrate was then dried under a stream of air and redissolved in 5 mL water, applied to a 9 cm x 1.3 cm column of Dowex-1-Cl^- and washed with 30 mL water. The water eluate from Dowex-1-Cl^- was then applied to a 4 cm x 1 cm Dowex-50-H^+ column, eluting $^{2}H_6$-DMSP with 12 mL 2.5 M HCl which was then dried and crystallized under a stream of air. Recovery of $^{2}H_6$-DMSP.HCl was 0.0965g which was then dissolved in water to give a final concentration of 100 mM and stored at -20°C.

Cell Culture

Growth medium was adapted from (4a) and contained 27 $g.L^{-1}$ NaCl, 5 $g.L^{-1}$ $MgSO_4.7H_20$, 0.9 $g.L^{-1}$ KCl, 1.5 $g.L^{-1}$ $CaCl_2.6H_20$, 0.02 $g.L^{-1}$ KH_2PO_4, 1.01 $g.L^{-1}$ KNO_3, 1 $g.L^{-1}$ Tris, and 1 $mL.L^{-1}$ of a stock mineral solution containing 1 $g.L^{-1}$ $FeCl_3.4H_20$, 20.8 $g.L^{-1}$ $Na_2EDTA.2H_2O$, 0.2 $g.L^{-1}$ $MnCl_2.4H_20$, 0.002 $g.L^{-1}$ $ZnSO_4.7H_20$, 0.002 $g.L^{-1}$ $CuSO_4.5H_20$ and 0.001 $g.L^{-1}$ $Na_2MoO_4.2H_20$. Medium was brought to pH 7.6 with concentrated HCl, and 50 mL aliquots dispensed into 200 mL flasks with foam stoppers, capped with aluminum foil, and then autoclaved at 121°C for 20 min. *Tetraselmis* cells were inoculated into fresh medium under sterile conditions, and cultures were grown on a rotary shaker under constant fluorescent lights (two 15W/118V/60hz fluorescent lamps which yielded a light intensity at the culture surface of 27.75 $\mu mol.m^{-2}.s^{-1}$) in a growth chamber maintained at either 5, 11, 19 or 23°C.

Determination of Cell Doubling Times

Cell number per mL of culture was monitored daily with a haemocytometer. Aliquots of 1 mL culture were removed under sterile conditions, cells were immobilized by adding 50 µL 25% glutaraldehyde, and then counted under a microscope at 100x magnification. Cell doubling times were determined by regression analysis of log cell number per mL versus time for the 3 d period preceding extraction of onium compounds at each growth temperature.

Extraction and Purification of Onium Compounds

Cultures (three replicate flasks at each growth temperature) were typically harvested at cell densities of 0.5 to 2 x 10^6 cells per mL. Aliquots of cell cultures were filtered through glass fiber filter paper and the filter paper plus cells was extracted in 10 mL methanol at 4°C in the dark. To the methanol extracts was added 5 mL chloroform and 6 mL distilled H_2O. The samples were vortexed, filtered through glass fiber paper, and the upper aqueous phase of the filtrate was then removed and dried at room temperature under a stream of air. After drying, the samples were redissolved in 2 mL water. For cells grown at 23°C, aliquots of the aqueous extracts were analyzed for DMSP by liberation of DMS under alkaline conditions, and gas chromatography analysis of DMS essentially as described in (7). For most PD-MS analyses, 1000 nmol of $^2H_9$-glycinebetaine and 2000 nmol $^2H_6$-DMSP.HCl were added as internal standards to the aqueous extracts, except for preliminary analyses of cells grown at 23°C in which internal standards were excluded in order to determine if ions of the same mass : charge ratio as the internal standards were detectable in the purified onium compound fractions *per se*. Onium compounds were purified from the aqueous extracts of *Tetraselmis* by passing the extract over a mixed bed resin (Dowex-1-OH^- and Biorex-70-H^+ (1:1 v/v))(5 cm x 1 cm) and washing with 6 mL H_2O. This effectively removed choline and amino acids (1). The aqueous eluate from the mixed bed resin was then applied to Dowex-50-H^+ (4 cm x 1 cm) columns. After washing with 6 mL H_2O, onium compounds were eluted with 6 mL 2.5 M HCl. Use of HCl as eluting solvent avoids possible losses of DMSP (1,8). The dried residue was extracted with 1 mL acetonitrile:methanol (20:1 v/v) (to remove inorganic ions) (27), and the acetonitrile : methanol extract was evaporated to dryness under a stream of air before PD-MS analysis.

Plasma Desorption Mass Spectrometry

PD-MS analyses were performed using a BIOION 20R Plasma Desorption Mass Spectrometer (BIOION KB, Uppsala, Sweden), essentially as described in (1,27). Sample targets were prepared by electrospraying ~50 μL of a 2 mg/mL nitrocellulose solution in acetone onto an aluminized mylar target. 50 μL of methanol was added to the dried onium sample and 1 to 2 μL of this solution was applied to the sample target and dried with a stream of nitrogen. Samples were inserted into the BIOION 20R carousel (capable of holding up to 8 samples) and spectra were collected for 15 minutes at an acceleration potential of 17 kV.

RESULTS AND DISCUSSION

PD-MS Spectra of Synthetic Internal Standards

The PD-MS spectrum of $^2H_9$-glycinebetaine (40 nmol applied to the sample target), displayed over the mass range 100 - 170 atomic mass units (amu), is given in Figure 1A. The major ion observed in the spectrum is the protonated molecule ($M + H^+$) = m/z 127 (Fig. 1A). The ion at m/z 149 corresponds to the $^2H_9$-glycinebetaine sodium adduct ion ($M + Na^+$). $^2H_9$-Glycinebetaine also gives rise to a methylation product at m/z 144 [$(C^2H_3)_3N^+$-CH_2-$COOC^2H_3$] ($M + C^2H_3^+$) (Fig. 1A). As shown previously, authentic unlabeled glycinebetaine gives equivalent ions at m/z 118 ($M + H^+$), m/z 140 ($M + Na^+$) and at m/z 132 ($M + CH_3^+$) in this method (1).

The PD-MS spectrum of $^2H_6$-DMSP (80 nmol applied to the sample target) displayed over the mass range 100 - 170 amu is shown in Figure 1B. The major ion observed in the spectrum is the protonated molecule ($M + H^+$) = m/z 141 (Fig. 1A). Authentic unlabeled DMSP gives an equivalent ion at m/z 135 ($M + H^+$)(1). The ion at m/z 158 observed in the spectrum of $^2H_6$-DMSP (Fig. 1B) may represent a deuteriomethylation product of $^2H_6$-DMSP ($M + C^2H_3^+$).

When equimolar mixtures of $^2H_6$-DMSP and $^2H_9$-glycinebetaine were tested, $^2H_6$-DMSP gave a signal at m/z 141 which was 6-fold less abundant than the signal from $^2H_9$-glycinebetaine at m/z 118 (not shown).

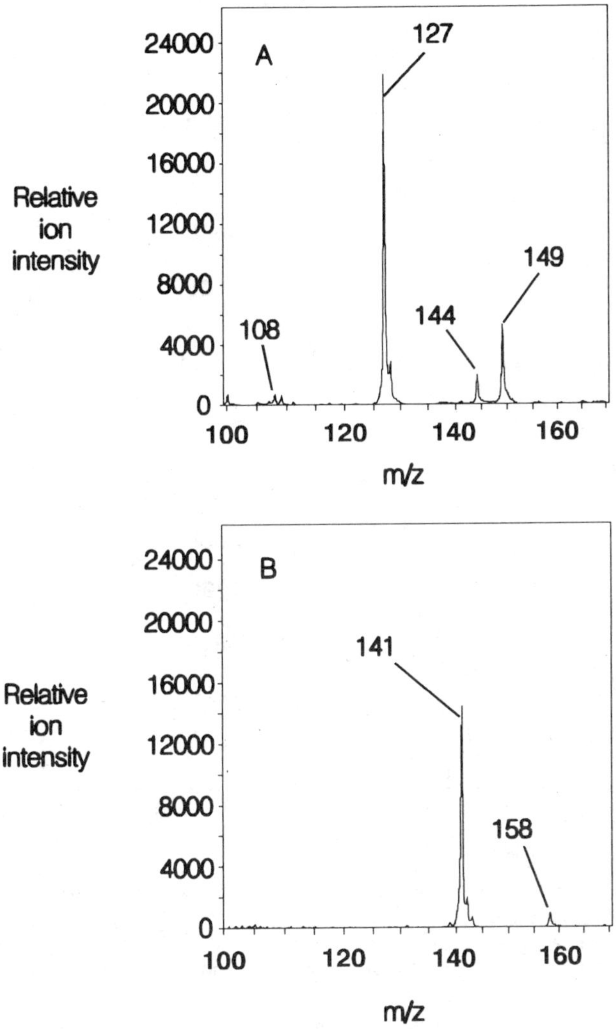

Figure 1. PD-MS spectra of $^{2}H_{9}$-glycinebetaine (40 nmol applied to sample target) (A), and $^{2}H_{6}$-DMSP.HCl (80 nmol applied to sample target) (B), displayed over the mass range 100 - 170 amu. See text for interpretation of spectra.

PD-MS Analyses of Onium Fractions from *Tetraselmis* Cells Grown at Different Temperatures

Preliminary studies showed that no ions at m/z 127 and 141 were detectable in PD-MS analyses of onium fractions purified from 23°C grown *Tetraselmis* cells when no internal standards were added (not shown). Therefore the signals at m/z 127 and m/z 141 (from $^{2}H_{9}$-glycinebetaine and $^{2}H_{6}$-DMSP, respectively) can be attributed exclusively to these internal standards when these were added to cell extracts.

A typical PD-MS spectrum of the purified onium fraction (containing the two internal standards) from *Tetraselmis* cells grown at 23°C is illustrated in Fig. 2A. Note the abundant signal at m/z 118 from unlabeled glycinebetaine (M + H^{+}), a signal at m/z 140 from the

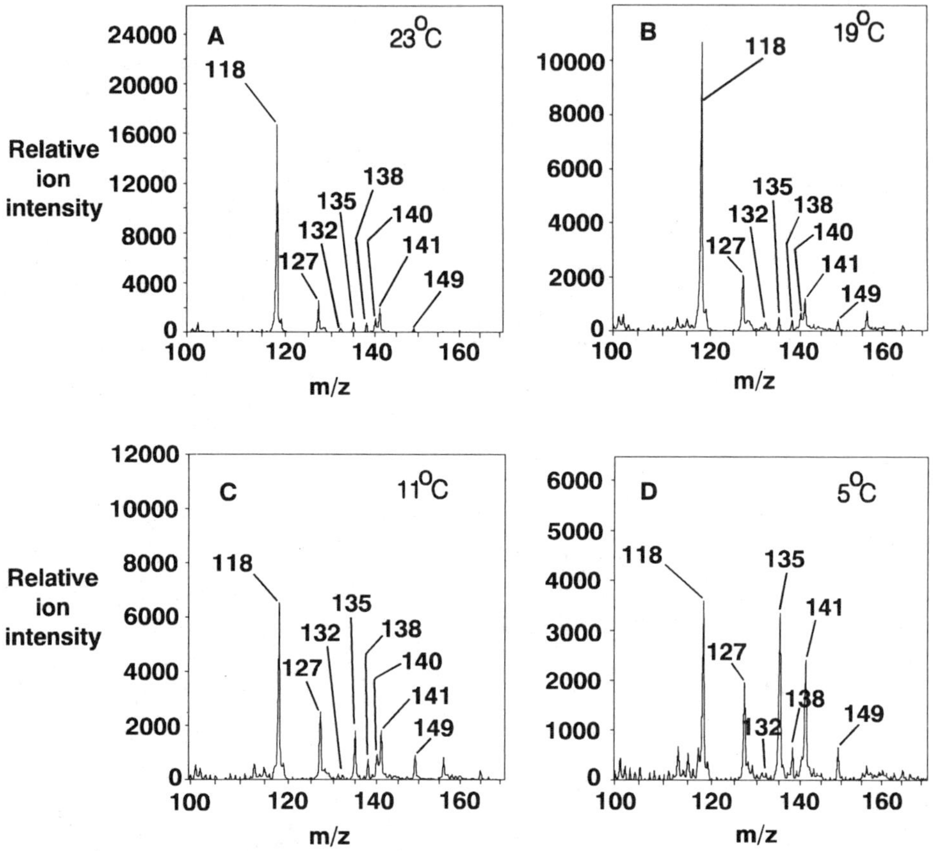

Figure 2. PD-MS spectra (displayed over the mass range 100 - 170 amu) of purified onium compound fractions from *Tetraselmis* cells grown at 23°C (A), 19°C (B), 11°C (C) and 5°C (D). Sample sizes were: 193 x 10^6 cells (A), 89 x 10^6 cells (B), 96 x 10^6 cells (C) and 60 x 10^6 cells (D). Each sample included 1000 nmol $^2H_9$-glycinebetaine and 2000 nmol $^2H_6$-DMSP.HCl as internal standards. See text for interpretation of spectra.

sodium adduct of unlabeled glycinebetaine ($M + Na^+$), and the methylation product of unlabeled glycinebetaine at m/z 132 ($M + CH_3^+$). Ions at m/z 127 and 141 correspond to the protonated molecules of the two internal standards, as discussed above. The ion at m/z 135 corresponds to unlabeled DMSP. The ion at m/z 138 (Fig. 2A) could correspond to either trigonelline (nicotinic acid betaine) ($M + H^+$)(1), or (more likely) homarine ($M + H^+$), which has been previously reported in *Tetraselmis* (4).

An advantage of this PD-MS method is that it can correct for losses of DMSP during sample purification, since any chemical breakdown of DMSP to DMS and acrylate should apply equally to the $^2H_6$-DMSP internal standard. DMSP can be quantified from the ratio of ions at m/z 135 : 141, the known amount of $^2H_6$-DMSP added as internal standard, and the known cell density of the extract. Likewise, glycinebetaine can be quantified from the ratio of ions at m/z 118 : 127. Mean glycinebetaine and DMSP levels (+/- standard deviations)[n = 3 samples] for *Tetraselmis* cells grown at 23°C determined by this PD-MS method were 31.62 (+/- 4.96) fmol/cell and 14.66 (+/- 5.35) fmol/cell, respectively (Table 1). The latter values are similar to values of 10 - 15 fmol DMSP/cell reported by Gröne and Kirst (1992) for *Tetraselmis subcor-*

Table 1. Effect of growth temperature on cell doubling time, glycinebetaine (Bet) and DMSP content of *Tetraselmis subcordiformis*

	Growth temperature (°C)			
	5	11	19	23
Doubling time (d)	4.55	3.61	2.42	1.51
Bet (fmol/cell) [a]	30.01	38.68	44.16	31.62
	(12.00)	(7.04)	(9.28)	(4.96)
DMSP (fmol/cell) [a]	121.23	49.88	14.83	14.66
	(30.61)	(9.49)	(2.32)	(5.35)
DMSP : Bet ratio	4.04	1.29	0.34	0.46

[a] Each value is the mean of three determinations; standard deviations are shown in parentheses.

diformis cells grown at 23°C on N-replete, artificial seawater medium, as measured by gas chromatography of DMS liberated from DMSP (7).

The latter gas chromatography method was applied to analysis of DMSP in aqueous extracts of *Tetraselmis* without added internal standards, and mean (+/- standard deviation) DMSP levels determined by this DMS liberation method were 12.33 +/- 3.47 fmol/cell (n = 5 samples); not significantly different from the values determined by stable isotope dilution PD-MS (above).

Dickson and Kirst (1986) reported glycinebetaine levels which are 1.64-fold greater than those of DMSP in *Tetraselmis* grown at 22°C and 500 mM NaCl, as determined by semi-quantitative TLC (4). This molar ratio of DMSP : glycinebetaine of 0.61 is of the same order of magnitude as observed in the present study for cells grown at a similar NaCl concentration and temperature (DMSP : glycinebetaine ratio = 0.46) (Table 1).

Because the above results obtained with PD-MS were in general agreement with other published values for *Tetraselmis* grown at 22 - 23°C, we applied this method to analysis of onium compounds from cells grown at different temperatures. Typical PD-MS spectra of purified onium compound fractions (containing internal standards) from cells grown at 19°C, 11°C and 5°C are illustrated in Figs. 2B, 2C and 2D, respectively. It should be noted that cell number per extract differed for each of the samples illustrated in Fig. 2 (see Fig. 2 legend) and this largely accounts for the observed differences in ratio of m/z 118 : 127. Thus, glycinebetaine level per cell remained fairly constant at the different growth temperatures (Table 1). This trend was not seen for the ion ratio m/z 135:141. As growth temperature was decreased, the intensity of the signal at m/z 135 corresponding to DMSP was greater, despite fewer cells harvested per extract at the lower growth temperatures (Fig. 2). Results summarized in Table 1 show that as growth temperature was decreased there was a substantial increase (8-fold) in DMSP content per cell which resembles that reported by Karsten et al (1992) for marine macroalgae (13).

The large increase in DMSP titer of *Tetraselmis* cells with low growth temperatures (Table 1) appears consistent with the proposed role for DMSP as a cryoprotectant in marine algae (13,15).

We have not yet fully resolved the problem of quantifying homarine (and/or trigonelline) in *Tetraselmis* extracts. The most obvious stable isotope labeled internal standard for homarine and/or trigonelline quantification would be $^2H_3$-homarine and/or $^2H_3$-trigonelline, prepared by reaction of C^2H_3I with picolinic acid or nicotinic acid, respectively. However, both of these internal standards would be expected to have the same mass : charge ratio as $^2H_6$-DMSP (i.e. m/z 141), and would therefore interfere with routine DMSP quantification.

The simplest solution to this problem would be to prepare separate samples for homarine (and/or trigonelline) quantification in which $^2H_6$-DMSP is excluded as internal standard. For such samples an NH_4OH eluate from Dowex-50-H^+ could be employed to prepare onium compound fractions which are essentially DMSP-free; i.e. containing only glycinebetaine and homarine (and/or trigonelline).

Naturally occurring and genetically engineered variation for quaternary ammonium and tertiary sulfonium compounds is of great interest in terms of osmotolerance of plants and photosynthetic microorganisms (16,19-21,27). Rapid and sensitive screening techniques for quantifying these compounds (such as the PD-MS method described here) have been instrumental in advancing our knowledge of genetic variability for glycinebetaine and trigonelline accumulation in maize (1,27). Such desorption mass spectrometry methods in conjunction with the use of stable isotope labeled precursors have also played a key role in the elucidation of biosynthetic pathways of these compounds (8,19,20).

The present studies have not addressed whether the increase of DMSP content of *Tetraselmis* in response to low growth temperatures is the result of increased synthesis from methionine (5,7) via *S*-methylmethionine (8), and/or decreased degradation of DMSP to DMS and acrylate (17,25). In principle, PD-MS, in combination with suitable stable isotope labeled DMSP precursors, could be employed to address the turnover rates of the DMSP pool of *Tetraselmis* as a function of growth temperature.

ACKNOWLEDGMENTS

We thank Marshall Overley (Biology Teacher at West Lafayette High School, West Lafayette, IN) for guidance and encouragement of E.B.S. in this project, and Connie C. Bonham (Biochemistry Dept., Purdue University, West Lafayette, IN) for the PD-MS analyses. This work was supported by a grant from the USDA National Research Initiative Competitive Grants Program (contract 93-37100-8870).

REFERENCES

1. Bonham, C.C., K.V. Wood, W.-J. Yang, A. Nadolska–Orczyk, Y. Samaras, D.A. Gage, J. Poupart, M. Burnet, A.D. Hanson and D. Rhodes. 1995. Identification of quaternary ammonium and tertiary sulfonium compounds by plasma desorption mass spectrometry. J. Mass. Spectrom. *30:* 1187-1194.
2. Charlson, R.J., J.E. Lovelock, M.O. Andreae and S.G. Warren. 1987. Oceanic phytoplankton, atmospheric sulphur, cloud albedo and climate. Nature *326*: 655-661.
3. Dacey, J.W.H., G.M. King and S.G. Wakeham. 1987. Factors controlling emission of dimethylsulphide from salt marshes. Nature *330*: 643-645.
4. Dickson, D.M.J. and G.O. Kirst. 1986. The role of β-dimethylsulphoniopropionate, glycine betaine and homarine in the osmoacclimation of *Platymonas subcordiformis*. Planta *167*: 536-543.

4a. Gooday, G. W. 1970. A physiological comparison of the symbiotic alga *Platymonas convolutae* and its free living relatives. J. Mar. Biol. Ass. U.K. *50*: 199-208.

5. Greene, R.C. 1962. Biosynthesis of dimethyl-β-propiothetin. J. Biol. Chem. *237*: 2251-2254.
6. Gröne, T. and G.O. Kirst. 1991. Aspects of dimethylsulfonio-propionate effects on enzymes isolated from the marine phytoplankter *Tetraselmis subcordiformis* (Stein). J. Plant Physiol. *138*: 85-91.
7. Gröne, T. and G.O. Kirst. 1992. The effect of nitrogen deficiency, methionine and inhibitors of methionine metabolism on the DMSP contents of *Tetraselmis subcordiformis* (Stein). Mar. Biol. *112*: 497-503.
8. Hanson, A.D., J. Rivoal, L. Paquet and D.A. Gage. 1994. Biosynthesis of 3-dimethylsulfoniopropionate in *Wollastonia biflora* (L.) DC.: Evidence that *S*-methylmethionine is an intermediate. Plant Physiol. *105*: 103-110.
9. Karsten, U., C. Wiencke and G.O. Kirst. 1990. The β-dimethylsulphoniopropionatę (DMSP) content of macroalgae from Antarctica and Southern Chile. Botanica Marina *33*: 143-146.

10. Karsten, U., C. Wiencke and G.O. Kirst. 1990. The effect of light intensity and daylength on the β-dimethylsulphonio-propionate (DMSP) content of green macroalgae at different irradiances. Plant, Cell & Environ. *13*: 989-993.
11. Karsten, U., C. Wiencke and G.O. Kirst. 1991. The effect of salinity changes upon the physiology of eulitoral green macroalgae from Antarctica and Southern Chile. II. Intracellular inorganic ions and organic compounds. J. Exp. Bot. *42*: 1533-1539.
12. Karsten, U., C. Wiencke and G.O. Kirst. 1991. Growth pattern and β-dimethylsulphoniopropionate (DMSP) content of green macroalgae at different irradiances. Mar. Biol. *108*: 151-155.
13. Karsten, U., C. Wiencke and G.O. Kirst. 1992. Dimethyl-sulphoniopropionate (DMSP) accumulation in green macroalgae from polar to temperate regions: interactive effects of light versus salinity and light versus temperature. Polar Biol. *12*: 603-607.
14. Malin, G., S.M. Turner and P.S. Liss. 1992. Sulfur: the phytoplankton/climate connection. J. Phycol. *28*: 590-597.
15. Nishiguchi, M.K. and G.N. Somero. 1992. Temperature- and concentration-dependence of compatibility of the organic osmolyte β-dimethylsulfoniopropionate. Cryobiol. *29*: 118-124.
16. Nomura, M., M. Ishitani, T. Takabe, A.K. Rai and T. Takabe. 1995. *Synechococcus* sp. PCC7942 transformed with *Escherichia coli bet* genes produces glycine betaine from choline and acquires resistance to salt stress. Plant Physiol. *107*: 703-708.
17. Pakulski, J.D. and R.P. Kiene. 1992. Foliar release of dimethylsulfoniopropionate from *Spartina alterniflora*. Mar. Ecol. Prog. Ser. *81*: 277-287.
18. Reed, R.H. 1983. Measurement and osmotic significance of β-dimethylsulphoniopropionate in marine macroalgae. Mar. Biol. Lett. *4*: 173-181.
19. Rhodes, D. and A.D. Hanson. 1993. Quaternary ammonium and tertiary sulfonium compounds in higher plants. Annu. Rev. Plant Physiol. Plant Mol. Biol. *44*: 357-384.
20. Rhodes, D., P.J. Rich, A.C. Myers, C.C. Reuter and G.C. Jamieson. 1987. Determination of betaines by fast atom bombardment mass spectrometry: Identification of glycine betaine deficient genotypes of *Zea mays*. Plant Physiol. *84*: 781-788.
21. Saneoka, H., C. Nagasaka, D.T. Hahn, W.-J. Yang, G.S. Premachandra, R.J. Joly and D. Rhodes. 1995. Salt tolerance of glycinebetaine-deficient and -containing maize lines. Plant Physiol. *107*: 631-638.
22. Steudler, P.A. and B.J. Peterson. 1984. Contribution of gaseous sulphur from salt marshes to the global sulphur cycle. Nature *311*: 455-457.
23. Steudler, P.A. and B.J. Peterson. 1985. Annual cycle of gaseous sulfur emissions from a New England *Spartina alterniflora* marsh. Atmos. Environ.*19*: 1411-1416.
24. Stewart, G.R., F. Larher, I. Ahmad and J.A. Lee. 1979. Nitrogen metabolism and salt-tolerance in higher plant halophytes, p. 211-227. *In* R.L. Jefferies and A.J. Davy (ed.), Ecological Processes in Coastal Environments. Blackwell, Oxford.
25. Vairavamurthy, A., M.O. Andreae and R.L. Iverson. 1985. Biosynthesis of dimethylsulfide and dimethylpropiothetin by *Hymenomonas carterae* in relation to sulfur source and salinity variations. Limnol. Oceanogr. *30*: 59-70.
26. van Diggelen, J., J. Rozema, D.M.J. Dickson and R. Broekman. 1986. β-3-dimethylsulphoniopropionate, proline and quaternary ammonium compounds in *Spartina anglica* in relation to sodium chloride, nitrogen and sulphur. New Phytol. *103*: 573-586.
27. Yang, W.-J., A. Nadolska–Orczyk, K.V. Wood, D.T. Hahn, P.J. Rich, A.J. Wood, H. Saneoka, G.S. Premachandra, C.C. Bonham, J.C. Rhodes, R.J. Joly, Y. Samaras, P.B. Goldsbrough and D. Rhodes. 1995. Near-isogenic lines of maize differing for glycinebetaine. Plant Physiol. *107*: 621-630.

6

AN ALTERNATIVE APPROACH TO THE MEASUREMENT OF β-DIMETHYLSULFONIOPROPIONATE (DMSP) AND OTHER PRECURSORS OF DIMETHYLSULFIDE

A. G. Howard and D. W. Russell

Chemistry Department
University of Southampton
Southampton, Hampshire, United Kingdom

SUMMARY

The restricted selectivity of conventional methods for the measurement of β-dimethylsulfoniopropionate (DMSP) has prompted the development of novel HPLC instrumentation tailored to the measurement of dimethylsulfide (DMS) precursors. By combining the separation and identification capabilities of ion-exchange HPLC with flame photometric detection after a post-column alkaline hydrolysis step, a high degree of selectivity is achieved towards compounds which readily yield dimethylsulfide on base hydrolysis. The selectivity of the new instrumentation to a number of dimethylsulfonium compounds is described. The detection limit of the prototype instrument for DMSP, based on three times the standard deviation of the blank and using a 200 μl injection volume, corresponds to a concentration of 200nM DMSP or an absolute injected mass of 6 ng of DMSP.

A novel means of response linearization has been achieved by the post-column addition of DMSP. Subsequent hydrolysis of this added DMSP gives a slow continuous bleed of sulfur into the flame photometric detector resulting in a reduction of the log-log calibration graph slope from typically 1.8 to 1.1. This permits essentially linear calibration to be achieved. Sensitivity is improved by such an addition, but only at the expense of increased noise. No significant practical improvement in signal-to-noise characteristics has as yet been attained from such an addition.

INTRODUCTION

Sulfate metabolism by marine algae results in the formation of a number of organosulfur compounds, the most commonly reported being dimethylsulfoniopropionate (DMSP)

(Compound A, Figure 1). DMSP is believed to have an osmoregulatory role, its level in algae responding to changes in salinity (7). Its chemical, bacterial and/or enzymatic breakdown releases volatile dimethylsulfide (DMS) into the atmosphere which is rapidly oxidized to methane sulfonic acid, sulfur dioxide, and sulfate. In turn these products cause the nucleation of cloud droplets and increased aerosol and precipitation acidity (2).

Since DMS was first identified as the "odoriferous principle" evolved from the macro-alga *Polysiphonia fastigiata* in 1935 (8) numerous papers have been published concerning DMS and one of its biological precursors, DMSP. DMSP itself was first extracted and characterized from *Polysiphonia fastigiata* by Challenger and Simpson in 1948 and its alkaline hydrolysis became the basis for the measurement of DMSP (4). Indirect methods employing flame photometric detection (1) have been used almost exclusively for DMSP measurement since that time and have contributed significantly to our understanding of the role of DMS-precursors (denoted by DMS-Pr) in the global sulfur cycle.

DMS-Pr are widely distributed through the plant kingdom but are most commonly found in saline environments. In the marine environment such compounds are present in macro-algae (4), marsh grasses and phytoplankton, particularly prymnesiophytes such as *Phaeocystis* and coccolithophores (12). In mid and high latitude regions DMS concentrations differ widely between summer and winter (18) as a result of changes in biological activity.

DMSP is not however the only potential DMS-precursor present in marine organisms. A number of other compounds have been isolated and characterized which could contribute to the release of DMS. These include 4-(dimethylsulfonio)-2-(methoxy)butyrate (Compound B) (3,17), S-methylmethionine (SMM)(13,19), gonyol (Compound D) (14), gonyauline (Compound C) (15) and less certainly 5-(dimethylsulfonio)pentanoate (Compound E) (9,11).

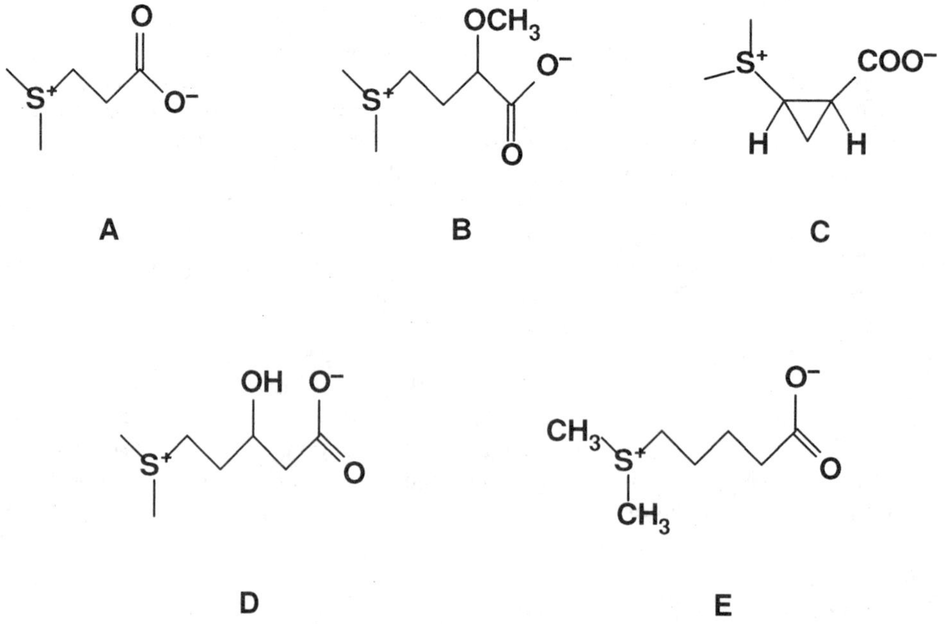

Figure 1. Some naturally occurring dimethylsulfonium compounds (A = DMSP; B = 4-(dimethylsulfonio)-2-(methoxy)butyrate; C = gonyauline; D = gonyol; E = 5-(dimethylsulfonio)pentanoate).

Although the literature contains many reports of measurements of DMSP, closer inspection reveals that in most cases DMSP has been measured by methods which quantitate DMS released on room temperature base hydrolysis. The extent to which other dimethylsulfonium compounds would contribute to such measurements has yet to be determined. Current DMSP measurements may therefore, for some sample types, be misleading. Improved methods which can distinguish between DMSP and the other DMS-precursors which might be present are therefore required. We have recently developed prototype instrumentation which, by linking HPLC separation capability with sulfur-specific detection (10), offers the potential for distinguishing between such DMS-precursors. This paper describes the selectivity of the apparatus and presents a novel approach to the linearization of its response.

MATERIALS AND METHODS

Standards and Reagents

Two main routes were employed for the synthesis of standard sulfonium compounds. DMSP and 2-(methyl)dimethylsulfoniopropionate (2Me-DMSP) were prepared by stirring dimethylsulfide with an unsaturated acid (acrylic acid or methacrylic acid respectively). The preparation of dimethylsulfonioacetate (DMS-Ac), 4-(dimethylsulfonio)butanoate (DMS-but) and 5-(dimethylsulfonio)pentanoate (DMS-pent) involved the reaction of the corresponding bromo-acid with dimethylsulfide. The synthesis of dimethylsulfoniocholine (DMS-Chol) followed a similar route involving the reaction of 2-bromoethanol with dimethylsulfide. S-methylmethionine (SMM) was prepared by the methylation of methionine with iodomethane. The products were characterized by NMR spectroscopy and carbon/hydrogen/nitrogen analysis. Analytical reagent grade materials were employed elsewhere unless otherwise stated.

Instrumentation

The common features of all the DMS-precursors which have been identified to date are that they are low molecular weight ionic, dimethylsulfonium compounds which are unstable with respect to base hydrolysis. The apparatus employed in this work was specially designed to exploit this instability to produce instrumentation which only responds to compounds, such as DMSP, which readily break down to give a volatile sulfur product. It incorporates isocratic ion-exchange HPLC of the DMSP followed by post-column base hydrolysis. This releases volatile DMS from the compounds, which after separation from the liquid stream can be detected using a custom-designed flame photometric detector (Figure 2).

HPLC Conditions

Depending on the nature of the DMSP precursors to be determined, chromatography can either be carried out on an anion or cation-exchange column. A Du Pont liquid chromatography pump was employed to deliver the eluent and the sample was injected onto the column using a Rheodyne 7125 injection valve fitted with a 200μl sampling loop. Cation-exchange chromatography employed a Spherisorb 5 SCX column (5μm packing, 25cm x 4 mm i.d.) eluted with 50 mM aqueous potassium dihydrogen orthophosphate (pH adjusted to 5.7 with 3mM sodium hydroxide solution) at a flow rate of 0.8 ml/min. Anion-exchange chromatography was carried out on a Spherisorb 5 SAX column (5μm

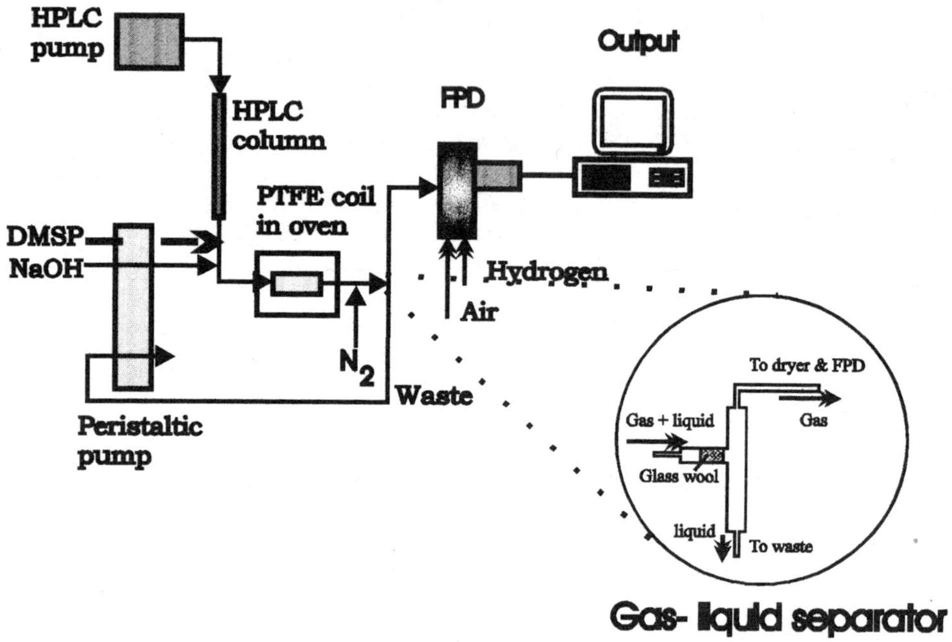

Figure 2. HPLC instrumentation for the measurement of DMS-precursors. Post-column DMSP addition is employed only when response linearization is required.

packing, 25cm x 4 mm i.d.) eluted with 50 mM aqueous potassium dihydrogen orthophosphate (pH adjusted to 6.7 with 20 mM sodium hydroxide solution) at a flow rate of 1.0 ml/min.

Post Column Hydrolysis

Post column hydrolysis was carried out by mixing the HPLC column eluent with a pumped stream (2.0 ml/min) of 4M sodium hydroxide solution. The rate of hydrolysis was enhanced by passing the mixture through a PTFE reaction tube (length: 6m, 0.74 mm i.d.) which was maintained at 85°C in a thermostatted oven. Although the apparent yield of DMS increased at temperatures > 85°C, increased levels of water vapour deteriorated the signal-to-noise characteristics of the detector.

DMS Detection

After the hydrolysis step the solution was mixed with nitrogen (flow rate: 50ml/min). The DMS was purged from solution in a gas-liquid separator (Figure 2) and the residual liquid was pumped to waste. The gas stream was then partially dried. In the first approach this was carried out using two traps; the first drying trap was empty and used to physically condense water vapour and spray, the second contained the chemical drying agent anhydrous magnesium perchlorate (14-22 mesh). In more recent instrumentation a Nafion dryer (12" x 0.06", held in molecular sieves) has replaced these traps to minimize band broadening of the DMS peaks.

The gas stream then passed into the air-hydrogen flame of a custom-built flame photometric detector. The S_2 band emission from the flame was monitored by a high sensitivity photomultiplier tube (EMI 6256B, run at 800V) viewing through a wide bandpass glass filter (Oriel BG-12). The signal was amplified, damped, and displayed on a chart recorder or computing integrator. Increased selectivity was achieved using a narrow band-pass interference filter, but with reduced sensitivity.

Calibration

The instrument was calibrated using peak height or area measurements obtained from standard solutions prepared in deionized water. In its standard configuration the FPD response depends on approximately the square of the sulfur concentration in the flame and calibration is therefore normally carried out using a log (concentration) / log (response) plot.

Response to Other Dimethylsulfonium Compounds

In order to assess the selectivity of the instrument towards dimethylsulfonium compounds the HPLC column was removed from the apparatus and 200µl aliquots of standard sulfonium solutions were injected. In this way column retention time dependent peak broadening and peak tailing effects were removed, permitting the response of each compound to be assessed under identical conditions. The response of the system depends approximately on the square of the concentration and the comparison was therefore made by assessing the concentration of each compound which was required to give a fixed instrument response.

For two compounds, 1 and 2, their responses, R_1 and R_2 respectively, are related to their concentrations (C_1 and C_2) by constants k_1 and k_2.

$$R_1 = k_1 C_1^2 \qquad R_2 = K_2 C_2^2$$

If the two compounds are to have equal responses:

$$R_1 = R_2$$

$$k_1 C_1^2 = k_2 C_2^2$$

$$K = \frac{C_1}{C_2} = \sqrt{\frac{k_2}{k_1}}$$

If one of the measured compounds is DMSP, K will therefore reflect the concentration of a compound required to give the same response as DMSP.

RESULTS AND DISCUSSION

Instrument Performance

Chromatography. The potential DMS-precursors which have been identified to date have in common the positively charged dimethylsulfonium group, which makes them amenable to cation-exchange chromatography. Compounds like DMSP, however are zwitterionic over a range of pH values around neutrality and can be satisfactorily chromatographed on both anion- and cation-exchange media, thereby enhancing confidence in the compound identification.

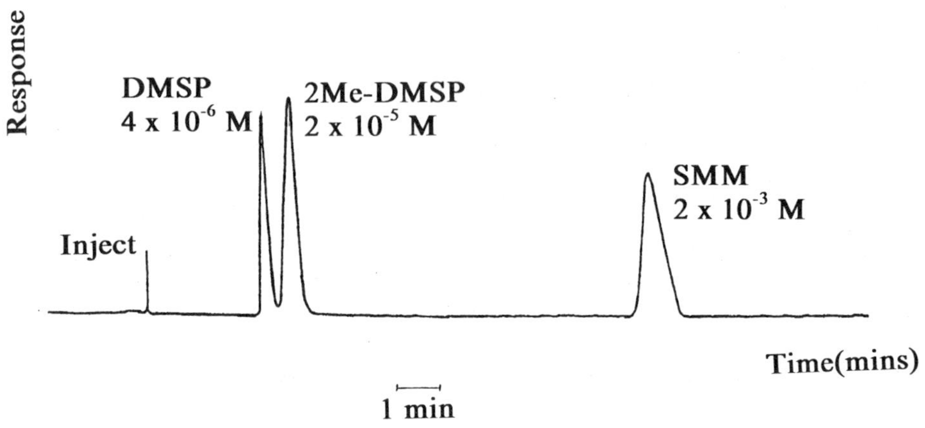

Figure 3. Sulfur-specific chromatography of three sulfonium compounds (2Me-DMSP = 2-(methyl)dimethylsulfoniopropionate, SMM = S-methylmethionine).

A typical HPLC trace, obtained from the system to show its sensitivity to three sulfonium compounds, is shown in Figure 3.

Response Characteristics. The $\log_{10}$ (concentration) v $\log_{10}$ (response) calibration plot for DMSP gives a slope of approximately two which is characteristic of the expected squared relationship obtained from emission resulting from the diatomic S_2 species (Figure 4).

The detection limit for DMSP, based on three times the standard deviation of the blank, and using a 200 μl injection volume, corresponds to a concentration of 200 nM DMSP

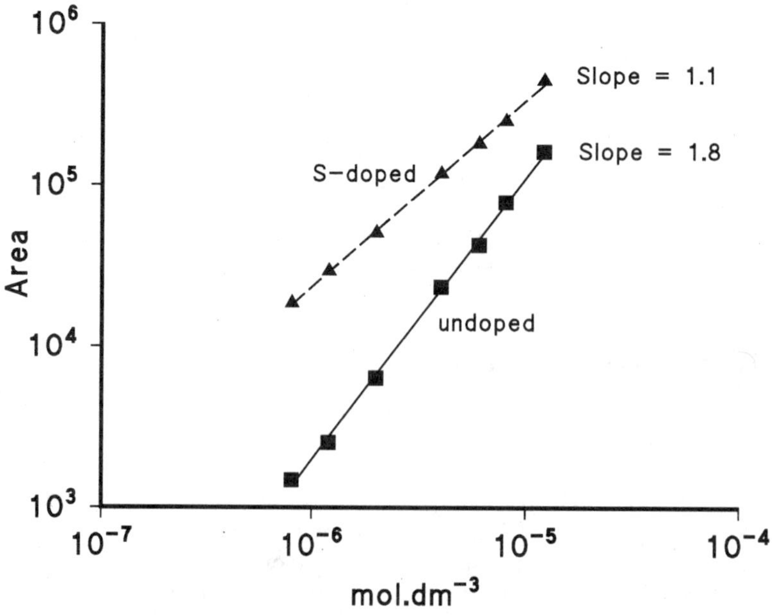

Figure 4. Linearized calibration obtained by bleeding a low concentration of DMSP into the column effluent.

or an absolute mass of 6 ng of DMSP. The presented instrumentation is in the early stages of development and significant improvements in sensitivity are anticipated from future versions of the instrument.

Selectivity in Response to Dimethylsulfonium Compounds. The relative response factors were determined for a number of available compounds that are structurally related to DMSP (Table 1).

Although it has not as yet been possible to obtain or synthesize samples of some of the potential DMS-precursors which have been isolated from marine flora, the preliminary results presented in Table 1 illustrate that the present instrumentation is highly selective towards DMSP and its methyl analogue. The chemical structure of DMSP makes it particularly susceptible to base hydrolysis and the rate of formation of dimethylsulfide is therefore greater than for other dimethylsulfonium compounds. As the hydrolysis reaction time in the HPLC instrumentation is short, a high yield of DMS is therefore only possible for DMSP and its structurally similar methyl analogue. All the studied compounds are thermodynamically capable of yielding DMS on base hydrolysis, but due to slow kinetics of hydrolysis only generate DMS at a low rate. Within the time allowed in the flowing stream for the reaction to proceed, only small quantities of DMS are therefore produced. To boost the responses from other dimethylsulfonium compounds would therefore require either longer reaction times or elevated hydrolysis temperatures. Neither of these approaches would however be viable for continuous flow instrumentation as these would both increase band-broadening and baseline noise. It is worth noting however that if gas chromatographic procedures which employ extended reaction times are used for the measurement of DMS-Pr, the yield of DMS from any other DMS-Pr might become significant.

Response Linearization. It has been reported that the response to sulfur compounds by flame photometric detectors can be linearized by bleeding low concentrations of sulfur into the flame gases (5). Such linearization is normally carried out by doping the flame gases with a volatile sulfur compound such as carbon disulfide or dimethylsulfide from a permeation source. With this HPLC system a simple modification of the apparatus can be used to supply the additional volatile sulfur to give a linearized detector response. This is achieved by the delivery of a small flow of DMSP directly into the effluent from the HPLC column (Figure 2). The subsequent hydrolysis then results in the release of a constant quantity of dimethylsulfide into the reagent stream giving rise to the required response linearization, taking, in the illustration, the log-log slope from 1.8 to 1.1 (Figure 4).

The concentration of added DMSP must be chosen with due regard to the concentrations of DMSP to be measured by the system and must be limited in order to restrict the baseline shift which occurs due to S_2 emission from the added sulfur. The increased sensitivity resulting from this addition is only however achieved with an increased noise

Table 1. Relative response factors for some dimethylsulfonium compounds (response factors relative to DMSP)

Compound	Relative response factor K
Dimethylsulfoniopropionate	1.0
2-(methyl)dimethylsulfoniopropionate	0.3
Dimethylsulfoniocholine	6×10^{-3}
S-methylmethionine	2×10^{-3}
3-(dimethylsulfonio)butanoate	8×10^{-4}
Dimethysulfonioacetate	4×10^{-5}
5-(dimethylsulfonio)pentanoate	3×10^{-5}

level and has not in our experience to date been favorable in terms of improved signal-to-noise characteristics. The ease with which the slow bleed of additional sulphur into the detector can be achieved and the degree of concentration control which is possible by simple dilution of the DMSP dopant solution, makes the liquid dopant system a simpler and easier to operate alternative to the permeation tubes employed in direct vapour or gas doping systems.

CONCLUSIONS

A detailed understanding of the origins and role of DMS-precursors in the production of atmospheric dimethylsulfide requires the application of rigorous analytical methods which are capable of distinguishing between a number of related compounds. The instrumentation described in this paper offers a high level of selectivity towards DMSP. The selectivity comes from three independent sources. In addition to the compound identification which is possible from characteristic retention time behaviour, few compounds produce volatile sulfur compounds on base hydrolysis. In terms of selectivity, few instruments are better tailored to a group of compounds as the release of volatile sulfur compounds is a prerequisite for the system to respond to a particular compound. DMS-Pr differ significantly in their rates of hydrolysis and with the instrumentation presented here the sensitivity towards DMSP and its methyl analogue are significantly higher than for the other compounds which have been investigated to date. Additional confidence in DMSP identification can be achieved by parallel analyses of DMSP on both cation and anion-exchange columns.

Current instrumentation is in prototype form and many design enhancements are possible to improve sensitivity. Whilst the ultimate goal of such developments must be the ability to measure DMS-Pr and in particular DMSP in seawater, current performance is more than adequate for the analysis of algal material. For the determination of natural levels of DMSP in seawater some further sensitivity enhancement will be necessary, but current performance would indicate that these should not be unattainable.

There are many applications for which the current instrumentation is too selective towards DMSP and variants of the technique are therefore under development for the measurement of a wider range of sulfonium compounds which are intermediaries in the marine and terrestrial sulfur cycles.

REFERENCES

1. Andreae, M.O. and W.R. Barnard. 1983. Determination of trace quantities of dimethylsulfide in aqueous solutions. Anal.Chem. *55:* 608-612
2. Andreae, M. O. 1990. Ocean-atmosphere interactions in the global biogeochemical sulfur cycle. Mar. Chem. *30:* 1-29
3. Blunden, G. and S.M. Gordon. 1986. Betaines and their sulfonio analogues in marine algae. Prog. in Phycol. Res.*4:* 39-80.
4. Challenger, F. and M.I. Simpson. 1948. Studies on biological methylation. Part XII A precursor of dimethylsulfide evolved by *Polysiphonia fastigiata*. Dimethyl-2-carboxyethylsulfonium hydroxide and its salts. J.Chem.Soc. *43:* 1591-1597.
5. Coburn, W.G. and R.E. Mulrooney. 1988. Linearized response of a doped FPD continuous monitor for low level sulphur measurements. Atmos. Environ. *22(9):* 1941-1947.
6. DeSouza, M.P. and D.C. Yoch. 1995. Purification and characterization of DMSP-lyase from an alcaligenes-like DMS-producing marine isolate. Appl. Environ. Microbiol. *61*: 21-26
7. Dickson, D.M., R.G. Wyn Jones and J. Davenport. 1980. Steady state osmotic adaption in Ulva lactuca. Planta *150*:158-165.
8. Haas, P. 1935. The liberation of dimethylsulfide in seaweed. J. Biochem. *29*: 1297-1299.

9. Hanson,A.D. Z-H Huang and D.A. Gage. 1993. Evidence that the putative compatible solute S-dimethyl-sulfoniopentanoate is an extraction artifact. Plant Physiol. *101:* 1391-1393.
10. Howard, A.G. and D.W. Russell. 1995. HPLC-FPD instrumentation for the measurement of the atmospheric DMS-precursor DMSP. Anal. Chem. *67:* 1293-1295.
11. Larher, F. and J. Hamelin. 1979. L'Acide dimethylsulfonium-S pentanoique de *Diplotaxis tenuifolia*. Phytochem. *18:* 1396-1397.
12. Malin, G., S. Turner, P. Liss, P. Holligan, and D. Harbour. 1993. Dimethylsulfide and dimethylsulfoniopropionate in the North East Atlantic during the summer coccolithophore bloom. Deep Sea Res. *40:* 1487-1508.
13. McRorie, R.A., G.L. Sutherland, M.S. Lewis, A.D. Barton, M.R. Glazener and W. Shive. 1954. Isolation and identification of a naturally occurring analog of methionine. J.Am.Chem.Soc. *76:* 115-118
14. Nakamura, H., K. Fujimaki, O. Sampei, and A. Murai. 1993. Gonyol: methionine-induced sulfonium accumulation in a dinoflagellate *Gonyaulex polyhedra*. Tet. Lett. *34*: 8481-8484.
15. Nakamura, H., M. Ohtoshi, O. Sampei, Y. Akashi and A. Murai. 1992. Synthesis and absolute configuration of (+)-gonyauline: a modulating substance of bioluminescent circadium rhythm in unicellular alga *Gonyaulex polyhedra*. Tet. Lett. *33*: 2821-2822.
16. Reed, R.H. 1983. Measurement and osmotic significance of ß-dimethylsulfoniopropionate in marine macroalgae. Mar. Biol. Lett. *4*: 173-181.
17. Sciuto, S., M. Piatelli and R. Chillemi. 1982. (-)-(S)-4-dimethylsulfonio-2-methoxybutyrate from the red alga *Rytiphloea tinctoria*. Phytochem. *21*: 227-228.
18. Turner, S.M., G. Malin, P. Liss, D.S. Harbour and P.M. Holligan. 1988. The seasonal variation of dimethylsulfide and dimethylsulfoniopropionate concentrations in nearshore waters. Limnol. Oceanogr. *33:* 364-375.
19. White, R.H. 1982. Analysis of dimethylsulfonium compounds in marine algae. J.Mar.Res. *40: 529-536.*

7

3-DIMETHYLSULFONIOPROPIONATE BIOSYNTHESIS AND USE BY FLOWERING PLANTS

A. D. Hanson[1] and D. A. Gage[2]

[1] Horticultural Sciences Department
University of Florida
Gainesville, Florida 32611

[2] Department of Biochemistry
Michigan State University
East Lansing, Michigan 48824

SUMMARY

3-Dimethylsulfoniopropionate (DMSP) is known to be an effective osmoprotectant. It is accumulated to high levels (5-50 μmol g^{-1} fresh weight) by several diverse flowering plants, including intertidal *Spartina* species and sugarcane (Poaceae), and the coastal plant *Wollastonia biflora* (Asteraceae). Many other species may have traces of DMSP (0.01-0.5 μmol g^{-1} fresh weight). The biosynthetic pathway has been investigated in *W. biflora* leaves. *In vivo* isotope labeling data are consistent with the following sequence of steps:

$$CH_3SCH_2CH_2CH(NH_2)COOH \rightarrow (CH_3)_2S^+CH_2CH_2CH(NH_2)COOH \rightarrow (CH_3)_2S^+CH_2CH_2CHO \rightarrow (CH_3)_2S^+CH_2CH_2COOH$$

Methionine — *S*-Methylmethionine — DMSP-aldehyde — DMSP

It is not clear whether *S*-methylmethionine is converted to DMSP-aldehyde directly or via an intermediate. The methyltransferase catalyzing the conversion of methionine to *S*-methylmethionine has been purified and shown to be a tetramer of 115-kD subunits; the other biosynthetic enzymes are not yet known. *In vivo* radiotracer studies of DMSP catabolism in *W. biflora* leaves indicate that demethylation to methylthiopropionate and breakdown to dimethyl sulfide are very slow (1% of the DMSP pool per day). In *W. biflora* leaves, both *S*-methylmethionine and DMSP-aldehyde are degraded to dimethyl sulfide far more rapidly than DMSP, which suggests that part of the dimethyl sulfide emitted by DMSP-producing plants might come from these compounds. The efficacy of DMSP as an osmoprotectant coupled with its simple biosynthetic pathway, metabolic stability and lack of nitrogen make DMSP a rational candidate for the genetic engineering of osmotic stress resistance in crop plants.

Biological and Environmental Chemistry of DMSP and Related Sulfonium Compounds
edited by Ronald P. Kiene et al., Plenum Press, New York, 1996

INTRODUCTION

Like many marine algae (4,33), certain flowering plants (angiosperms) contain 3-dimethylsulfoniopropionate (DMSP). This is of both environmental and agricultural interest. The environmental interest arises because DMSP in angiosperms is a potential source of the dimethyl sulfide (DMS) emitted from terrestrial ecosystems (45). The agricultural interest stems from the effectiveness of DMSP as an osmoprotectant and cryoprotectant (53) and from the possibility of enhancing the stress resistance of crops by genetic engineering of the accumulation of such compounds (50). With these interests in mind, we will first review the occurrence, biosynthesis and catabolism of DMSP in angiosperms, making comparisons to algae where appropriate. We will then briefly discuss the regulation of DMSP levels and the prospects for the genetic engineering of DMSP accumulation in crop plants.

OCCURRENCE OF DMSP IN ANGIOSPERMS

Only a few hundred angiosperms have been analyzed for DMSP and, as these are mostly mesophytic or halophytic herbs of temperate regions, they do not adequately represent the group (46). In general, only leaves have been examined. It is therefore hard at this point to make general statements about taxonomic or ecological patterns in the occurrence of DMSP. Further, some reports of DMSP need cautious interpretation. Those based on chromatography or on DMS release from samples treated with cold base do not prove that DMSP is present (60). Also, DMSP in marine angiosperms may be due, at least in part, to epiphytic algae (10). In view of these problems, the data cited below establish only (a) that a few, taxonomically disparate angiosperms certainly accumulate high levels of DMSP, and (b) that many others probably contain traces.

DMSP-Accumulating Species. High levels of DMSP (about 5 to 50 μmol g^{-1} fresh weight) have been found in leaves of *Spartina anglica* (36), *S. alterniflora* (11,15), *S. foliosa* (38) and *Saccharum* spp. (41) from the family Poaceae, and *Wollastonia biflora* (= *Melanthera biflora*) (25,56) from the family Asteraceae. The *Spartina* species are salt marsh halophytes and *W. biflora* is a salt-tolerant coastal strand plant, whereas *Saccharum* (sugarcane) is a moderately salt-sensitive crop. In these plants there is no doubt that DMSP is present as it has been identified by spectroscopic methods. Large quantities of DMSP have also been reported from the marine angiosperms *Posidonia* sp. (61) and *Zostera* sp. (60). However, both these reports require confirmation as the analyses were based on thin layer chromatography or DMS release, and contributions from epiphytic algae were not excluded.

Reports for *W. biflora* and *S. alterniflora* show DMSP levels proportional to the external salt concentration, as observed in various marine algae (Fig. 1). However, other studies with *S. alterniflora* (38), *S. anglica* (54,59) and sugarcane (39) show no relationship between DMSP levels and salinity. The discrepant findings for *Spartina* species might reflect differing experimental designs or problems in quantifying DMSP (31,41). In *S. alterniflora* and *W. biflora*, DMSP levels increase when N supply is limiting (Fig. 2, and ref. 38), again as has been observed in algae. Conversely, S-deficiency depresses DMSP accumulation (56); high concentrations of external sulfide may (59) or may not (38) promote it. The parallels in DMSP response to environmental conditions between flowering plants and algae (Figs. 1 and 2) are noteworthy since they suggest similar regulatory mechanisms exist in both groups.

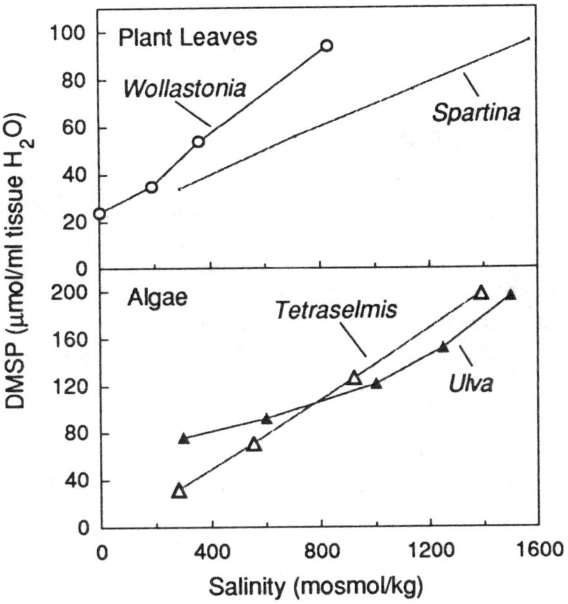

Figure 1. Responses of DMSP levels to salinity in two flowering plants and two marine algae. Data for *Wollastonia biflora* (56), *Spartina alterniflora* (11), *Tetraselmis subcordiformis* (12) and *Ulva lactuca* (13) have been expressed in the same units for ease of comparison. The data for *S. alterniflora* were derived by linear regression (r = 0.62) from a scatter plot of DMSP level *versus* porewater salinity in the range 1-6%; other data are actual experimental values. *T. subcordiformis* is a unicellular flagellate alga of coastal and marine waters; *U. lactuca* is a littoral macroalga.

Plants with Low DMSP Levels. Several species of cultivated Poaceae (about half those surveyed) were reported to have small amounts of DMSP (0.01 to 0.3 µmol g^{-1} fresh weight) (7,41). These species include maize, wheat and other cereals. Another survey of 177 species representing 90 angiosperm families from 55 orders indicated that about 16% of the sample (29 species, from 22 orders) had DMSP levels of 0.01 to 0.5 µmol g^{-1} fresh weight (40). These studies all used DMSP assays based on release of DMS by cold base treatment; although this DMS is unlikely to originate from *S*-methylmethionine (SMM), it could possibly come from other (unknown) dimethylsulfonium compounds (10,40,60).

Artifactual Origin of the DMSP Homolog 5-Dimethylsulfoniopentanoate. Apart from DMSP and SMM, the only dimethylsulfonium compound so far reported from angiosperms is 5-dimethylsulfoniopentanoate, from flowers of *Diplotaxis tenuifolia* (Brassicaceae) (35).

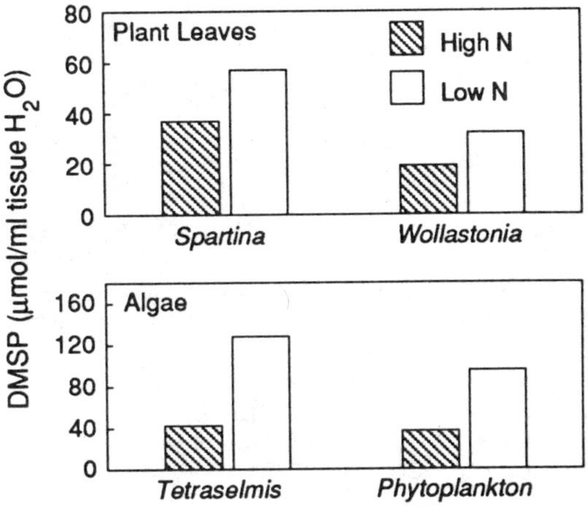

Figure 2. Responses of DMSP levels to environmental N supply in flowering plants and marine microalgae. Data for *Wollastonia biflora* (25), *Spartina alterniflora* (11), *Tetraselmis subcordiformis* (22) and phytoplankton assemblages (58) have been expressed in the same units for ease of comparison. In all cases, the salinity level was at or near that of normal seawater and S supply was not limiting. Note that the responses of the microalgae may be larger because their low N treatments were more severe than for the angiosperms.

However, this compound was recently shown to be an artifact generated from the glucosinolate glucoerucin by hot acid treatment (23).

DMSP BIOSYNTHESIS

Early radiotracer experiments with the alga *Ulva lactuca* showed that the carbon skeleton, the S atom and the methyl groups of DMSP are derived from methionine (Met) (20,30). Subsequent radiolabeling studies with other algae confirmed that Met is the precursor of DMSP (6,44). While this work established that DMSP originates from Met via methylation, deamination, decarboxylation and oxidation, it did not show the order of these steps or provide information about the intermediates and enzymes involved. Progress in these

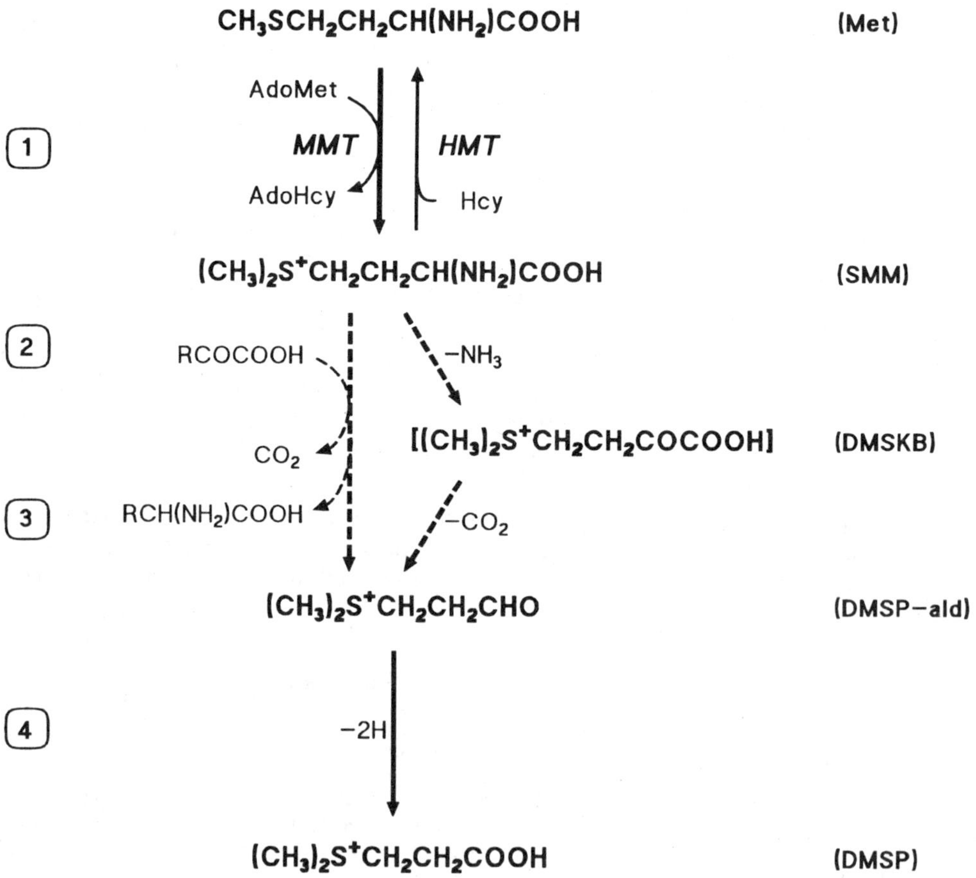

Figure 3. Biosynthesis of DMSP in *Wollastonia biflora* leaves. Bold arrows denote steps that have been demonstrated experimentally. Steps shown with dashed arrows are hypothetical, as is the intermediate DMSKB. The reverse step (thin arrow) appears to be a minor reaction. Numbers on the left correspond to the following steps: 1, *S*-methylation; 2, transamination or deamination; 3, decarboxylation; 4, oxidation. Abbreviations: Met, methionine; SMM, *S*-methylmethionine; DMSKB, 4-dimethylsulfonio-2-ketobutyrate; DMSP-ald, 3-dimethylsulfoniopropionaldehyde; DMSP, 3-dimethylsulfoniopropionate; *MMT*, AdoMet:Met *S*-methyltransferase; *HMT*, SMM:Hcy *S*-methyltransferase.

areas has recently been made with *Wollastonia biflora* (25,27,28), as summarized below and in Figure 3.

Step 1: S-Methylation. Four lines of evidence from isotope labeling studies with *W. biflora* leaf disks indicate that methylation of Met to SMM is the first step in DMSP biosynthesis (25). (a) In pulse-chase experiments with [^{14}C]Met, SMM had the labeling pattern expected of a pathway intermediate, acquiring label rapidly during the [^{14}C]Met pulse and then losing it during the chase period when [^{14}C]Met was replaced with unlabeled Met. (b) [^{14}C]SMM was efficiently converted to DMSP. (c) Supplying unlabeled SMM reduced ^{14}C incorporation into DMSP from [^{14}C]Met and caused accumulation of [^{14}C]SMM. (d) The dimethylsulfonium group from [$^{13}CH_3$,C^2H_3]SMM was incorporated as a unit into DMSP. Detailed studies have not been made of other angiosperms, but leaf disks of sugarcane readily converted supplied [^{35}S]SMM to [^{35}S]DMSP, suggesting that SMM is an intermediate in DMSP synthesis in this species also (40). The situation in algae is unclear. *Ulva lactuca* (44) and *Chondria coerulescens* (6) metabolized [$^{14}CH_3$]Met to both SMM and DMSP, and excess unlabeled SMM reduced incorporation of label from Met into DMSP in *U. lactuca* (30). However, neither species converted exogenous radiotracer SMM to DMSP (6,20). The negative results with supplied SMM might be explained by poor absorption (26), or by compartmentation of SMM of exogenous origin away from the metabolic pool of SMM.

The enzyme catalyzing the *S*-adenosylmethionine (AdoMet) dependent methylation of Met has been purified from *W. biflora* leaves and characterized (27). This enzyme (AdoMet:Met *S*-methyltransferase, EC 2.1.1.12, MMT) is a homotetramer of 115-kDa subunits. Like other methyltransferases that utilize AdoMet (5), it is strongly inhibited by *S*-adenosylhomocysteine (AdoHcy). It is also inhibited by SMM. SMM and MMT occur very widely in plants (19), so are not unique to the DMSP biosynthesis pathway. Apart from conversion to DMSP, SMM has only two known metabolic fates: cleavage to homoserine and DMS, and reaction with homocysteine (Hcy), catalyzed by SMM:Hcy *S*-methyltransferase, to give two molecules of Met (19). The tandem action of MMT and SMM:Hcy *S*-methyltransferase, together with the hydrolysis of AdoHcy, can interconvert SMM and Met; this sequence of reactions has been termed the SMM cycle (37). This cycle appears to function only at a low level in *W. biflora* leaves (25).

Steps 2 and 3: Trans- or Deamination and Decarboxylation. Three lines of evidence indicate that SMM is converted to DMSP via 3-dimethylsulfoniopropionaldehyde (DMSP-ald) in *W. biflora* leaf disks (28). (a) In pulse-chase experiments with [^{35}S]SMM, DMSP-ald labeled as expected for an intermediate, whereas three other possible intermediates (3-dimethylsulfoniopropylamine, 3-dimethylsulfoniopropionamide and 4-dimethylsulfonio-2-hydroxybutyrate) did not. (b) When [^{35}S]SMM was supplied along with unlabeled compounds, only DMSP-ald promoted [^{35}S]DMSP-ald accumulation while the three other possible intermediates had no such trapping effects. (c) Plants that do not accumulate DMSP did not form [^{35}S]DMSP-ald from [^{35}S]SMM. DMSP-ald is rather unstable in aqueous solution, and is present only in trace amounts in *W. biflora*. The DMSP-ald content estimated from radiolabeling data was only 0.3-0.6 nmol g^{-1} fresh weight (28), which may be compared with the contents of SMM and DMSP, respectively around 300 and 12,000 nmol g^{-1} fresh weight (25). The radiolabeling data also indicated that the DMSP-ald pool turns over once every 15-30 sec.

Conversion of SMM to DMSP-ald involves loss of the amino and carboxyl groups, but the enzyme reactions involved are not known. It seems improbable that the first reaction is decarboxylation because this would produce 3-dimethylsulfoniopropylamine, and radiotracer data indicated this compound was not an intermediate (see above). Two more likely possibilities are shown in Figure 3. One is analogous to a Strecker degradation (49): SMM might be converted directly to DMSP-ald, in a reaction catalyzed by a decarboxylation-dependent transaminase (52,57). This would be unusual inasmuch as there appear to be no other cases in which an amino acid with an -H undergoes such a reaction. Alternatively, transamination or oxidative deamination could produce the -keto acid 4-dimethylsulfonio-2-ketobutyrate (DMSKB), which could then undergo decarboxylation to DMSP-ald. DMSKB is not a known natural product and appears never to have been chemically synthesized, although there is indirect evidence that its adenosine analog (*S*-adenosyl-2-keto-4-methylthiobutyrate) is a natural metabolite in *Escherichia coli* (55). Amino acids with a good leaving group in the position are known to undergo ß, α-elimination upon conversion to the corresponding keto (or imino) acid (9). Since SMM has a good leaving group, $(CH_3)_2S^+$ (48), its α-keto analog DMSKB would be expected to be unstable and to undergo a fragmentation reaction.

Step 4: Oxidation. If DMSP-ald is an intermediate in DMSP biosynthesis, then the last step in the pathway must be the oxidation of the aldehyde group. It was not possible to confirm this directly by supplying [^{14}C]DMSP-ald to *W. biflora* leaf disks, due to the chemical instability of DMSP-ald and its rapid catabolism to DMS (28). However, preliminary tests with leaf extracts have indicated the presence of pyridine nucleotide-dependent DMSP-ald dehydrogenase activity (James and Hanson, unpublished data). This suggests a parallel with the glycine betaine biosynthesis pathway, where the last step is oxidation of betaine aldehyde, catalyzed by an NAD^+-dependent betaine aldehyde dehydrogenase (46). As DMSP-ald is structurally similar to betaine aldehyde, and as *W. biflora* (genotype B) can produce glycine betaine as well as DMSP (25), the enzymes involved may be closely related or perhaps identical. In this connection it is noteworthy that the glycine betaine and DMSP biosynthesis pathways occur together in other DMSP-rich angiosperms and algae (4,46).

DMSP CATABOLISM

DMSP catabolism has been studied very little in angiosperms. Two catabolic routes are known for DMSP in bacteria and algae (34): cleavage by a lyase enzyme to DMS and acrylate, and demethylation or transmethylation to give 3-methylthiopropionate (MTP). There is fragmentary evidence to suggest that angiosperms may have both routes, neither of them very active. This evidence is discussed below.

Release of DMS. When *W. biflora* leaf disks were incubated for 4 h with tracer [^{35}S]DMSP about 0.1% of the absorbed label was released as [^{35}S]DMS (28). Supplied [^{35}S]SMM and [$^{14}CH_3$]DMSP-ald were degraded to DMS much more rapidly, by 60- and 2,500-fold respectively (28). The enzymes involved were not investigated, but SMM lyases catalyzing conversion of SMM to DMS and homoserine are widely distributed in plants (17,19), and it is possible that they can also attack DMSP and DMSP-ald. A low rate of DMS emission has also been observed from intact *S. alterniflora* leaves, equivalent to around 0.8% of the endogenous DMSP content per day (11). As this rate was 100-fold higher than those for grasses that do not accumulate DMSP it seems likely that DMSP was the precursor. However, the possibility that endogenous SMM or DMSP-ald (or perhaps DMSKB) was the source cannot be excluded.

Conversion to MTP. A partially purified SMM:Hcy *S*-methyltransferase preparation from Jack bean seeds was found to use DMSP as a methyl donor (1), showing that angiosperms may have the potential to demethylate DMSP, as has been demonstrated for the alga *Chondria coerulescens* (6). However, following incubation of *W. biflora* leaf disks with tracer [^{35}S]DMSP for 24 h, 99% of the absorbed ^{35}S was recovered as [^{35}S]DMSP, and [^{35}S]MTP was not detected (28). These data indicate that there is little net degradation *in vivo* of DMSP via MTP or any other route. Although they do not rule out turnover of the DMSP methyl groups via a transmethylation-remethylation cycle, this seems unlikely to be highly active in *W. biflora* because supplied [^{14}C]MTP was not extensively converted to DMSP (25).

REGULATION OF DMSP LEVEL

The slow rates of DMSP breakdown in *W. biflora* and *S. alterniflora* (1% per day) make it seem likely that DMSP levels are regulated mainly by the rate of synthesis in relation to growth (11,28). Little is yet known about how flux through the DMSP biosynthesis pathway is controlled, particularly by environmental factors such as N-nutrition and salinity. The following is a synopsis of the data available. They are consistent with control being shared among several pathway steps, as is generally the case (16).

The conversion of SMM to DMSP-ald is the committing step in DMSP biosynthesis, and so would seem one likely control point. There is indirect evidence for this: conversion of supplied [^{35}S]SMM to DMSP was proportional to the endogenous DMSP contents of *W. biflora* and three other species (40). The Met SMM step in *W. biflora* appears to be modulated by the intracellular concentrations of both SMM and AdoHcy (25,27). The supply of Met can also exert control, as large doses of Met raised DMSP production in *W. biflora* to about 4 μmol g^{-1} fresh weight day^{-1}, well above the likely range of endogenous synthesis rates (25). Exogenous Met also enhanced DMSP accumulation in the alga *Tetraselmis subcordiformis* (22). The physiological significance of control via Met supply is uncertain, as Met levels *in vivo* are probably stabilized by feedback mechanisms (18,19).

METABOLIC ENGINEERING PROSPECTS

Figure 4 sets DMSP accumulation in the framework of overall sulfur and methyl group metabolism in leaves. Although the data are approximations, they bring out two essential points: engineering a DMSP-free plant to produce as much DMSP as *W. biflora* could triple the amount of reduced sulfur required, and would also increase the demand for methyl groups. Because lignin and pectin methylation are minor activities in the mature leaf tissues that produce DMSP, the methyl group demand in these tissues could also easily triple. These considerations show that DMSP accumulation has major repercussions on the metabolism of sulfur amino acids and on methyl group biogenesis and transfer, and hence that engineering it requires attention to these areas as well to the DMSP biosynthesis pathway itself (2,29). We expand upon this briefly below and in Figure 5, which shows how DMSP and Met biosynthesis are related. The essential point is that DMSP biosynthesis requires two molecules of Met: one for the methylthiopropionate moiety, the other to donate a methyl group. The Hcy moiety left after methyl transfer is recycled to Met.

DMSP Biosynthesis Enzymes. In the simplest case it might not be necessary to increase MMT expression, since plants that do not accumulate DMSP can have MMT activities and SMM levels comparable to *W. biflora* (3,27). It would almost certainly be necessary to introduce the enzyme(s) responsible for converting SMM to DMSP-ald, as this is the committing step in DMSP biosynthesis and no DMSP-ald was found in plants lacking DMSP (28). It is probable that an enzyme to oxidize DMSP-ald would also be required, although perhaps not in glycine betaine-accumulating plants if betaine aldehyde dehydrogenase can catalyze DMSP-ald oxidation (see above).

Enzymes of Methyl Group Biogenesis and Transfer. These include methylene tetrahydrofolate reductase, Met synthase, AdoMet synthetase and AdoHcy hydrolase. The activi-

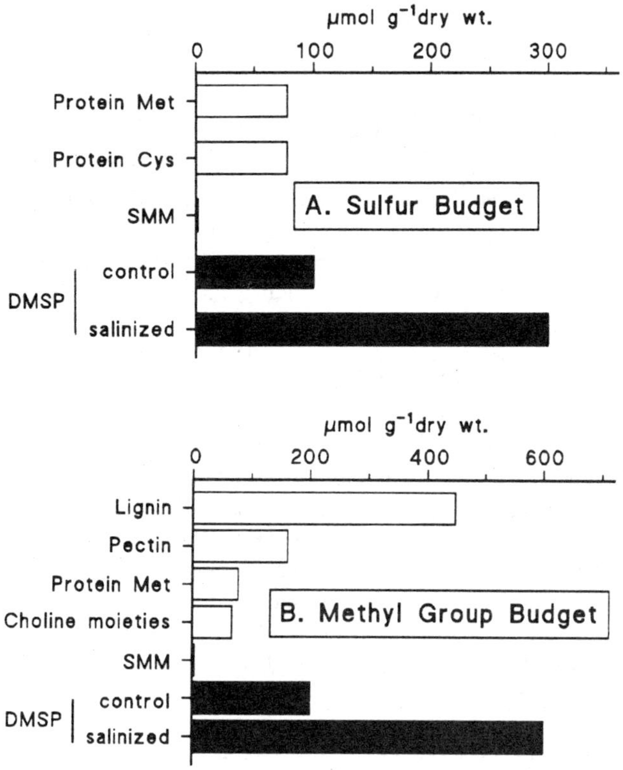

Figure 4. Theoretical sulfur (A) and methyl group (B) budgets for leaves of a plant metabolically engineered to accumulate DMSP. This plant is assumed to contain no betaines or other methyl-rich small molecules. Open bars show sulfur and methyl group demands before engineering; solid bars show the new demands created by two levels of DMSP, equal to those in *W. biflora* grown without (control) or with (salinized) 450 mM NaCl (56). SMM values are also from *W. biflora* (25). Sulfur and methyl group demands for proteins were estimated from typical biomass composition data (43), assuming mole abundances of methionine (Met) and cysteine (Cys) of 5%; the pools of free sulfur amino acids and glutathione were assumed to be negligible (19). The methyl group demand for choline moieties was estimated by summing typical values for free choline (21) and phosphatidylcholine (47). Some assumptions made in calculating methyl demands were: that lignin is polyconiferyl alcohol (43); that pectin is fully carboxymethylated polygalacturonic acid; that there is no turnover of methyl groups; and that 10% of nucleic acid bases are methylated (making this demand negligible). Note that the theoretical sulfur budget is consistent with experimental data for *S. alterniflora*, in which as much as 50 to 86% of the total sulfur can be allocated to DMSP (38,59).

ties of these enzymes in a plant lacking DMSP might not suffice to cope with a large additional methyl group demand for DMSP synthesis. This inference is strengthened by two types of observations: (a) high levels of AdoMet synthetase and AdoHcy hydrolase are constitutively expressed or induced in tissues with a large requirement for methyl groups for lignin synthesis (14,32,42); and (b) the methyl group supply can limit growth of cotyledons, which have a high methyl group demand for pectin synthesis (8). It is possible that the entire methyl group biogenesis and transfer subsystem in leaves would respond adaptively when presented with a larger load (29). If it did not, then inadequate AdoHcy hydrolase activity might be the most serious problem (51). This enzyme is responsible for removing AdoHcy, which strongly inhibits AdoMet-dependent methyltransferases (5). If inadequate AdoHcy hydrolase activity led to AdoHcy accumulation, the consequent suppression of methylation reactions

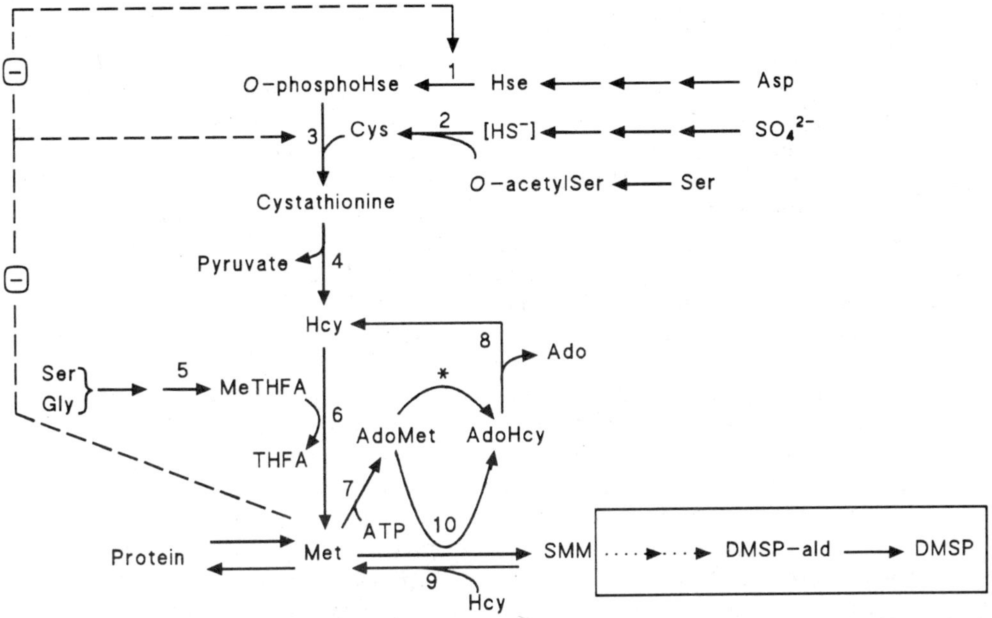

Figure 5. Methionine and DMSP biosynthesis. The reactions inside the box are unique to DMSP biosynthesis. The broken lines represents negative feedback control by methionine of assimilation of sulfate into cystathionine and of *O*-phosphohomoserine synthesis (18,19). Ado, adenosine; AdoHcy, *S*-adenosylhomocysteine; AdoMet, *S*-adenosylmethionine; THFA, tetrahydrofolate; MeTHFA, N^5-methyltetrahydrofolate; Asp, aspartate; Cys, cysteine; Gly, glycine; Hcy, homocysteine; Hse, homoserine; Met, methionine; Ser, serine; SMM, *S*-methylmethionine. Enzymes: 1, Hse kinase; 2, Cys synthase; 3, cystathionine- -synthase; 4, ß-cystathionase; 5, methylenetetrahydrofolate reductase; 6, Met synthase; 7, AdoMet synthetase; 8, AdoHcy hydrolase; 9, SMM:Met *S*-methyltransferase; 10, AdoMet:Met *S*-methyltransferase; *, other methyl transfer reactions.

could have many undesirable effects on the rest of metabolism. Engineering higher levels of this enzyme may therefore be required. Alternatively, if the plant to be engineered already had the capacity to accumulate a betaine, it could have a pre-existing capacity for methyl group synthesis and transfer high enough to cope with an additional demand from DMSP (24).

Enzymes of Sulfate Assimilation and Hcy Biosynthesis. DMSP accumulation would greatly increase the requirement for assimilation of sulfate into cystathionine and its product Hcy, the precursor of Met. Met (or a close metabolic product thereof) may exercise strong negative feedback control over the sulfur and carbon metabolism involved (18,19), so that increased consumption of Met might elicit a matching increase in its production. In this case it would not be necessary to increase expression of the many enzymes upstream from Hcy (29).

DMS Emissions from Metabolically Engineered Crops. The prospect of engineering DMSP accumulation in crop plants raises the issue of a potential increase in terrestrial DMS emissions, via catabolism of DMSP in the crop or by microbial action on crop residues. The significance of such hypothetical emissions is obviously hard to evaluate. However, as a point of comparison, the DMSP-accumulating crop sugarcane is presently grown on a land area of about 17 x 10^6 hectares. As it is unlikely that crops genetically engineered to produce DMSP would ever occupy so large an area, their contributions to terrestrial DMS emissions would most probably be small in relation to those currently made by the world's sugarcane crop.

CONCLUSION

Although some progress has been made in elucidating the DMSP biosynthesis pathway in *W. biflora*, many issues remain for future research. These can be summarized as follows. With respect to the pathway in *W. biflora*, we do not know (a) whether SMM is converted to DMSP-ald directly or via an intermediate, (b) where in the cell the pathway is localized or (c) what types of enzymes catalyze the steps after SMM. We know very little about how flux through the pathway is controlled. More generally, it is uncertain whether the DMSP biosynthesis pathway in other angiosperms or in algae is the same as in *W. biflora*.

Many questions remain about the catabolism of DMSP in angiosperms. In *W. biflora* and *S. alterniflora* leaves it seems to be slow, but we do not know if this applies to all developmental stages and all environmental conditions, or to other species. Nor do we know whether DMSP is translocated from leaves to roots or flowers, or whether these organs can degrade it. Except for preliminary reports on SMM lyase activity, nothing is known about the angiosperm enzyme(s) that catalyze release of DMS from dimethylsulfonium compounds.

Lastly, it must be emphasized that our present picture of the occurrence of DMSP in angiosperms is deficient in two respects. First, too few species (particularly tropical and woody plants) have been surveyed for DMSP to give a clear picture of its taxonomic distribution or its ecological significance. Second, the reports of low levels of DMSP (0.5 $\mu mol\ g^{-1}$ fresh weight) in various angiosperms cannot be interpreted rigorously until the identity of DMSP is established by spectroscopic methods (mass spectrometry, NMR).

ACKNOWLEDGMENTS

The authors' research was supported by grants from the Natural Sciences and Engineering Research Council of Canada and by grant RR 00484 from the National Institute of Health, National Center for Research Resources to the Michigan State University-National Institutes of Health Mass Spectrometry Facility.

REFERENCES

1. Abrahamson, L. and S. K. Shapiro. 1965. The biosynthesis of methionine: partial purification and properties of homocysteine methyltransferase of Jack bean meal. Arch. Biochem. Biophys. *109:* 376-382.
2. Bailey, J.E. 1991. Toward a science of metabolic engineering. Science *252:* 1668-1675.
3. Bezzubov, A.A. and N.N. Gessler. 1992. Plant sources of *S*-methylmethionine. Prikl. Biokhim. Mikrobiol. *28:* 423-429.
4. Blunden, G. and S.M. Gordon. 1986. Betaines and their sulphonio analogues in marine algae. Prog. Phycol. Res. *4:* 39-80.
5. Cantoni, G.L., H.H. Richards and P.K. Chiang. 1979. Inhibitors of *S*-adenosylhomocysteine hydrolase and their role in the regulation of biological methylation, p. 155-164. *In* E. Usdin, R.T. Borchardt and C.R. Creveling (eds.), Transmethylation. Elsevier/North Holland, New York.
6. Chillemi, R., A. Patti, R. Morrone, M. Piatelli and S. Sciuto. 1990. The role of methylsulfonium compounds in the biosynthesis of *N*-methylated metabolites in *Chondria coerulescens*. J. Nat. Prod. *53:* 87-93.
7. Chrominski, A., D.J. Weber, B.N. Smith and D.F. Hegerhorst. 1989. Is dimethylsulfonium propionate an osmoprotectant of terrestrial glycophytes? Naturwissenschaften *76:* 473-475.
8. Coker, G.T., J.R. Garbow, and J. Schaefer. 1987. ^{15}N and ^{13}C NMR determination of methionine metabolism in developing soybean cotyledons. Plant Physiol. *83:* 698-702.
9. Cooper, A.J.L., M.M. Hollander, and M.W. Anders. 1989. Formation of highly reactive vinylglyoxylate (2-oxo-3-butenoate) from amino acids with good leaving groups in the α-position. Biochem. Pharmacol. *38:* 3895-3901.

10. Dacey, J.W.H., G.M. King, and P.S. Lobel. 1994. Herbivory by reef fishes and the production of dimethylsulfide and acrylic acid. Mar. Ecol. Prog. Ser. *112:* 67-74.
11. Dacey, J.W.H., G.M. King and S.G. Wakeham. 1987. Factors controlling emission of dimethylsulphide from salt marshes. Nature *330:* 643-645.
12. Dickson, D.M.J. and G.O. Kirst, G.O. 1986. The role of ß-dimethylsulfoniopropionate, glycine betaine and homarine in the osmoacclimation of *Platymonas subcordiformis*. Planta *167:* 536-543.
13. Dickson, D.M., R.G. Wyn Jones and J. Davenport. 1980. Steady state osmotic adaptation in *Ulva lactuca*. Planta *150:* 158-165.
14. Espartero, J., J.A. Pintor-Toro and J.M. Pardo. 1994. Differential accumulation of *S*-adenosylmethionine synthetase transcripts in response to salt stress. Plant Mol. Biol. *25:* 217-227.
15. Fan, T.W.-M., T.D. Colmer, A.N. Lane and R.M. Higashi. 1993. Determination of metabolites by ^{1}H NMR and GC: analysis for organic osmolytes in crude tissue extracts. Anal. Biochem. *214:* 260-271.
16. Fell, D.A. 1992. Metabolic control analysis: a survey of its theoretical and experimental development. Biochem. J. *286:* 313-330.
17. Gessler, N.N. and A.A. Bezzubov. 1988. *S*-Methylmethionine sulfonium hydrolase activity in plant and animal tissues. Prikl. Biokhim. Mikrobiol. *24:* 240-246.
18. Giovanelli, J. 1990. Regulatory aspects of cysteine and methionine biosynthesis, p. 33-48. *In* H. Rennenberg, C. Brunold, L.J. De Kok and I. Stulen (eds.), Sulfur Nutrition and Sulfur Assimilation in Higher Plants. SPB Academic Publishing, The Hague.
19. Giovanelli, J., S.H. Mudd and A.H. Datko. 1980. Sulfur amino acids in plants, p. 453-595. *In* B.J. Miflin (ed.), Plant Biochemistry, Vol. 5. Academic Press, New York.
20. Greene, R.C. 1962. Biosynthesis of dimethyl-ß-propiothetin. J. Biol. Chem. *237:* 2251-2254.
21. Grieve, C.M. and E.V. Maas. 1984. Betaine accumulation in salt-stressed sorghum, Physiol. Plant. *61:* 167-171.
22. Gröne, T. and G.O. Kirst. 1992. The effect of nitrogen deficiency, methionine and inhibitors of methionine metabolism on the DMSP contents of *Tetraselmis subcordiformis* (Stein), Mar. Biol. *112:* 497-503.
23. Hanson, A.D., Z.-H. Huang and D.A. Gage. 1993. Evidence that the putative compatible solute 5-dimethylsulfoniopentanoate is an extraction artifact. Plant Physiol. *101:* 1391-1393.
24. Hanson, A.D., J. Rivoal, M. Burnet and B. Rathinasabapathi. 1995. Biosynthesis of quaternary ammonium and tertiary sulfonium compounds in response to water deficit, p. 189-198. *In* N. Smirnoff (ed.), Environment and Plant Metabolism. Bios Scientific, Oxford.
25. Hanson, A.D., J. Rivoal, L. Paquet and D.A. Gage. 1994. Biosynthesis of 3-dimethylsulfoniopropionate in *Wollastonia biflora* (L.) DC., Plant Physiol. *105:* 103-110.
26. Hellebust, J.A. and R.R.L. Guillard. 1967. Uptake specificity for organic substrates by the marine diatom *Melosira nummuloides*. J. Phycol. *3:* 132-136.
27. James, F., K.D. Nolte and A.D. Hanson. 1995. Purification and properties of *S*-adenosyl-L-methionine:L-methionine *S*-methyltransferase from *Wollastonia biflora* leaves. J. Biol. Chem. (in press).
28. James, F., L. Paquet, S.A. Sparace, D.A. Gage and A.D. Hanson. 1995. Evidence implicating dimethylsulfoniopropionaldehyde as an intermediate in dimethylsulfoniopropionate biosynthesis. Plant Physiol. *108:* 1439-1448.
29. Kacser, H. and L. Acerenza. 1993. A universal method for achieving increases in metabolite production. Eur. J. Biochem. *216:* 361-367.
30. Kahn, V. 1964. Glycine as a methyl donor in dimethyl-ß-propiothetin synthesis. J. Exp. Bot. *15:* 225-231.
31. Karsten, U., K. Kück, C. Daniel, C. Wiencke and G.O. Kirst. 1994. A method for complete determination of dimethylsulfoniopropionate (DMSP) in marine macroalgae from different geographical regions. Phycologia *33:* 171-176.
32. Kawalleck, P., G. Plesch, K. Hahlbrock and I.E. Somssich. 1992. Induction by fungal elicitor of *S*-adenosyl-L-methionine synthetase and *S*-adenosyl-L-homocysteine hydrolase mRNAs in cultured cells and leaves of *Petroselinum crispum*. Proc. Natl. Acad. Sci. USA *89:* 4713-4717.
33. Keller, M.D., W.K. Bellows and R.R.L. Guillard. 1989. Dimethyl sulfide production in marine phytoplankton, p. 167-182. *In* E.S. Saltzman and W.J. Cooper (eds.), Biogenic Sulfur in the Environment. American Chemical Society, Washington DC.
34. Kiene, R.P. 1993. Microbial sources and sinks for methylated sulfur compounds in the marine environment, p. 15-33. *In* D.P. Kelly and J.C. Murrell (eds.), Microbial Growth on C_1 Compounds. Intercept, Andover.
35. Larher, F. and J. Hamelin. 1979. L'acide diméthylsulfonium-5 pentanoïque de *Diplotaxis tenuifolia*. Phytochemistry *18:* 1396-1397.
36. Larher, F., J. Hamelin and G.R. Stewart. 1977. L'acide diméthylsulfonium-3 propanoïque de *Spartina anglica*. Phytochemistry *16:* 2019-2020.

37. Mudd, S.H. and A.H. Datko. 1990. The *S*-methylmethionine cycle in *Lemna paucicostata*. Plant Physiol. *93:* 623-630.
38. Otte, M.L. and J.T. Morris. 1994. Dimethylsulfoniopropionate (DMSP) in *Spartina alterniflora* Loisel. Aquatic Botany *48:* 239-259.
39. Paquet, L. 1995. Répartition et quelques aspects du métabolisme du 3-diméthylsulfoniopropionate chez les angiospermes. M.Sc. Thesis, Université de Montréal.
40. Paquet, L., P.J. Lafontaine, H.S. Saini, F. James and A.D. Hanson. 1995. Évidence en faveur de la présence du 3-diméthylsulfoniopropionate (DMSP) chez une large gamme d'angiospermes. Can. J. Bot. (in press).
41. Paquet, L., B. Rathinasabapathi, H. Saini, L. Zamir, D.A. Gage, Z.-H. Huang and A.D. Hanson. 1994. Accumulation of the compatible solute 3-dimethylsulfoniopropionate in sugarcane and its relatives, but not other gramineous crops. Aust. J. Plant Physiol. *21:* 37-48.
42. Peleman, J., W. Boerjan, G. Engler, J. Seurinck, J. Botterman, T. Alliotte, M. van Montagu and D. Inze. 1989. Strong cellular preference in the expression of a housekeeping gene of *Arabidopsis thaliana* encoding *S*-adenosylmethionine synthetase. Plant Cell *1:* 81-93.
43. Penning de Vries, F.W.T., A.H.M. Brunsting and H.H. van Laar. 1974. Products, requirements and efficiency of biosynthesis: a quantitative aproach. J. Theor. Biol. *45:* 339-377.
44. Pokorny, M., E. Marcenko and D. Keglevic. 1970. Comparative studies of L- and D-methionine metabolism in lower and higher plants. Phytochemistry *9:* 2175-2188.
45. Rennenberg, H. 1991. The significance of higher plants in the emission of sulfur compounds from terrestrial ecosystems, p. 217-260. *In* T.D. Sharkey, E.A. Holland and H.A. Mooney (eds.), Trace Gas Emissions by Plants. Academic Press, San Diego.
46. Rhodes, D. and A.D. Hanson. 1993. Quaternary ammonium and tertiary sulfonium compounds in higher plants. Annu. Rev. Plant Physiol. Plant Mol. Biol. *44:* 357-384.
47. Roughan, P.G. and R.D. Batt. 1969. The glycerolipid composition of leaves. Phytochemistry *8:* 363-369.
48. Schlenk, F. 1965. The chemistry of biological sulfonium compounds. Fortschr. Chem. Org. Naturst. *23:* 61-112.
49. Schönberg, A. and R. Moubacher. 1952. The Strecker degradation of α-amino acids. Chem. Reviews *50:* 261-277.
50. Serrano, R. and R. Gaxiola. 1994. Microbial models and salt stress tolerance in plants. Critical Reviews in Plant Sciences *13:* 121-138.
51. Sibley, M.H. and J.H. Yopp. 1987. Regulation of *S*-adenosylhomocysteine hydrolase in the halophilic cyanobacterium *Aphanothece halophytica*: a possible role in glycine betaine biosynthesis. Arch. Microbiol. *149:* 43-46.
52. Snell, E.E. and S.J. Di Mari. 1970. Schiff base intermediates in enzyme catalysis, p. 335-370. *In* P.D. Boyer (ed.), The Enzymes, Third Edition, Vol. II. Academic Press, New York.
53. Somero, G.N. 1992. Adapting to water stress: convergence on common solutions, p. 3-18. *In* G.N. Somero, C.B. Osmond and C.L. Bolis (eds.), Water and Life. Springer, Berlin.
54. Stewart, G.R., F. Larher, I. Ahmad and J.A. Lee. 1979. Nitrogen metabolism and salt-tolerance in higher plants, p. 211-227. *In* R.L. Jefferies and A.J. Davy (eds.), Ecological Processes in Coastal Environments. Blackwell Scientific Publishers, Oxford.
55. Stoner, G.L. and M.A. Eisenberg. 1975. Purification and properties of 7,8-diaminopelargonic acid aminotransferase. J. Biol. Chem. *260:* 4029-4036.
56. Storey, R., J. Gorham, M.G. Pitman, A.D. Hanson and D.A. Gage. 1993. Response of *Melanthera biflora* to salinity and water stress. J. Exp. Bot. *44:* 1551-1560.
57. Toney, M.D., E. Hohenester, S.W. Cowan and J.N. Jansonius. 1993. Dialkylglycine decarboxylase: bifunctional active site and alkali metal sites. Science *261:* 756-759.
58. Turner, S.M., G. Malin, P.S. Liss, D.S. Harbour and P.M. Holligan. 1988. The seasonal variation of dimethyl sulfide and dimethylsulfoniopropionate concentrations in nearshore waters. Limnol. Oceanogr. *33:* 364-375.
59. Van Diggelen, J., J. Rozema, D.M.J. Dickson and R. Broekman. 1986. ß-3-Dimethylsulphoniopropionate, proline and quaternary ammonium compounds in *Spartina anglica* in relation to sodium chloride, nitrogen and sulphur. New Phytol. *103:*573-586.
60. White, R.H. 1982. Analysis of dimethyl sulfonium compounds in marine algae. J. Mar. Res. *40:* 529-536.
61. Wyn Jones, R.G. and R. Storey. 1981. Betaines, p. 171-204. *In* L.G. Paleg and D. Aspinall (eds.), The Physiology and Biochemistry of Drought Resistance in Plants. Academic Press, Sydney.

8

EFFECTS OF SULFIDE ON GROWTH AND DIMETHYLSULFONIOPROPIONATE (DMSP) CONCENTRATION IN *SPARTINA ALTERNIFLORA*

J. T. Morris, C. Haley, and R. Krest

Department of Biological Sciences
University of South Carolina
Columbia, South Carolina 29208

SUMMARY

The effects of sulfide on growth and dimethylsulfoniopropionate (DMSP) concentration in *Spartina alterniflora* were studied in greenhouse cultures. Spartina plants were maintained in sand-filled pots and supplemented with a balanced nutrient mixture and 1 g/l of NaCl. After they were well established, plants were separated into four treatment groups that were maintained in different sulfide concentrations ranging from 0 to 2 mM Na_2S for 5 weeks in March and April, 1994. Relative growth rates, determined from weekly plant heights, final dry weights, and final DMSP concentrations in tissues were measured. The results indicated that sulfide concentrations in the neighborhood of 1 mM were optimal for growth. Sulfide treatment affected leaf growth, but not root growth. The maximum relative growth rate (2.0±0.3% day^{-1}) was found in the 1 mM treatment, while the minimum growth rate (0.5±0.2% day^{-1}) was found in the 0 mM treatment. Mean stem height reached 64 cm in the 1 mM treatment and was 47 cm in the 0 mM treatment. Final stem weights were 1.4±1.2 and 0.8±0.65 g/pot in the 2 and 0 mM treatments, respectively. In general, growth at 2 mM was equivalent to that in the 1 mM treatment. The sulfide treatments had marginally significant effects on the DMSP concentration in leaf tissues and no effect in roots. Data on sulfide distributions in pore water in several marsh sites suggest that there is no consistent relationship between Spartina production and sulfides. The results demonstrate that the relationship between Spartina growth and sulfides is complex.

INTRODUCTION

Previous research suggests that growth (19, 1) and nutrient uptake (2) by marsh grasses (*Spartina* spp.) are negatively affected by sulfides in controlled experiments. Sulfide toxicity has

Biological and Environmental Chemistry of DMSP and Related Sulfonium Compounds
edited by Ronald P. Kiene et al., Plenum Press, New York, 1996

been reported to be the cause of dieback of *Spartina alterniflora* in the field (14). It has also been hypothesized that DMSP is produced in Spartina in response to sulfides. In greenhouse experiments, van Diggelen et al. (19) found a positive relationship between DMSP concentration in *S. anglica* tissue and the concentration of sulfide in nutrient solutions. They speculated that DMSP and the subsequent formation of DMS may be a detoxification mechanism that allows this species to tolerate toxic sulfides in anoxic marine soils. However, there is contradictory evidence about the effect of sulfides on Spartina growth and DMSP production. A recent greenhouse study of *Spartina alterniflora* (16) failed to demonstrate a relationship between sulfides and DMSP. Furthermore, this study showed that the concentration of DMSP in *S. alterniflora* tissues was not related to either salinity or sulfide concentration in field populations growing along a natural salinity gradient. Finally, the stable sulfur isotope composition of *S. alterniflora* taken from the field indicates that the sulfur contained in this plant is derived from sulfides, not from sulfate, despite an overwhelming abundance of sulfate in marine sediments (3). The collective data indicate that the traditional view of sulfide as a toxin may be an over simplification.

This paper reports results of a study that was undertaken to provide additional information about the effects of sulfides on the growth and DMSP production of *Spartina alterniflora*. A greenhouse experiment was conducted in which Spartina plants were cultured at different levels of sulfide. Field data on concentrations of soluble sulfides in pore water in several healthy marshes are reported to provide a context for judging the ecological relevance of the greenhouse study.

MATERIALS AND METHODS

Single newly emerged stems of *Spartina alterniflora* with their rhizomes and roots were collected in the field from a salt marsh near Georgetown, SC and cultured in sand-filled pots in the greenhouse. Single ramets were maintained, initially one per pot, for several weeks to establish the plants before beginning the experiment. The pots were randomly divided into eight groups and kept in tubs containing a nutrient solution. All treatments contained 1 g/l of NaCl in their nutrient solutions. In addition, the nutrient solution was composed of a balanced mixture of macro and micronutrients (Table 1) patterned after a modified Hoagland's solution (7). The level of solution was kept even with the surface of the sand in the pots. Each tub contained 12 pots. At the start of the experiment in March the tubs were organized into two replicates each of four treatments of sulfide concentration. The sulfide concentrations used in the experiment were 0, 0.5, 1.0, and 2.0

Table 1. Composition of nutrient solution used in growth experiments, exclusive of NaCl which was maintained at a final concentration of 1 g/l

Compound	Conc. of stock solution	Volume of stock soln. (ml) per liter of final soln.
KH_2PO_4	1 M	2
$Ca(NO_3)_2 4H_2O$	1 M	4
$MgSO_4 7H_2O$	1 M	1
NH_4Cl	1 M	8
Fe-EDTA	20 mM	2
H_3BO_3	25 mM	1
$MnSO_4 H_2O$	2 mM	1
$ZnSO_4 7H_2O$	2 mM	1
$CuSO_4 5H_2O$	0.5 mM	1
H_2MoO_4	0.5 mM	1

mmol/liter of Na_2S. The nutrient solutions were replaced weekly after the experiment was started, while Na_2S was added daily and the pH adjusted as necessary to maintain neutrality. Sulfide concentrations decreased following each renewal largely because of the volatilization of H_2S and were close to zero after 24 h. The experimental treatments continued for five weeks in the greenhouse under natural light and ambient temperature during spring 1994 at a time when the plants were in log-phase growth.

Every week the total number of shoots and the length of the tallest plant in each pot were recorded. At the conclusion of the experiment all of the plants were harvested, belowground and aboveground parts were divided, and the parts dried and weighed. Subsamples of fresh root and leaf tissues were collected for DMSP analysis following a modified procedure (16) of van Diggelen et al. (19). About 0.3 g fresh plant tissue was transferred first to liquid nitrogen then to 25 ml vials containing 5 ml 4.25 M NaOH. The vials were immediately sealed with Teflon-lined grey butyl septa with crimp caps. Standards were made using known amounts of pure DMSP. After incubating the vials for 24 hr, DMS in the headspace gas was analyzed by gas chromatography (Carle AGC with flame ionization detector and 2 m glass column packed with 0.1% SP-1000 on 80/100 Carbopack C). The concentration of DMS in solution was calculated by Henry's Law, and the total quantity of DMSP expressed as a concentration per unit weight of fresh tissue.

Pore water sulfides were sampled in the field using diffusion samplers screened with 45 μm Nitex membranes. Samplers were deployed in triplicate and were collected monthly from several salt marsh locations in Charleston Harbor and North Inlet (Fig. 1) for 1.5-yr. The sampler

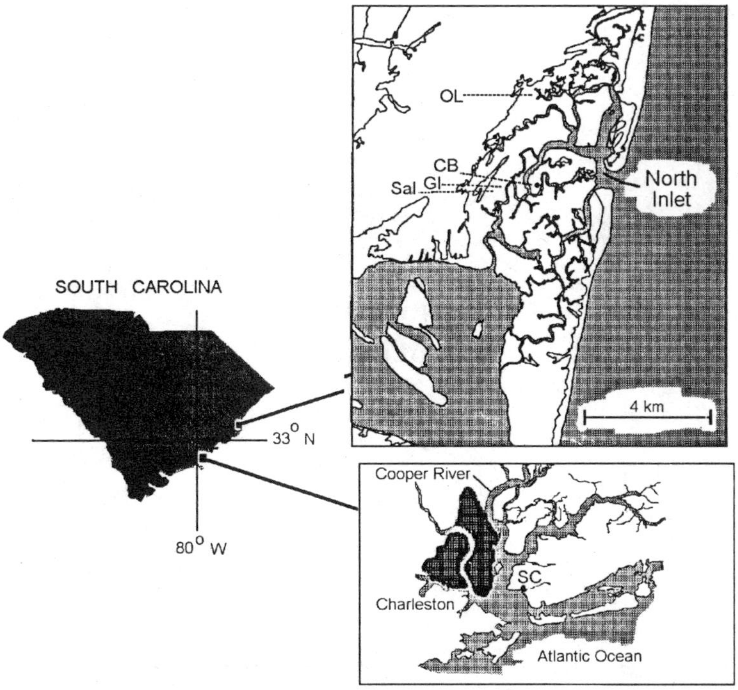

Figure 1. Locations of study sites in Charleston Harbor and North Inlet, SC. The Charleston Harbor and North Inlet maps are not drawn to the same scale. The site designated SC in Charleston Harbor is the location of four sampling stations including creek bank and interior sites. At North Inlet, there are creek bank and interior sampling stations at OL and GI. Sal is a high marsh site dominated by *Salicornia* spp., CB is an interior site 15 m from a creekbank. With the exception of Sal, all sampling stations are in areas vegetated by *Spartina alterniflora*.

bottles (30 ml each) were placed in notches cut into PVC pipes such that when the pipes were inserted into the sediment the pore water at regular depth intervals was sampled. The samplers were left in the sediment to equilibrate for 1 month. As they were collected, the Nitex membranes were replaced by Teflon septa. Sulfides were stabilized in the field immediately with ZnAc and analyzed the following day. ZnAc-treated pore water samples were analyzed colorimetrically for sulfide using a modification of the Cline procedure (5) as described by Otte and Morris (16).

Statistical analyses were performed using the SAS statistical package. The GLM procedure was used to test for treatment effects both by simple linear regression (treatment variable treated as continuous) and by one-way ANOVA (treatment treated as a class variable). Tukey's studentized range test on all main effects was used to test for significant differences among pairs of treatments. Pore water sulfide concentrations are reported as geometric means to correct for skewness, and ranges of their 95% confidence limits are given in the text.

RESULTS

Pore water sulfide concentrations showed greater differences between marshes (Fig. 2) than within marshes. Within these marshes sulfide concentrations generally increase from

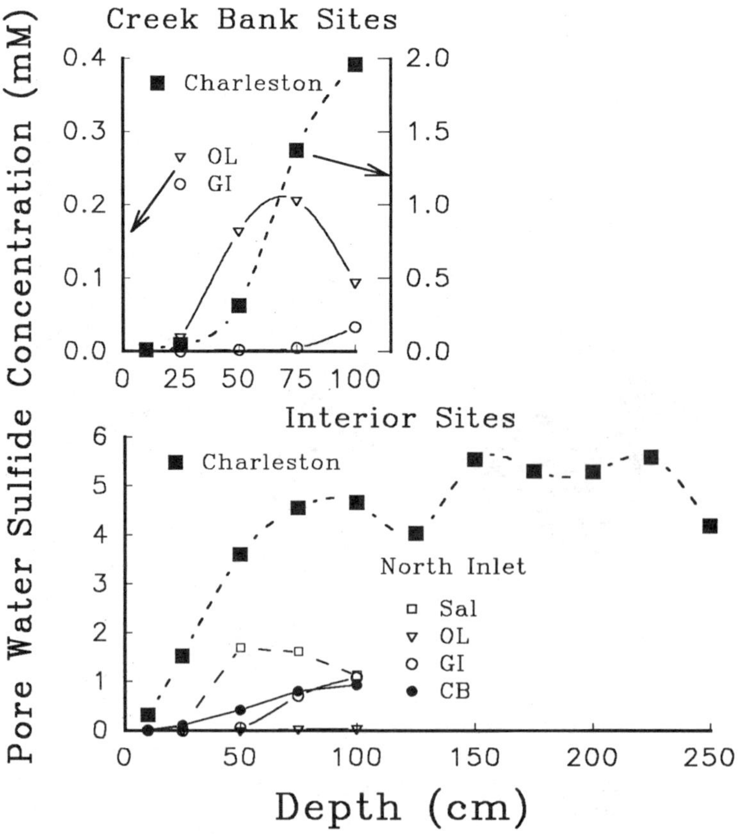

Figure 2. Depth profiles of geometric mean concentrations of soluble sulfide in pore water from several marsh sites in Charleston Harbor and North Inlet, SC. Geometric means were computed from 1.5-yr time series of monthly data.

relatively low values along the creek banks toward higher levels in the interior marsh sites. Past studies have shown that spatial gradients in sulfide concentration correlate negatively with Spartina production within a marsh (10, 6). However, this trend breaks down between sites. For instance, grand geometric means over all interior sites at North Inlet at 10 and 25 cm depth (within the root zone) were 3.9 μM (range of 95% confidence limits is 2.7-5.8 μM) and 14.1 (8.8-22.6) μM, which is less than sulfide concentrations at creek bank locations in Charleston Harbor. In Charleston Harbor, geometric mean concentrations of sulfide along the creek bank were 13.3 (9.9-17.8) and 49.4 (31.8-76.7) μM at 10 and 25 cm depth respectively. At the 100 cm depth, the sulfide concentration at the Charleston Harbor creekbank sites exceeds that of North Inlet sites by a factor of 25 (1957 vs 76 μM). In both of these marshes the production of Spartina is greatest along the creekbanks where sulfide concentrations are generally lower. Yet, despite their elevated sulfide levels, Charleston Harbor marshes are very productive, and the productivity of their creekbank areas is certainly greater than interior North Inlet marshes (data not shown). Interior marshes in Charleston Harbor, which have healthy stands of Spartina, had sulfide concentrations of 324 (253-414) μM at 10 cm and 1528 (1301-1795) μM at 25 cm.

In the greenhouse experiments, sulfide treatment had significant effects on several growth parameters, and there was a consistent trend toward increased growth at higher sulfide concentrations. The maximum relative growth rate (RGR, 2.0±0.3% day^{-1}) was found in the 1 mM treatment, while the minimum RGR (0.5±0.2% day^{-1}) was found in the 0 mM treatment (Table 2). Simple regression analysis indicated that sulfide had a significant effect on RGR at the 0.004 level. Linear regression analyses likewise indicated that there were significant treatment effects on mean stem density and weight (Table 2). Final density per pot was greatest in the 2 mM treatment (2.75±1.7) and least (2.0±1) in the 0 and 0.5 mM treatments. Final stem weight was greatest in the 2 mM treatment (1.7±1.2) and least (0.8±0.6) in the 0 mM treatment. The height of the tallest plant from each pot, averaged by treatment, reached 67.6±22.3 cm in the 2 mM treatment and was 47.0±13.0 cm in the 0 mM treatment. The treatment effect was significant at the 0.0005 level. There was no significant effect of sulfide treatment on final root weight (Table 2).

Sulfide level had a marginally significant effect on the concentration of DMSP in leaf tissues (significance level for non-zero slope=0.01) and no effect on DMSP concentration in root tissues (Table 3). In leaf tissues DMSP concentration averaged 9.5±3.4 μmol/g (fresh weight) in the 2 mM sulfide treatment and 6.3±1.7 μmol/g in the 0 mM treatment. DMSP

Table 2. Summary growth characteristics of *Spartina alterniflora* from a greenhouse experiment in which potted plants were grown at different levels of sulfide (n=24 per treatment). Data are means ±1 SD (or ±1 SE of the estimated regression slope in the case of relative growth rate). Mean final plant heights with different superscript letters are significantly different at the 0.05 level according to Tukey's studentized range test. Significance of the treatment effect using a simple linear regression model is given by the value of p (ns=not significant)

	Sulfide treatment (mM)				
	0	0.5	1.0	2.0	p
RGR (%/day)	0.5±0.2	0.8±0.26	2.0±0.3	1.27±0.22	0.004
Final height (cm)	47.0±13.0^{a}	55.3±15.3^{ab}	64.1±18.9^{b}	67.6±22.3^{b}	0.0005
Stem density/pot	2.0±1.2	2.0±1.1	2.3±1.5	2.75±1.7	0.05
Stem weight (g/pot)	0.8±0.65	0.96±0.8	1.18±0.9	1.4±1.17	0.02
Root weight (g/pot)	0.33±0.3	0.33±0.3	0.45±0.3	0.42±0.3	ns

Table 3. Mean DMSP concentrations (μmol/gfw) in *Spartina alterniflora* grown at different levels of sulfide. Data are means ±1 SD (n=6 per treatment). Plants were analyzed after 5 weeks of treatment. Significance of the treatment effect using a simple linear regression model is given by the value of p (ns=not significant)

	Sulfide treatment (mM)				
	0	0.5	1.0	2.0	p
Root tissue	1.1±0.5	1.3±0.7	1.1±0.9	1.4±1.0	ns
Leaf tissue	6.3±1.7	5.6±1.8	7.3±3.0	9.5±3.4	0.01

concentrations in roots were lower and averaged between 1.1±0.5 and 1.4±1.0 μmol/g (fresh weight) in the 0 and 2 mM treatments, respectively (Table 3).

DISCUSSION

The DMSP concentrations that we have observed in greenhouse cultures of *S. alterniflora* (Table 3) are consistent with other reports for this and closely related species. In specimens of *S. alterniflora* collected in the field along an estuarine salinity gradient, DMSP concentrations in roots varied from 0.4 to 5.6 μmol/g (fresh weight) and in leaves they varied from 9 to 48 μmol/g (fresh weight), depending on time of year and collection site (16). Weber et al. (20) reported concentrations in leaves of 7.5 to 26.1 μmol/g (fresh weight) depending on time of year and collection site. Van Diggelen et al. (19) reported concentrations of DMSP in leaves of *S. anglica* between 5 and 22 μmol/g (fresh weight) in the greenhouse. In every example where leaf and root values have been reported, the concentration of DMSP in leaves exceeds that in roots.

DMSP concentrations in *S. alterniflora* leaves responded marginally to the sulfide concentration in culture solutions (Table 3). There was a significant concentration of DMSP in tissues even in the 0 mM sulfide treatment. Sulfide had no significant effect on DMSP concentrations in roots. These results are consistent with earlier reports by van Diggelen et al. (18) and Otte and Morris (16). These results do not rule out the possibility that DMSP has significance as an adaptation to toxic sulfides, though they do indicate that DMSP production is not induced by and is independent of external sulfide, at least at external sulfide concentrations of less than 2 mM.

Our results challenge the long held view that any concentration of sulfide is toxic to *S. alterniflora* under any circumstance. There are negative correlations between Spartina production and sulfide within marshes (10, 6), and there are reports that sulfide levels in marsh sediments of approximately 1 mM cause the dieback of Spartina (15), but we have seen sulfide levels exceed 1 mM in healthy marsh sites in Charleston Harbor. In the interior marsh sites in Charleston Harbor, the time-averaged geometric mean sulfide concentration in the upper 50 cm of sediment was 958 μM (810-1132 μM confidence limits). Sulfide concentrations vary seasonally at all the sites we have examined. In interior Charleston Harbor sites they reached a seasonal maximum of 3.6 mM (3.0-4.4 mM) in the upper 50 cm during November with no visible affect on the health of the vegetation. Furthermore, the commonly observed trend toward decreasing production with increasing sulfide concentration within a marsh breaks down when different marshes are compared as in the case of North Inlet and Charleston Harbor. Creek bank areas in Charleston Harbor that support highly productive stands of Spartina have sulfide concentrations that exceed those in interior sites at North Inlet (Fig. 2), and these interior North Inlet sites support the typical short form of Spartina.

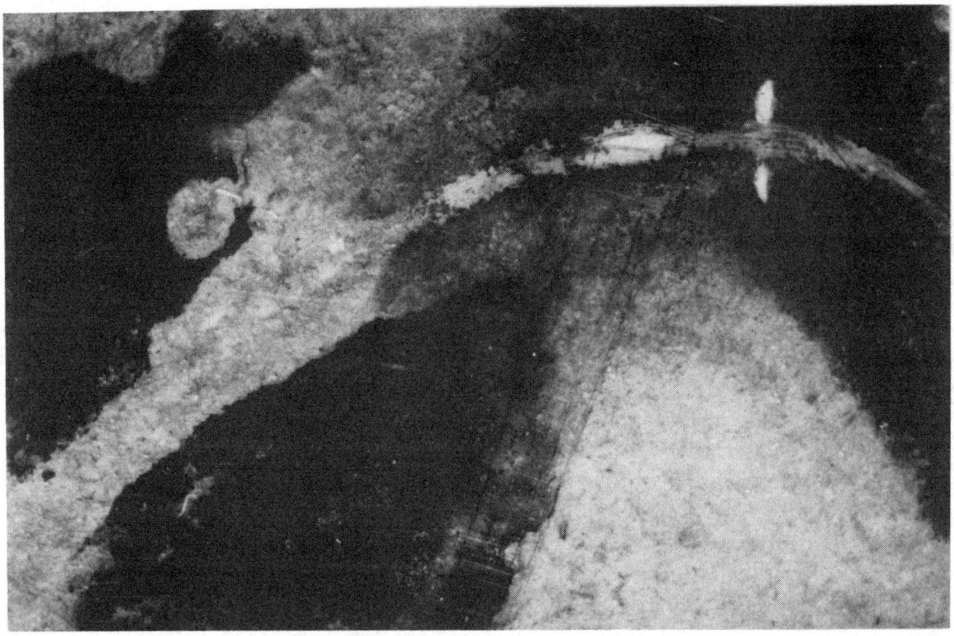

Figure 3. Photograph of a rhizome, the large structure, and root of *Spartina alterniflora* growing along the window of a sand-filled tank. The photo is of an area of approximately 10 × 15 cm. The large black area around the rhizome is an anoxic zone with significant sulfate reduction as indicated by the dark coloration that can be attributed to iron monosulfides. The light area immediately around the root and part of the rhizome (right-center of photo) is an oxidized rhizosphere. (A color version of this figure can be found in the color insert following p. 94.)

The experimental evidence for toxic sulfide effects comes largely from water cultures. Bradley and Dunn (1) found that sulfide concentrations of 0.5 and 1.0 mM decreased the growth of *S. alterniflora* in water culture, while 2 mM sulfide was lethal. Koch et al. (12) also saw declines in growth rate and a suppression of alcohol dehydrogenase activity in roots with increasing sulfide levels up to 4 mM. Similarly, van Diggelen et al. (19) saw a depression of growth of *S. anglica* at sulfide concentrations of 0.5 and 1.0 mM in water culture. However, in a second experiment in water culture, but at lower sulfide levels and pH buffered to 6.5, sulfide significantly stimulated growth at 0.2 and 0.5 mM (18).

Results from sand cultures indicate that low levels of sulfide stimulate growth of *S. alterniflora*. In the present study we observed a 4-fold increase in RGR as the sulfide treatment was raised from 0 to 1 mM, while the RGR at 1 mM and 2 mM did not differ significantly. Except for root production, every metric of growth, including stem density, showed an increasing trend with higher sulfide concentration up to 2 mM, though it appeared that the 1 mM level was nearly optimal. Otte and Morris (16) saw a trend of increased production of Spartina in sand cultures treated with as much as 2 mM sulfide. However, a negative effect of 1 mM sulfide added to permanently flooded sediment cores containing *S. alterniflora* was observed by Koch and Mendelssohn (11).

One important difference between the sand and water cultures is the presence and absence of a stable rhizosphere. There is clear evidence that *S. alterniflora* is able to maintain an oxidized rhizosphere, in situ, around at least a portion of its root system. The evidence is the formation of a plaque of ferric iron that is visible on the root surfaces (15). In sand culture it is possible for an oxidized rhizosphere to develop (Fig. 3), and this interface between the root and its edaphic environment is probably important for the plant in anoxic sediments. This photo (Fig. 3) of a *Spartina alterniflora* rhizome and root, taken through the side of the tank in which the plants were growing in sand, demonstrates the spatially complex environment of the rhizosphere. The large dark area around the root and rhizome is a reduced zone colored by iron monosulfides. Sulfate reducing bacteria in this zone are most likely utilizing organic substrates diffusing away from the root and rhizome. Inside this reduced zone is a small pocket of oxidized space surrounding the root and part of the rhizome. A plaque of orange ferric iron is visible on the surface of the root. It is unlikely that such a rhizosphere could develop in water culture because of the unimpeded diffusion, and possibly advection, of solutes around the root surfaces.

The oxidized rhizosphere may protect the plants from toxins in the surrounding sediment. However, if the sulfides were simply oxidized in the rhizosphere prior to their absorption as sulfate, then the sulfur isotope composition of the plant would be closer to that of sulfate. In this case, the plant sulfur should have the isotopic composition of sulfate, because sulfate concentrations are normally far greater than sulfide concentrations. Sea water has a sulfate concentration of about 29 mM (17). In fact, the isotopic composition of sulfur in Spartina is like that of sulfide (3). The rhizosphere may have a regulatory role by limiting the influx of sulfide. Or, these small oxidized zones may help scavenge nutrients, like phosphate and iron, that will precipitate onto the root where they can be assimilated after solubilization by root exudates. Root exudates from other species have been identified that dissolve ferric phosphate and that are induced by iron deficiency (13).

Questions remain about the role of DMSP in *Spartina alterniflora*, and the nature of the sulfide effect. Sulfide may in fact be the preferred sulfur source, in low concentrations, because of favorable energetics, requirements of bacterial symbionts, or secondary effects involving other nutrients. Sulfide is used by chemolithotrophic bacteria as an energy source. Perhaps Spartina, which is known to fix CO_2 in its root system (8), is able to utilize this energy source directly or indirectly by hosting bacterial symbionts. The rhizosphere bacterial community may also influence plant mineral nutrition in a variety of ways. Sulfide may have secondary effects on plant mineral nutrition because of its effect on the solubility and availability of nutrients. Van Diggelen et al. (18) speculated that the greater growth of *S. anglica* that they observed at low sulfide concentrations could have been caused by an iron deficiency in the 0 mM sulfide treatment. Plants grown without sulfide were chlorotic. Plants grown with sulfide were green and had higher concentrations of Fe and P in their tissues.

The function of DMSP in Spartina is unclear. The evidence that it functions as an osmolyte is weak (19, 16). DMSP certainly does not respond to changes in salinity (19) in the same way as proline and glycinebetaine (4), which are believed to function as osmolytes, though they share the same curious characteristic that concentrations of all of these compounds are considerably lower in roots than in leaves. Unlike proline and glycinebetaine, which respond positively to nitrogen (4), DMSP concentration in Spartina decreases with increasing nitrogen supply (16), which raises the possibility that DMSP is an alternative osmolyte when the production of nitrogen-based proline and glycinebetaine are limited. DMSP-S accounts for a major fraction of the total S in Spartina (19, 16) and, thus, it is linked to sulfide metabolism. Perhaps DMSP functions as a storage compound for sulfur or for methyl groups. As a potential methyl donor, DMSP may be involved in the methylation and translocation of metals (20), though more recent work does not support this (9). The prevalence of DMSP in marine algae and the fact that its production is confined to a relatively few halophytic higher plants strongly suggests that its adaptive significance is marine-related. Unfortunately the work on Spartina to date has not resulted in the definitive answer about its function.

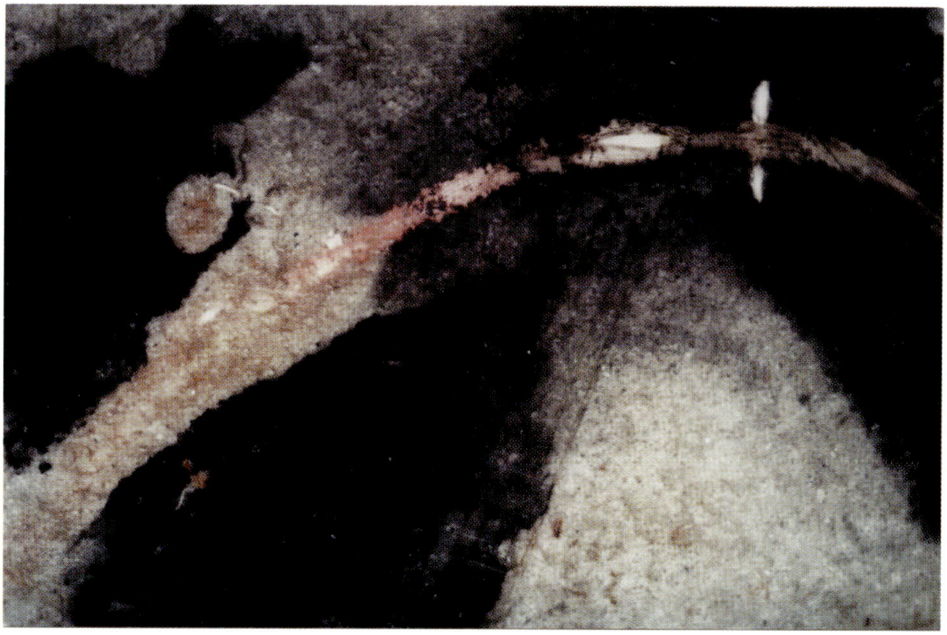

Figure 8.3. Photograph of a rhizome, the large structure, and root of *Spartina alterniflora* growing along the window of a sand-filled tank. The photo is of an area of approximately 10 × 15 cm. The large black area around the rhizome is an anoxic zone with significant sulfate reduction as indicated by the dark coloration that can be attributed to iron monosulfides. The light area immediately around the root and part of the rhizome (right-center of photo) is an oxidized rhizosphere. Notice the orange deposits of ferric iron on the root surface. The red coloration on the rhizome (center of photo) is most likely an anthocyanidin pigment.

ACKNOWLEDGMENTS

Aspects of this work were supported by an Undergraduate Science Education Grant from the Howard Hughes Medical Institute, the National Science Foundation (LTREB program), and the South Carolina Sea Grant Consortium.

REFERENCES

1. Bradley, P. M. and E. L. Dunn. 1989. Effects of sulfide on the growth of three salt marsh halophytes of the southeastern United States. Amer. J. Bot. 76:1707-1713
2. Bradley, P. M. and J. T. Morris. 1990. Influence of oxygen and sulfide concentration on nitrogen uptake kinetics in *Spartina alterniflora*. Ecol. 71:282-287
3. Carlson, Jr., P. R. and J. Forrest. 1982. Uptake of dissolved sulfide by *Spartina alterniflora*: evidence from natural sulfur isotope abundance ratios. Science 216:633-635
4. Cavalieri, A. J. 1983. Proline and glycinebetaine accumulation by *Spartina alterniflora* Loisel. in response to NaCl and nitrogen in a controlled environment. Oecologia 57:20-24
5. Cline, J. D. 1969. Spectrophotometric determination of hydrogen sulfide in natural waters. Limnol. Oceanogr. 14:454-458
6. DeLaune, R. D., C. J. Smith and W. H. Patrick Jr. 1983. Relationship of marsh elevation, redox potential and sulfide to *Spartina alterniflora* productivity. Soil Sci. Soc. Am. J. 47:930-935
7. Epstein, E. 1972. Mineral nutrition of plants: principles and perspectives. Wiley, New York.
8. Hwang, Y. H. and J. T. Morris. 1992. Fixation of inorganic carbon from different sources and its translocation in *Spartina alterniflora* Loisel. Aquat. Bot. 43:137-147
9. Fake, A. M. and J. H. Weber. 1993. Variations in concentrations of methyltin compounds and inorganic tin in Spartina alterniflora and porewater in the Great Bay estuary (NH) during the 1991 growing season. Environ. Technol. 14:851-859
10. King, G. M. and M. Klug. 1982. Relation of soil water movement and sulfide concentration to *Spartina alterniflora* production in a Georgia salt marsh. Science 218:6-163
11. Koch, M. S. and I. A. Mendelssohn. 1989. Sulphide as a soil phytotoxin: differential responses in two marsh species. J. Ecol. 77:565-578
12. Koch, M. S., I. A. Mendelssohn and K. L. McKee. 1990. Mechanism for the hydrogen sulfide-induced growth limitation in wetland macrophytes. Limnol. Oceanogr. 35:399-408
13. Masaoka, Y., M. Kojima, S. Sugihara, T. Yoshihara, M. Koshino and A. Ichihara. 1993. Dissolution of ferric phosphate by alfalfa (*Medicago sativa* L.) root exudates. Plant and Soil 155/156:75-78
14. Mendelssohn, I. A. and K. L. McKee. 1988. *Spartina alterniflora* dieback in Louisiana: timecourse investigation of soil waterlogging effects. J. Ecol. 76:509-521
15. Mendelssohn, I. A. and M. T. Postek. 1982. Elemental analysis of deposits on the roots of *Spartina alterniflora* Loisel. Amer. J. Bot. 69:904-912
16. Otte, M. L. and J. T. Morris. 1994. Dimethylsulphoniopropionate (DMSP) in *Spartina alterniflora* Loisel. Aquat. Bot. 48:239-259.
17. Stumm, W. and J. J. Morgan. 1981. Aquatic Chemistry. 2nd ed. Wiley, New York.
18. van Diggelen, J., J. Rozema, and R. Broekman. 1987. Growth and mineral relations of saltmarsh species on nutrient solutions containing various sodium sulphide concentrations, p. 260-268. *In* A.H.I. Huiskes, C. W. P. M. Blom, and J. Rozema (eds), Vegetation between the land and sea. Dr. W. Junk, Dordrecht.
19. van Diggelen, J., J. Rozema, D. M. J. Dickson and R. Broekman. 1986. 3-dimethylsulphoniopropionate, proline and quaternary ammonium compounds in *Spartina anglica* in relation to sodium chloride, nitrogen and sulphur. New Phytol. 103:573-586
20. Weber, J. H., M. R. Billings and A. M. Fake. 1991. Seasonal methyltin and (3dimethylsulphonio)propionate concentrations in leaf tissue of *Spartina alterniflora* of the Great Bay Estuary (NH). Est. Coast. Shelf Sci. 33:549-557

9

BIOSYNTHESIS OF DIMETHYLSULFONIOPROPIONATE IN *CRYPTHECODINIUM COHNII* (DINOPHYCEAE)

Aritsune Uchida, Tomoaki Ooguri, Takehiro Ishida, Hirotaka Kitaguchi, and Yuzaburo Ishida

Laboratory of Microbiology, Department of Fisheries
Faculty of Agriculture, Kyoto University
Oiwake-cho, Kitashirakawa, Sakyo-ku, Kyoto 606-01, Japan

SUMMARY

Dimethylsulfoniopropionate(DMSP) is the precursor of dimethylsulfide (DMS) which is the most abundant volatile sulfur compound in sea water. Various marine macroalgae and microalgae contain DMSP in high concentration and release DMS. *Crypthecodinium cohnii* (Dinophyceae) is one of the DMSP producers. In *C. cohnii*, the sulfur, the methyl carbon and carbons 2 and 3 of the methionine molecule are efficiently incorporated into DMSP. The location of label in the radioactive DMSP is consistent with the conversion of methionine to DMSP by decarboxylation, deamination, and methylation. Methionine incorporation into DMSP was not inhibited in the presence of methionine-related compounds such as S-methylmethionine and S-adenosylmethionine. Methionine incorporation into DMSP was completely inhibited in the presence of methylmercaptopropionic acid but was not inhibited by methylthiooxybutanoic acid. These data are consistent with a biosynthetic pathway in which methionine is first decarboxylated to methylthiopropylamine. This is followed by loss of the amino group and oxidation to give methylmercaptopropionic acid, which is subsequently methylated. Methionine decarboxylase, which may be the key enzyme of biosynthesis of DMSP, was purified from *C. cohnii* and characterized. This enzyme is pyridoxalphosphate-dependent. The N-terminal amino acid sequence of this enzyme is Ala-Leu-Cys-Trp-Ser-Asp-Ile-Ser-Pro—.

INTRODUCTION

Recent concern over acid precipitation has increased interest in the relative importance of the various sources of sulfur to the atmosphere (20,21). Volatile sulfur compounds are emitted both by terrestrial and marine biota (3). Recently it has been suggested that in the marine environment, most volatile sulfur is emitted in the form of DMS (22), which is excreted by living macroalgae (12), especially from members of the Rhodophyceae (3) and Chlorophyceae (14), and also by microalgae including members of the Dinophyceae (13,23) and Haptophyceae (21). DMS volatilization is of great interest because in the atmosphere it acts as a precursor for cloud condensation nuclei, which can increase cloud reflectance and so reduce the effect of solar heating (4).

It is generally thought that DMSP is the major precursor of DMS and that it is produced by many algae. This DMSP is important in osmoregulation in a number of phytoplankton species (5-8,25) and may also participate in the methylation reactions. DMSP is excreted by algae, and its enzymatic and nonenzymatic breakdown in sea water releases DMS.

Data on the biosynthesis of DMSP in algae are lacking. Although Green (10) first reported that the green alga *Ulva lactuca* synthesized DMSP from methionine, the mechanism has not yet been clarified. It was therefore of interest to determine the biosynthesis pathway of DMSP in algae, especially in microalgae such as dinoflagellates.

MATERIALS AND METHODS

Organism

The dinoflagellate *Crypthecodinium cohnii* (ATCC e32001), which is able to grow heterotrophically, was cultured in an ESW medium (2% glucose and 0.2% yeast extract in natural sea water) or A_2E_6 medium (9) at 25 °C in dark conditions. In the growth experiments, cells were incubated in 50 ml of ESW medium in which sulfate was replaced with 2.5 mM L-cysteine or L-methionine. Cell growth was determined by measuring optical density at 660 nm.

Chemicals

Sodium[^{35}S]sulfate, L-[^{35}S]cysteine, L-[^{35}S]methionine, L-[methyl-^{14}C]methionine, L-[3,4-^{14}C]methionine,and L-[1-^{14}C]methionine were obtained from Ammersham Life Science, Ltd. DMSP bromide was obtained from Shiono Koryo Kaisha, Ltd. S-methylmethionine, L-methionine, methylcysteine, L-cysteine, taurine, betaine, sulfate, 3-methyl mercaptopropionic acid (MMPA), and methylthiooxybutanoic acid(MTOB) were obtained from Nakarai Tesque. Ltd., Waken Chem. Ltd. and Tokyo Kasei Kogyo Co.,Ltd.

Determination of DMS and DMSP in the Culture Medium

DMS was analyzed using a Shimadzu Model GC-9A gas chromatograph equipped with a specially designed, sulfur specific, highly sensitive, linearized flame photometric detector (FPD) as described previously (23). DMSP was quantified by alkaline decomposition to yield DMS(1). Sub-samples (1 ml) of culture media, or cell free extracts were placed in 10 ml vials and the vials were then sealed with rubber caps. Two ml of 1 N NaOH were then injected through rubber caps to decompose the DMSP to DMS. After 90 min, 50 μl of the gas phase was sampled and analyzed by gas chromatography as described above.

Counting Procedure

Radioactivity measurements on thin layer chromatograms were made with an Aloka Radiochromanyzer.

Isolation and Determination of DMSP

Cells were incubated in 50 ml of ESW medium containing 1480 kBq of labeled compound at 25 °C in dark conditions. Cultured cells collected by centrifugation were suspended in 4% perchloric acid (PCA) and the suspension was centrifuged at 12,000 g for 10 min. The supernatant was neutralized with 5 N KOH and centrifuged to remove the potassium salt of perchloric acid. The aliquot of supernatant fraction was separated by two-dimensional cellulose thin layer chromatography (TLC) (Merck) with two different developing solutions (developing solution 1; n-butanol:acetic acid:water = 4:1:1 and developing solution 2; phenol:water = 7:3). The air-dried plates were sprayed with Dragendorff reagent to visualise spots of onium compounds. Only one spot corresponding to the authentic DMSP was observed.

Purification and Characterization of Methionine Decarboxylase

After being treated with acetone, the precipitates were concentrated on a paper filter (Advantec No.2) with vacuum aspirator. The precipitates were resuspended in buffer A (0.01 M potassium-phosphate buffer, pH 7.2). The cell debris was removed by centrifugation. The supernatant fraction was treated with solid ammonium sulfate at 4 °C to give a final concentration of 80%. After centrifugation, the sediment was resuspended in buffer A and dialyzed against the same buffer. The sample was then applied to a DEAE-cellulose column equilibrated with Buffer A; the column was washed with the same buffer and eluted with a linear gradient on NaCl (0 - 0.5 M NaCl in Buffer A). The fractions containing methionine decarboxylase activity were pooled and concentrated to a small volume. The concentrated preparation was applied to a Sephadex G-100 column equilibrated with buffer A; the elution was carried out with the same buffer. The active fractions were pooled. The active fraction was further purified using Mono-Q and Superose-200HR FPLC columns.

Enzyme Assay

The L-methionine decarboxylase assay was based on measurement of the release of $^{14}CO_2$ from L-[1-^{14}C]methionine. The reaction mixture contained L-methionine (20 mM containing L-[1-^{14}C]methionine (2.2 x 10^6 dpm)), pyridoxal phosphate (1 mM), and enzyme in a total volume of 1 ml of buffer A. Incubation was carried out in a test tube with a butyl cap. After incubation at 30 °C for 10 min, the reaction mixture was acidified with 0.2 ml of 10% trichloroacetic acid and bubbled with air. The $^{14}CO_2$ released from the reaction mixture was absorbed in hyamine solution. Scintillation cocktail was added to the hyamine solution and the radioactivity was determined by using Aloka Liquid Scintillation System LCS-3050.

RESULTS

Sulfur Sources and Biosynthesis of DMSP

C. cohnii was able to grow in an A_2E_6 medium in which sulfate was replaced with L-cysteine. From the growth pattern, it appeared that *C. cohnii* was able to utilize L-cysteine as well as sulfate, and that the final cell yields reached almost the same level. However, there was an initial delay period in the medium supplemented with L-cysteine, indicating a possible difference in the uptake mechanism from that of sulfate. In the medium supplemented with L-methionine, growth of *C. cohnii* was very poor. This indicated that L-methionine was not able to act as the sole source of sulfur. Taurine and methylcysteine were also not used as sulfur sources. The DMS and DMSP levels in the culture (mg/*l*) increased in parallel with cell numbers during the logarithmic phase, but the levels stopped increasing in the stationary phase. The DMSP levels per cells (mg/*l*/O.D.$_{660nm}$) did not vary so much with incubation time. The DMSP level per cells was highest in the cells incubated with L-methionine. These results indicated that the L-methionine incorporated was not efficiently utilized as a sulfur source of growth, but was utilized in the biosynthesis of DMSP, and that L-methionine could be a direct or close precursor of DMSP.

Incorporation of Labeled Sulfur-Containing Compounds into DMSP

The cells were incubated in the ESW medium with sodium-[^{35}S]sulfate, L-[^{35}S]cysteine and L-[^{35}S]methionine, respectively, for 10 days, and the harvested cells were extracted with cold PCA. The PCA extract was analyzed by TLC. These three labeled sulfur-containing compounds were efficiently incorporated into the cells; among them methionine showed the highest incorporation. In a short period (135-min) incubation experiment, among the low molecular weight sulfur-containing materials, DMSP was also the most heavily labeled with L-[^{35}S]methionine (Fig. 1).

In the next experiment, incorporation of radioactivity from methionine labeled with ^{14}C at various positions or with ^{35}S was examined (Fig. 2). The methyl, C-3 and C-4 carbons, and the sulfur atom of methionine were efficiently incorporated into DMSP. However, the C-1 carbon of methionine was not incorporated into DMSP. From the relationship between the structures of methionine and DMSP, these data show that the methyl groups and the sulfur of DMSP are derived from the methyl group and the sulfur of methionine, respectively, and that the carboxyl group of

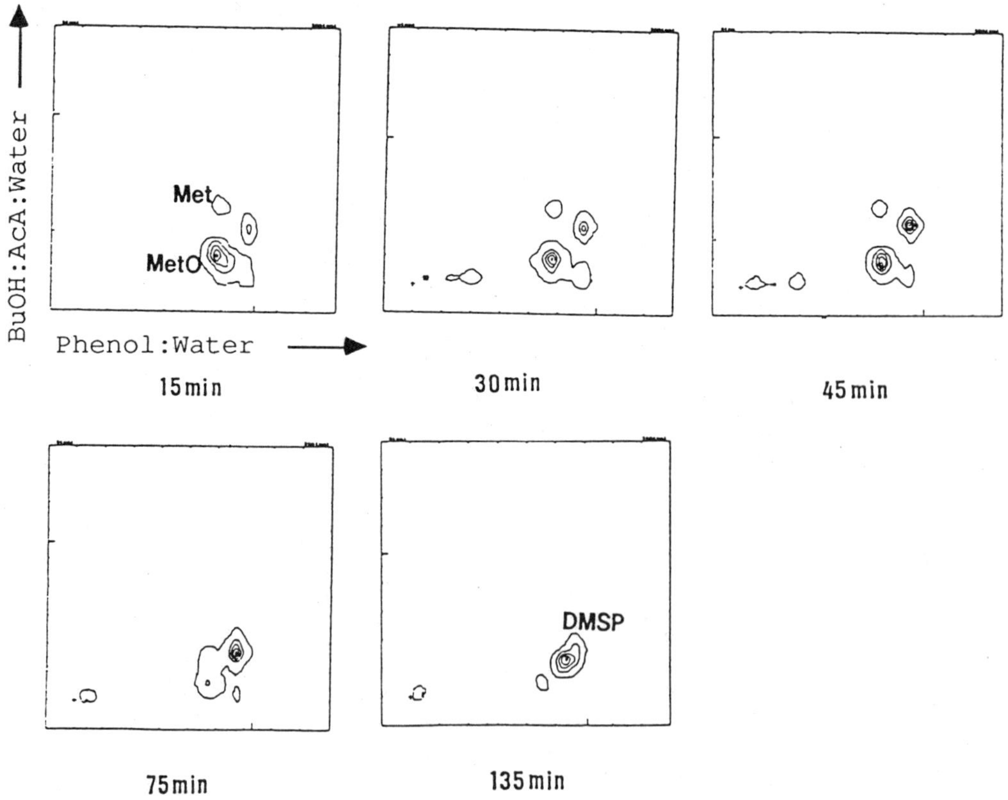

Figure 1. TLC distribution pattern of labeled sulfur compounds in the PCA soluble fraction of intact cells of *C. cohnii*. The intact cells were incubated in A_2E_6 medium containing L-[^{35}S]methionine for 135 min. Met: methionine; MetO: methionine sulfoxide; DMSP:dimethyl-sulfoniopropionate.

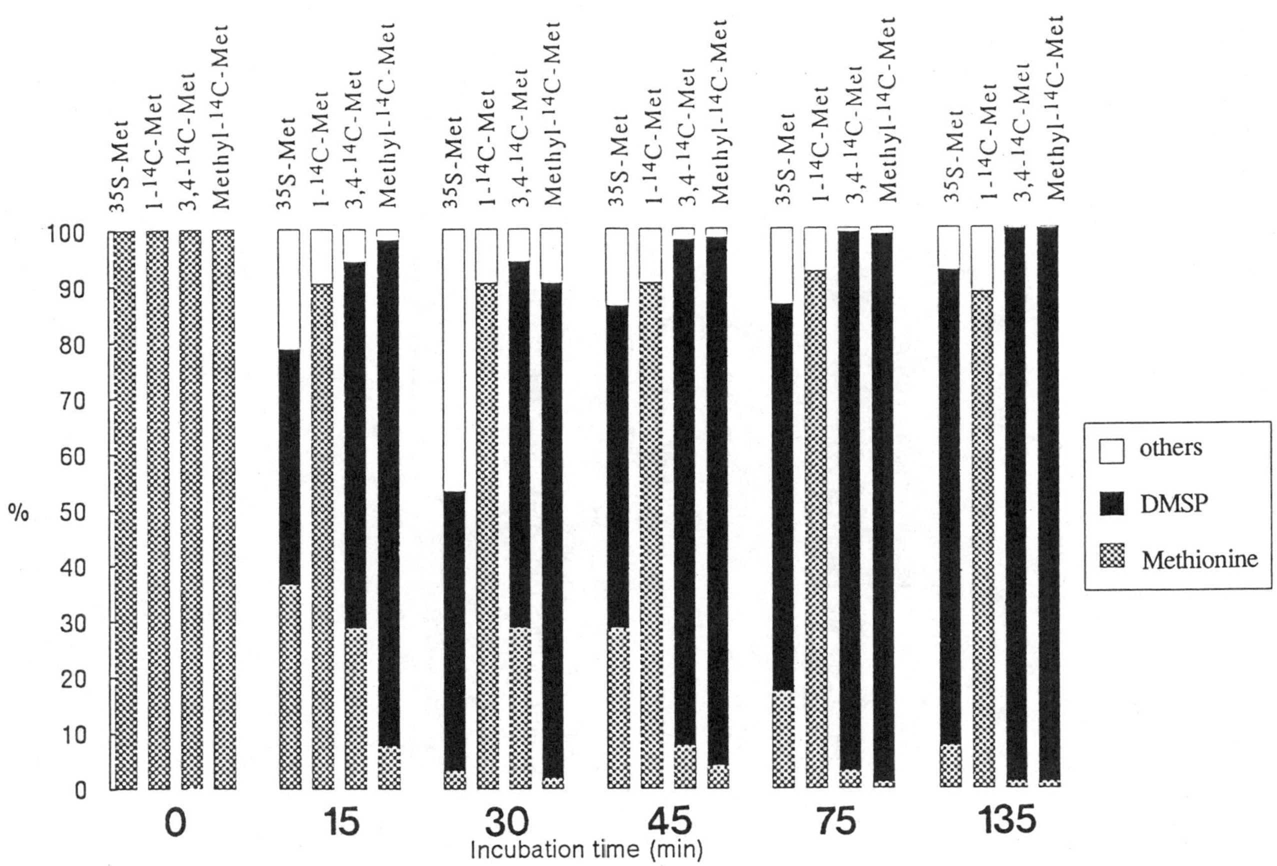

Figure 2. Incorporation of labeled atoms of L-methionine into DMSP in the intact cells of *C. cohnii*. The cells were incubated in A_2E_6 medium containing L-[^{35}S]methionine, L-[methyl-^{14}C], L-[3,4-^{14}C methionine and L-[1-^{14}C]methionine, respectively, for 135 min. The PCA soluble fraction of the cells was analyzed by TLC as described in Fig. 1.

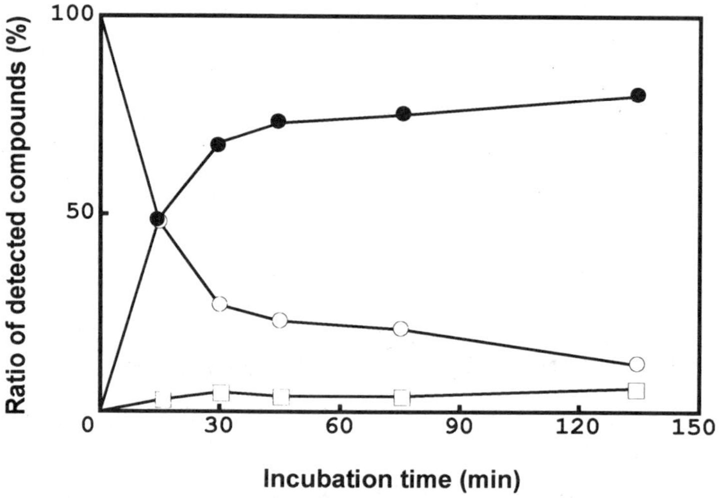

Figure 3. Effect of 1 mM S-methylmethionine on the incorporation of L-[^{35}S]methionine into DMSP in intact cells of *C. cohnii*. S-methylmethionine was added at time zero of the incubation. Symbols: ❍, methionine + methionine sulfoxide; ●, DMSP; ❑, others. The labeling pattern in the absence of S-methylmethionine is shown in Fig. 2.

DMSP may arise from the C-2 carbon of methionine. These observations are consistent with the hypothesis that methionine is converted to DMSP by decarboxylation, deamination, and methylation. In the *in vivo* system the incorporation of L-[^{35}S]methionine into DMSP was not inhibited by addition of 1 mM S-methylmethionine (SMM) to the medium (Fig 3). The incorporation of L-[^{35}S]methionine into DMSP was not inhibited by 1 mM betaine, but interestingly a spot corresponding to S-adenosylmethionine (SAM) appeared after 15 min incubation and then it

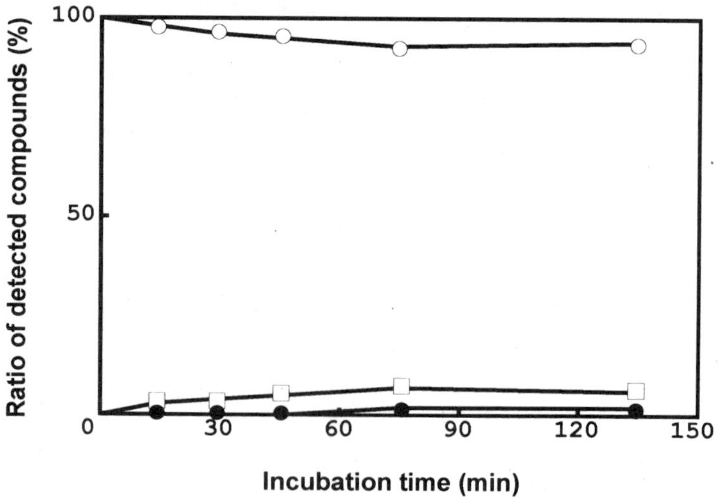

Figure 4. Effect of 1 mM betaine on the incorporation of L-[^{35}S]-methionine into DMSP in the intact cells of *C. cohnii*. Betaine was added at time zero of the incubation. Symbols: ❍, methionine + methionine sulfoxide; ●, DMSP; ❑, others. The labeling pattern in the absence of betaine is shown in Fig. 2.

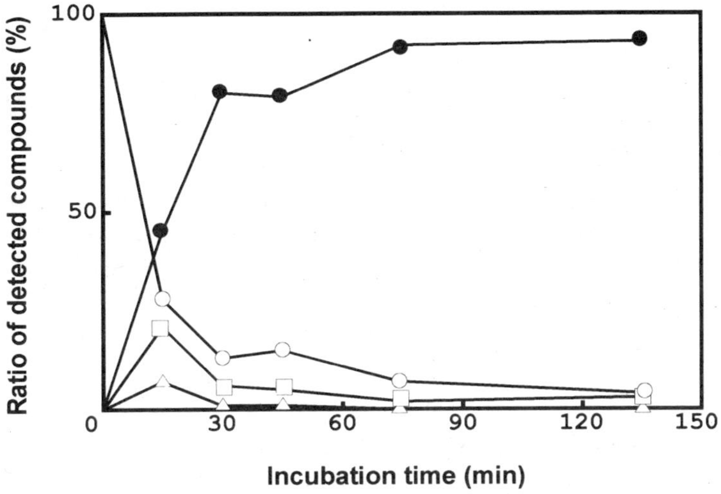

Figure 5. Inhibitory effect of 1 mM methylmercaptopropionic acid on the incorporation of L-[^{35}S]methionine into DMSP in the intact cells of *C. cohnii*. Methylmercaptopropionic acid was added at time zero of the incubation. Symbols: ○, methionine + methionine sulfoxide; ●, DMSP; Δ, S-adenosylmethionine;❑,others. The labeling pattern in the absence of methylmercapto-propionic acid is shown in Fig. 2.

disappeared immediately (Fig. 4). The incorporation of L-[^{35}S]methionine was completely inhibited by inclusion of 1 mM MMPA in the growth medium but not by MTOB (Figs. 5 and 6). These results are consistent with a biosynthetic pathway in which the first step was not methylation or deamination but decarboxylation of methionine.

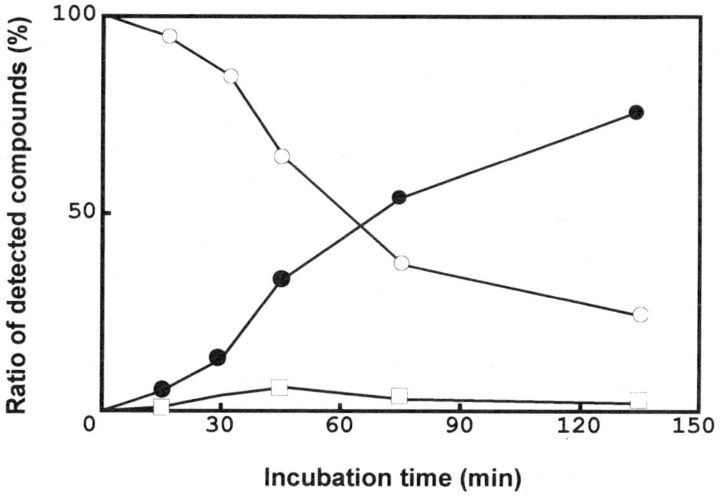

Figure 6. Effect of 1 mM methylthiobutanoic acid on the incorporation of L-[^{35}S]methionine into DMSP in the intact cells of *C. cohnii*. Methylthiobutanoic acid was added at time zero of the incubation. Symbols: ○, methionine + methionine sulfoxide; ●, DMSP; ❑, others. The labeling pattern in the absence of methylthiobutanoic acid is shown in Fig. 2.

Purification and Characterization of Methionine Decarboxylase of C. Cohnii

Mazelis and Ingraham (15,16) first reported the decarboxylation of methionine by an enzyme system from cabbage leaf. Therefore, we tested whether L-methionine decarboxylase activity is present in crude cell extracts of *C. cohnii* which were obtained by sonication or by grinding cells with glass beads. The enzyme assay consisted of measuring the amount of $^{14}CO_2$ released from L-[1-^{14}C]methionine after incubation for 30 min at 30°C. As *C. cohnii* had methionine decarboxylase activity in crude cell extracts, we purified the enzyme by means of acetone precipitation, ammonium sulfate precipitation, and chromatography on DEAE-cellulose, Mono-Q, and Superose-200HR. After these procedures, purity was determined by native polyacrylamide gel electrophoresis (PAGE). PAGE revealed a single protein band which corresponded to the enzyme activity (Fig. 7). The molecular weight of this enzyme was estimated by Superose-200HR. In comparison with the marker proteins, the molecular weight of this enzyme was calculated to be 230 kDa. Sodium dodecyl sulfate (SDS)- PAGE revealed that the purified enzyme was composed of a single polypeptide. The molecular weight of this polypeptide was estimated to be about 103 kDa by comparison with the SDS-PAGE mobility of standard proteins. Therefore, it is probable that this enzyme consists of two identical 103 kDa subunits. This enzyme requires pyridoxal phosphate as a co-enzyme and is stabilized by 1 mM Mg^{2+}. The optimum pH condition was 7.3. This enzyme activity was not inhibited in the presence of SAM in the reaction mixture. No other amino acids had significant inhibitory effect on the enzyme activity. The N-terminal amino acid sequence was determined up to the 9th step by automatic Edman degradation. It was Ala-Leu-Cys-Trp-Ser-Asp-Ile-Ser-Pro—.

DISCUSSION

From the relationship between the structures of methionine and DMSP, the data presented above show that the methyl groups and the sulfur of DMSP are derived from the methyl group and the sulfur of methionine, respectively, and that the carboxyl group of

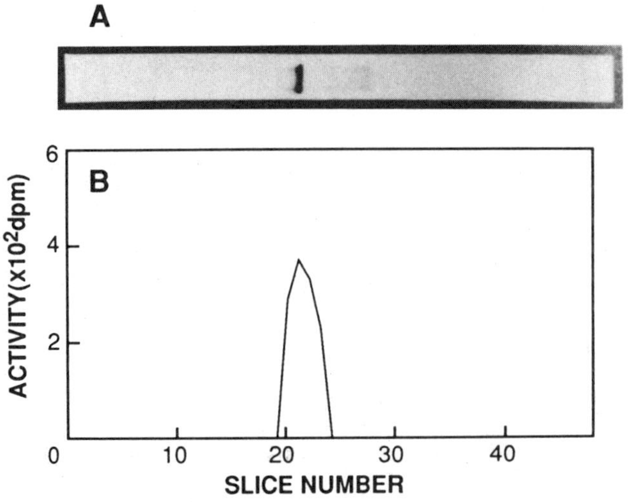

Figure 7. Native polyacrylamide gel electrophoresis of completely purified methionine decarboxylase of *C. cohnii*. A: a photograph of the gel stained with the Silver Stain Kit Wako; B: enzyme activity of methionine decarboxylase.

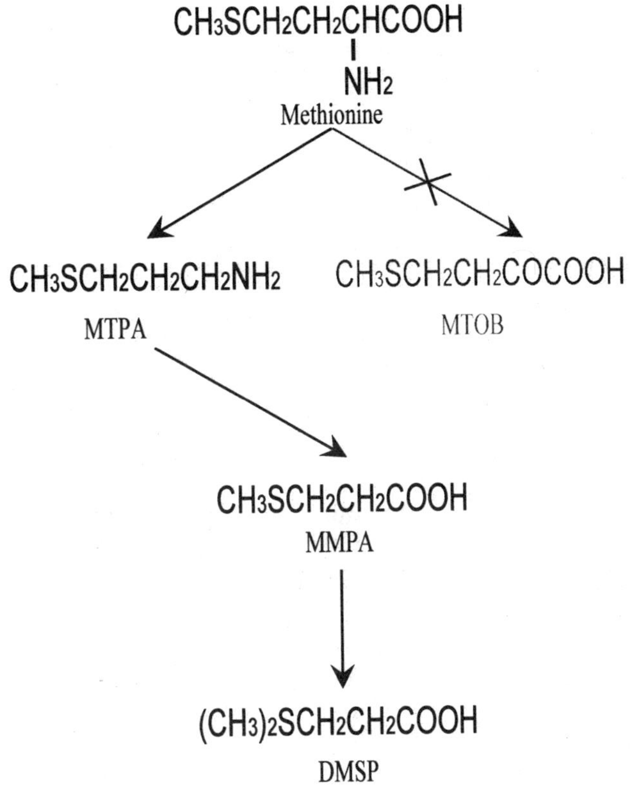

Figure 8. Hypothetical pathway from methionine to DMSP in *Crypthe codinium cohnii.* MTPA:methylthiopropylamine; MTOB:methylthio-oxyobutanoic acid; MMPA;methylmercapto-propionic acid; DMSP:dimethyl-sulfonio propionate.

DMSP may arise from the C-2 carbon of methionine. These observations are consistent with the hypothesis that methionine is converted to DMSP by decarboxylation, deamination, and methylation. The same conversion pathway has been reported by Green in *Ulva lactuca* thallus. On the other hand, Hanson et al. (11) reported that in the coastal strand plant *Wollastonia biflora,* the methylation of methionine was the first step.

In our experiment, the incorporation of L-[^{35}S]methionine into DMSP was hardly inhibited by betaine, and a spot corresponding to SAM was detected after 15 min incubation. As shown in Fig. 2, in the *in vivo* system L-[methyl-^{14}C]methionine was quickly incorporated into DMSP. These results suggest that SAM may not be an intermediate but may act only as a methyl donor in the final methylation of MMPA. There was no dilution effect of SMM on the incorporation of labeled methionine into DMSP, indicating that SMM may not be an intermediate. In other words, the evidence suggests that methylation of L-methionine may not be the first step of biosynthesis of DMSP in *C. cohnii.* The incorporation of methionine into DMSP was not inhibited by MTOB but was inhibited by MMPA. This means that the first step from L-methionine to DMSP is not deamination but decarboxylation. The methylation of MMPA may occur at the final step (Fig. 8). Since these experiments were performed with intact cells of *C. cohnii*, SAM, SMM and MTOB are not conclusively ruled out as intermediate, because we did not check the permeability of these compounds into the cells. Nor it is possible to rule out an effect of MMPA on methionine uptake as the explanation of the observed effect of MMPA on DMSP synthesis. In our experiments, very small amounts of SMM and SAM have been detected in the cell extracts. Green (10) also reported that incorporation of radioactivity from labeled SMM into DMSP was not detected and suggested that this compound was not an intermediate in *U. lactuca*. The results indicated above

Table 1. Comparison of methionine decarboxylases

Source	***Crypthecodinium cohnii***	*Dryopteris filix-mas*	*Streptomyces*	*Streptomyces*
M.W.	**230,000**	100,000	100,000	130,000
Subunit	**103,000 (homodimeric)**	57,000 (homodimeric)	59,000 (homodimeric)	-
opt.pH	**7.3**	5.0	6.9	6.4
Coenzyme	**PLP**	PLP	PLP	PLP
Inhibitors	**Zn^{2+} (Mn^{2+},Cu^{2+},Fe^{3+})**			Ag^{+},Hg^{2+},Cu^{2+}
reference	**(This study)**	(Stevenson et.al 1990)	(Stevenson et.al 1990)	(Misono et.al 1980)

strongly suggest that, in the dinoflagellate *C. cohnii,* methionine is directly converted to DMSP by decarboxylation, deamination, oxidation, and methylation.

There have been some reports concerning decarboxylation of methionine. Mazelis and Ingraham (15,16) have shown that horseradish peroxidase catalyzes oxidative decarboxylation of methionine to yield 3-methylthiopropionamide ($CH_3SCH_2CH_2CONH_2$) as a product. Methionine decarboxylases in which the reaction product is methylthiopropylamine (MTPA) have been purified from *Streptomyces* sp.(17,18) and from the fern *Dryopteris filix-mas* (19,2). The characteristics of our enzyme are summarized and compared with these two methionine decarboxylases (Table 1). This methionine decarboxylase which has been purified may be a key enzyme in DMSP synthesis.

ACKNOWLEDGMENTS

This research was supported by Grants in Aid for Scientific Research (No. 04660225 and 04304020) from the Ministry of Education, Science, and Culture, Japan and Grant Pioneering Research Project in Biotechnology from the Ministry of Agriculture, Forestry and Fisheries, Japan. The authors wish to thank Dr. T. Kawai, Shiono Koryo Kaisha, Ltd., for synthesizing DMSP.

REFERENCES

1. Ackman, R.G., C.S. Tocher, and J.McLachlan. 1966. Occurrence of dimethyl-β-propiothetin in marine phytoplankton. J. Fish. Res. Bd. Can. *23*:357-364.
2. Akhtar, M., D.E. Stevenson, and D. Gani. 1990. Fern L-methionine dedcarboxylase: kinetics and mechanism of decarboxylation and abortive transamination. Biochemistry. *29*: 7648-7660.
3. Challenger, F. and M.I. Simpson. 1948. Studies on biological methylation. Part 12. A precursor of the dimethyl sulphide evolved by P*olysiphonia fasigata*. Dimethyl-2-carboxyethyl-sulfonium hydroxide and its salts. J. Chem. Soc. *3*: 1591-1597.
4. Charlson, R.J., J.E. Lovelock, M.O. Andreae, and S.G. Warren. 1987. Oceanic phytoplankton, atmospheric sulphur, cloud albedo and climate. Nature. *326*: 655-661.
5. Dickson, D.J.J., and G.O. Kirst. 1986. The role of dimethylsulphoniopropionate, glycine betaine and homarine in the osmoacclimation of *Platymonas subcordiformis*. Planta. *167*: 536-543.

6. Dickson,D.J.J., and G.O. Kirst. 1987. Osmotic adjustment in marine eukaryotic algae:the role of inorganic ions, quaternary ammonium, tertiary sulfonium and cabohydrate solutes: I. Diatoms and a rhodophyte. New Phytol. *106*: 645-655.
7. Dickson, D.J.J., and G.O. Kirst. 1987. Osmotic adjustment in marine eukryotic algae:the role of inorganic ions, quaternary ammonium, tertiary sulfonium and carbohydrate solutes: II. Prasinophytes and haptophytes. New Phytol. *106*: 645-655.
8. Edwards, D.M., R.H. Reed and W.D.P. Stewart. 1988. Osmoacclimation in *Enteromorpha inter stinalis*: long-term effects of osmotic stress on organic solute accumulation. Mar. Biol. *98*: 467-476.
9. Gold, K. and C.F. Baren. 1966. Growth requirements of *Gymnodinium cohnii*. J. Protozool. *13*: 255-257.
10. Green, R.C. 1962. Biosynthesis of dimethyl-β-propiothetin. J. Biol. Chem. *98*: 467-476.
11. Hanson, A.D., J.Rivoal, L. Paquet, and D.A. Gage. 1994. Biosynthesis of 3-Dimethyl-sulfoniopropionate in *Wollastonia biflora* (L.) DC. Evidence that S-methylmethionine is an intermediate. 1994. Plant Physiol.*105*:103-110.
12. Iida, H., K. Nakamura and T. Tokunaga. 1985. Dimethyl sulfide and dimethyl-β-propiothetin in sea algae. Nippon Suisan Gakkaishi. *51*: 1145-1150.
13. Ishida, Y. 1968. Physiological studies on evolution of dimethyl sulfide from unicellular marine algae. Mem. Col. Agric. Kyoto Univ. *94*: 47-82.
14. Karsten, U., C. Wiencke, and G.O. Kirst. 1990. The β-dimethyl-sulphoniopropionate (DMSP) content of macroalgae from Antarctica and southern Chile. Bot. Mar. *33*: 143-146.
15. Mazelis, M. 1962.: The pyridoxal phosphate-dependent oxidative decarboxylation of methionine by peroxide. I. Characteristics and properties of the reaction. J. Biol. Chem. *237*: 104-108.
16. Mazelis, M. and L.L. Ingraham. 1962. The pyridoxal phosphate-dependent oxidative decarboxylation of methionine by peroxide. II.Identification of 3-methylthiopropionamide as a product of the reaction. J. Biol. Chem. *237*: 109-112.
17. Misono, H., Y. Kawabata, M. Toyosato, T. Yamamoto and K. Soda. 1980. Purification and properties of L-methionine decarboxylase of *Streptomyces* sp. Bull. Inst. Chem. Res. Kyoto Univ. *58*: 323-333.
18. Stevenson, D.E., M. Akhtar, and D. Gani. 1990. *Streptomyces* L-methionine decarboxylase: purification and properties of the enzyme and stereochemical course of substrate decarboxylation. Biochemistry. *29*: 7660-7666.
19. Stevenson, D.E., M. Akhtar, and D. Gani. 1990. L-methionine decarboxylase from *Dryopteris filix-mas*: purification, characterization, substrate specificity, abortive transamination of the coenzyme, and stere ochemical courses of substrate decarboxylation and coenzyme transamination. Biochemistry. *29*: 7631-7647.
20. Turner, S.M. and P.S. Liss. 1985. Measurement of various sulphur gases in a coastal marine environment. J. Atoms. Chem. *2*: 223-233.
21. Turner, S.M., G. Malin, and P.S. Liss 1988. The seasonal variation in nearshore waters. Limnol. Oceanogr. *11*: 364-375.
22. Uchida, A., T. Ooguri, and Y. Ishida. 1992. The distribution of dimethylsulphide in the water off Japan and in the subtropical and tropical Pacific Ocean. Nippon Suisan Gakkaishi. *58*: 261-265.
23. Uchida, A., T. Ooguri, T. Ishida, and Y. Ishida. 1992. Seasonal variation in dimethylsulfide in the water of Maizuru Bay. Nippon Suisan Gakkaishi. *58*: 255-259.
24. Uchida, A., T. Ooguri, T. Ishida, and Y. Ishida. 1993. Incorporation of methionine into dimethylthiopropanoic acid in the dinoflagellate *Crypthecodinium cohnii*. Nippon Suisan Gakkaishi.*59*: 851-855.
25. Vairavamurthy, A., M.O. Andreae, and R.L. Iverson. 1985. Biosynthesis of dimethylsulfide and dimethylpropiothetin by *Hymenomonas carterae* in relation to sulfur source and salinity variations. Limnol. Oceanogr. *30*: 59-70.

10

BIOSYNTHETIC PATHWAYS FOR PHOSPHATIDYLSULFOCHOLINE, THE SULFONIUM ANALOGUE OF PHOSPHATIDYLCHOLINE, IN DIATOMS

Morris Kates[1] and Benjamin E. Volcani[2]

[1] Department of Biochemistry
University of Ottawa
40 Marie Curie (Priv.)
Ottawa, Ontario, K1N 6N5, Canada
[2] Scripps Institution of Oceanography
University of California San Diego
La Jolla, California 92093

SUMMARY

The biosynthetic pathways for phosphatidylsulfocholine (PSC) and phosphatidylcholine (PC) have been studied in intact cells of the non-photosynthetic diatom *Nitzschia alba* and the photosynthetic diatom *Nitzschia angularis*. Sulfocholine and choline were incorporated intact into PSC and PC, respectively, in *N. alba* and presumably also in *N. angularis*. In both diatoms, incorporation of sulfocholine into PSC occurred at a greater rate than that of choline into PC. However, choline strongly inhibited the incorporation of sulfocholine into PSC in both diatoms, while sulfocholine weakly inhibited the incorporation of choline into PC in *N. alba* and more strongly in *N. angularis*. Serine supplied the S-methyl carbon atoms (presumably via CH_2-tetrahydrofolate) of PSC in both *N. alba* and *N. angularis*, but not the sulfocholine carbons 1 & 2 which are presumably derived from homoserine/aspartate. Carbon from serine was also incorporated into phosphatidylserine and phosphatidylethanolamine (PE) in both diatoms but was incorporated into PC only in *N. angularis*. Ethanolamine was incorporated into both PE and PC in *N. angularis* but only into PE in *N. alba*. Thus, in both diatoms, synthesis of PSC proceeds from serine to cysteine to methionine, then presumably to dimethylsulfoniopropionate (DMSP) and thence to sulfocholine, which is incorporated into PSC by the phosphocholine transferase (nucleotide) pathway. This enzyme can transfer either phosphocholine or phosphosulfocholine to diacylglycerol. However, in *N. angularis* but not in *N. alba*, PC appears to be formed also from PE by the N-methylation pathway. These results may explain the presence of both PSC and PC in *N. angularis* but largely PSC in *N. alba*.

Biological and Environmental Chemistry of DMSP and Related Sulfonium Compounds
edited by Ronald P. Kiene et al., Plenum Press, New York, 1996

INTRODUCTION

The non-photosynthetic diatom, *Nitzschia alba*, contains phosphatidylsulfocholine (PSC)(phosphatidyl-S,S-dimethyl mercaptoethanol), the sulfonium analogue of phosphatidylcholine (PC), as the major membrane phospholipid (2,3a,3b,6,15) (Fig.1). This sulfonium analogue completely replaces PC in *N. alba*, but occurs only in minor amounts in photosynthetic diatoms, such as *Nitzschia angularis*, the major phospholipid being PC (6,11,15). In addition, all diatoms studied contain the plant sulfolipid, sulfoquinovosyldiacylglycerol (SQD) and two other novel sulfolipids: 1-deoxyceramide-1-sulfonic acid (DCS)(N-acyl-1-deoxysphingenine-1-sulfonate) and the sterol sulfate (SS), 24-methylene cholesterol sulfate (3a,3b,11,15) (Fig.1).

Previous studies have shown that both [^{35}S]cysteine and [^{35}S]methionine serve as precursors in the biosynthesis of PSC (4). Cysteine provided only the sulfur atom to methionine by the well-known cystathionine pathway (17), while methionine was shown to supply both the sulfur and the S-methyl groups in PSC (4) suggesting that methionine was converted to dimethylsulfoniopropionate (DMSP)(dimethyl-β-propiothetin) (1,9) and then to sulfocholine:

$$\text{cysteine} \xrightarrow[\text{ATP}]{+\text{ homoserine}} [\text{cystathionine}] \xrightarrow{-\text{N-CH}_2\text{-THF}} \text{homocysteine} \longrightarrow$$

$$\text{methionine} \xrightarrow{-\text{NH}_2,\ -\text{CO}_2,\ \text{O}_2,\ \text{SAM}} \text{dimethylsulfoniopropionate} \xrightarrow{-\text{CO}_2,\ \text{O}_2} \text{sulfocholine} \qquad (1)$$

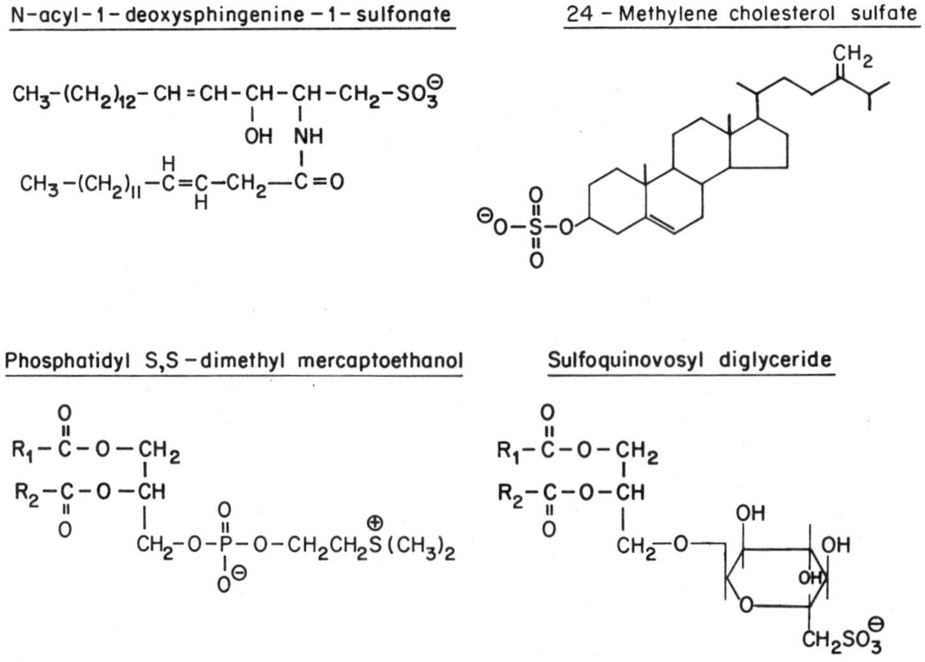

Figure 1. Structures of the major sulfolipids in diatoms.

Sulfocholine was then presumably incorporated into PSC by a pathway analogous to the nucleotide pathway for PC biosynthesis in animals (16) and plants (12):

$$\text{sulfocholine} \xrightarrow{\text{ATP}} \text{P-sulfocholine} \xrightarrow{\text{CTP}}$$

$$\text{CDP-sulfocholine} \xrightarrow{sn\text{-1,2-diacylglycerol}} \text{phosphatidylsulfocholine} \qquad (2)$$

To test for the presence of the nucleotide pathway, we have studied the incorporation of doubly-labelled [^{3}H-Me]/[^{35}S]sulfocholine into phosphatidylsulfocholine in intact cells of *N.alba*. To determine whether the CDP-base transferase shows any preference for sulfocholine over choline and whether the sulfo-analogue inhibits the incorporation of choline and vice-versa, we have also studied the incorporation into PSC and PC of labelled sulfocholine or choline, separately, together, or separately in the presence of cold choline or sulfocholine, respectively, in both *N. alba* and *N. angularis*.

An alternative pathway for PC biosynthesis, the N-methylation pathway, is known in animals (16) and plants (12):

$$sn\text{-1,2-diacylglycerol} \xrightarrow{\text{CTP}} \text{CDP-diacylglycerol} \xrightarrow{\text{serine}} \text{phosphatidylserine}$$

$$\xrightarrow{-CO_2} \text{phosphatidylethanolamine} \xrightarrow{3 \times \text{SAM}} \text{phosphatidylcholine} \qquad (3)$$

To help explain why *N. alba* contains virtually no nitrogenous phospholipids while photosynthetic diatoms, like *Nitzschia angularis*, contain phosphatidylcholine and phosphatidylethanolamine (PE) (11,15), the incorporation of radiolabeled serine and ethanolamine into the phospholipids of these two diatoms was compared to determine whether they contain an active N-methylation pathway.

The present paper is a report of these exploratory radioisotopic experiments with intact cells of *N. alba* and *N. angularis* which have revealed the existence of novel pathways for the biosynthesis of PSC and PC and their regulation.

MATERIALS AND METHODS

Radioactive Precursors

L-[^{35}S]methionine (specific activity, 1.45 Ci/μmol; 54 GBq/μmol), *L*-[^{35}S]cysteine (spec. act., 1.19 Ci/μmol; 44 GBq/μmol), [^{35}S]S-methylthioethanol (spec. act., 1-10 mCi/mmol; 37-370 MBq/mmol), and [2-^{14}C] ethanolamine (spec. act., 50 mCi/mmol; 1.85 GBq/mmol) were purchased from Amersham (Boston, MA); [^{14}C]methyl iodide (spec. act., 50 mCi/mmol; 1.85 GBq/mmol), L-[^{14}C]serine (spec. act., 135 mCi/mmol; 5 GBq/mmol) and [Me-^{14}C]choline-Cl (spec. act. 50 mCi/mmol; 1.85 GBq/mmol) from ICN (Irvine, CA); and *L*-[U-^{14}C]cysteine (spec. act., 306 mCi/mmol; 11.3 GBq/mmol) and [^{3}H]methyl iodide (spec. act., 72 mCi/mmol; 2.66 GBq/mmol) from New England Nuclear (du Pont, Boston, MA). Unlabelled S-methylthioethanol was synthesized as described elsewhere (18).

Synthesis of [Me-^{3}H], [Me-^{14}C], or [^{35}S]sulfocholine

Synthesis of radioactively-labeled sulfocholine was carried out as described for the unlabelled sulfocholine (18), with the following modifications. A mixture of unlabelled S-methylthioethanol (127 mg; 1.38 mmol) and [^{3}H]methyl iodide (0.5 ml; 8.4 mmol; specific activity, 3.0 mCi/mmol or 111 MBq/mmol) or [^{14}C]methyl iodide (0.33 ml; 0.42 mmol; spec.act., 2.38 mCi/mmol; 88 MBq/mmol) was vortexed for several minutes and left overnight at room temperature. After evaporation of the excess methyl iodide under a stream of nitrogen, the residual dark brown oil was dissolved in acetone (1 ml) and left at 4 C for 2-3 hr; the crystalline sulfocholine iodide was washed with three 1 ml portions of cold acetone and dried in vacuo; yield of [Me-^{3}H]sulfocholine iodide, 195 mg (0.83 mmol; 60%; 4.8 mCi; 178 MBq). The iodide was converted to the chloride by ion-exchange chromatography on a column of Dowex-50 cation exchange resin (1 g) by elution with 1N HCl [18]; yield of [Me-^{3}H]sulfocholine chloride, 114 mg (0.8 mmol; 3.33 mCi or 123 MBq ; specific activity, 4.16 mCi/mmol or 154 MBq/mmol); yield of [Me-^{14}C]sulfocholine chloride, 34.6 mg (0.24 mmol; 0.58 mCi or 22 MBq; spec. act., 2.4 mCi/mmol or 89 MBq/mmol).

[^{35}S]Sulfocholine chloride was synthesized as described for [^{3}H]sulfocholine using [^{35}S]S-methylthioethanol instead of unlabelled methylthioethanol and unlabelled methyl iodide instead of [^{3}H]methyl iodide. The final product had spec. act., 58 μCi/mmole or 2.2 MBq/mmol.

Biosynthesis of Phosphatidylsulfocholine (PSC) from [Me-^{3}H]/[^{35}S]Sulfocholine (Long Term Labeling)

Cells of *N. alba* were grown in 250 ml Erlenmeyer flasks containing 75 ml of synthetic sea water medium (modified Rila marine mixture (10) to which was added 3 ml of 0.5 M NaCl containing 32 mg (225 μmol) of [Me-^{3}H]/[^{35}S]sulfocholine [144 μCi (5.3 MBq) ^{3}H and 13.1 μCi (0.49 MBq) ^{35}S; spec. act., 0.058 μCi (2.2kBq) ^{35}S/μmol and 0.64 μCi (24 kBq) ^{3}H/μmol; ratio [^{3}H]/[^{35}S], 11.0]; incubation was carried out with stirring at 30° C for 44 hr (late log phase). Cells were harvested by centrifugation, washed with unlabelled culture medium, lyophilized, and the dried cells (35 mg) were extracted (14) with hot (60° C) isopropanol (16 ml) and with chloroform/methanol (1:1, v/v, 8 ml). The combined extracts were brought to dryness in vacuo. The residue was dissolved in chloroform-methanol (1:1, v/v, 8 ml) and two phases were formed by addition of water (3.6 ml). The lower chloroform phase was concentrated to a small volume (1 ml), counted and subjected to TLC on silica gel G plates (Brinkmann) in a solvent system of chloroform/methanol/water (65:35:5, v/v). After autoradiography on X-ray film (Fuji Film Corp.), the PSC and lyso-PSC spots were scraped into vials and counted; ^{35}S counts were corrected for background, counting efficiency and decay to the beginning of the incubation.

Biosynthesis of PSC and PC from [^{35}S]Methionine, [Me-^{3}H]sulfocholine, [^{14}C]choline, [^{14}C]serine or [^{14}C]ethanolamine (Long Term Labeling)

Cells of *N. alba* were grown in 125 ml Erlenmyer flasks in 50 ml of a modified synthetic sea water medium (10) in which the "marine-mix" salts were substituted with the following (per liter): NaCl, 27.5 g; $MgCl_2 . 6 H_2O$, 5.38 g; $MgSO_4 . 7 H_2O$, 6.77 g; KCl, 0.72 g; $CaCl_2$, 1.35 g (final pH, 7.5). Cells of *N. angularis* were grown in 50 ml of artificial sea water + 0.1% Bacto-Tryptone (Difco) (8) supplemented with trace elements and vitamins (11). The growth media for both cells contained either of the following labelled precursors: 1) [^{35}S]methionine [0.40 mCi(15 kBq); 8 nmol; spec. act., 50.3 mCi(1.86 GBq)/μmol]; 2)

[Me-^{3}H]sulfocholine [0.32 mCi (12 MBq); 75 umol]; 3) [Me-^{14}C]choline [0.25 mCi (9.3 MBq); 75 umol]; 4) [Me-^{3}H]sulfocholine [317 μCi (11.7 MBq); 75 μmol] + [Me-^{14}C]choline [250 μCi (9.3 MBq);75 μmol]; 5) [^{14}C]serine [125 μCi (4.6 MBq); 10 μmol]; or 6) [2-^{14}C]ethanolamine [125 μCi (4.6 MBq); 10 μmol). Incubation was carried out with magnetic stirring at 30° C for *N. alba* and at 18° C for *N. angularis* ; growth was allowed to proceed to the stationary phase followed by a further 52 and 110 hr for *N. alba* and *N. angularis*, respectively.

Cells were harvested by centrifugation, resuspended in 1 ml of water and extracted (14) with 3.75 ml of methanol/chloroform (2:1, v/v). The mixture was centrifuged, then the supernatant was removed by Pasteur pipet, and the residue extracted again with the same amounts of water/methanol/chloroform (0.8:2:1, v/v); the combined supernatants were then diluted with 2.5 ml of chloroform and water, and the chloroform phase was recovered by Pasteur pipet. The upper methanol/water phase was washed twice with 1 ml portions of chloroform and the combined chloroform phases were counted and brought to a small volume (0.1 ml). The residual lipids were separated by TLC on silica gel G analytical plates (Brinkmann) in the solvent system: chloroform/methanol/conc. ammonium hydroxide (65:35:5, v/v), followed by autoradiography on X-ray film. The radioactive spots were then scraped into scintilation vials and counted in Aquasol-II (NEN) for ^{3}H, ^{14}C or ^{35}S; all counts are corrected for efficiency of counting, background and decay.

Biosynthesis of PSC from [Me-^{14}C]Sulfocholine (with or without Cold Choline) and PC from [Me-^{14}C]Choline (with or without cold sulfocholine) in *N. alba* and *N. angularis* cultures (Short Term Labeling)

N. alba cells grown in medium (8) for 20 h (log phase) were harvested by centrifugation and resuspended in spent medium to give a total cell count of 5 X 10^6 cells/ml; 6 ml aliquots were then incubated, in 25 ml erlenmeyer flasks, with stirring for 4 h at 29° C in the presence of radiolabeled substrates with or without inhibitors, as given in Table 3. Duplicate 2 ml portions of each mixture were then extracted (14) with 5 ml of methanol + 2.5 ml of chloroform, followed by addition of 2.5 ml of chloroform and 2.5 ml of water to form two phases. After centrifugation, the chloroform phase was removed, diluted with benzene, brought to a small volume (0.1 ml) under a stream of nitrogen, and subjected to TLC, autoradiography and counting as described above.

This experiment was repeated with a log-phase culture of *N. angularis*, as described for *N. alba*, except that incubation was carried out at 18-19° C in the light.

Deacylation of labeled material was carried out as described elsewhere (14); water-soluble degradation products were chromatographed on cellulose TLC plates (Whatman) in butanol/propionic acid/water (15:7:10 v/v) and chloroform-soluble products were run on silica gel TLC plates in chloroform/methanol/28% NH_4OH (65:35:5, v/v) and in petroleum ether/ethyl ether/acetic acid (70:30:1, v/v/). Radioactive deacylated phospholipid and glycolipid spots were detected by autoradiography and identified by their R_f values (14).

RESULTS AND DISCUSSION

Biosynthesis of PSC from Doubly-Labeled Sulfocholine

After incubation of *N. alba* cells for 44 hr with [Me-^{3}H]/[^{35}S]sulfocholine (initial average $^3H/^{35}S$ ratio, 8.9 ± 1.4), a low incorporation of radioisotopes (0.5%) into the total

Table 1. Incorporation of [Me-^{3}H]/^{35}S]Sulfocholine into Lipids of *Nitzschia alba*

	Radioactivity, nCi		*Ratio,
Sample	^{3}H	^{35}S	^{3}H/^{35}S
Sulfocholine precursor**	(1) 143x10^3 (2)47.8x10^3	13.0x103 6.62x10^3	11.0 7.2
Culture medium at zero time (+precursor)	64.0	7.7	8.4 Ave$^+$ = 8.9 ± 1.4
Total lipids	247	16.1	15.3
Isolated PSC	20.8	2.5	8.3
Isolated lysoPSC†	0.248	0.025	9.8

*^{35}S counts are corrected for decay to 0 time.
**Values are given for two independent measurements of ^{3}H/^{35}S ratio.
$^+$ Average (± mean deviation) of the ^{3}H/^{35}S ratio of the precursor.
†The "lysoPSC" is a monoacylPSC, probably formed by phospholipase A degradation of PSC.

chloroform-soluble lipid fraction was observed (Table 1), probably due to inhibition of growth of diatom cells by sulfocholine. However, TLC/autoradiography of the extracted lipids showed the presence of one major spot corresponding to PSC with ^{3}H/^{35}S ratio 8.3 and a minor spot corresponding to lyso-PSC with ^{3}H/^{35}S ratio 9.8. The ^{3}H/^{35}S ratios for PSC and lysoPSC do not differ significantly from that of the sulfocholine precursor (Table 1)(note that small amounts of lyso-PSC were always observed even after extraction with hot isopropanol, which would inactivate any degradative enzymes (4)). The fact that the ^{3}H/^{35}S ratio of the precursor was the same as that of PSC thus shows that sulfocholine is most likely incorporated intact into PSC, presumably by a pathway analogous to the nucleotide pathway (see eq.1 in Introduction) for PC biosynthesis in plants (12). The alternative pathway involving methylation of phosphatidyl-2-mercaptoethanol, suggested previously (2), is

Table 2. Incorporation of sulfocholine and choline into phosphatidylsulfocholine (PSC) and phosphatidylcholine (PC) in diatoms*

	Incorporation (nmol) into		% of Control	
Radioactive precursor	PSC (+ lyso PSC †)	PC (+ lysoPC†)	PSC	PC$^+$
Nitzschia alba				
[Me-^{3}H]Sulfocholine	145		100	
[Me-^{14}C]Choline		219		100(150)
[Me-^{3}H]Sulfocholine	154		106	
+ [Me-^{14}C]Choline		62		28
Nitzschia angularis				
[Me-^{3}H]Sulfocholine	90		100	
[Me-^{14}C]Choline		291		100(320)
[Me-^{3}H]Sulfocholine	144		160	
+ [ME^{14}C]Choline		39		13

*Cultures of diatoms (50 ml; 1x10^6 cells/ml) were grown in the presence of 1.5 mM concentration of all labelled precursors for 52 hr (*N. alba*) or 110 hr (N. angularis).
$^+$Values in parentheses are % of PSC control
†Lyso PSC and lysoPC are the monoacyl derivatives of PSC and PC, respectively.

highly unlikely since the sulfur atom is not derived from mercaptoethanol but from cysteine or methionine, the latter also supplying both methyl groups (4).

It is of interest that virtually complete replacement of PC by PSC has been achieved by growth of mouse LM fibroblast cells (5) or the yeast *Saccharomyces cerevisia* (13) in the presence of sulfocholine. Incorporation of [Me-^{3}H]/[^{35}S]sulfocholine into PSC in animal tissues has also been reported (7). These results are consistent with the concept that the phosphocholine transferase has a wide substrate specificity and can utilize sulfocholine, as well as choline, in animal and yeast cells and in diatoms.

Biosynthesis of PSC and PC from Labeled Sulfocholine and Choline

To obtain further information on the pathway for PSC biosynthesis and to answer the question whether *N. alba* is capable of synthesizing PC, as does *N. angularis*, the incorporation of labelled sulfocholine, choline, and 1:1 mixtures of the two bases into the lipids of these two diatoms was examined in a long-term incubation study (Table 2). The results with labeled choline and sulfocholine, separately, showed that in *N. alba,* choline incorporation into PC was 1.5 times that of sulfocholine into PSC; however, in 1:1 mixture, sulfocholine incorporation into PSC remained essentially the same while that of choline into PC was reduced by about 70% (Table 2). In *N. angularis*, however, incorporation of choline into PC was about 3 times that of sulfocholine into PSC, but in 1:1 mixture, sulfocholine incorporation was increased 1.6 fold while that of choline was reduced by 86% (Table 2).

These results are consistent with the hypothesis that sulfocholine is incorporated intact into PSC, presumably as the CDP-sulfocholine derivative, by the phospho-base transferase enzyme system (see eq. 2), which acts on either CDP-base in both diatoms, although somewhat better with choline in *N. angularis*. However, in a competitive situation, this transferase system appears to have a preference for sulfocholine in both *N. alba* and *N. angularis*. Furthermore, sulfocholine appears to inhibit the incorporation of choline into PC to a greater extent in *N. angularis* than in *N. alba* (Table 2).

Further study of the inhibition of PC synthesis by sulfocholine and of PSC synthesis by choline was carried out by short-term (4 h) incubation of each diatom with labelled choline and sulfocholine in the absence or presence of cold sulfocholine or choline, respectively (Table 3). The results show that in both diatoms, sulfocholine alone is incorporated into PSC at more than twice the rate of choline alone into PC; also, in *N. alba,* choline strongly inhibits the incorporation of sulfocholine into PSC and sulfocholine relatively weakly inhibits choline incorporation into PC, while in *N. angularis* these bases are equally effective as moderate inhibitors (Table 3).

Biosynthesis of PSC and PC from Labeled Methionine, Serine and Ethanolamine

In long term labeling studies, *N. alba* was found to incorporate ^{35}S from [^{35}S]methionine exclusively into PSC and lyso-PSC, as was reported previously (8), but *N. angularis* incorporated ^{35}S from methionine into PSC + lyso-PSC only to the extent of about 70%, the remainder appearing in deoxyceramide sufonate (DCS) and its precursors, and also in sterol sulfate (SS) (Table 4). This suggests that in *N. angularis* but not in *N. alba* methionine is degraded to sulfate (or sulfide) which is incorporated into DCS and SS (4,8).

In both diatoms, serine was incorporated at high levels into total lipids, of which PSC (+ lyso PSC) accounted for 15% and PC (+lyso-PC) for about 6 % in *N. alba*, while the corresponding values in *N. angularis* were 13% and 1%, respectively. Serine was also incorporated, in both diatoms, into phosphatidylserine (PS) and phosphatidylethanolamine

Table 3. Inhibition of Biosynthesis Phosphatidylsulfocholine by Choline and Phosphatidylcholine by sulfocholine*

Precursor (mM)	Inhibitor (mM)	Incorporation, nmol		% Inhibition	
		PSC+lyso	PC+lyso	PSC+lyso	PC+lyso
N. alba					+
[Me-14C]Sulfocholine (1.33mM)	Choline (0 mM)	29.1	–	0	–
[Me-14C]Sulfocholine (1.33mM)	Choline (1.33 mM)	7.9	–	73	–
[Me-14C]Sulfocholine (1.33mM)	Choline (6.66 mM)	2.9	–	90	–
[Me-14C]Choline (1.33 mM)	Sulfocholine (0 mM)	–	12.7	–	0
[Me-14C]Choline (1.33 mM)	Sulfocholine (1.33 mM)	–	11.1	–	13
[Me-14C]Choline (1.33 mM)	Sulfocholine (6.66 mM)	–	9.2	–	28
N. angularis					
[Me-14C]Sulfocholine (1.33mM)	Choline (0 mM)	21.2	-	0	–
[Me-14C]Sulfocholine (1.33mM)	Choline (1.33 mM)	10.2	-	46	–
[Me-14C]Sulfocholine 1.33mM)	Choline (6.66 mM)	7.1	-	60	–
[Me-14C]Choline (1.33 mM)	Sulfocholine (0 mM)	–	7.4	–	0
[Me-14C]Choline (1.33 mM)	Sulfocholine (4.33 mM)	–	2.9	–	60
[Me-14C]Choline (1.33 mM)	Sulfocholine (6.66 mM)	–	3.2	–	57

* Log-phase cells were incubated for 4 h in presence of the indicated precursors and inhibitors

Table 4. Synthesis of PSC and PC from various labeled precursors

Precursor (μM)	Total ncorp. (nmol)	% Incorporation						
		PSC	lysoPSC	PC +lysoPC	PS	PE	DCS	X*
N. alba								
[^{35}S]Methionine (160μM)	0.51	51.2	42.9	–	–	–	–	–
[^{14}C]Serine (200μM)	333	11.0	3.7	5.8	2.1	1.1	5.2	62.1
[^{14}C]Ethanolamine (200μM)	38.4	tr^{+}	tr	tr	tr	1.6	3.6	79.1
N. angularis								
[^{35}S]Methionine (140μM)	0.012	65.2	5.5	–	–	–	4.4	18.9
[^{14}C]Serine (170μM)	1744	12.4	0.7	0.7	0.7	4.7	2.7	51.3
[^{14}C]Ethanolamine (190μM	132	3.8	1.4	3.0	tr	7.6	2.4	59.1

*X = three fast- moving unidentified precursors of DCS; in addition a small proportion of the label also appeared in the sterol sulfate (SS) component.$^{+}$"tr" = trace.

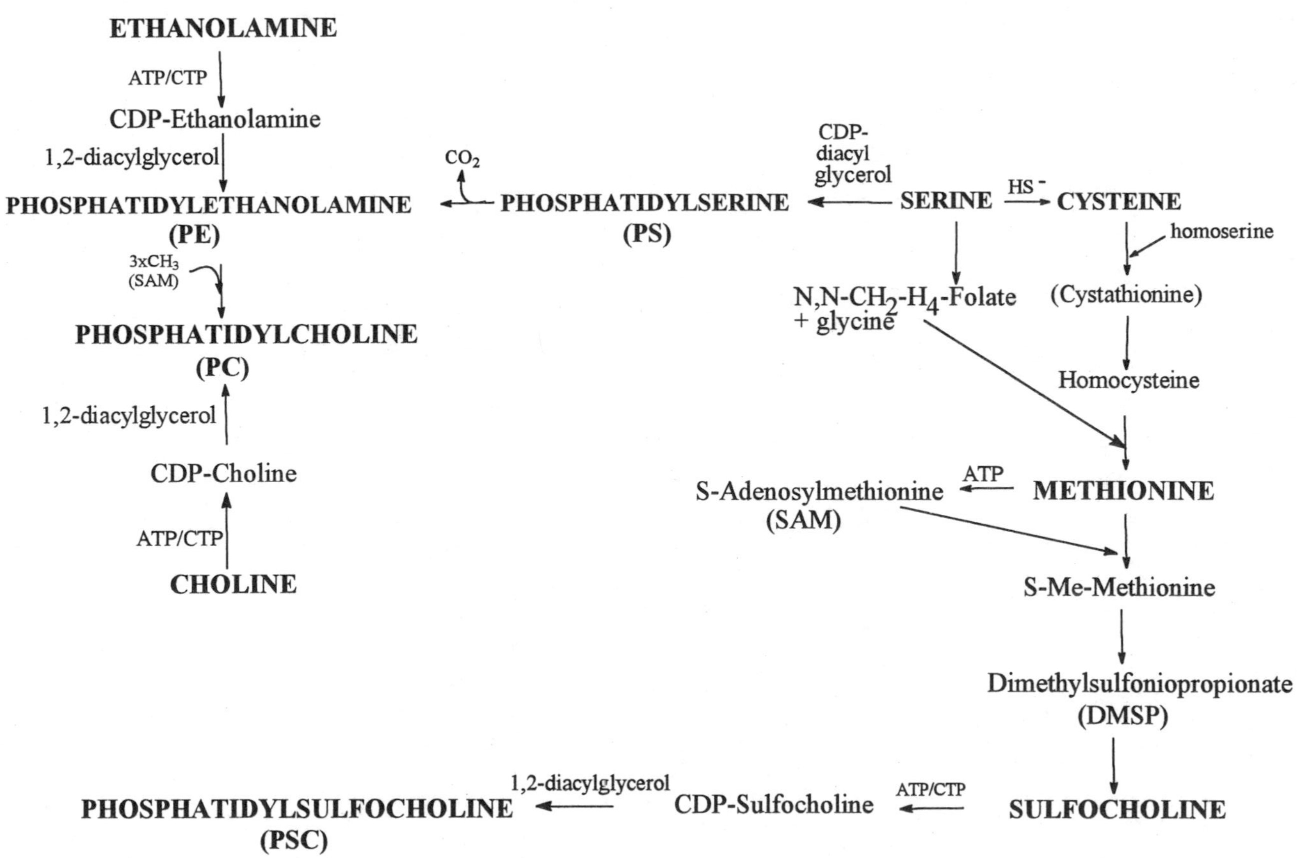

Figure 2. Proposed pathways for biosynthesis of phosphatidylsulfocholine, phosphatidylcholine, phosphatidylserine, and phosphatidyl ethanolamine in the diatoms *N. alba* and *N. angularis*.

(PE), but further incorporation into PC was significant only in *N. angularis* (Table 4). Note that in both diatoms the remaining ^{14}C-activity (>50%) appeared in DCS and its precursors (Table 4).

Ten-fold lower incorporation of ethanolamine into total lipids was observed in both diatoms, with only trace incorporation into PSC (+ lyso-PSC) and PC (+lyso-PC) and significant incorporation into PE in *N. alba*, but higher incorporation into these lipids in *N. angularis* (Table 4). Again, most of the ^{14}C-activity was found in DCS and its precursors (Table 4).

The results of the present study suggest that in both diatoms serine was converted to cysteine and thence to methionine by the cystathionine pathway, well established in plants (17); methionine was presumably then converted via DMSP to sulfocholine which was incorporated into PSC by the Kennedy nucleotide pathway (see eq. 1 and 2 and Fig. 2). The biosynthesis of PC in both diatoms may thus occur by the nucleotide pathway (Tables 2 and 3) but also by the methylation pathway (eq. 3 and Fig. 2) in *N. angularis* and to a much lesser extent, or not at all, in *N. alba*. The present findings thus might explain why *N. alba* contains PSC and virtually no PC or PE while *N. angularis* contains all three of these phospholipids (15). It seems reasonable to suppose that the absence of PC in *N. alba* may be due to the absence of the enzymes for methylation of PE to form PC. However, further studies with cell-free systems or isolated enzymes are required to establish the pathways for PSC biosynthesis as well as that of PS, PE and PC in these diatoms unambiguously.

In conclusion, it appears that serine is the key precursor of the novel sulfonium phospholipid PSC in diatoms (as well as of the ceramide sulfonate (DCS)) (Table 4). However, it should be noted that serine provides only the CH_3-carbon atoms of sulfocholine in PSC, the sulfur atom arising from cysteine/methionine or sulfate or H_2S (Fig. 2). The new pathways outlined in Fig. 2 should provide a basis for further studies of PSC and DCS biosynthesis using cell-free systems.

ACKNOWLEDGMENT

This work has supported in part by grants from NSERC of Canada (A-5324) (M.K.) and by USA Public Health Service, NIH Grant GM-08229 (B.E.V.). The authors are grateful to Dr. Paul Tremblay for assistance in the sulfocholine double-labelling experiment.

REFERENCES

1. Ackman, R.G., C.S. Tocher and J. McLachlan. 1966. Occurrence of dimethyl- β-propiothetin in marine phytoplankton. J. Fish. Res. Bd. Can. *23*: 357-364.
2. Anderson, R., M. Kates and B.E. Volcani. 1976. Sulfonium analogue of lecithin in diatoms. Nature *263*: 51-53.

3a. Anderson, R., B.P. Livermore, M. Kates and B.E. Volcani. 1978. The lipid composition of the non-photosynthetic diatom *Nitzschia alba*. Biochim. Biophys. Acta *528*: 77-88.

3b. Anderson, R., M. Kates and B.E. Volcani. 1978. Identification of the sulfolipids in the non-photosynthetic diatom *Nitzschia alba*. Biochim. Biophys. Acta *528*: 89-106.

4. Anderson, R., M. Kates, and B.E. Volcani. 1979. Studies on the biosynthesis of sulfolipids in the diatom *Nitzschia alba*. Biochim. Biophys. Acta *573*: 557-561.
5. Anderson, R. and P. Bilan. 1981. Replacement of mouse LM fibroblast choline phospholipids by a sulfonium analog. Biochim. Biophys. Acta *640*: 91-99.
6. Bisseret, P., S. Ito, P.-A. Tremblay, B.E. Volcani, D. Dessort and M. Kates. 1984. Occurrence of phosphatidylsulfocholine, the sulfonium analog of phosphatidylcholine in some diatoms and algae. Biochim. Biophys. Acta *796*: 320-327.

7. Bjerve, K.S. and J.Bremer. 1969. Sulfocholine (dimethylhydroxyethylsulfonium chloride) and choline metabolism in the rat. Biochim. Biophys. Acta *176*: 570-58.
8. Darley, W.M. and Volcani, B.E. 1969. Role of silicon in diatom metabolism. A silicon requirement for deoxyribonucleic acid synthesis in the diatom *Cylindrotheca fusiformis* Reimann and Lewin. Exp. Cell. Res. 58, 334-342.
9. Greene, R.C. 1962. Biosynthesis of dimethyl- β-propiothetin. J. Biol. Chem. *237*: 2251-2254.
10. Hemmingsen, B.B. 1971. A mono-silicic acid stimulated adenosinetriphosphatase from protoplasts of the apochlorotic diatom *Nitzschia alba*. Ph.D. thesis, University of California, San Diego.
11. Kates, M. and B.E. Volcani. 1966. Lipids of diatoms. Biochim. Biophys. Acta *116*: 264-278.
12. Kates, M. and M.O. Marshall. 1975. Biosynthesis of Phosphoglycerides in Plants. pp. 115-159. In T.Galliard and E.I. Mercer (eds.), Recent Advances in the Chemistry and Biochemistry of Plant Lipids. Academic Press, London, New York.
13. Kates, M. and P.A. Tremblay. 1981. L'analogue sulfonium de la lecithin, le phosphatidylsulfocholine, peut-il remplacer efficacement la lecithine dans les membranes naturelles? Rev. Can. Biol. *40*: 343-349.
14. Kates, M. (1986) Techniques of Lipidology. Isolation, Analysis and Identification of Lipids, 2nd edition, pp. 396-404, Elsevier, Amsterdam, New York.
15. Kates, M. 1989. The sulfolipids of diatoms. p. 389-427. *In* R.G. Ackman (ed.), Marine Lipids. CRC Press, Boca Raton, Florida.
16. Kennedy, E.P. 1962. Biosynthetic pathways for phospholipids. The Harvey Lectures, Series *57*, pp. 143-171, Academic Press, New York.
17. Maw, G.A. 1972. *In* A.Senning (ed.), Sulfur in Organic and Inorganic Chemistry, Vol. 2, p. 113-142, M. Dekker, New York.
18. Tremblay, P. A. and M. Kates. 1979. Chemical synthesis of *sn*-3-phosphatidylsulfocholine, a sulfonium analogue of lecithin. Can. J. Biochem. *57:* 595-604.

11

OSMOTIC ADJUSTMENT IN PHYTOPLANKTON AND MACROALGAE

The Use of Dimethylsulfoniopropionate (DMSP)

G. O. Kirst

Marine Botany (FB 2: Biology)
University of Bremen, P.O. Box 330 440
D-28334 Bremen, Germany

SUMMARY

Salinity, next to light, temperature and nutrients, is one of the major abiotic factors affecting algal growth and distribution in various habitats. The osmotic balance is maintained by adjusting the cellular solute concentrations. Besides the ions K^+, Na^+ and Cl^-, which are controlled by selective ion transport, the synthesis, accumulation and degradation of organic osmolytes such as polyols, sugars, tertiary sulfonium and quaternary ammonium compounds are involved in regulation of osmotic pressure. Organic osmolytes act as compatible solutes, i.e. they generate a low water potential in the cytoplasm without incurring metabolic or membrane damage. DMSP is one of those compounds abundant in many algal classes, particularly in phytoplankton species containing chlorophyll *a / c*, and in macroalgae, mostly belonging to the chlorophyll *a / b* type. Although DMSP plays a role in osmotic adjustment, the change of its concentration according to osmotic demands is very slow. Usually, DMSP is already present in high concentrations thus providing a buffer capacity which may help to bridge transient stresses under salinity shock. One of the precursors in DMSP biosynthesis appears to be methionine, however, many steps in the synthesis still remain obscure. A DMSP lyase enzyme, which cleaves the compound into DMS and acrylic acid, has been shown to be present, and to exhibit high activities, in an increasing number of micro- and macroalgae.

INTRODUCTION

Growth and distribution of algae are controlled by environmental factors such as light, temperature, nutrient availability and salinity. Salinity typically is a local factor, highly variable in tidal areas including rock pools and lagoons, as well as river inlets and estuaries, and in special habitats, such as sea ice in the polar regions (35). This puts an additional selection pressure on organisms living in those areas. Although large salinity fluctuations

Biological and Environmental Chemistry of DMSP and Related Sulfonium Compounds
edited by Ronald P. Kiene et al., Plenum Press, New York, 1996

occur in limited areas, the ability to tolerate changes in salinity appears to be present in all algae, even those growing in fresh water or in the open ocean, environments which typically have constant salinities (34).

Changes of salinity affect organisms in three ways: (i) by osmotic stress with direct impact on the cellular water potential, (ii) by ion (salt) stress, a result of the inevitable uptake or loss of ions, which in turn is part of the transient acclimation after an abrupt change of salinity, and (iii) by a shift in the ratios of cellular ions because of the acclimation to the new conditions.

RESPONSE TO SALINITY CHANGES: OSMOTIC ACCLIMATION

The immediate responses to salinity changes are rapid water fluxes according to the osmotic gradients established. These result either in volume changes: swelling or shrinking, if naked organisms such as flagellates are considered, or in enhanced turgor pressure and plasmolysis, observed in organisms possessing a proper cell wall (refs. in 34). Subsequently, these fast reactions, which cause severe disturbance of the metabolism by affecting the cellular water potential, are replaced by processes of osmotic acclimation to restore cell volume or turgor pressure. This is achieved by adjusting cellular osmolyte concentrations such as ions and low molecular weight organic compounds accordingly. Osmotic acclimation cannot be separated from general ionic and metabolic regulation. The osmolarity of the cytoplasm and the other cell compartments (e.g. vacuole) is controlled by a variety of feedback mechanisms acting on active ion transport, permeability of membranes (pores / channels) and turnover rates of the metabolite pools (see 4 for further details). However, there is an increasing number of examples of incomplete regulation in species which are nevertheless able to thrive under variable salinities (refs. in 34).

The major ions involved in osmotic acclimation are the monovalent cations K^+ and Na^+ and the anion Cl^- (4, 23, 25, 34). The cellular concentrations of divalent ions such as Mg^{2+}, Ca^{2+} and sulfate usually are not affected by salinity changes, but Ca^{2+} may play an important role in metabolic regulation (32). Nitrate and phosphate are metabolized solutes although they might be accumulated in osmotically significant amounts. The high concentrations of ions necessary to balance external hyperosmotic potentials may exert adverse effects on cellular metabolism (8, 18, 34, 56). Therefore, it is generally assumed that ionic osmolytes are sequestered mainly in the vacuole while a combination of ions and organic osmolytes accounts for the osmotic potential in the cytoplasm.

ORGANIC OSMOLYTES AND DMSP

Low molecular weight organic compounds are involved in osmotic acclimation in all species investigated so far, except for some extremely halophilic archaea which employ high concentrations of monovalent cations. Because of the fairly slow accumulation rates compared to ionic adjustments, organic osmolytes seem to be of importance mostly under long term hyperosmotic conditions. In many cases, the organic osmolytes are identical to the main photosynthetic products of the algae (3, 8, 26, 34, 44) such as sugars (chlorophytes: sucrose), polyols (chlorophytes and phaeophytes: glycerol, mannitol) and heterosides (rhodophytes and cyanobacteria). Other osmolytes are not immediate photosynthates: N-organic compounds (proline, betaines) and the tertiary sulfonium compound dimethylsulfoniopropionate (DMSP). These compounds play an important role not only in algae but also in higher plants and in a wide variety of organisms, including prokaryotes (cyanobacteria, bacteria) and animals (7, 11, 18, 56). Usually most of these compounds are accumulated with increasing salinity or degraded and transferred into osmotically non-active polymeric storage products under decreasing salinity. From these polymers, organic osmolytes may be remobilized again under hyperosmotic stress, or in the dark when photosynthesis is not possible.

In the following, we will focus mainly on DMSP. It has long been suggested that DMSP might be involved in osmotic acclimation in those organisms in which it is present. However, it is only in the last decade that substantial evidence has been provided to prove that DMSP acts as an organic osmolyte in micro- and macroalgae (13, 15, 16, 43, 51, 53). Through this research we have come to know more about the special properties of DMSP.

In most cases, DMSP is not the only osmolyte present in a given algal species; it is typically accumulated in combination with other organic compounds. For example in *Tetraselmis (Platymonas) subcordiformis,* it occurs in addition to mannitol, glycine betaine and homarine (14). It is interesting to note, that except for mannitol, these compounds are accumulated after a substantial lag phase of about three hours. Hence, DMSP seems to be one of the typical long term osmolytes especially involved in algae under severe hyperosmotic stresses. In contrast, with decreasing salinities, DMSP is released almost instantaneously into the medium, presumably by controlled leakage of the cell membranes. This may be of importance when considering the question of whether it is possible that healthy algal cells release DMSP directly into the environment, rather than degrading it themselves to compounds such as DMS and acrylic acid.

In macroalgae, the abundance of DMSP in various taxa is very variable. In some groups, there are examples of a single genus or species with very high concentrations of DMSP, such as the rhodophyte, *Polysiphonia* sp. Other rhodophytes and phaeophytes do not contain DMSP at all or just traces and minor amounts, respectively (5, 6, 43, 55). Many of the benthic green algae are high DMSP producers (13, 29, for review see also 5, 34). With respect to the abundance of DMSP in relation to taxa, we can derive a rule of thumb from an ever increasing number of screening investigations: In microalgae (phytoplankton), many of the chlorophyll *a / c* algae are high DMSP producers and within the macroalgae, many of the chlorophyll *a / b* algae produce much DMSP (in addition to the above mentioned papers e.g. 6, 27, 33; see also the contributions by Karsten et al., this volume).

DMSP AS "COMPATIBLE SOLUTE"

What is the advantage of accumulating organic osmolytes such as DMSP in addition to ions? Organic compounds act not only as osmolytes but also as compatible solutes (9). Unlike ions, they exert a stabilizing effect on proteins (enzymes) and membranes and protect them against inactivation, inhibition and denaturation under conditions of low water potentials. Even in high concentrations, organic osmolytes do not have adverse effects on metabolism, or the disturbing effect is at least much less than that of equivalent levels of inorganic ions.

Three mechanisms are under discussion at present to explain the protective effect of compatible solutes by means of solute interactions. They might act singly or in combination. It is suggested that the binding of the solute to the macromolecule replaces and mimics water to mitigate the loss of the "water sphere" at low water potential. This may be valid especially for polyols. Similarly, it was postulated that proline acts as an amphiphilic (detergent like) compound to mask hydrophobic sites of a protein and convert them into hydrophilic parts (45). This stabilizes the hydration shell, as lowering the water potential affects hydrophobic moieties of macromolecules first, because they loose their hydration water more easily. The third mechanism recognizes the very good solubility of compatible solutes. According to the theory of "preferential exclusion", compatible solutes are excluded from direct contact with the surface of the macromolecules (2, 18). They bind water in their own large hydration shell

Table 1. Characteristic properties of compatible solutes compared with those of DMSP. Data on the solubility of DMSP in water are not available.

General properties	With respect to DMSP
High solubility (proline, betaines: ca. 14 osmolal sucrose: ca. 5.8 osmolal)	Reasonably good solubility, much less than proline
Neutral compounds no net charges at physiological pH	Yes ("Zwitterion")
Controlled permeability / compartmentation	Yes
Non toxic at high concentrations	Moderate (acrylic acid ?)
Rapid synthesis or degradation (turnover)	Moderate
Separated from major "maintenance" metabolic pathways	Yes

leaving the bulk water for the macromolecule. This is entropically an unfavorable condition that results in a reduction of the volume and stabilizes the conformation of the macromolecule. More simply, the macromolecule is stabilized, entrapped in its "waistcoat" of bulk water, surrounded by highly hydrated compatible solutes.

The characteristic properties of compatible solutes are summarized in Table 1. With respect to the general properties of high solubility, charge neutrality at physiological pH, non toxicity, synthesis and degradation according to the demand under osmotic stress, DMSP does not seem to be as suitable as e.g. betaines, proline or glycerol. This is supported by the results of an investigation on the effects of DMSP as a compatible solute in *in-vitro* enzyme test systems (21). A comparison of the inhibitory effect of isosmolal concentrations of NaCl, DMSP and mannitol in cell free enzyme preparations of glucose-6-phosphatedehydrogenase, glutamate dehydrogenase and malate dehydrogenase (MDH) obtained from *Tetraselmis* revealed that DMSP was not as "compatible" as mannitol (Table 2). In fact, at high concentrations, it was as inhibitory as NaCl. However, under substrate limitation, the protection of MDH was clearly enhanced by DMSP.

SYNTHESIS AND DEGRADATION OF DMSP

Methionine is the immediate precursor in DMSP biosynthesis (20, 24, 41, see also the contribution by Hanson and Gage, this Volume). Because of the reduction of sulfate, it is generally

Table 2. Effect of isosmolal concentrations of NaCl, DMSP(hydrochloride), and mannitol in *in-vitro* tests of malate dehydrogenase (MDH). Modified from (21)

Osm. potential (mosmol / kg)	Remaining MDH	Activity	(Contr. = 100%)
	NaCl	DMSP-Cl	mannitol
330	70	85	87
500	60	60	80
950	35	- -	- -

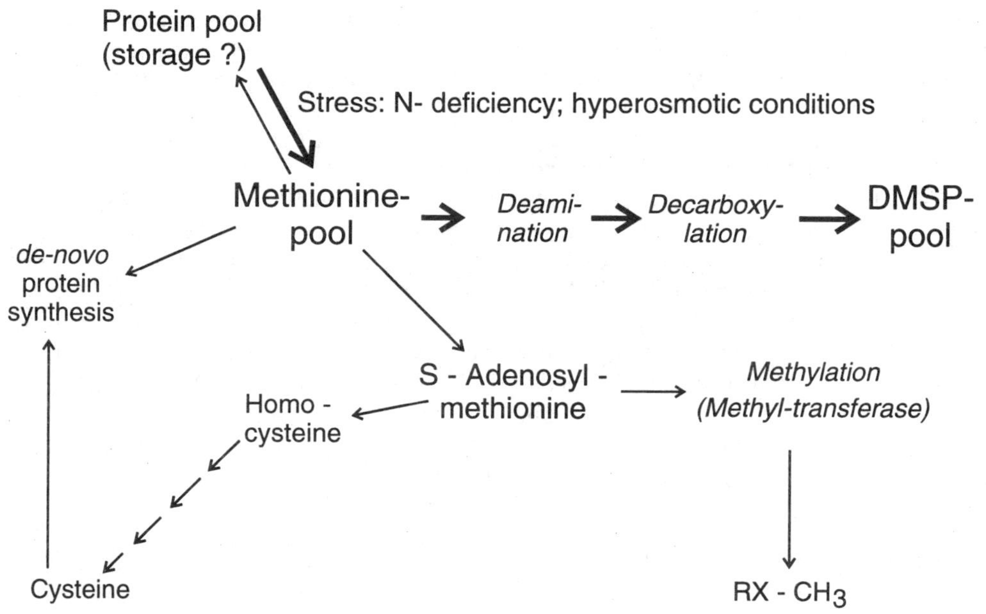

Figure 1. Biosynthesis of DMSP and related methionine metabolic pathways. RX-CH3: methylated compound. Summarized from 20, 22, 41, 51.

accepted that photosynthesis is a necessary prerequisite to enable organisms to synthesize methionine and hence DMSP, in large amounts under aerobic conditions. The biosynthesis of DMSP, and the related pathways of methionine metabolism (summarized from 19, 20, 22, 41), are given in Fig. 1 . From experiments with S-labeled precursors, it was concluded that the availability of methionine, via degradation of proteins, controls the size of the DMSP pool. Stress conditions, such as N-deficiency or hyperosmotic conditions, result in the degradation of protein which liberates methionine and in turn increases the synthesis of DMSP. The synthesis steps from methionine to DMSP involve decarboxylation, deamination, oxidation and methylation. However, it is not yet clear in what order the reaction sequences proceed (51).

The mechanism of DMSP release from organisms has attracted great interest because of the importance of DMSP as the precursor of DMS, the most important biogenic volatile sulfur compound (e.g. 1). There is substantial evidence that DMSP is released via lysis after cell death, cell rupture by grazing, large hypo-osmotic shocks or leakage during excess of synthesis (e.g. 36, 37). There is increasing evidence to support the idea that many algae also possess an enzyme system capable of cleaving DMSP into DMS and acrylic acid (27, 48, 49). DMSP lyase was partially enriched by ammoniumsulfate precipitation (0-35% fraction) from the macroalga, *Enteromorpha* sp., and its activity measured in cell free extracts. In the presence of detergents during preparation of the enzyme extract, the activity was increased, suggesting that the enzyme system may be located in membranes (49). DMSP lyase is also present in marine bacteria (12).

OTHER FACTORS AFFECTING CELLULAR LEVELS OF DMSP

There is clearly a positive correlation between DMSP content and increasing salinity in all algae tested in this respect (14, 34, 43, 52, 56). However, a variety of other factors also influence DMSP accumulation in algae, indicating additional biological roles. Numerous examples demonstrate that increasing light intensity stimulates DMSP production and accumulation (30, 31, 52). The duration of irradiance, e.g. long-day or short-day conditions, increase and decrease DMSP concentrations, respectively (29).

Nutrient supply also affects the DMSP concentrations, e.g. N deficiency has been reported to increase DMSP levels in phytoplankton and *Spartina*, (50; see also Keller and Korjeff-Bellows, this Volume, respectively) while Fe deficiency results in a decrease in DMSP in *Tetraselmis* (Bolt and Kirst, unpublished). Reducing sulfate to less than 2.5 % of its average sea water level causes a decrease in DMS production, presumably as a result of an indirect effect on the synthesis of DMSP (52).

There are reports of extracellular accumulation of DMSP in senescent algal cultures, increasing with cell age, and in decaying plankton blooms (39, 42). In late stationary phases of cultures of dinoflagellates and diatoms, increasing leakage of DMSP from the cells was observed (40). From our own observations, we know that with increasing population densities or changes in the species composition of algal culture mixtures, some microalgae increase their DMSP levels, e.g. *Tetraselmis* sp.

CONCLUSIONS

There is common agreement on the importance of DMSP as a biogenic precursor for DMS. At the organismic level, what makes DMSP so special compared to other compatible solutes?

DMSP does not appear to be as effective a compatible solute as betaines, proline, glycerol or mannitol. However, in those organisms which produce it, it is present at high levels and hence may be suitable as a buffer during the initial phases after a hyperosmotic shock. In this case it can act as a moderate compatible solute under substrate or metabolite limitation. DMSP increases with rising salinity, but especially under long term stress conditions.

Unlike betaines, no additional nitrogen is needed for the biosynthesis of DMSP. However, this advantage may not be very significant since organisms can produce polyols, which are usually better compatible solutes than DMSP, without the need to reduce the sulfate that is necessary for DMSP synthesis.

Other biological roles for DMSP have been suggested, including the activity of its byproduct, acrylic acid, as a grazing repellent or antibiotic (46). However, concentrations of DMSP or acrylic acid in the sea water are usually much too low to be effective for these purposes. Only under special conditions, when phytoplankton forms aggregates ("marine snow"), or in proximity to, e.g. *Phaeocystis* colonies, could acrylic acid retard bacterial activity (47).

In Antarctic macroalgae, DMSP content was observed to increase with decreasing temperature (28). It was suggested that DMSP may also act as a cryoprotectant. Freezing point depression is part of the colligative properties of osmolytes, yet on a cellular level, the concentrations (usually between 50 and 300 mM) are much too low to account for a reasonable freezing point depression. Therefore, we assume that it is the protective effect, i.e. the feature of a compatible solute, rather than the colligative properties which distinguishes DMSP as a cryoprotectant.

DMSP appears to be a typical "secondary metabolite"; potentially handling an excess of reducing power under favorable photosynthetic conditions and hence, channeling overcapacities. Alternatively, DMSP may serve to detoxify sulfide present in large amounts in the anoxic zones of mud flats. It may act as a transient sulfur store or sink for sulfur released out of the cell via its volatile byproduct, DMS (refs. in 34).

In algae and especially in bacteria, DMSP may serve as a methyl donor in metabolism (10, 17, 38, 54). Although this is frequently discussed, our knowledge of these pathways remains scarce.

All in all, there seems to be no single clear cut "benefit" of possessing DMSP. However, it may be of some selective advantage, taking part in all or some of the above mentioned biological roles.

ACKNOWLEDGMENTS

The author gratefully acknowledges the support of his research on salinity tolerance and DMSP by the Deutsche Forschungsgemeinschaft (DFG). I thank two anonymous referees for valuble suggestions and brushing up the language.

REFERENCES

1. Andreae, M.O., and H. Raemdonk. 1983. Dimethyl sulfide in the surface ocean and the marine atmosphere: a global view. Science 221, 744-747
2. Arakawa, T. and S.N. Timasheff. 1985. The stabilization of proteins by osmolytes. Biophys. J. *47:* 411-414.
3. Ben-Amotz, A. and M. Avron. 1983. Accumulation of metabolites by halotolerant algae and its industrial potential. Annu. Rev. Microbiol. *37:* 95-119.
4. Bisson, M.A. and G.O. Kirst. 1995. Osmotic acclimation and turgor pressure regulation in algae. Naturwissenschaften (in press).
5. Blunden, G. and S.M. Gordon. 1986. Betaines and their sulphonio analogues in marine algae. Progress in Phycological Research *4:* 39-80
6. Blunden, G., B.E. Smith, M.W. Irons, M. Yang, O.G. Roch and A.V. Patel. 1992. Betaines and tertiary sulphonium compounds from 62 species of marine algae. Biochemical Systematics and Ecology *20:* 373-388
7. Borowitzka, L.J. 1986. Osmoregulation in blue-green algae. Progress in Phycological Research *4:* 243-256.
8. Borowitzka, L.J. 1981. Solute accumulation and regulation of cell water activity. p. 97-130. *In* L.G.Paleg and D. Aspinall (eds.), Physiology and Biochemistry of Drought Resistance in Plants. Academic Press Melbourne.
9. Brown, A.D. and J.R. Simpson. 1972. Water relations of sugar-tolerant yeasts: the role of intracellular polyols. J General Microbiology *72:* 589-591.
10. Challenger, F., and M.I. Simpson. Studies on biological methylation 12. J. Chem. Soc. *43:* 1591-1597.
11. Csonka, L.N., and A.D. Hanson. 1991. Prokaryotic osmoregulation: Genetics and physiology. Annu. Rev. Microbiol. *45:* 569-606.
12. De Souza, M.P. and D.C. Yoch. 1995. Purification and characterization of dimethylsulfoniopropionate lyase from an *Alcaligenes*-like dimethyl sulfide-producing marine isolate. Applied and Environmental Microbiology *61:* 21-26.
13. Dickson, D.M.J., R.G. Wyn Jones and J. Davenport. 1980. Steady state osmotic adaptation in *Ulva lactuca*. Planta *150:* 158-165.
14. Dickson, D.M.J. and G. O. Kirst. 1986. The role of ß-dimethylsulfoniopropionate, glycine betaine and homarine in the osmoacclimation of *Platymonas subcordiformis*. Planta *167:* 536-543.
15. Dickson, D.M.J. and G. O. Kirst. 1987a. Osmotic adjustment in marine eukaryotic algae: The role of inorganic ions, quaternary ammonium, tertiary sulphonium and carbohydrate solutes. I. Diatoms and a rhodophyte. New Phytologist *106:* 645-655.
16. Dickson, D.M.J. and G. O. Kirst. 1987b. Osmotic adjustment in marine eukaryotic algae: The role of inorganic ions, quaternary ammonium, tertiary sulphonium and carbohydrate solutes. II. Prasinophytes and haptophytes. New Phytologist *106:* 657-666.
17. Dubnoff, J.W., and H. Borsook. 1948. Dimethylthetin and dimethyl-ß-propiothetin in methionine synthesis. J. Biol. Chem. *176:* 789-798.

18. Galinski, E.A. 1993. Compatible solutes of halophilic eubacteria: molecular principles, water-solute interaction, stress protection", Experientia *49:* 487-496.
19. Giovanelli, J., S.H. Mudd and A.H. Datko. 1985. Quantitative analysis of pathways of methionine metabolism and their regulation in *Lemna*. Plant Physiol. *78:* 555-560.
20. Greene, R.C. 1962. Biosynthesis of dimethyl-ß-propiothetin. J. Biol. Chem. *237:* 2251-2254.
21. Gröne, T. and G. O Kirst. 1991. Aspects of dimethylsulfoniopropionate effects on enzymes isolated from the marine phytoplankter *Tetraselmis subcordiformis* (Stein). J. Plant Physiol. *138:* 85-91.
22. Gröne, T. and G. O. Kirst. 1992. The effect of nitrogen deficiency, methionine and inhibitors of methionine metabolism on the DMSP contents of *Tetraselmis subcordiformis*. Marine Biology *112:* 497-503.
23. Gutknecht, J., D.F. Hastings and M.A. Bisson. 1978. Ion transport and turgor pressure regulation in giant algal cells. p. 125-174. *In* G. Giebisch, D.C. Tosteson and H.H. Ussing (eds.) Membrane Transport in Biology III. Springer Berlin Heidelberg.
24. Hanson, A.D., J. Rivoal, L. Paquet and D.A. Gage. 1994. Biosynthesis of 3-dimethylsulfoniopropionate in *Wollastonia biflora*. Plant Physiol. *105:* 103-110.
25. Hellebust, J.A. 1985. Mechanisms of response to salinity in halotolerant microalgae. Plant and Soil *89:* 69-81.
26. Hellebust, J.A. 1976. Osmoregulation. Annu. Rev. Plant Physiol. *27:* 485-505.
27. Ishida, Y. and H. Kadota. 1967. Production of dimethyl sulfide from unicellular algae. Bulletin of the Japanese Society of Scientific Fisheries *33:* 782-787.
28. Karsten, U., C. Wiencke and G. O. Kirst. 1990 a. ß-Dimethylsulphoniopropionate (DMSP) content of macroalgae from Antarctica and Southern Chile. Bot. Mar. *33:* 143-146.
29. Karsten, U., C. Wiencke and G. O. Kirst. 1990 b. The effect of light intensity and daylength on the ß-dimethylsulphoniopropionate (DMSP) content of marine macroalgae from Antarctica," Plant, Cell and Environment *13:* 989-993.
30. Karsten, U., C. Wiencke and G. O. Kirst. 1991. Growth pattern and ß-dimethylsulphoniopropionate (DMSP) content of green macroalgae at different irradiances. Marine Biology *108:* 151-155.
31. Karsten, U., C. Wiencke and G. O. Kirst. 1992. Dimethylsulphoniopropionate (DMSP) accumulation in green macroalgae from polar to temperate regions: interactive effects of light versus salinity and light versus temperature. Polar Biol. *12:* 603-607.
32. Kauss, H. 1987. Some aspects of calcium-dependent regulation in plant metabolism," Annu. Rev. Plant Physiol. *38:* 47-72.
33. Keller, M.D., W.D. Bellows and R.R.L. Guillard. 1989. Dimethyl sulfide production in marine phytoplankton. p. 167-182. *In* E.S. Saltzmann and W.J.Cooper (eds.), Biogenic Sulfur in the Environment. A.C.S. Sympsoium Series 393.
34. Kirst, G. O. 1990. Salinity tolerance in eukaryotic marine algae. Annu. Rev. Plant Physiol. Mol. Biol. *41:* 21-53.
35. Kirst, G. O. and C. Wiencke. 1995. Ecophysiology of polar algae. J. Phycol. *31:* 181-199.
36. Leck, C., U. Larsson, L.E. Bagander, S. Johansson and S. Hadju. 1990. DMS in the Baltic Sea - annual variability in relation to biological activity. J. Geophys. Res. *95:* 3353-3363.
37. Liss, P.S., G. Malin and S.M. Turner. 1993. Production of DMS by phytoplankton. p. 1-14. In G. Restelli and G. Angeletti (eds.), Dimethylsulphide: Oceans, Atmosphere and Climate. Kluwer Academic Publishers Dordrecht (NL).
38. Marc, J.E., C. van der Maarel, M. Jansen and T.A. Hansen. 1995. Methagonetic conversion of 3-S-Methylmercaptopropionate to 3-Mercaptopropionate. Appl. Environ. Microbiol. *61:* 48-51.
39. Matrai, P.M. and M.D. Keller. 1993. Dimethylsulfide in a large-scale coccolithophore bloom in the Gulf of Maine. Continental Shelf Research *13:* 831-843.
40. Matrai, P.M. and M.D. Keller. 1994. Total organic sulfur and dimethylsulfoniopropionate in marine phytoplankton: intracellular variations. Marine Biology *119:* 61-68.
41. Maw, G.A. 1981. The biochemistry of sulphonium salts. p. 703-770. *In* C.J.M. Stirling and S. Patai (eds.), The chemistry of the sulphonium group. J. Wiley & Sons
42. Nguyen, C., S. Belviso, N. Mihalopoulos, J. Gostan and P. Nival. 1988. Dimethyl sulfide production during natural phytoplanktonic blooms. Marine Chem. *24:* 133-141.
43. Reed, R.H. 1983 Measurement and osmotic significance of ß-dimethyl-sulphoniumpropionate in marine macroalgae. Marine Biology Letters *4:* 173-181.
44. Reed, R.H. 1990. Solute accumulation and osmotic adjustment. p. 147-170. In K.M. Cole and R.G. Sheath (eds.), Biology of the Red Algae. Cambridge Univ. Press.
45. Schobert, B. and H. Tschesche. 1978. Unusual solution properties of proline and its interaction with proteins. Biochim. Biophys. Acta *541:* 270-277.
46. Sieburth, J.M. 1959. Antibacterial activity of antarctic marine phytoplankton. Limnol. Ocean. *4:* 419-424

47. Slezak, D.M., S. Puskaric and G.J. Herndl. 1994. Potential role of acrylic acid in bacterioplankton communities in the sea. Mar. Ecol. Prog. Ser. *105:* 191-197.
48. Stefels, J. and W.H.M. van Boekel. 1993. Production of DMS from dissolved DMSP in axenic cultures of the marine phytoplankton species *Phaeocystis* sp. Mar. Ecol. Progr. Ser. *97:* 11-18.
49. Steinke, M. and G.O. Kirst. (submitted). Enzymatic cleavage of DMSP in cell free extracts of the marine macroalga, enteromorpha clathrata (Ulvales, Chlorophyta). J. Exp. Mar. Ecol.
50. Turner, S.M., G. Malin and P.S. Liss. 1989. Dimethyl sulfide and (dimethylsulfonio)-propionate in European coastal and shelf waters. p. 183-200. *In* E.S. Saltzman and W.J. Cooper (eds.), Biogenic Sulfur in the Environment. Am. Chem. Soc. 393.
51. Uchida, A., T. Ooguri, T. Ishida and Y. Ishida. 1993. Incorporation of methionine into dimethylthio-propanoic acid in the dinoflagellate *Crypthecodinium cohnii.* Nippon Suisan Gakkaishi *59:* 851-855.
52. Vairavamurthy, A., M.O. Andreae and R.L. Iverson. 1985. Biosynthesis of dimethyl sulfide and dimethyl-propiothetin by *Hymenomonas carterae* in relation to sulfur source and salinity variations. Limnol. Oceanogr. *30:* 59-70.
53. Vetter, Y.A. and J.H. Sharp. 1993. The influence of light intensity on dimethylsulfide production by a marine diatom. Limnol. Oceanogr. *38:* 419-425.
54. Visscher, P.T., R.P. Kiene and B.F. Taylor. 1994. Demethylation and cleavage of dimethylsulfoniopropion-ate in marine intertidal sediments. FEMS Microbiol. Ecol. *14:* 179-190.
55. White, H. 1982. Analysis of dimethyl sulfonium compounds in marine algae. J. Marine Research *40:* 529-536.
56. Yancey, P.H., M.E. Clark, S.C. Hand, R.D. Bowlus and G.N Somero. 1982. Living with water stress: evolution of osmolyte systems. Science *217:* 1214-1222.

12

PHYSIOLOGICAL ASPECTS OF THE PRODUCTION OF DIMETHYLSULFONIOPROPIONATE (DMSP) BY MARINE PHYTOPLANKTON

Maureen D. Keller and Wendy Korjeff–Bellows

Bigelow Laboratory for Ocean Sciences
McKown Point
West Boothbay Harbor, Maine 04575

SUMMARY

Marine phytoplankters are the primary source of dimethylsulfoniopropionate (DMSP). Only certain groups of phytoplankton, notably the prymnesiophytes and the dinoflagellates, produce significant amounts of these compounds on a per cell basis. However, even within these groups, there is considerable variability in concentration, and the function of DMSP within the cells is not fully resolved. In macroalgae, there is a clear relationship between DMSP and osmotic adaptation. Most planktonic species, however, do not experience significant salinity variation; other factors must be responsible for any observed shifts in intracellular DMSP content. We have examined the effects of light intensity and nitrogen availability on the production of DMSP by several isolates of marine phytoplankton. An inverse relationship appears to exist between intracellular DMSP levels and nitrogen availability. The algae examined in this study produced more DMSP per cell under nitrogen-deplete conditions. There was no common response to light variations. It is important to understand the factors that control the production of DMSP in the oceans, as DMSP is the major precursor of the atmospherically important gas, dimethyl sulfide (DMS).

INTRODUCTION

Dimethyl sulfide (DMS) is the primary volatile sulfur compound involved in the global sulfur cycle (2,6). DMS production in the oceans is associated with phytoplankton, but correlations between DMS concentrations and the principal indicator of phytoplankton biomass, chlorophyll *a,* have been surprisingly poor (4,5). This can be attributed, at least in part, to the fact that not all phytoplankton produce DMS, or more properly, the precursor of DMS in algae, dimethylsulfoniopropionate (DMSP). In addition, it is clear that factors

Biological and Environmental Chemistry of DMSP and Related Sulfonium Compounds
edited by Ronald P. Kiene et al., Plenum Press, New York, 1996

external to the algal cells are important in the production and distribution of DMS in the water column. DMSP is cleaved to DMS and acrylic acid by the action of lyase enzymes. It is still not entirely clear if phytoplankton release DMS from DMSP as part of their normal physiology, or if mechanical breakage of the cells, via grazing or senescence, is necessary for its release and the release of dissolved DMSP. There is evidence to support both conclusions. DMS release is certainly enhanced in the presence of zooplankton (11). DMS and DMSP consumption by bacteria are also important controls on water column concentrations of DMS and its eventual flux to the atmosphere (see Ledyard and Dacey; Kiene; and Taylor and Visscher, all this volume). Thus, trophodynamics is an important aspect of the DMSP/DMS cycle.

Many member of the classes Dinophyceae and Prymnesiophyceae, in particular, produce copious amounts of DMSP (intracellular concentrations of several hundred mmol DMSP . $liter^{-1}$ cell volume). Other classes, like the Chlorophyceae, Cryptophyceae and Cyanophyceae, produce very little. Members of the classes Bacillariophyceae (diatoms) and Prasinophyceae produce intermediate amounts (3,19). However, even among those algae known to produce DMSP, variations in intracellular content are substantial. This appears to be related to growth stage and physiology. Within microalgae, DMSP can serve as an osmoticum, changing intracellularly in response to salinity fluctuations (29). DMSP is one of a suite of compounds that can serve an osmotic function and are categorized as "compatible solutes," so-called because they do not interfere with cellular function, as some inorganic salts do. Kirst (pers. comm.) estimates that DMSP contributes up to 15% of the cells' osmolarity in species with high intracellular concentrations. All organic osmolytes can contribute up to 50% of the cellular osmotic pressure with the remaining 50% comprised of inorganic ions, mainly potassium. In marine waters, planktonic algae seldom experience significant changes in salinity, unless they are in an estuarine system. This does not eliminate the need for osmolytes, as marine organisms must continually balance internal and external osmotic pressure. However, it is unlikely that fluctuations in DMSP content in marine phytoplankton occur in response to any experienced salinity change. Therefore, other environmental factors must be involved in observed differences in DMSP within algae. Phytoplankton do experience frequent and large changes in their light and nutrient levels. Changes in DMSP might be attributed to changes in these parameters

MATERIALS AND METHODS

Cultures were acquired from the Provasoli–Guillard National Center for the Culture of Marine Phytoplankton (CCMP) (Bigelow Laboratory for Ocean Sciences, W. Boothbay Harbor, ME 04575, USA). The algae were grown at 18°C in 500 ml batch cultures in autoclaved K medium (18) made with GF/F filtered local coastal seawater. Inocula for each experiment were taken from cultures in late exponential or early stationary stages of growth. The inocula were preadapted to conditions replicating the experimental intermediate light level and high nitrate medium (5×10^{15} quanta. $m^{-2}.sec^{-1}$ and K levels of nitrate). Lighting was provided on a 14:10 light: dark cycle with a combination of "cool white" (GE) and "power twist"(Duro test) fluorescent bulbs, screened with neutral density mylar sheeting to produce light levels of 10^{16}, 5×10^{15} and 10^{15} quanta.$m^{-2}.sec^{-1}$. These light levels correspond to ca. 170 , 85 and 17 umol.$m^{-2}.sec^{-1}$ or 10, 5 and 1% incident sunlight, respectively. The low and high nitrate levels at the time of inoculation corresponded to 44 (N:P ratio of 1.2:1) and 883 (N:P ratio of 24.3:1) μM respectively. The low nitrate level was chosen to ensure that the depletion of N would limit growth in batch cultures, as demonstrated previously (21,24).

Cultures used for the light intensity and nitrate experiments (4 clones: CCMP 288, 376, 495, 1322; two prymnesiophytes, a diatom and a dinoflagellate, respectively) were sampled at five points during their growth cycle: twice in exponential phase, twice in stationary phase and once in senescence, as determined by daily measurements of *in vivo* fluorescence (8). Triplicate cultures of 26 clones were used to test for the intraclonal variation of intracellular DMSP content and DMSP:chl ratio comparisons. These were grown under conditions identical to the experimental inocula cultures and were sampled in late exponential growth as determined above. Aliquots of these cultures were taken for measurements of DMS, total DMSP [DMSP(T)], dissolved DMSP [DMSP(d)] (by gentle GF/F filtration), chlorophyll *a* and cell counts and volumes. In some cases, only samples for DMSP(T) were taken. Particulate DMSP [DMSP(p)] was determined by difference between the DMSP(T) and the DMSP(d) values. DMS analyses were conducted immediately. DMSP samples were treated with base (5 M KOH), sealed into serum vials, and held in the dark for at least 24 hours prior to analysis. DMS was assayed, following collection in a cryogenic purge and trap system, on a Varian 3300 Gas Chromatograph with a Chromosil 330 column (Supelco). Standards of DMSP (Research Plus), dissolved in DMSP/DMS free seawater, were treated similarly and used for calibration. Cell samples were preserved with Lugol's solution and counted using a Speirs-Levy counting chamber. Cell volumes were calculated from live cultures. Chlorophyll analyses followed the protocol of Strickland and Parsons (25).

RESULTS

Culture Observations

All cultures were grown in batch mode and monitored daily with measurements of *in vivo* fluorescence. When fluorescence values suggested certain stages of growth, samples were taken for all parameters, including cell counts and volumes. The batch cultures exhibited typical growth, with defined exponential and stationary phases in all cases. The timing of the onset of stationary phase was different for the different algae and was initiated by an unknown limiting factor. Much higher biomass was achieved in the high nitrogen medium. There was little difference in growth cycle pattern or biomass at the different irradiances (Figure 1 A,B; Figure 2 A,B). Cell volume was determined at the same time as cell counts and no consistent changes in cell volumes were observed in the different conditions or at different points in the growth cycle.

Since we were examining changes in DMSP production in response to different growth conditions, it was first necessary to determine the inherent variation in intracellular DMSP content in any one experimental clone. To do this, we examined 26 clones of phytoplankton representing 6 algal classes (Tables 1 and 2). The cultures were grown in triplicate, under the same growth conditions, and all were sampled in late exponential stage of growth. Coefficients of variations for DMSP(p) ranged from 1-112%, with the majority under 30%. Greater than 90% of the DMSP (T) was present in the particulate fraction in the majority of the isolates. The dissolved fraction, DMSP(d), was dominant in only one clone of a cryptophyte, *Rhodomonas salina*. The values for DMSP(d) were more variable (C.V's 13-115%), with the majority having CV's greater than 50%. The Prymnesiophytes, as a group, have a sizeable percentage of the DMSP(T) in the dissolved fraction.

Chlorophyll *a* concentrations were also determined for this same group of cultures, and the ratio of nM DMSP: μg chl/l for the cultures was computed (Table 3). The "non-DMSP-producing" classes of algae, diatoms, chlorophytes and cryptomonads, had low ratios, ranging from 1-10, although the diatom, *Melosira nummuloides* had a ratio of 13.4; the prymnesiophytes had ratios of 15+ (range ca. 15-44); dinoflagellates of 30+ (range ca. 33-124).

Only DMSP(T) concentrations were measured for most of the light and nutrient experiments. These concentrations, normalized on a per cell basis (Tables 4 and 5), were similar to those previously reported (17,19,20,22). Where DMSP(d) was also measured, most of the DMSP(T) was found in particulate form throughout the growth cycle until late stationary growth or senescence, when the dissolved component increased. Dissolved DMSP became particularly high in the low light conditions (up to 80% of the total DMSP in some of the cultures). DMS represented a very small percentage (1-3%) of the total DMSP. In many cases, DMSP(T) is much higher on a per cell basis in early exponential growth than at other points in the growth cycle (Figure 1C; 2C,D). Concentrations often rise again in senescence, but this is probably an artifact caused by high DMSP(d) values in combination with declining cell numbers. Because of the uncertainty associated with interpreting data from the early exponential and late stationary stages of growth, the observed concentrations of DMSP (T) at two points in the growth cycle, late exponential and late stationary, were selected for comparative purposes (Tables 4 and 5). Cell numbers are fairly stable during this period and concentrations of DMSP(d) are typically at a minimum.

Light Effects

Replicate cultures of four clones of microalgae were grown at three different irradiances (Figure 1 and Table 4). Levels of DMSP (T) in *Chrysochromulina* were highest at the lowest light level. The other prymnesiophyte, *Emiliania huxleyi*, had the lowest levels at low light. The diatom, *Minidiscus trioculatus*, also had lower levels in low light. There was no pattern in the dinoflagellate *Heterocapsa pygmaea.* None of the differences are statistically significant.

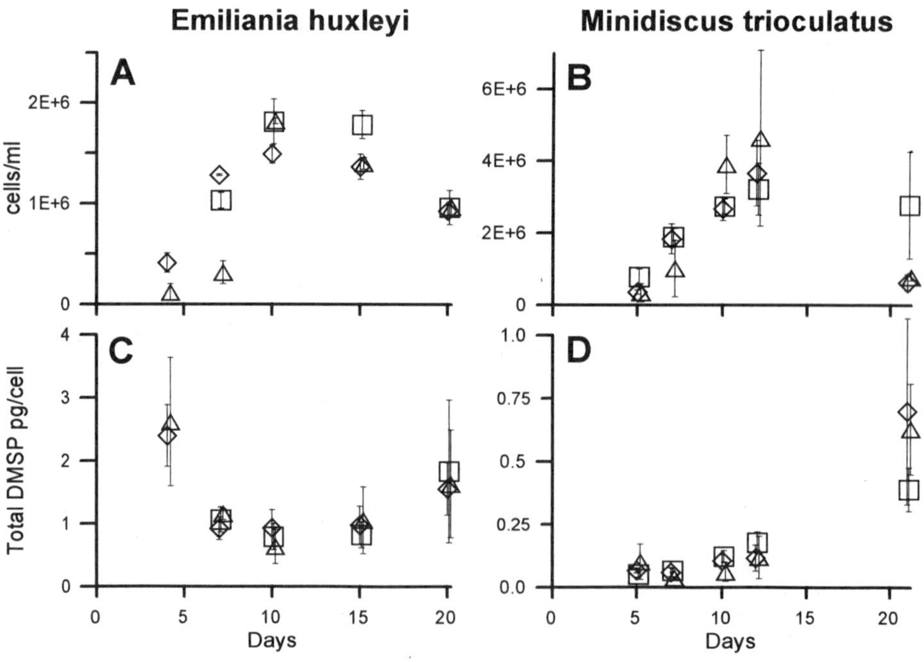

Figure 1. Cell counts (A and B) and DMSP(T) concentrations (C and D) over the growth cycle in two clones of marine phytoplankton, *Emiliania huxleyi* and *Minidiscus trioculatus* grown at three different irradiances. Low light (Δ) = 10^{15} quanta.m^{-2}.sec^{-1}, equivalent to the 1% light level. Medium light (❑) = 5 x 10^{15}quanta.m^{-2}.sec^{-1}, equivalent to the 5% light level. High light (◇)= 10^{16}quanta.m^{-2}.sec^{-1}, equivalent to the 10% light level. Symbols represent mean values; lines standard deviation.

Nutrients Effects

Replicate clones of the same four microalgae used in the light experiments were grown in high and low nitrate media (Figure 2 and Table 5). In three of the algae, DMSP(T) levels increased in response to low nitrate conditions and from exponential to stationary growth, when cells were also presumably nitrogen-limited. The prymnesiophyte, *Emiliania huxleyi*, did not exhibit changes in response to nitrate, but we believe that this alga routinely becomes carbon-limited in batch cultures, thus making the data difficult to interprete. In spite of some large standard deviations in these data, the trend is clear that intracellular levels of DMSP do increase with nitrogen limitation. The standard deviations are much larger in the low nitrate samples, perhaps a function of cellular stress.

DISCUSSION

Culture Observations

We have attempted in this study to identify some physiological aspects of DMSP production in marine phytoplankton that may be critical in our understanding of the marine DMSP/DMS cycle. Although conditions in laboratory cultures are simplistic and can't hope to duplicate the complex conditions encountered in the field, they can be used effectively to identify potentially important parameters. It is possible to examine the effects of individual parameters in a way not possible in field situations. The use of batch cultures is problematic,

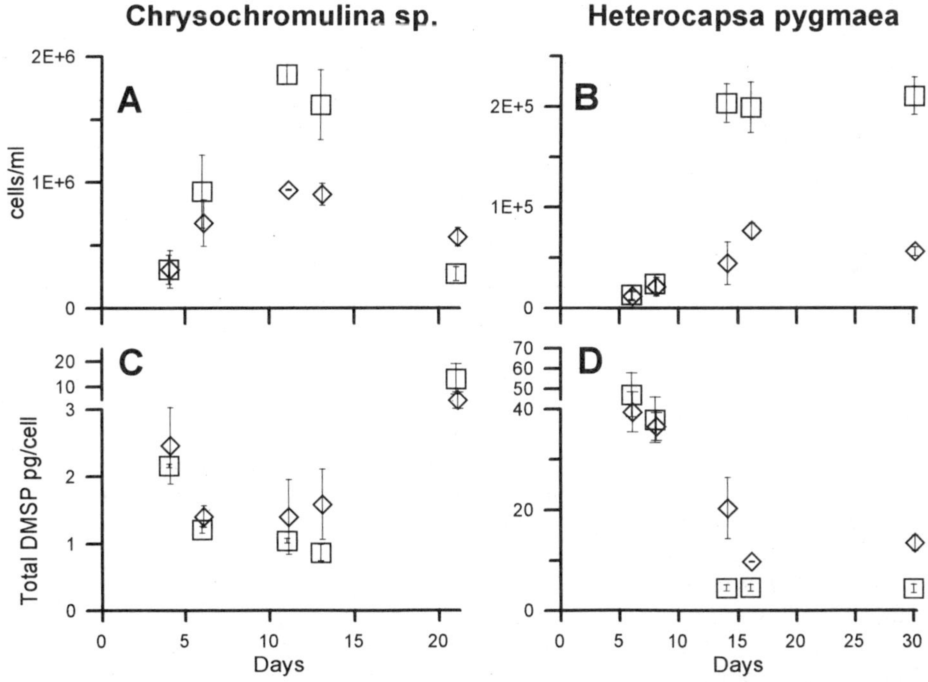

Figure 2. Cell counts (A and B) and DMSP(T) concentrations (C and D) over the growth cycle in two clones of marine phytoplankton, *Chrysochromulina sp.* and *Heterocapsa pygmaea* grown at two different nitrate (N) levels. High N (❑) = 883μM. Low N (◇) = 44 μM. Symbols represent mean values; lines standard deviation.

Table 1. Intraclonal variation in DMSP(p) content in 26 clones of marine phytoplankton. N=3 for each clone. Average = pg DMSP(p) cell^{-1}

Class/species	Clone	Average	S. D.	% Coeff. of Variation	DMSP(p) as % of DMSP(T)
Bacillariophyceae					
Minidiscus trioculatus	CCMP495	0.101	0.002	3	98
Chaetoceros simplex	CCMP199	0.025	0.001	4	87
Cylindrotheca closterium	CCMP342	0.306	0.003	1	97
Thalassiosira pseudonana	CCMP1335	0.063	0.030	48	97
Melosira nummuloides	CCMP482	6.970	4.269	61	99
Skeletonema costatum	CCMP1332	0.712	0.039	5	90
Nitzschia laevis	CCMP560	0.485	0.099	20	96
Chlorophyceae					
Chlamydomonas sp.	CCMP231	0.003	0.001	26	100
Chlorella capsulata	CCMP246	0.006	0.001	21	77
Chrysophyceae					
Rhizochromulina sp.	CCMP1150	1.396	0.185	13	96
Cryptophyceae					
unidentified cryptophyte	CCMP1178	0.445	0.151	34	95
Rhodomonas salina	CCMP1319	0.018	0.020	112	20
Dinophyceae					
Prorocentrumminimum	CCMP1329	38.116	3.512	9	96
Amphidinium carterae	CCMP1314	16.547	3.813	23	97
Crypthecodinium cohnii	CCMP316	4.853	2.370	49	91
Prorocentrum micans	CCMP691	239.085	80.327	34	91
Gymnodinium simplex	CCMP419	46.544	7.435	16	94
Prasinophycea					
Micromonas pusilla	CCMP490	0.051	0.015	30	99
Tetraselmis levis	CCMP896	1.473	0.445	30	98
Prymnesiophyceae					
Coccolithus neohelis	CCMP298	7.474	1.144	15	87
Emiliania huxleyi	CCMP378	0.792	0.264	33	80
Emiliania huxleyi	CCMP376	1.130	0.146	13	80
Chrysochromulina sp.	CCMP288	1.299	0.264	20	57
Phaeocystis sp.	CCMP628	4.090	1.061	26	76
Pavlova pinguis	CCMP609	1.066	0.157	15	93
Prymnesium parvum	CCMP708	1.831	0.499	27	70

especially for studies of this kind, but large numbers of samples can be surveyed in a way that is not possible with more exacting chemostat cultures. It is necessary to control as many variables as possible and sample at similar points in the growth cycle for all batch cultures. Even then, it is unlikely that growth conditions are identical.

No nutrient analyses were conducted from these cultures, although similar batch cultures became nitrogen- limited at the onset of stationary growth (Kiene, Keller and Matrai, unpublished data), as previously observed in other studies (21, 24). Dortch et al (12), using seven different microalgae grown in batch cultures, examined cellular changes from N sufficiency to N deficiency. External N became depleted early on and there was a 2-6 day lapse between this depletion and the minimum in cellular N content. Growth continued for only 3 days after external nitrogen depletion. By day 9, nitrogen starvation was occurring. Thus, it seems plausible that the batch cultures used herein did become nitrogen limited

Table 2. Intraclonal variation in dissolved DMSP (DMSP(d)) concentrations in culture medium of 26 clones of marine phytoplankton. N=3 for each clone. Average = nM. L^{-1}

Class/species	Clone	Average	S. D.	% Coeff. of variation	DMSP(d)as % of DMSP(T)
Bacillariophyceae					
Minidiscus trioculatus	CCMP495	12.0	7.989	66	2
Chaetoceros simplex	CCMP199	57.4	41.111	72	13
Cylindrotheca closterium	CCMP342	23.2	18.2	78	3
Thalassiosira pseudonana	CCMP1335	25.3	4.38	17	3
Melosira nummuloides	CCMP482	43.1	16.8	39	1
Skeletonemacostatum	CCMP1332	149.9	19.7	13	10
Nitzschia laevis	CCMP560	9.8	3.8	39	4
Chlorophyceae					
Chlamydomonas sp.	CCMP231	0.59	0.23	38	0
Chlorella capsulata	CCMP246	2.344	2.518	107	23
Chrysophyceae					
Rhizochromulina sp.	CCMP1150	66.5	22.8	34	4
Cryptophyceae					
unidentified cryptophyte	CCMP1178	15.1	8.71	58	5
Rhodomonas salina	CCMP1319	61.1	25	41	80
Dinophyceae					
Prorocentrum minimum	CCMP1329	266.4	145.7	55	4
Amphidinium carterae	CCMP1314	249.6	239.3	96	3
Crypthecodinium cohnii	CCMP316	7.499	5.514	74	9
Prorocentrum micans	CCMP691	454.5	389.8	86	9
Gymnodinium simplex	CCMP419	52.199	33.754	65	6
Prasinophyceae					
Micromonas pusilla	CCMP490	19.3	22.1	115	1
Tetraselmis levis	CCMP896	146.4	120.9	83	2
Prymnesiophyceae					
Coccolithus neohelis	CCMP298	148.3	52.8	36	13
Emiliania huxleyi	CCMP378	1022.3	573	56	20
Emiliania huxleyi	CCMP376	899.3	471.9	52	20
Chrysochromulina sp.	CCMP288	2098.1	657.1	31	43
Phaeocystis sp.	CCMP628	312.7	87.9	28	24
Pavlova pinguis	CCMP609	299.4	38.0	13	7
Prymnesium parvum	CCMP708	1797.8	810.2	45	30

during stationary growth, and nitrogen limitation did occur early on in the low nitrate experimental cultures.

On a per cell basis, the DMSP concentrations reported herein for cultures in exponential growth are generally comparable, though somewhat higher, than those previously measured (19). This may be a function of changing methodologies; the earlier study used headspace analysis, while this study used a cryogenic trap and purge method, or it may be a function of growth stage. The species-specific nature of production was the same. Because we are reporting concentrations of DMSP on a per cell basis, potential changes in cell volume were a concern. Although we observed no changes, there are reports of increases in cell volume in cultures of a dinoflagellate (*Heterocapsa* sp.) clone under nitrogen or phosphorus limitation (9). Other studies have found no difference or a slight decrease with nitrogen starvation in a variety of cultures (12).

Intraclonal variations in DMSP(p) concentrations in phytoplankton cultures in late exponential stages of growth were relatively minor (Table 1). Variation in DMSP(d) values (Table 2) was higher but since DMSP(d) is typically a very small percentage of the DMSP (T), such variation may be of minor concern. The size of the dissolved pool has been previously observed to increase in later stages of growth (17), but it is unknown if this is active release by the algae or simply cell leakage with senescence. It is also possible that changes in the dissolved pool are related to filtration, as previously demonstrated to be problematic with DMS sampling (27).

The high initial DMSP(T) value that was observed in early exponential growth (Figures 1C and 2C,D) is a common occurrence. We have seen this high initial time point in 31 of 60

Table 3. Concentrations of DMSP(P) and chlorophyll *a* and DMSP/Chl *a* ratios (nM/μg/ L) in a variety of phytoplankton cultures. DMSP and Chl *a* values represent means of triplicate samples.

Class/species	Clone	nM DMSP	μg Chl *a*	DMSPChl *a*:
Bacillariophyceae				
Minidiscus trioculatus	CCMP495	677.7	261.5	2.59
Chaetoceros simplex	CCMP199	196.7	766.8	0.26
Cylindrotheca closteriu	CCMP342	763.5	683.1	1.12
Thalassiosira pseudonana	CCMP1335	835.8	936.9	0.89
Melosira nummuloides	CCMP482	4794.8	357.8	13.40
Skeletonema costatum	CCMP1332	1204.9	347.0	3.91
Nitzschia laevis	CCMP560	217.6	163.2	1.33
Chlorophyceae				
Chlamydomonas sp.	CCMP231	12.8	77.8	0.17
Chlorella capsulata	CCMP246	7.1	30.5	0.23
Chrysophyceae				
Rhizochromulina sp.	CCMP1150	1557.0	32.5	47.95
Cryptophyceae				
unidentified cryptophyte	CCMP1178	306.1	493.5	0.62
Rhodomonas salina	CCMP1319	5.7	144.6	0.04
Dinophyceae				
Prorocentrum minimum	CCMP1329	6787.9	184.8	36.74
Amphidinium carterae	CCMP1314	7536.0	178.4	42.25
Prorocentrum micans	CCMP691	10300.1	307.8	33.46
Gymnodinium simplex	CCMP419	817.6	6.6	124.25
Prasinophyceae				
Micromonas pusilla	CCMP490	987.9	124.1	7.96
Tetraselmis levis	CCMP896	4634.8	234.6	19.76
Prymnesiophyceae				
Coccolithus neohelis	CCMP298	1020.0	28.2	36.19
Emiliania huxleyi	CCMP378	3984.6	258.6	15.41
Emiliania huxleyi	CCMP376	3174.8	177.5	17.88
Chrysochromulina sp.	CCMP288	2716.4	109.6	24.79
Phaeocystis sp.	CCMP628	1145.8	26.1	43.97
Pavlova pinguis	CCMP609	3880.8	255.2	15.21
Prymnesium parvum	CCMP708	4062.8	235.0	17.29

similar batch cultures, representing a wide variety of algal species (Keller and Bellows, unpubl). Originally thought to be an artifact of carryover from the inoculum or an imprecise cell count, either of which could produce artificially high values, it now appears to us to be accurate, if unexplained. This same high initial concentration has also been observed for particulate organic carbon and sulfur in phytoplankton cultures as well as DMS (22, 30). It is possible that DMSP, as well as these other compounds, accumulate rapidly in early exponential growth as precursors. The values equilibrate as growth becomes more balanced. The values often rise again in senescence, but this, we believe, is an artifact caused by high extracellular DMSP values being included in intracellular calculations. Because of this, we believe that data collected in late exponential through stationary growth are the most reliable.

The taxonomic relationship between certain groups of algae and DMSP production is well established, with the chromophyte microalgae clearly the most significant producers (2,17,19,20). "Significance" here is defined as having substantial, albeit somewhat arbitary, intracellular concentrations of DMSP. When calculated on a per unit cell volume basis, members of the Prymnesiophytes and Dinoflagellates stand out, with concentrations in excess of several hundred mmol DMSP. liter $^{-1}$ cell volume. Members of other groups rarely contain concentrations in excess of 50 mM. However, this does not mean that species composition of the phytoplankton source is the sole determinant of "significance." Exceptional biomass, such as that seen in diatoms in sea ice or in certain bloom situations, can result in very high DMSP concentrations, even though diatoms are thought of as less significant sources of DMSP (e.g 17, 19, 26, 28). When presented on a per cell basis, the DMSP content of diatoms is typically quite small.

In this regard, the ratios of DMSP:Chl calculated in this study (Table 3) support the previously noted taxonomic relationship, with significant DMSP producing groups with high ratios and "non-producers" with low ratios. The range in ratios in fact reinforces previous reports of the poor correlation between DMS/DMSP and chlorophyll (2,4,5). No culture examined here had values as high as those reported in the literature (in excess of 200+) (e.g. 28). For comparative purposes, the use of this ratio has become widespread, but it is unclear how to interpret any field ratio, since both mixed populations and the effects of photoadaptation or nutrient sufficiency will alter DMSP and/or chlorophyll concentrations. Although high ratios may be indicative of certain groups of microalgae, they may also be attributed to heterotrophic organisms with accumulated pools of DMSP. Some consideration of the population composition must be included to aid interpretation.

Even measurements of plant biomass in combination with species information will not always yield predictable DMSP concentrations. Measurements of the production of DMSP by any particular group of algae are undoubtably complicated by a variety of factors, including physiology and sampling methodology. DMSP does appear to function as an osmoticum in phytoplankton, with DMSP concentrations responding to salinity variations (29). We have attempted in this paper to identify other parameters that affect DMSP in marine phytoplankton.

Light Effects

The effect of light on DMSP production by microalgae is not consistent nor is it apparent in our experiments (Table 4, Figure 1). This is not to say that DMSP production is not a light dependent process, as suggested by Karsten et al (15,16), but light intensity did not appear to have a significant effect on DMSP production in the algae examined in this study. The highest light condition used represents only ca. 10% of incident sunlight values. Thus, the high light conditions are not really that high, although they are typical of surface waters and chlorophyll maxima in coastal areas. The low light conditions are ca. 1% incident sunlight, that are typical of deep chlorophyll maxima in more oligotrophic areas. In macroalgae, high DMSP levels were found only under long day conditions and high irradiances (16). Additional observations suggested that

Table 4. Mean total DMSP (DMSP(T)) content, calculated on a per cell basis, at different stages of growth in nutrient-replete phytoplankton cultures grown in varying light conditions. DMSP concentration = pg DMSP(T)/cell. Light intensities: low = 10^{15}; intermediate=5 x 10^{15}; high=10^{16} quanta.$m^{-2}sec^{-1}$

			Stages of growth					
			Late exponential			Late stationary		
Species	Clone	Light levels	X	s.d.	N	X	s.d.	N
Chrysochromulina sp.	CCMP288	Low	1.35	0.52	2	1.59	0.92	2
		Interm.	0.89	0.27	3	1.07	0.07	2
		High	0.75	0.31	2	0.82	0.47	2
Emiliania huxleyi	CCMP376	Low	0.64	0.38	2	1.06	0.75	2
		Interm.	0.95	0.43	4	0.89	0.32	4
		High	0.93	0.41	2	0.97	0.41	2
Minidiscus trioculatus	CCMP495	Low	0.06	0.05	2	0.12	0.12	2
		Interm	0.12	0.06	5	0.23	0.11	5
		High	0.11	0.03	2	0.12	0.07	2
Heterocapsa pygmaea	CCMP1322	Low	6.62	–	1	6.81	–	1
		Interm.	5.09	0.62	3	5.27	0.27	3
		High	6.39	–	1	6.30	–	1

DMSP production occurs only in the light; species kept in the dark degraded DMSP. We observed higher DMSP concentrations, especially in the dissolved fraction, in the prymnesiophyte *Chrysochromulina* grown in low light. This is consistent with higher DMS release previously observed at low light conditions in the prymnesiophyte *Phaeocystis* (7). Higher levels of DMS release have been observed in high light in several diatoms (7, 30). There was no consistent pattern of DMSP production in response to varying light intensity in the diatom used in this study.

Nutrient Effects

Three of the four algal clones examined in this study did exhibit increases in DMSP content when nitrogen-limited. This trend is evident not only between cultures grown in nitrogen

Table 5. Mean total DMSP (DMSP(T)) content, calculated on a per cell basis, at different stages of growth in phytoplankton cultures grown at two different nitrate concentrations. Low and high nitrate = 44 and 883 uM respectively, equivalent to K/20 and K levels of nitrogen (18). All other nutrients were at K levels.

			Stage of growth					
			Late exponential			Late stationary		
		Nitrate level	X	s.d.	N	X	s.d.	N
Chrysochromulina sp.	CCMP288	Low	1.40	0.78	2	1.60	0.74	2
		High	0.89	0.27	3	0.97	0.18	3
Emiliania huxleyi	CCMP376	Low	0.88	0.41	2	0.67	0.15	2
		High	0.82	0.47	5	0.78	0.37	5
Minidiscus trioculatus	CCMP495	Low	0.23	0.12	2	0.40	–	1
		High	0.12	0.05	5	0.24	0.11	5
Heterocapsa pygmaea	CCMP1322	Low	9.85	0.23	3	11.44	2.15	3
		High	4.63	1.28	4	3.94	1.89	2

replete and nitrogen deplete conditions, but also between samples collected in late exponential (nitrogen still available) and late stationary (nitrogen exhausted) stages of growth. There is a small amount of information which supports a relationship between DMSP and available nitrogen in phytoplankton. Turner et al (26) reported that in cultures of *Emiliania huxleyi* supplemented with extra nitrate, the per cell levels of DMSP were about 20% lower than in cultures maintained with low nitrate. Recently it has been shown that DMSP levels in *Tetraselmis subcordiformis* are 75-100% higher in N-limited batch cultures (13). This relationship has also been observed in *Spartina alterniflora* (11a). Here, we add to the evidence supporting the importance of nitrogen in intracellular DMSP dynamics. One role nitrogen may play in DMSP synthesis appears to be connected to another compatible solute in algae, the nitrogen analog of DMSP, glycine betaine (GBT). It was suggested a decade ago (1) that DMSP and GBT may be related in marine phytoplankton, with GBT being produced when nitrogen is available and DMSP when nitrogen is limiting. In the oceans, this may be very important, as most areas of the worlds' oceans appear to be nitrogen-limited. It is unclear why GBT would be preferentially produced, although Kirst (pers. comm.) believes that GBT is a more compatible solute, providing better protein protection than DMSP. Also, sulfate assimilation is more energetically expensive than nitrate assimilation, and thus GBT production may be favored. It is clear that nitrogen plays a critical role but more definitive work is needed to characterize it.

DMSP is a significant compound in the biochemistry of marine algae. Because of its role as the source compound of the atmospherically-important gas DMS, it deserves additional consideration in the areas of general marine ecology and chemistry. We now know that, in many cases, DMSP production is related to taxonomic position and cell physiology. Variations in light and nutrient levels, as well as salinity, may affect its accumulation..

ACKNOWLEDGMENTS

This research was supported by EPA grants R81-5077-010 and R81-9515-010. Thanks to Paty Matrai, Gunter Kirst and Ron Kiene for discussions and review of this manuscript.

REFERENCES

1. Andreae, M.O. 1986. The ocean as a source of atmospheric sulfur compounds, p. 331-362. *In* P. Buat-Menard (ed.) The role of air-sea exchange in geochemical cycling, Reidel, Dordrecht.
2. Andreae, M.O. 1990. Ocean-atmosphere interactions in the global biogeochemical sulfur cycle. Mar. Chem. *30:* 1-29.
3. Andreae, M.O., W.R. Barnard, and J.M. Ammons. 1983. The biological production of dimethylsulfide in the ocean and its role in the global atmospheric sulfur budget. Ecol. Bull. *35*: 167-177.
4. Barnard, W.R., M.O. Andreae, W.E. Watkins, H. Bingemer and H.W. Georgii. 1982. The flux of dimethylsulfide from the oceans to the atmosphere. J. Geophys. Res. *87:* 8787-8793.
5. Bates, T.S. and J.D. Cline. 1985. The role of the ocean in a regional sulfur cycle. J. Geophys. Res. *90:* 9168-9172.
6. Bates, T.S., B.K. Lamb, A. Guenther, J. Dignon, and R.E. Stoiber. 1992. Sulfur emissions to the atmosphere from natural sources. J. Atmos. Chem. *14:* 315-337.
7. Baumann, M.E.M., F.P. Brandini and R. Staubes. 1994. The influence of light and temperature on carbon-specific DMS release by cultures of *Phaeocystis antarctica* and three antarctic diatoms. Mar. Chem. *45:* 129-136.
8. Brand, L.E., R.R.L. Guillard. and L.S. Murphy. 1982. A method for the rapid and precise determination of acclimated phytoplankton reproduction rates. J. Plankton Res. *3:* 193-207.
9. Berdalet, E., M. Latasa and M. Estrada. 1994. Effects of nitrogen and phosphorus starvation on nucleic acid and protein content of *Heterocapsa* sp. J. Plankton Res. *16*: 303-316.

10. Charlson, R. J., J.E. Lovelock, M.O. Andreae, and S.G. Warren. 1987. Oceanic phytoplankton, atmospheric sulfur, cloud albedo and climate. Nature. *326:* 655-661.
11. Dacey, J.W.H. and S. G. Wakeham. 1986. Oceanic dimethyl sulfide: production during zooplankton grazing on phytoplankton. Science. *233:* 1314-1316.
11a. Dacey, J. W. H., G. M. King, and S. G. Wakeham. 1987. Factors controlling emission of dimethylsulphide from salt marshes. Nature. *330:*643-645.
12. Dortch, Q., J.R. Clayton, Jr., S.S. Thoresen and S.I. Ahmed. 1984. Species differences in accumulation of nitrogen pools in phytoplankton. Mar.Biol. *81*: 237-250.
13. Grone, T. and G.O. Kirst. 1992. The effect of nitrogen deficiency, methionine and inhibitors of methionine metabolism on the DMSP contents of *Tetraselmis subcordiformis* (Stein). Mar. Biol. *112*: 497-503.
14. Guillard, R.R.L. 1975. Culture of phytoplankton for feeding marine invertebrates, p. 29-60. *In* W.L. Smith and M.H. Chanley (eds.), Culture of marine invertebrate animals, Plenum Press, New York.
15. Karsten, U., C. Wiencke and G.O. Kirst. 1990. The -dimethylsulphoniopropionate (DMSP) content of macroalgae from Antarctica and Southern Chile. Bot. Mar. *33:* 143-146.
16. Karsten, U., C. Wiencke and G.O. Kirst. 1991. Growth pattern and -dimethylsulphoniopropionate (DMSP) content of green macroalgae at different irradiances. Mar. Biol. *108:* 151-55.
17. Keller, M.D. 1991. Dimethyl sulfide production and marine phytoplankton: the importance of species composition and cell size. Biol. Oceanogr. *6:* 375-382.
18. Keller, M.D., R.C. Selvin,W. Claus and R.R.L. Guillard. 1987. Media for the culture of oceanic ultraphytoplankton. J. Phycol. *23:* 633-638.
19. Keller, M.D., W. K. Bellows and R.R.L. Guillard. 1989a. Dimethyl sulfide production in marine phytoplankton, p. 167-182. *In* E.S. Saltzmann and W.J. Cooper (eds.), Biogenic sulfur in the environment. ACS Symposium Series No. 393.
20. Keller, M.D., W.K. Bellows and R.R. L. Guillard. 1989b. Dimethyl sulfide production and marine phytoplankton: An additional impact of unusual blooms, p. 101-115. *In* E.M. Cosper, V.M. Bricelj and E.J. Carpenter (eds.), Novel phytoplankton blooms: Causes and impacts of recurrent brown tides and other unusual blooms. Springer-Verlag, Berlin.
21. Lewitus, A.L. and D.A. Caron. 1990. Relative effects of nitrogen or phosphorus depletion and light intensity on the pigmentation, chemical composition, and volume of *Pyrenomonas salina* (Cryptophyceae). Mar. Ecol. Prog. Ser. *61:* 171-181.
22. Matrai, P.A. and M.D. Keller. 1994. Total organic sulfur and dimethylsulfoniopropionate in marine phytoplankton: intracellular variations. Mar. Biol. *119:* 61-68.
23. Nishiguchi, M.K. and G.N. Somero. 1992. Temperature- and concentration dependence of compatibility of the organic osmolyte -dimethylsulfoniopropionate. Cryobiol *29*: 118-124.
24. Sakshaug, E. and O. Holm-Hansen. 1977. Chemical composition of *Skeletonema costatum* (Grev.) and *Pavlova (Monochrysis) lutheri* (Droop) Green as a function of nitrate-, phosphate-, and iron-limited growth. J. Exp. Mar. Biol. Ecol. *29*: 1-34.
25. Strickland, J.H.D. and T.R. Parsons. 1972. A practical handbook of seawater analysis. 2nd edn. Bull.Fish.Res.Bd.Can. *167:* 1-310.
26. Turner, S.M., G. Malin, P.S. Liss, D.S. Harbour and P.M. Holligan. 1988. The seasonal variation of dimethyl sulfide and dimethylsulfoniopropionate concentrations in nearshore waters. Limnol. Oceanogr. *33*: 364-375.
27. Turner, S.M., G. Malin, L.E. Bagander and C. Leck. 1990. Interlaboratory calibration and sample analysis of dimethyl sulphide in water. Mar.Chem. *29:* 47-62.
28. Turner, S.M. , P.D. Nightingale and P.S. Liss. In press. The distribution of dimethyl sulphide and dimethylsulphoniopropionate in Antarctic waters and sea ice. Deep Sea Res. Special Issue: Biogeochemical cycling in the Bellingshansen Sea.
29. Vairavamurthy, A., M.O. Andreae, and R.L. Iverson. 1985. Biosynthesis of dimethylsulfide and dimethylpropiothetin by *Hymenomonas carterae* in relation to sulfur source and salinity variations. Limnol. Oceanogr. *30:* 59-70.
30. Vetter, Y. and J.H. Sharp. 1993. The influence of light intensity on dimethyl sulfide production by a marine diatom. Limnol. Oceanogr. *38:* 419-425.

13

DIMETHYLSULFONIOPROPIONATE PRODUCTION IN PHOTOTROPHIC ORGANISMS AND ITS PHYSIOLOGICAL FUNCTION AS A CRYOPROTECTANT

U. Karsten,[1,4] K. Kück,[2] C. Vogt,[3] and G. O. Kirst[2]

[1] Max-Planck Institute for Marine Microbiology
Fahrenheitstrasse 1
D-28359 Bremen, Germany
[2] University of Bremen
FB 2, Marine Botany
D-28334 Bremen, Germany
[3] University of Bremen
FB 2, Marine Microbiology
D-28334 Bremen, Germany
[4] School of Biological Science
University of New South Wales
Sydney, NSW 2052, Australia

SUMMARY

Dimethylsulfoniopropionate (DMSP) was detected in many strains of phototrophic bacteria and cyanobacteria. These data, together with results from earlier studies and the literature, clearly indicate that mainly photosynthetic micro- and macroorganisms have the metabolic capability to synthesize this compound. A relationship with temperature has also been identified. Polar macroalgae contain significantly higher DMSP contents compared to temperate or tropical species (5, 17). In this study, the cryoprotective function of DMSP was investigated on the activity of the model enzyme lactate dehydrogenase (LDH) as well as on that of malate dehydrogenase (MDH), extracted from the polar alga *Acrosiphonia arcta* (Chlorophyta). LDH activity was stabilized during freezing and thawing by the presence of DMSP concentrations up to 230 mM. The addition of higher amounts of DMSP to the enzyme assay even led to a 1.5 fold stimulation in the activity. DMSP stabilized MDH activity at the extremely low temperature of -2°C. In the presence of high solute concentrations, up to 300 mM, the enzyme activity rose to 165% of the control.

Biological and Environmental Chemistry of DMSP and Related Sulfonium Compounds
edited by Ronald P. Kiene et al., Plenum Press, New York, 1996

INTRODUCTION

In recent years, the tertiary sulfonium compound, DMSP, has attracted particular interest as the biogenic precursor of the predominant sulfur gas, dimethylsulfide, that is dissolved in oceanic, coastal and estuarine waters (1, 16). The biogeochemical cycling of dimethylsulfide has important implications for the chemistry of the marine atmosphere, and subsequently in the process of cloud formation, that may influence weather and climate (8).

DMSP is widely distributed in marine micro- and macroalgae, as well as in some coastal higher plants and crops (5, 17, 21, 26, 28, 36). While reports in the literature on the occurrence of this compound in cyanobacteria are contradictory (4, 21, 34), data on other phototrophic prokaryotes are to our knowledge still lacking. To fill this gap, we determined the DMSP concentration in different phototrophic bacteria and compared the results with previously published data on macroalgae and other photosynthetic organisms. One aim of the present study was to get a broad overview of the groups of DMSP-containing organisms, which should indicate whether the physiological capability to form this compound is linked to photosynthesis and light conditions. Previous studies demonstrated stimulating effects of increasing irradiances on the intracellular DMSP pool in eulitoral green macroalgae (19). Other factors such as senescence (24), grazing (9), salinity (11) and temperature (20) also control the intracellular DMSP pool in algae.

DMSP is known to act as an organic osmolyte, accumulating in micro- as well as in macroalgae under hypersaline salt stress (10, 11, 19, 35). Other biological functions of DMSP, such as an antibiotic compound (3), methyl group donor (36), storage pool for excess sulfur (32) or cryoprotectant (17) have been suggested rather than experimentally proven. Green macroalgae from polar latitudes contain significantly higher DMSP concentrations compared to related species from temperate to tropical regions (5). This points to a connection between accumulated DMSP content and average habitat temperature. Therefore, another goal of the present study was to investigate the possible cryoprotective role of DMSP in Antarctic green macroalgae. This was tested on the activity of enzymes extracted from polar seaweeds, as well as on the stability of the cold-inactivated model enzyme lactate dehydrogenase.

MATERIALS AND METHODS

Phototrophic Organisms and Culture

The purple sulfur and purple non-sulfur bacteria were obtained from the DSM (Deutsche Sammlung von Mikroorganismen und Zellkulturen, Braunschweig, FRG) and cultivated according to the DSM instructions. The cyanobacterial strains of *Microcoleus chthonoplastes* MPI EBD-1, MPI GNL-1, MPI MEL-1 and MPI TOW-1 were recently isolated at the Max-Planck-Institute for Marine Microbiology from marine microbial mats collected in Ebre Delta, Spain, Guerrero Negro Saltern, Mexico, Mellum Island, Germany and Towra Point Sydney, Australia, respectively. The strains were grown in membrane-filtered North Sea water (32 PSU; Sartorius Sartobran II, 0.2 μm), which was enriched with nutrients (PES/2) according to Starr and Zeikus (30). A growth incubator that was kept at 25°C and provided 25-40 μmol photons m^{-2} s^{-1} from cool-white fluorescent tubes under a 18/6 h light/dark cycle was used for cultivation. The other cyanobacterial species were originally isolated in the Baltic Sea and grown under estuarine conditions at 15 PSU (Seawater diluted with distilled water and enriched with PES/2). Temperature and light conditions were very similar to those used for *Microcoleus*.

The eulitoral green macroalgae, *Acrosiphonia arcta*, *Enteromorpha bulbosa* and *Ulothrix subflaccida*, used in the present study, were isolated at King George Island

(Antarctica) by Dr. C. Wiencke, Alfred Wegener Institute for Polar and Marine Research, Bremerhaven, FRG (37). The water temperature in the Antarctic habitat seasonally fluctuates only between -1.9°C and +1.7°C, while the air temperature may change from -30°C up to +10°C (20). For the enzyme study, an Arctic isolate of *Acrosiphonia arcta* collected at Disko Island (Greenland) was utilized. The water temperature in this habitat is in the range from -2.0°C to +6.0°C. All species were grown from spores and subsequently cultivated in 0.5 l beakers containing membrane-filtered North Sea water (see above) enriched with Provasoli's ES nutrients (27). The cultures were grown at 0 C, and illuminated with cool-white fluorescent neon tubes at 20-30 µmol photons m^{-2} s^{-1} under a 18/6 h light/dark rhythm.

Temperature Experiments

For testing the influence of different temperatures on the DMSP pool, *A. arcta*, *E. bulbosa* and *U. subflaccida* were cultivated for one year at 10° and 0°C, respectively and at a photon fluence rate of 55 µmol m^{-2} s^{-1} under a 18/6 h light/dark cycle. Media were changed in intervals of 2 weeks to avoid nutrient limitation.

Enzyme Experiments

The physiological function of DMSP as a cryoprotectant was tested on the activity of malate dehydrogenase (MDH, E.C. 1.1.1.37) extracted from the Arctic *A. arcta*, and on the stability of the commercially available cold-labile enzyme lactate dehydrogenase (LDH, E.C. 1.1.1.28) (Type XXX-S, porcine muscle, Sigma). Approximately 1g fresh weight of *A. arcta* were homogenized under liquid nitrogen in a mortar with quartz sand. 3-5 ml of an ice-cold extraction buffer were added to the homogenate and kept for 5 min at 0°C. The extraction buffer (50 mM HEPES, pH 7.4) contained 4% Polyclar AT, 2 mM EDTA (Na^+ salt), 2mM L-ascorbic acid (Na^+ salt), 5 mM $MgCl_2$, 5 mM dithiothreitol, 13 mM $CaCl_2$, 0.5% Tween 80 (w/v) and 20% glycerol (w/v). All manipulations were conducted at 0°C. The homogenates were centrifuged at 20000 x g for 15 min, and then to the supernatants, $(NH_4)_2SO_4$ was added step-wise to 70% saturation and stirred for several min. Insoluble material was recovered by centrifuging at 20000 x *g* for 15 min and supernatants were discarded. The pellets were resuspended in extraction buffer without Polyclar AT, L-ascorbic acid, $CaCl_2$ and Tween 80, and directly used for determination of MDH activity. For each estimate, the protein content was kept constant. MDH activity was measured photometrically by following the NADH decrease at 340 nm in 1 ml cuvettes incubated for exactly 20 min in a waterbath filled with 50% EtOH (v/v) at 30° and -2°C, respectively. The temperature of 30°C was chosen because this temperature is strongly recommended as a standard parameter for all enzyme assays for inter-comparision by the International Union of Biochemistry Enzyme Commission. Moreover, preliminary studies also demonstrated an optimum activity at 30°C for the MDH of *A. arcta* (Kück et al., unpublished results). MDH was assayed in 100 mM HEPES, pH 7.5, 5 mM $MgCl_2$, 2 mM dithiothreitol, 0.15mM NADH, and 0.25 mM of the substrate oxalacetate. For testing possible protective effects of DMSP, this compound (Cl-salt) was dissolved and buffered in 10 mM HEPES, pH 6.5, followed by addition of increasing concentrations to the enzyme assay. MDH was incubated in the presence of DMSP for 20 min at the temperatures mentioned above. The LDH- solution was centrifuged at 8000 xg, and the pellet resuspended in 10 mM HEPES buffer, pH 6.5, to remove most of the $(NH_4)_2SO_4$ that may disturb the enzyme activity. The enzyme was shock-frozen for several min in liquid nitrogen in the presence of various concentrations of proline, sucrose and DMSP. After thawing, the LDH activity was photometrically measured by following the NADH decrease at 340 nm in 1-ml cuvettes at 25°C

Table 1. The occurrence of dimethylsulfoniopropionate in different pro- and eukaryotic phototrophic organisms

Organisms	Concentration	Reference
μmol cm^{-3} cell volume		
Microalgae		
Bacillariophycea	0.2 - 264	(21)
Chrysophyceae	200 - 596	(21)
Dinophyceae	0.1 - 2201	(21)
Prasinophyceae	0.5 - 484	(10, 21)
Prymnesiophyceae	3.3 - 413	(21)
Organisms	**Concentration**	**Reference**
mmol kg^{-1} fresh weight		
Purple sulfur bacteria		
Chromatium gracile DSM 203	0.4	this study
Chromatium minus DSM 178	0.6	this study
Chromatium minutissimum DSM 1376	0.07	this study
Chromatium violascens DSM 198	0.04	this study
Thiocapsa roseopersicina DSM 5653	0.1	this study
Thiocapsa pfennigii DSM 226	0.2	this study
Purple non-sulfur bacteria		
Rhodopseudomonas sulfoviridis DSM 729	2.2-12.5	this study
Rhodopseudomonas salexigens DSM 2132	0.6	this study
Rhodovulum euryhalinum DSM 4868	0.3	this study
Rhodovulum sulfidophilum DSM 1374	0.06	this study
Cyanobacteria		
Microcoleus chthonoplastes MPI EBD-1	0.16	this study
M. chthonoplastes MPI MEL-1	0.06	this study
M. chthonoplastes MPI GNL-1	0.319	this study
M. chthonoplastes MPI TOW-1	0.10	this study
Synechocystis sp. Bo79	0.14 - 0.75	this study
Lyngbya aestuarii Bo 9	4.9	this study
Nostoc sp. Bo 84	1.17 - 2.39	this study
Anabaena sp. Bo 70	6.28	this study
Macroalgae		
Chlorophyceae		
temperate	2.5 - 48.7	(14, 28, 36)
tropical	7.1 - 63.1	(5)
polar	53.6 - 290	(17)
Rhodophyceae (*Polysiphonia* 151)	0.01 - 1	(17, 28)
Phaeophyceae	0.01 - 0.2	(17, 28)
Higher plants		
Spartina anglica	5 - 15	(32)
Saccharum sp.	0.1 - 6	(26)
Cereals (*Zea mays*, *Hordem vulgare*, *Sorghum vulgare*, *Oryza sativa*)	0.01 - 0.1	(26)

according to Nishiguchi and Somero (25), but modified as follows: 80 mM HEPES, pH 6.5, 10 mM KCl, 0.15 mM NADH and 1 mM pyruvate.

Dimethylsulfoniopropionate Analysis

The DMSP content of the phototrophic organisms was determined by gas-liquid chromatography according to previous reports (17).

RESULTS

The DMSP concentrations of the different pro- and eukaryotic phototrophic organisms are shown in Table 1. Within the purple sulfur and purple non-sulfur bacteria, the contents ranged from 0.06 mmol kg^{-1} FW in *Rhodovulum sulfidophilum* up to 12.5 mmol kg^{-1} FW in *Rhodopseudomonas sulfoviridis*. The cyanobacteria exhibited very similar DMSP concentrations between 0.06 and 6.28 mmol kg^{-1} FW. In contrast to the prokaryotes, most micro- and macroalgae contained significantly higher contents, e.g. polar Chlorophyceae showed up to 290 mmol kg^{-1} FW. Within the green macroalgae, all isolates collected in Antarctica and the Arctic exhibited the highest DMSP concentrations compared to related species that were collected in temperate to tropical regions (Table 1). Except for the red alga *Polysiphonia*, all other Rhodophyceae, as well as Phaeophyceae, showed only traces of

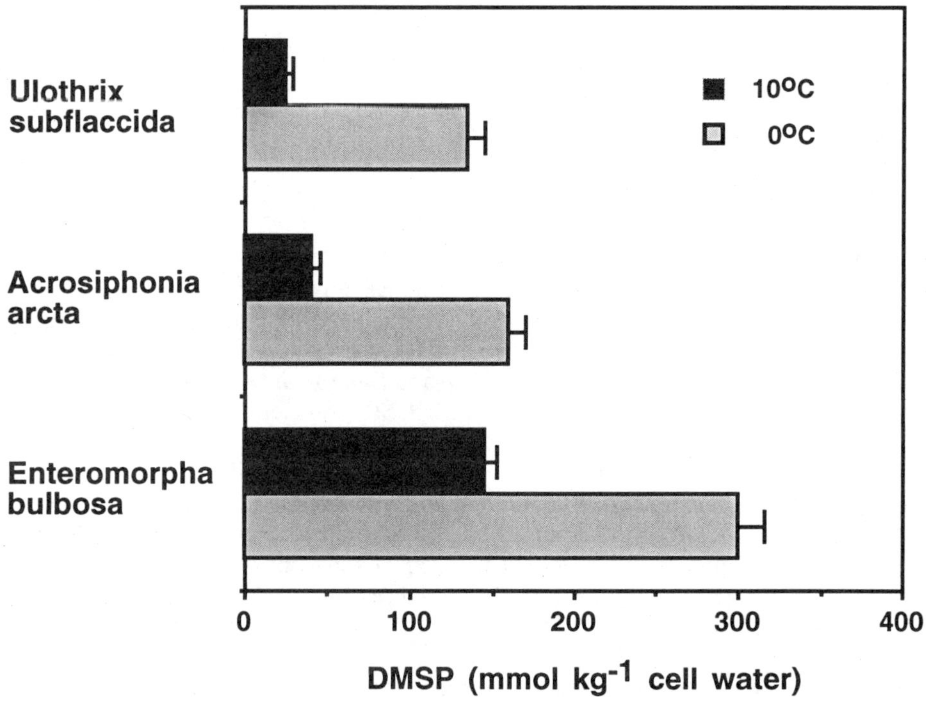

Figure 1. The effect of long-term acclimation at 10° and 0°C on the intracellular dimethylsulfoniopropionate (DMSP) content of the Antarctic green macroalgae *Acrosiphonia arcta*, *Enteromorpha bulbosa* and *Ulothrix subflaccida*. Data are presented as mean ± standard deviation (n=4).

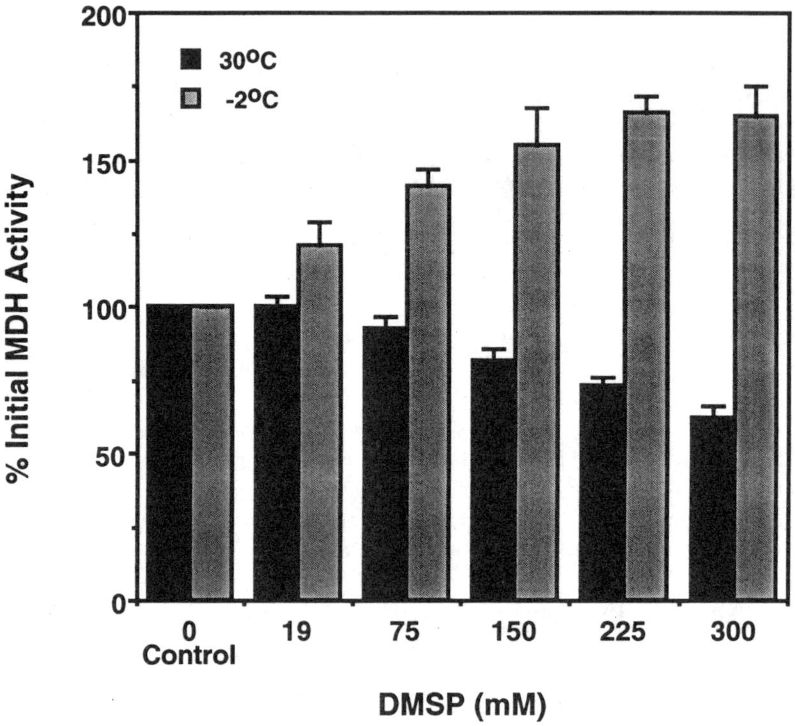

Figure 2. The effect of increasing dimethylsulfoniopropionate (DMSP) concentrations on the activity of the enzyme malate dehydrogenase (MDH) extracted from the Antarctic green macroalga *Acrosiphonia arcta*. The activity was tested at 30° and -2°C, and expressed as percentage of the control without any addition of DMSP. The absolute control activity of MDH at -2°C was 80.0 nkat mg^{-1} protein, that for 30°C 255.6 nkat mg^{-1} protein.Data are given as mean ± standard deviation (n=4). The control activity at 30°C was about 10 fold higher compared with the activity at -2°C.

DMSP (<1 mmol kg^{-1} FW). In higher plants DMSP has only been found in members of the Gramineae. The contents are generally low in the range between 0.01 and 15 mmol kg^{-1} FW.

The influence of long-term acclimation to 10° and 0 °C on the intracellular DMSP pool in *Ulothrix subflaccida*, *Acrosiphonia arcta* and *Enteromorpha bulbosa* is summarized in Fig. 1. All species contained 2-5.5 fold higher DMSP concentrations at 0° C compared with plants maintained at 10°C.

The effect of increasing DMSP concentrations on the activity of the enzyme malate dehydrogenase (MDH) extracted from *A. arcta* and measured at 30 ° and -2° C is shown in Fig. 2. In the presence of DMSP from 19 to 300 mM (final concentration) at 30° C, the activity of MDH continuously decreased to 62% of the control. Just the opposite was observed under sub-zero conditions (-2° C), i.e. increasing DMSP concentrations led to a stimulation of the enzyme activity up to 165% of the control (Fig. 2). However, it has to be taken into account that the control MDH activity at 30° C was about 3.5-10 fold higher compared to the assay at -2° C because of the general temperature dependency of enzyme activities.

The influence of increasing concentrations of DMSP, proline and sucrose on the stability of the cold-labile enzyme lactate dehydrogenase (LDH) is given in Fig. 3. Freezing and thawing of LDH without the presence of any organic solute was accompanied by a strong loss of activity down to 15% of the control. The addition of increasing concentrations of

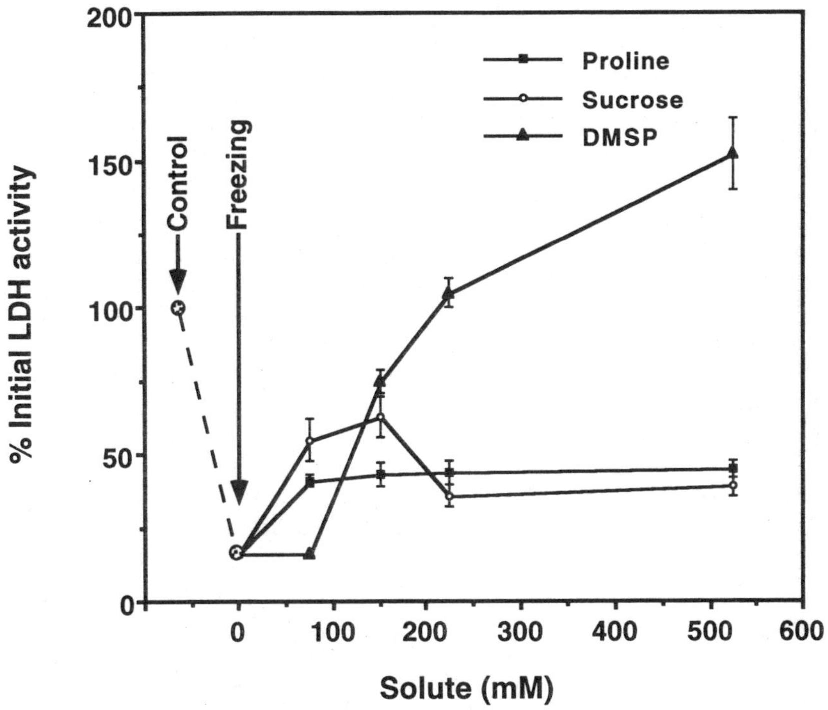

Figure 3. The effect of increasing concentrations of the organic solutes proline, sucrose and dimethylsulfonio-propionate (DMSP) on the stability of the cold-labile enzyme lactate dehydrogenase (LDH). LDH was frozen in the presence of the solutes followed by thawing and an activity determination. Data are given as mean ± standard deviation (n=4).

proline up to 530 mM led only to a small recovery of the LDH activity in the range of 40-45% of the activity without any solute. While at sucrose additions up to 150 mM increased the activity of the LDH to about 60% of controls, higher sucrose concentrations up to 530 mM again inhibited the enzyme activity to about 35% of the control activity. In contrast to proline and sucrose, increasing concentrations of DMSP stabilized the LDH. The addition of 230 mM DMSP before the freezing-thawing procedure led to full recovery of the enzyme activity. The presence of higher DMSP concentrations up to 530 mM resulted in a stimulation of the LDH activity of approximately 150% of the control (Fig. 3).

DISCUSSION

The distribution patterns of DMSP within different phototrophic organisms (Table 1) as well as a literature survey clearly demonstrate that mainly photosynthetic organisms are able to synthesize and accumulate this compound. The only exception known from the literature to synthesize and accumulate DMSP is the heterotrophic dinoflagellate *Crypthecodinium cohnii* (as *Gymnodinium cohnii*) (15). The opposite conclusion that all phototrophic organisms are able to synthesize DMSP is not valid. Some microalgal groups, for example, do not contain any trace of this compound (21). The data on the benthic and planktonic cyanobacteria clearly prove that many species of this group are able to form DMSP which is in agreement with other reports

(4, 34). Highest DMSP concentrations are usually found in many marine micro- and macroalgae. This strong correlation between DMSP metabolism and photosynthetic organisms indicates that light may be of great importance for the accumulation of this compound. The effect of light on the DMSP content was previously investigated in various green macroalgae. Species grown under fluctuating daylengths and various light intensities accumulated DMSP only under long day conditions and high irradiances (17-19). In addition, species kept in the dark degraded DMSP, and hypersaline treated species formed DMSP as an osmolyte only in the light (20, 22). All data point to a light-dependent DMSP biosynthesis. Recently, the influence of light intensity on the production of dimethylsulfide by the diatom *Skeletonema costatum* was studied (33). These authors measured significantly reduced production rates under low light conditions. Although DMSP biosynthesis is not yet fully understood, it is reasonable to conclude that parts of this pathway are directly light-dependent via light-activated enzymes or at least indirectly via the supply of NADH/NADPH and ATP by photosynthesis. The hypothesis of light stimulation of enzymes in the DMSP pathway is strengthened by the fact that the sulfur containing amino acid methionine is a precursor in the DMSP biosynthesis (12), and that several enzymes of the amino acid metabolism are light-activated in algae (31). Enzymes involved in sulfate uptake and assimilatory sulfate reduction are also very often activitated by light (29). In conclusion, light is an important environmental factor influencing the ability of algae and most probably also of other phototrophic organisms to regulate the intracellular DMSP pool. The finding that a heterotrophic dinoflagellate is also capable to synthesize DMSP (15) is difficult to interprete. Dinoflagellates have been regarded both as plants and as animals due to the simultaneous occurrence of photosynthesis and heterotrophy within this group. Most photosynthetic species of dinoflagellates have extremely high DMSP concentrations (21). It is possible that *Crypthecodinium* evolved from a photosynthetic ancestor and therefore still carries the biochemical capability to synthesize DMSP. This, however, would indicate that photosynthesis is not essential for the biosynthesis. Another explanation could be the development of an alternative pathway for DMSP synthesis in *C. cohnii* during evolution which is light-independent. However, before any final conclusions can be drawn, much more detailed studies on the regulation of the DMSP biosynthesis in *C. cohnii* have to be conducted.

The physiological function of DMSP as an organic osmolyte that is synthesized and accumulated under increasing salinities is well proven (10, 22). It has been speculated that the ability to accumulate this compound was evolved during the last ice age, when the salinity of the oceans was higher (8). Moreover, DMSP acts not only as an osmolyte but also as a compatible solute, as demonstrated with enzyme preparations of the unicellular alga *Tetraselmis subcordiformis* (13). Compatible solutes generally have protecting and stabilizing effects on metabolic pathways and membrane-dependent processes against adverse effects of high salt concentrations (6, 22). Other physiological roles of DMSP in metabolism are still obscure. However, the significantly higher DMSP concentrations in many algae grown under low temperatures, compared with those maintained in temperate conditions, suggest another biological function of this compound as a cryoprotectant (Fig. 1; 20, 25). While this response pattern has been demonstrated for all Arctic and Antarctic Ulvales, as well as for all polar isolates of *A. arcta* so far tested, some temperate populations of the latter species do not accumulate DMSP under low temperatures (Kück et al., unpublished data). Studies to explain these differences in the response patterns are in progress. Our experiments on the enzymes MDH from *A. arcta* and on the cold-labile LDH clearly illustrate the role of DMSP as a cryoprotectant. In both assays, i.e. in the case of MDH at -2° C, the presence of increasing DMSP concentrations positively affected enzyme activities in such a way that even significantly higher rates could be measured compared to the control. As already mentioned in Materials and Methods, the MDH of the Arctic *A. arcta* showed optimum activity at 30°C (Kück et al., unpublished data). Even at 0° C, this enzyme exhibited an activity of about 30% of the maximum. It seems to be a general phenomenon in polar algae, that many of their physiological processes, such as growth or photosynthesis, show temperature optima very similar to those of temperate species (23). Instead of the development of specific genetic adaptations to the cold-water environment, algae from Antarctica and the Arctic seem rather to exhibit a broad physiological potential for acclimation. The increase in DMSP under low temperatures in polar algae can be attributed to such an acclimation strategy.

Nishiguchi and Somero (25) found in their physicochemical study on the compatibility of DMSP with the structural stability of three enzymes that this compound protected at low but not high temperatures. These authors clearly proved the function of DMSP as an effective cryoprotectant for phosphofructokinase under conditions of cold-induced denaturatuion. The opposite was not valid, i.e. DMSP did not stabilize the protein structure under conditions of heat denaturation. Consequently, Nishiguchi and Somero (25) concluded that organisms that experience high temperatures in their habitat generally lack DMSP due to its perturbing effects under these conditions.

The mechanism of cryoprotection of LDH by 28 different solutes such as sugars, polyols, amino acids, methylamines and lyotropic salts was studied by Carpenter and Crowe (7). Most of the organic compounds protected the enzyme to varying degrees from damage during freezing-thawing. However, none of them led to a stimulation in LDH activity as DMSP did in the present study. Organic solutes exert their protective function by preferentially excluding contact with the enzyme's surface (7). This suggests that compensation for temperature at the molecular level may be achieved by changes in thermodynamic activation parameters (enthalpy of activation) and a reduction in thermal stability. Thus, the complex nature of the interaction between enzyme activity, enzyme stability and temperature in the presence of DMSP is emphasized. However, under elevated temperatures organic solutes that are effective cryoprotective compounds are often destabilizing to enzymes and under some circumstances even toxic to cells (2, 23). This observation is also explained by the preferential hydration model (2, 7), i.e. especially compounds with hydrophobic components such as DMSP tend to bind preferentially to proteins which then causes denaturation. This binding is temperature-dependent, i.e. the higher the temperature the stronger the effects.

It is also important to mention that the DMSP concentrations found in Antarctic green macroalgae are high enough to protect/stabilize enzymes after periods of freezing. This will have important physiological and ecological implications for the survival of these organisms under the extreme environmental conditions of Antarctica. Finally, the data presented call for further careful studies on the activity of many different enzymes isolated from polar algae. The experiments have to be conducted at various temperatures and in the presence of different DMSP concentrations to better understand the cryoprotective function of this compound.

ACKNOWLEDGMENTS

This project was funded by the Max-Planck Society, the Deutsche Forschungsgemeinschaft and the Bundesministerium für Forschung und Technologie, FRG. We deeply thank Christian Wiencke, Alfred Wegener Institute for Polar and Marine Research, for providing the Antarctic and Arctic macroalgae, as well as Jan Kuever, Max-Planck-Institute for Marine Mikrobiology, for providing most of the phototrophic bacteria. We also acknowledge the excellent technical help of Christina Langreder, Claudia Daniel and Anja Eggers.

REFERENCES

1. Andreae, M.O. 1990. Ocean-atmosphere interactions in the global biogeochemical sulfur cycle. Mar. Chem. *30:* 1-29.
2. Arakawa, T., J.F. Carpenter, Y.A. Kita and J.H. Crowe. 1990. The basis for toxicity of certain cryoprotectants: a hypothesis. Cryobiol. *27:* 401-415.
3. Barnard, W.R., M.O. Andreae and R.L. Iverson. 1984. Dimethylsulfide and *Phaeocystis pouchetii* in the southeastern Bering Sea. Cont. Shelf Res. *3:* 103-113.
4. Bechard, M.J. and W.R. Rayburn. 1979. Volatile organic sulfides from freshwater algae. J. Phycol. *15:* 379-383.
5. Bischoff, B., U. Karsten, C. Daniel, K. Kück, B. Xia and C. Wiencke. 1994. Preliminary assessment of the β-dimethylsulfoniopropionate (DMSP) content of macroalgae from the tropical island Hainan (People's Republic of China). Aust. J. Mar. Freshwater Res. *45:* 1329-1336.
6. Brown, A.D. and J.R. Simpson. 1972. Water relations of sugar-tolerant yeasts: the role of intracellular polyols. J. Gen. Microbiol. *72:* 589-591.
7. Carpenter, J.D. and J.H. Crowe. 1988. The mechanism of cryoprotection of proteins by solutes. Cryobiol. *25:* 244-255.
8. Charlson, R.J., J.E. Lovelock, M.O. Andreae and S.G. Warren. 1987. Oceanic phytoplankton, atmospheric sulphur, cloud albedo, and climate. Nature *326:* 655-661.
9. Dacey, J.W.H. and S.G. Wakeham. 1986. Oceanic dimethylsulfide: production during zooplankton grazing on phytoplankton. Science *233:* 1314-1316.
10. Dickson, D.M.J. and G.O. Kirst. 1986. The role of dimethylsulphoniopropionate, glycine betaine and homarine in the osmoacclimation of *Platymonas subcordiformis*. Planta *155:* 409-415.
11. Edwards, D.M., R.H. Reed, J.A. Chudek, R. Foster and W.D.P. Stewart. 1987. Organic solute accumulation in osmotically-stressed *Enteromorpha intestinalis*. Mar. Biol. *95:* 583-592.
12. Greene, R.C. 1962. Biosynthesis of dimethyl- β-propiothetin. J. Biol. Chem. *237:* 2251-2254.
13. Gröne, T. and G.O. Kirst. 1991. Aspects of dimethylsulfoniopropionate effects on enzymes isolated from the marine phytoplankter *Tetraselmis subcordiformis* (Stein). J. Plant Physiol. *138:* 85-91.
14. Iida, H., K. Nakamura and T. Tokunaga. 1985. Dimethyl sulfide and dimethyl-β -propiothetin in sea algae. Bull. Jap. Soc. Sci. Fish. *51:* 1145-1150.
15. Ishida, Y. and H. Kadota. 1967. Isolation and identification of dimethyl-β-propiothetin from *Gymnodinium cohnii*. Agr. Biol. Chem. *31:* 756-757.
16. Iverson, R.L., F.L. Nearhoff and M.O. Andreae. 1989. Production of dimethylsulfonium propionate and dimethylsulfide by phytoplankton in estuarine and coastal waters. Limnol. Oceanogr. *34:* 53-67.
17. Karsten, U., C. Wiencke and G.O.Kirst. 1990a. The β-dimethylsulphoniopropionate (DMSP) content of macroalgae from Antarctica and southern Chile. Bot. Mar. *32:* 143-146.
18. Karsten, U., C. Wiencke and G.O. Kirst. 1990b. The effect of light intensity and daylength on the β-dimethylsulfoniopropionate (DMSP) content of marine green macroalgae from Antarctica. Pl. Cell Envir. *13:* 989-993.
19. Karsten, U., C. Wiencke and G.O. Kirst. 1991. Growth pattern and β-dimethylsulphoniopropionate (DMSP) content of green macroalgae at different irradiances. Mar. Biol. *108:* 151-155.
20. Karsten, U., C. Wiencke and G.O. Kirst. 1992. Dimethylsulphoniopropionate (DMSP) accumulation in green macroalgae from polar to temperate regions: interactive effects of light versus salinity and light versus temperature. Pol. Biol. *12:* 603-607.
21. Keller, M.D., W.K. Bellows and R.R.L. Guillard. 1989. Dimethyl sulfide production by marine phytoplankton, p. 167-182. *In* E.S. Saltzman and W.J. Cooper (eds.), Biogenic Sulfur in the Environment. American Chemical Society, Washington, DC.
22. Kirst, G.O. 1990. Salinity tolerance of eukaryotic marine algae. Ann. Rev. Plant Physiol. Plant Mol. Biol. *41:* 21-53.
23. Kirst, G.O. and C. Wiencke. 1995. Ecophysiology of polar algae. J. Phycol. *31:* 181-199.
24. Matrai, P.A. and M.D. Keller. 1993. Dimethylsulfide in a large-scale coccolithophore bloom in the Gulf of Maine. Cont. Shelf Res. *13:* 831-843.
25. Nishiguchi, M.K. and G.N. Somero. 1992. Temperature- and concentration-dependence of compatibility of the organic osmolyte -dimethylsulfoniopropionate. Cryobiol. *29:* 118-124.
26. Paquet, L., B. Rathinasabapathi, H. Saini, L. Zamir, D.A. Gage, Z.H. Huang and A.D. Hanson. 1994. Accumulation of the compatible solute 3-dimethylsulfoniopropionate in sugarcane and its relatives, but not other gramineous crops. Aust. J. Plant Physiol. *21:* 37-48.
27. Provasoli, L. 1968. Media and prospects for cultivation of marine algae, p. 47-74. *In* A. Watanabe and A. Hattori (eds.), Cultures and Collections. Jap. Soc. Plant Physiol., Tokyo.

28. Reed, R.H. 1983. Measurement and osmotic significance of β-dimethylsulfoniopropionate in marine macroalgae. Mar. Biol. Lett. *4:* 173-181.
29. Schmidt, A. 1983. Thioredoxins in the regulation of nitrogen and sulfur metabolism, p. 124-128. *In* R. Scheibe (ed.), Light dark modulation of plant enzymes. L. Ellwanger Press, Bayreuth.
30. Starr, R. and J. Zeikus. 1987. UTEX - the culture collection of algae at the University of Texas at Austin. J. Phycol. *23* (Supplement)*:* 1-47.
31. Tischner, R. and K. Hüttermann. 1980. Regulation of glutamine synthetase by light during nitrogen deficiency in synchronous *Chlorella sorokiniana*. Plant Physiol. *66:* 805-808.
32. Van Diggelen, J., J. Rozema, D.M.J. Dickson and R. Broekman. 1986. 3-dimethylsulphoniopropionate, proline and quaternary ammonium compounds in *Spartina anglica* in relation to sodium chloride, nitrogen and sulphur. New Phytol. *103:* 573-586.
33. Vetter, Y.A. and J.H. Sharp. 1993. The influence of light intensity on dimethylsulfide production by a marine diatom. Limnol. Oceanogr. *38:* 419-425.
34. Visscher, P.T. and H. van Gemerden. 1991. Production and consumption of dimethylsulfoniopropionate in marine microbial mats. Appl. Environ. Microbiol. *57:* 3237-3242.
35. Young, A.J., J.C. Collins and G. Russell. 1987. Solute regulation in the euryhaline marine alga *Enteromorpha prolifera* (O.F.Müll.). J.Exp.Bot. *38:* 1298-1308.
36. White, R.H. 1982. Analysis of dimethyl sulfonium compounds in marine algae. J. Mar. Res. *40:* 529-536.
37. Wiencke, C. 1988. Notes on the development of some benthic marine macroalgae of King George Island, Antarctica. Ser.cient. Inst. Ant. Chil. *37: 23-47.*

14

THE DETERMINATION OF DMSP IN MARINE ALGAE AND SALT MARSH PLANTS

D. W. Russell and A. G. Howard

Chemistry Department
University of Southampton
Southampton, Hampshire, SO17 1BJ, United Kingdom

SUMMARY

Factors influencing the analysis of dimethylsulfoniopropionate (DMSP) are discussed with particular reference to the measurement of DMSP using HPLC-FPD instrumentation. Significant losses of DMSP can occur during sampling, sample storage and DMSP extraction. During sampling and sample transport, care must be taken to minimize stresses such as changes in temperature, light level or salinity. During the extraction procedure cell rupture releases DMSP-lyase into the extract leading to the breakdown of DMSP to dimethylsulfide (DMS). This can be minimized by maintaining acidic extraction conditions. A similar DMSP loss due to the release of DMSP-lyase occurs when samples are frozen and then rethawed prior to the extraction step.

DMSP has been measured in a number of macro-algae, salt marsh plants and cultured phytoplankton from Southern England. DMSP has been confirmed by compound-specific HPLC-FPD instrumentation to be present in three red, two green and three brown macro-algae. The levels of this compound vary greatly, ranging from 0.0010 mmol DMSP. kg^{-1} (fresh weight basis) in *Laminaria saccharina* to 161 mmol DMSP. kg^{-1} (fresh weight basis) in *Enteromorpha linza*. DMSP was also confirmed as being present in the higher plants *Halimione portulacoides* and *Spartina alterniflora*. DMSP was shown to be present in a number of phytoplankton. Of the studied species, levels ranged from 0.82 µmol DMSP/mg chlorophyll 'a' in *Tetraselmis suercica* to 157 µmol DMSP/ mg Chl 'a' in *Emiliania huxleyi*.

INTRODUCTION

In 1948 Challenger and Simpson (10) extracted β-dimethylsulfoniopropionate (DMSP) from the red alga *Polysiphonia fastigiata* and precipitated it from a crude alcoholic extract as a reineckate. It was then confirmed as DMSP by mixed melting point determination with a synthetic sample of the compound. Later DMSP was also isolated from the green alga *Enteromorpha intestinalis* (9) and a 'presumptive' test was proposed for the identification

Biological and Environmental Chemistry of DMSP and Related Sulfonium Compounds
edited by Ronald P. Kiene et al., Plenum Press, New York, 1996

of DMSP based on the cold alkaline hydrolysis of DMSP to DMS and acrylic acid. Variants of this test have been used almost exclusively since that time; the evolved DMS being used to infer the DMSP content of a sample. This method has been applied to a wide range of macro-algae (5,11,25,42,43,49,53), micro-algae (3,5,19,27,47,50), grasses (38,52), sediments (51), fish (1,11), sugar cane (39), bacteria (33), shellfish (2), and sea water (7,22,28,29,35,54).

On few occasions since the initial work has the presence of DMSP been recorded directly. In 1962, Greene (18) identified DMSP in *Ulva lactuca* by thin-layer chromatography. Ackman (3), four years later, used a similar process but followed it with the precipitation of the reineckate to confirm the presence of DMSP. Larher *et al* (31) subjected an alcoholic extract of *Spartina anglica* to electrophoresis and then precipitated the DMSP as its reineckate, liberated it from the counter ion and finally purified it by further electrophoresis. Thin-layer electrophoresis has also been used with scanning reflectance densitometry (12,13,16) as a means of quantifying DMSP.

More recently proton NMR, FAB-MS (39,48) and GC-MS of S-demethylated silyl derivatives (20) of DMSP have been used to study extracts from the salt-tolerant plants *Melanthera biflora* and *Diplotaxis tenuifolia*. Cation-exchange HPLC has been used to separate dimethylsulfonium compounds and betaines (15,48) with direct UV detection of the eluted compounds (190-220nm) or of their UV absorbing esters (17).

These direct methods of DMSP detection are not sufficiently sensitive to measure DMSP in a wide range of samples without prior preconcentration, and normally require quite extensive sample preparation to ensure that the DMSP is sufficiently enriched to be distinguished from co-extracted material. Thus, the way is open for a simple, thorough, direct DMSP analysis procedure that does not rely solely on the release of DMS. This is especially the case as the basic assumption of the test, that only DMSP will yield DMS upon base hydrolysis, is becoming increasingly uncertain as more potential DMS precursors (DMS-Pr) are being found in the environment. Examples of such compounds include: S-methyl methionine (SMM) (34), gonyol (36), gonyauline (37), (S)-4-dimethylsulfonio-2-methoxybutyrate (46), (R)-3-dimethylsulfonio-2-methoxypropanoate (41), and 3-phosphatidyl sulfocholine (4)). With the exception of SMM, which requires boiling in alkali to release DMS (14), and the dimethylsulfonium analogue of choline, which evolves DMS at 85°C, the behaviour of these compounds under basic conditions is largely unreported. Indeed experimental evidence suggests that there are other sources of methylated sulfur species in the environment. When Roberts (44) attempted to measure DMSP in water samples from an Antarctic lake that contained dimethylsulfide (DMS), dimethyl disulfide (DMDS), dimethyltrisulfide, and dimethyltetrasulfide using base hydrolysis, the concentration of DMDS increased five-fold. There was however, no increase in the level of DMS implying that no DMSP was present in the samples.

DMSP has been specifically identified in very few cases and whilst it is likely to be a major natural source of DMS, the compounds listed above and probably many others, could lead to a positive interference in the DMSP content inferred from DMS evolution. The development of simple HPLC-FPD instrumentation which is capable of separating potential DMS precursors by HPLC before on-line post column base hydrolysis to DMS (21), has opened up the opportunity to confirm the presence of DMSP.

The extraction of DMSP has in the past been carried out under either acidic (18,32,49) or alcoholic (10,31) conditions or using a mixture of solvents (most commonly methanol:chloroform:water (15,17,20,39,48)). A purification procedure has always then been necessary before the final identification could be made. This has involved either its precipitation as a reineckate, electrophoresis, or most commonly by ion-exchange chromatography using Dowex 50.

The use of very strong acids, such as hot 6N HCl (32), is to be avoided as such treatment has been shown to lead to the erroneous identification of dimethylsulfoniopentanoate (32), a compound which is produced by the breakdown of the glucosinolate glucoerucin (20). The HPLC-FPD apparatus employed in this work is exceptionally selective as compound identification arises from both characteristic chromatographic retention behaviour and specific reaction chemistry. Such selectivity obviates the need for time consuming sample clean-up procedures which might lead to the preferential concentration of DMSP over other potential DMS precursors.

MATERIALS AND METHODS

Apparatus

The apparatus employed for the measurement of DMSP is a custom-built instrument employing compound separation and identification by HPLC, rapid post-column hydrolysis and flame photometric detection of evolved volatile sulphur compounds (21).

Extraction Procedure

The extraction procedure developed for this work was chosen to minimize the decomposition of DMSP precursors for reasons that will be discussed in subsequent sections. The finally adopted procedure for the analysis of plant materials was:

i) ca. 3g of plant material was carefully blotted dry and weighed.
ii) the sample was homogenized to a fine powder in liquid nitrogen.
iii) without allowing the sample to thaw before the addition, 2ml of 10% HCl and 8ml of water were added and then ground together.
iv) the mixture and aqueous washings were centrifuged for 10 min.
v) the supernatant liquid was filtered (Whatman GF/C, 2.5cm)
vi) immediately prior to the chromatographic step, the aqueous extract was buffered to column eluent pH and diluted with column eluent.

Plankton was collected for analysis by gentle filtration (< 1.5 atmos.) through a Whatman GF/C filter. This filter was then subjected to the extraction procedure steps ii) to vi) (see above).

Plant biomass is reported on a fresh weight basis. For micro-algae, results are related to the chlorophyll ‘a’ content (measured spectrophotometrically (40)).

Gas Chromatographic Measurement of DMS and DMS-Pr

Headspace analyses of DMS were carried out using gas-solid chromatography with flame photometric detection. The PTFE column (1/8″ i.d. x 350 cm) contained Chromosorb 101 and was operated under isothermal conditions at 150°C with a carrier gas (N_2) flow rate of 50 ml/min. DMSP-Pr were measured by room temperature base hydrolysis in 50ml headspace bottles.

RESULTS

DMSP Breakdown during Extraction

Initial experiments involving the extraction of DMSP into water gave poor and erratic recoveries of added DMSP. Challenger and Simpson first suggested that an enzyme was

Table 1. Recovery of spiked DMSP (n=4)

Sample	Average % recovery of DMSP spike
Fucus (no acid)	94 ± 14
Enteromorpha (no acid)	25 ± 35
Enteromorpha (with acid)	73 ± 5

present in the red alga *Polysiphonia fastigiata* (10) that could be involved in the evolution of DMS. This was later confirmed by Cantoni and Anderson (8) and it was demonstrated that whilst the enzyme activity was highest at a pH of 5.1, it fell off sharply under more acidic conditions.

The effect of pH on DMSP-lyase activity was tested by the measurement of the DMS released during the extraction procedure. Subsamples (ca. 8g) of *Enteromorpha linza* (ground in liquid nitrogen) were placed in 60ml bottles. To half of these bottles 5ml of 4% sulfuric acid was added. All bottles were then made up to 50ml with distilled water and sealed with silicone rubber septa. The headspace in each bottle was periodically analyzed by gas solid chromatography. DMS was identified in the headspace and its level increased up to two hours. The quantity of DMS released corresponded to a loss of 20±1 mmol kg^{-1} (fresh weight) from the green alga when it was extracted with water. When acidified, however, the sample lost just 1.14±0.04 mmol of DMSP kg^{-1}.

In view of the enhanced stability brought about by the addition of a small amount of acid during the extraction, an experiment was carried out to assess the recovery of added DMSP. Ca. 3g of algae (liquid nitrogen ground) was spiked with DMSP during its extraction into water. Acid (2ml of 10% HCl) was simultaneously added to some of the extracts.

A significant improvement in both the recovery and reproducibility was achieved in the presence of acid. The DMSP-lyase activity varies significantly between algal species and this is reflected in the excellent spike recovery of DMSP from *Fucus* even in the absence of acid. The effectiveness of the acidification is evident in both the absolute DMSP recovery from *Enteromorpha* and its variability.

Sample Collection and Storage

The presence of DMSP-lyase (8) and the changes in DMSP content observed in response to a wide range of external physical stresses such as those brought about by changes in salinity (14,30,43), day length (25), temperature (26,30), and light intensity (25), indicates that there is great potential for changes to occur in the DMSP content of samples prior to their extraction and analysis. This must be addressed when considering sample collection and storage methods.

To test the effects of sampling and sample storage, *Enteromorpha linza* was collected from below the water line at the mouth of Langstone Harbour at Eastney, Portsmouth, U.K.. The algal samples were brought back to the laboratory in sea water from the sampling site and returned to the laboratory within one hour of collection. The algal material was divided into three sub-samples before being subjected to three different regimes of overnight storage:

Treatment 1: The alga was aerated and illuminated (150W) in natural unfiltered seawater.
Treatment 2: After blotting dry, the alga was oven dried at 80°C.
Treatment 3: The alga was blotted dry and frozen at -20°C.

After 18 hours these were all extracted and analyzed for DMSP. The samples stored in sea water, oven dried, and frozen contained 164 ± 20, 46 ± 3, and 55.1 ± 0.3 mmol kg^{-1} (dry wt) of DMSP (n=4) respectively.

Table 2. Quantification of DMSP in various taxa using the HPLC/FPD instrumentation (data represented as mean ± stdev (n=4))

Sample	DMSP content
Phaeophyceae	mmol kg^{-1} (fresh weight)
Fucus vesiculosus	0.48 ± 0.08
Sargassum muticum	0.045 ± 0.016
Laminaria saccharina	0.0010 ± 0.0003
Chlorophyceae	
Ulva lactuca	29.4 ± 2.2
Enteromorpha linza	161 ± 2
Rhodophyceae	
Palmeria palmata	0.028 ± 0.003
Chondrus crispus	0.047 ± 0.021
Polysiphonia urceolata	10.7 ± 1.7
Flowering plants	μmol kg^{-1} (fresh weight)
Halimione portulacoides (roots)	17 ± 2
Halimione portulacoides (leaves)	3.1 ± 0.4
Spartina alterniflora (rhizomes & roots)	2.9 ± 0.3
Phytoplankton	μmol mg^{-1} (Chl 'a')
Phaedactylum tricornutum	1.1 ± 0.1
Isochrysis galbana	6.3 ± 0.2
Tetraselmis suercica	0.82 ± 0.02
Emiliania huxleyi	158 ± *16*

After overnight storage at -20°C there was a strong smell of DMS in the container and as the sample thawed over a period of 90 minutes, further DMS evolution was evident. The same sample, which had contained 55 mmol DMSP. kg^{-1} when extracted directly from the frozen state, when left to thaw on the bench before extraction, only contained 5.05 mmol DMSP kg^{-1}. The enhanced decay which occurs when a sample has undergone a freeze/thaw cycle implies the mixing of DMSP and DMSP-lyase as a result of cell rupture.

Although these experiments are of a preliminary nature, and the illumination of the sub-sample stored in sea water could conceivably lead to an increase in DMSP content, it vividly illustrates the unusual severity of DMSP losses which can occur in all stages of the analysis protocol. Subsequent analyses were therefore carried out on samples which had been returned to the laboratory in water from the sampling site. These were analysed immediately on return to the laboratory or if necessary stored for the minimum possible time in aerated sea water with illumination for 18 hours a day.

Oven Drying of Samples

The oven drying of samples is a convenient means of preparing macroalgae for storage, but as has been shown by Karsten *et al* (23), it can significantly alter the levels of DMS-Pr which are measured. Interestingly, with some green and especially red seaweeds, the result was not the expected loss of DMS-Pr but an increase in the measured levels. To explain this rise he postulated that there might be a second DMS precursor such as S-methylmethionine. Bishoff *et al*, on the other hand (6), found that oven dried green and red macro algae contained less DMSP than fresh cultured samples from the same sampling site. In the case of *Ulva lactuca*, Karsten found a 24% increase in DMSP after oven drying. Using identical drying conditions and the same GC headspace method, we found only an

8% increase (45). More significantly, when the dried sample was extracted and analysed using the HPLC/FPD system, just 30% of the DMSP originally in the sample was still present.

Analysis of Flora

The techniques outlined previously were applied to the analysis of a number of samples of micro- and macro-flora. These were either collected from the coast of Southern England, or were in the form of non-axenic phytoplankton cultures.

Comparison of the HPLC and Headspace GC Methods

To obtain a direct comparison of the effectiveness of the HPLC procedure and the base hydrolysis headspace GC method, a sample of *Enteromorpha linza* was analysed by both methods. The DMSP result by HPLC (116 ± 10 mmol kg^{-1} (n=6)) was not significantly different from the DMS-Pr level measured by GC (125 ± 10 mmol kg^{-1} (n=4)).

DISCUSSION AND CONCLUSIONS

The determination of DMSP in marine flora requires particular attention to be paid to the sampling and sample pretreatment steps. It is fortunately rare that it is necessary to be as stringent in sampling procedure and in sample storage as it is with this compound. It is necessary to adopt a strict protocol in the collection of samples and to preserve their condition during the return journey to the laboratory and the time up to the analysis. For this reason we have adopted the procedure of returning samples in water from their site and minimizing changes in temperature, degree of oxygenation, light exposure etc. The rate at which the DMSP content of some samples adapts to environmental changes makes it impossible to use conventional sampling and storage procedures.

The modified procedures have been applied to the identification and quantification of DMS-Pr in a number of marine plants. Whilst to date it has only been possible to analyze comparatively few samples, in many cases these represent the first identifications of DMSP in these flora; previously reported values having been of DMS-Pr obtained by base hydrolysis.

The salt marsh plants were sampled during the winter months and this could explain the low DMSP content of *Spartina alterniflora* compared to those of Otte (38); 0.3-7.9 mmol kg^{-1} (fresh weight). There is some variability in the literature regarding the levels of DMSP in the same algal species. Taking for example the green seaweed *Ulva lactuca* from different localities, reported values are of the same magnitude as the 29.4 mmol.kg^{-1} given in this work (Greenland (23): 70.4±7.2; China (6): 11.04±1.23; Scotland (42): 23.4-38.7 and California (53): 15 mmol kg^{-1} (fresh weight)). Such variability is to be expected in view of the susceptibility of the DMSP levels to environmental conditions and analysis protocols.

The comparison of reported levels of DMSP in phytoplankton is clouded by the different means of reporting biomass. The level of DMSP in *Emiliania huxleyi* (2.014±0.007 (n=5) pg DMSP $cell^{-1}$) reported here compares relatively well with that reported by Keller *et al* (0.75pg DMS-Pr (as DMSP) $cell^{-1}$).

The results confirm the widespread presence of DMSP and significant differences in DMSP levels which are present in such flora.

ACKNOWLEDGMENTS

The authors would like to thank the University of Southampton for the provision of a University Studentship to D.W. Russell for this work. They would also like thank members of the Southampton University Department of Oceanography for supplying the phytoplankton cultures.

REFERENCES

1. Ackman, R.G. and J. Dale. 1965. Reactor for determination of dimethyl- β-propiothetin in tissue of marine origin by gas-liquid chromatography. J. Fish. Res. 22 (4): 875-883.
2. Ackman, R.G. and H.J. Hingley. 1968. The occurrence and retention of dimethyl- β-propiothetin in some filter feeding organisms. J. Fish. Res. Bd. Canada. *25*: 267-284.
3. Ackman, R.G., C.S. Tocher and J. McLachlan. 1966. Occurrence of dimethyl- β-propiothetin in marine phytoplankton. J. Fish. Res. 23: 357-364.
4. Anderson, R., M. Kates, and B.E. Volcani. 1976. Sulphonium analogue of lecithin in diatoms. Nature. 263, 51-53.
5. Belviso, S. 1990. Production of DMSP and DMS by a microbial food web. Limnol. Oceanogr. 35 (8): 1810-1821.
6. Bischoff, B., U. Karsten, C. Daniel, K. Kuck, B. Xia and C. Wienke. 1994. Preliminary assessment of the DMSP content of macroalgae from the tropical island of Hainan, China. Aust. J. Mar. Freshwater Res. 45: 1329-1336.
7. Burgermeister, S., R.C. Zimmerman, H-W. Georgii, H.G. Bingmer, G.O. Kirst, M. Janssen and W. Ernst. 1990. On the biogenic origin of DMS: relation between chlorophyll, ATP, organismic DMSP, phytoplankton species, and DMS distribution in Atlantic surface water and atmosphere. J. Geophys. Res. 95: 20607-20615.
8. Cantoni, G.C. and D.G. Anderson. 1956. Enzymatic cleavage of dimethylpropiothetin by *Polysiphonia lanosa*. J. Biochem. 222: 171-177.
9. Challenger, F., R. Bywood, P. Thomas and B.J. Hayward. 1957. Studies on biological methylation, part XVII. The natural occurrence and chemical reactions of some thetins. Arch. Biochem. Biophys. 69: 514-523.
10. Challenger, F. and M.I. Simpson. 1948. Studies on biological methylation. Part XII A precursor of dimethylsulfide evolved by *Polysiphonia fastigiata*. Dimethyl-2-carboxyethylsulfonium hydroxide and its salts. J. Chem. Soc. 43: 1591-1597.
11. Dacey, J.W.H., G.M. King and P.S. Lobel. 1994. Herbivory by reef fishes and the production of DMS and acrylic acid. Mar. Ecol. Prog. Ser. 112: 67-74.
12. Dickson, D.M.J. and G.O. Kirst. 1986. The role of β-dimethylsulphoniopropionate, glycine betaine and homarine in the osmoacclimation of *Platymonas subcordiformis*. Planta. 167: 536-543.
13. Dickson, D.M.J., R.G. Wyn Jones and J. Davenport. 1980. Steady state osmotic adaption in *Ulva lactuca*. Planta. 150: 158-165.
14. Dickson, D.M., R.G. Wyn Jones and J. Davenport. 1982. Osmotic adaption in *Ulva lactuca* under fluctuating salinity regimes. Planta. 155: 409-415.
15. Gorham, J. 1984. Separation of plant betaines and their sulphur analogues by cation exchange high-performance liquid chromatography. J. Chrom. 287: 345-351.
16. Gorham, J., S.J. Coughlan, R. Storey and R.G. Wyn Jones. 1981. Estimation of quaternary ammonium and tertiary sulphonium compounds by thin-layer electrophoresis and scanning reflectance densitometry. J. Chrom. 210: 550-554.
17. Gorham, J., E. McDonnell and R.G. Wyn Jones. 1982. Determination of betaines as ultra violet-absorbing esters. Anal. Chim. Acta. 138: 277-283.
18. Greene, R.C. 1962. Biosynthesis of Dimethyl- β-propiothetin. J. Biol. Chem. 237: 2251-2254.
19. Grone, T. and G.O. Kirst. 1992. The effect of nitrogen deficiency, methionine and inhibitors of methionine metabolism on the DMSP contents of *Tetraselmis subcordiformis*. Mar. Biol. 112: 497-503.
20. Hanson, A.D., Z-H. Huang, and D.A. Gage. 1993. Evidence that the putative compatible solute S-dimethylsulfoniopentanoate is an extraction artifact. Plant Physiol. 101: 1391-1393.
21. Howard, A.G. and D.W. Russell. 1995. HPLC-FPD instrumentation for the measurement of the atmospheric DMS-precursor DMSP. Anal. Chem. 67 (7): 1293-1295.

22. Iverson, R.L., F.L. Nearhoof and M.O. Andreae. 1989. Production of DMSP and DMS by phytoplankton in estuarine and coastal waters. Limnol. Oceanogr. 34 (1): 53-67.
23. Karsten, U., K. Kuck, C. Daniel, C. Wiencke and G.O. Kirst. 1994. A method for complete determination of DMSP in marine macroalgae from different geographical regions. Phycologia 33: 171-176.
24. Karsten, U., C. Wiencke and G.O. Kirst. 1990. The DMSP content of macroalgae from Antarctica and Southern Chile. Bot. Mar. 33: 143-146.
25. Karsten, U., C. Wienke and G.O. Kirst. 1990. The effect of light intensity and day length on DMSP content of marine green macroalgae from Antarctica. Plant, Cell Environ. 13: 989-993.
26. Karsten, U., C. Wienke and G.O. Kirst. 1992. DMSP accumulation in green macroalgae from polar to temperate regions: interactive effects of light versus salinity and light versus temperature. Polar Biol. 12: 603-607.
27. Keller, M.D., W.K. Bellows and R.R.L Guillard. 1989. Dimethyl sulfide production in marine phytoplankton, p.167-182. *In* Saltzman and Cooper (eds.), Biogenic sulfur in the environment. A.C.S. Symp. Ser. 393, Washington.
28. Kiene, R.P. 1992. Dynamics of dimethylsulfide and dimethylsulfoniopropionate in oceanic water samples. Mar. Chem. 37: 29-52.
29. Kiene, R.P. and S.K. Service. 1991. Decomposition of dissolved DMSP and DMS in estuarine waters: dependence on temperature and substrate comcentration. Mar. Ecol. Prog. Ser. 76: 1-11.
30. Kirst, G.O., C. Theil, H. Wolff, J. Nothnagel, M. Wanzek and R. Ulmke. 1991. Dimethylsulfoniopropionate in ice-algae and its possible biological role. Mar. Chem. 35: 381-388.
31. Larher, F., J. Hamelin and G.R. Stewart. 1977. L'Acide dimethylsulfonium-3 propanoique de *Spartina anglica*. Phytochem. 16: 2019-2020.
32. Larher, F. and J. Hamelin. 1979. L'Acide dimethylsulfonium-5 pentanoique de *Diplotaxis tenuifolia*. Phytochem. 18: 1396-1397.
33. Ledyard, K.M. and J.W.H. Dacey. 1994. Dimethylsulfide production from dimethylsulfoniopropionate by a marine bacterium. Mar. Ecol. Prog. Ser. 110: 95-103.
34. McRorie, R.A., G.L. Sutherland, M.S. Lewis, A.D. Barton, M.R. Glazener and W. Shive. 1954. Isolation and identification of a naturally occurring analog of methionine. J.A.C.S. 76: 115-118.
35. Matrai, P.A. and M.D. Keller. 1993. Dimethyl sulfide in a large scale Coccolithophore bloom in the Gulf of Maine. Cont. Shelf Res. 13: 831-843.
36. Nakamura, H., K. Fujimaki, O. Sampei and A. Murai. 1993. Gonyol: methionine-induced sulfonium accumulation in a dinoflagellate *Gonyaulax polyedra*. Tett. Letts. 34: 8481-8484.
37. Nakamura, H., M. Ohtoshi, O. Sampei, Y. Akashi and A. Murai. 1992. Synthesis and absolute configuration of (+)-gonyauline: a modulating substance of bioluminescent circadian rhythm in unicellular alga *Gonyaulax polyedra*. Tett. Letts. 33: 2821-2822
38. Otte, M.L. and J.T.Morris. 1994. Dimethylsulfoniopropionate (DMSP) in *Spartina alterniflora* Loisel. Aquatic Bot. 48: 239-259.
39. Paquet, L., B. Rathinasabapathi, H. Saini, L. Zamir, D.A. Gage, Z-H. Huang and A.D.Hansen. 1994. Accumulation of the compatible solute DMSP in sugarcane and its relatives, but not other graminous crops. Aust. J. Plant Physiol. 21: 37-48.
40. Parsons, T.R., Y. Maita and C.M. Lalli. 1984. A manual of chemical and biological methods for seawater analysis. Pergamon.
41. Patti, A., R. Morrone, R. Chillemi, M. Piatelli and S. Sciuto. 1993. Thetins and betaines of the red alga *Digenea simplex*. J. Nat. Prods. 56: 432-435.
42. Reed, R.H. 1983. Measurement and osmotic significance of β-dimethylsulphonio propionate in marine macroalgae. Mar. Biol. Letts. 4: 173-181.
43. Reed, R.H. 1983. The osmotic responses of *Polysiphonia lanosa* from marine and estuarine sites: evidence for incomplete recovery from turgor. J. Exp. Mar. Biol. Ecol. 68: 169-193.
44. Roberts, N.J., H.R. Burton and G.A.Pitson. 1993. Volatile organic compounds from Organic Lake, an Antarctic hypersaline, meromictic lake. Antarctic Sci. 5: 361-366.
45. Russell, D.W. unpublished data.
46. Sciuto, S., M. Piatelli and R. Chillemi. 1982. (-)- (S)-4-dimethylsulfonio-2-methoxybutyrate from the red alga *Rytiphloea tinctoria*. Phytochem. 21 (1): 227-228.
47. Stefels, J. and W.H.M. van Boekel. 1993. Production of DMS from dissolved DMSP in axenic cultures of the marine phytoplankton *Pheocystis* sp. Mar. Ecol. Prog. Ser. 97: 11-18.
48. Storey, R., J. Gorham, M.G. Pitman, A.D. Hanson and D. Gage. 1993. Response of *Melanthera biflora* to salinity and water stress. J. Exp. Bot. 44: 1551-1560.
49. Tocher, C.S., R.G. Ackman and J. McLachlan. 1966. The identification of dimethylpropiothetin in the algae *Syracosphaera carterae* and *Ulva lactuca*. Can. J. Biochem. 44: 519.

50. Vairavamurthy, A., M.O. Andreae and R.L. Iverson. 1985. Biosynthesis of DMS and DMSP by *Hymenomons carterae* in relation to sulfur source and salinity variations. Limnol. Oceanogr. 30: 59-70.
51. Visscher, P.T., R.P. Kiene and B.F. Taylor. 1994. Demethylation and cleavage of dimethylsulfoniopropionate in marine intertidal sediments. FEMS Microbiol. Ecol. 14: 179-190.
52. Weber, J., M.R. Billings and A.M. Falke. 1991. Seasonal methyltin and DMSP concentrations in leaf tissue of *Spartina alterniflora* of the Great Bay estuary. Estuarine Coast. Shelf Sci. 33: 549-557.
53. White, R. 1982. Analysis of dimethyl sulphonium compounds in marine algae. J. Mar. Res. 40: 529-535.
54. Wolfe, G.V., E.B. Sherr and B.F. Sherr. 1994. Release and consumption of DMSP from *Emiliania huxleyi* grazing by *Oxyrrhis marina. Mar. Ecol. Prog. Ser. 111: 111-119.*

15

EFFECTS OF DMSP AND RELATED COMPOUNDS ON BEHAVIOR, GROWTH AND STRESS RESISTANCE OF FISH, AMPHIBIANS AND CRUSTACEANS

Kenji Nakajima

Laboratory of Biochemistry, Department of Nutrition
Koshien University, Momijigaoka 10-1
Takarazuka, Hyogo 665, Japan

SUMMARY

Recording methods for determining the striking behavior of fish were developed. Among a number of sulfonium compounds tested, dimethylsulfoniopropionate (DMSP) proved to elicit the strongest effect on striking behavior of goldfish. Electrophysiological experiments revealed that DMSP stimulated the olfactory sense of carp more strongly than did glutamine. The effects of DMSP on growth and body movement of fish, amphibians, crustaceans, rats and chickens were examined. The results indicated that DMSP stimulated the growth and body movements of test organisms to varying degrees. Moreover, DMSP proved to accelerate the metamorphosis of amphibians and the molting of crustaceans. It also strengthened the resistance of fish against physical and chemical stresses and against stress-induced gastric ulcers in rats. The distribution of dimethylacetothetin (DMT)-homocysteine methyltransferase enzyme in various viscera of carp and substrate specificity of the enzyme were investigated. DMSP gave the highest specific activity as a methyl donor, as compared to other sulfonium compounds, and the highest activity of the enzyme was found in the hepotopancreas. Furthermore, dietary supplemented DMSP proved to accumulate rapidly in hepatopancreas tissues as compared to various viscera of carp. The addition of various concentrations of methionine to the diet or to the rearing water was found to promote the growth of carp and the molt of prawns but not the growth and metamorphosis of amphibians.

INTRODUCTION

Dimethylsulfoniopropionate (DMSP) is present in marine algae (10,41), crustaceans (44), shell fish (2, 12), and fish (1, 3,11), and also in some non-marine plants (17,40,45).

Biological and Environmental Chemistry of DMSP and Related Sulfonium Compounds
edited by Ronald P. Kiene et al., Plenum Press, New York, 1996

With the exception of algae, it is uncertain whether (or not) DMSP is biosynthesized by these organisms. Furthermore, its role in these organisms is poorly understood. The report that DMSP in fish may come from their diet (11), which may include DMSP-containing algae (10,41), suggests that only algae and/or microorganisms can biosynthesize it. Large amounts of DMSP in algae may act as a compatible solute (6, 20,42), or a cryoprotectant (14, 15). However, there are few reports on the effects of DMSP on other organisms. Because various sulfur-containing compounds occur in aquatic animals (13,43), this study examined the effects of DMSP and related compounds on the growth, body movements, stress resistance, molt or metamorphosis of fish, amphibians and crustaceans.

MATERIALS AND METHODS

Chemicals

DMSP and related sulfonium compounds were synthesized by refluxing equimolar amounts of the appropriate dialkylsulfide and corresponding bromoorganic acid at around 40°C for 3 h to 24 h (5). Crude products were washed with dry ether and crystallized from methanol, if necessary. Products were identified by NMR, IR, elemental and mass analysis or by x-ray or thermal-degradation. Vitamin U (3-amino-3-carboxypropyldimethylsulfoniumchloride), glycine betaine hydrochloride and choline chloride, purchased from Wako Pure Chemicals Co., Ltd., were washed separately with chilled dry ether, and purified by crystallizing from methanol. Other chemicals used were of the highest quality available.

Organisms

The test organisms used were: goldfish *Carassiuss auratus*, (n=27 individuals, body weight about 0.92-1.31 g); rainbow trout, *Oncorhynchus mikiss* (n=32, body weight about 0.91-4.85 g); carp, *Cyprinus carpio*, (n=8, average body weight 8.0 g); red sea bream, *Pagrus major* (n=12, average body weight 11.4 g); yellowtail, *Seriola quinqeradiata* (n=5, average body weight 3.3 g); and flounder *Paralichthys olivaceus* (n=10, average body weight 14.2 g).

Commercial fish diets used for determining striking strength or growth of test organisms were “Swimmy mini” (Nippon Pet Food Co., Ltd.) for goldfish and carp, “Nijimasu Chigyo C” (Nihon Haigo Shiryo, Co., Ltd.) for rainbow trout, “Ajikuranburu” (Nissin Shiryo Co., Ltd.) for red sea bream, flounder and striped prawn, “Buri-Mojako No. 1" (Nisshin Shiryo Co., Ltd.) for yellowtail and commercially available spinach leaves boiled for 10 min for tadpole.

Experimental

For determining striking strength of test fish, the test vessel (13x20x14 cm) with 2 liters of tap water or commercial synthetic sea water (Japan Biochemical Co., Ltd.) contained five fish. Test paste was made by mixing cellulose powder (0.5 g) in a mortar with either distilled water (or commercial sea water) or the test solution (0.7 ml). The paste was formed into a round ball and fixed 2.5 cm from the bottom of a thread which was suspended vertically from a bar across the vessel. Fish striking the test paste vibrated the thread; these vibrations were transferred to the bar, and were recorded by smoked paper on a running drum of a Kimographyone (Shimano Seisakusho Co., Ltd.). Striking strength of a trial was calculated from striking peak numbers by five test fish for 170 sec, which was one revolution of the drum. Control experiments with distilled or commercial sea water were performed at the start and end of each experiment. The experimental process was repeated five times and test paste was freshly prepared for each trial. Striking numbers obtained from five experiments with the same test solution were totaled and expressed in terms of striking frequency of the test solution (21). The striking frequency measured with the Kimographyone method gave the same results as electric techniques (22).

To estimate growth of the test organisms, diet paste was freshly prepared by mixing ground commercial pellets with a 1 mM DMSP solution at a ratio of 5:7 (w/v) for freshwater fish and crustaceans, 1 mM for yellowtail, and at 5 mM for red sea bream and flounder at a ratio of 2:1 (w/v). Feeding was *ad libitum* at 10:00 a.m. and 4:00 p.m. for fish and at 10:00 a.m. for striped prawn and tadpole. The total body weights were measured *en masse* for freshwater and marine fish, striped prawn and tadpole at specified times. They were then expressed as percentage of their initial values.

The molting of striped prawn was estimated every day from the numbers of sheds left off. The metamorphosis of tadpole was measured by counting the numbers of newly appearing hind and fore limbs every day.

For determining thrust power of test fish, fresh or commercial sea water (19 l) was put in a polyacrylate vessel (450 l). A polyacrylate test chamber (40x14x15 cm height) was placed in the center part of the vessel. Water was removed from the side part of the vessel by a pump (Iwaki, model MO-30R), and pumped into the test chamber. Three lattices were placed at the inlet, in front and rear of the test chamber. Aeration was carried out through the side of the vessel. All the test fish were placed between the front and rear lattices of the test chamber. In a trial, the flow speed was adjusted to a constant speed of 34 $cm \cdot s^{-1}$ just after 3 min, and the swimming time was recorded when the tails of half of the total number of the fish touched the rear lattice.

Resistance to oxygen deficiency was determined by exposing five fish of each group to air (12°C) on a wet soft cloth for specified times. After exposure, the fish were put back into the tank, and the time at which each fish began to swim normally was recorded. For estimating resistance of test fish against high water temperature, five fish in each group were placed into a round net basket (19.0 x 23.5 cm) in a vessel (44 x 53 x 18 cm height) with 31.2 *l* of water. Then, the water temperature in the vessel was raised at a rate of 0.54°C/min from the initial temperature of 20°C (Uni Cool Bath, NCB-130 (EYELA) Tokyo Rikakikai Co., Ltd.). The fish began to swim abnormally with the rising temperature and finally turned on their side. The water temperature and time at which each fish turned on its side were recorded.

Electrical responses from the carp olfactory tract were measured as follows: The fish (average weight 318 g) were anesthetized by injection of galamine triethiodide (2 mg/kg body weight). A portion of the upper skull of the sedated carp was excised by a drill (Leutor Mini Ace 200, Hihon Seimitsukikai Kousaku Co., Ltd.) and the inner layers of lipid and connective tissue were gently swabbed off. The olfactory tract was exposed to air, allowing the insertion of two insulated bipolar stainless steel electrodes (diam 0.2 mm) about 1 mm apart. Another electrode was introduced into the muscle above the eye as a ground. The gills of the fish were perfused with tap water for respiration. A microdispensor (Drumond Scientific Co., Ltd., USA) was used to deliver 50 µl of the test solution into the nasal cavity of the fish through a special glass inlet. Electric responses were recorded by an electroencephalograph (Model Me-135D, Nihon Koden Kogyo Co., Ltd.) and on a cassette tape recorder (Model RMG-5304, Nihon Koden Kogyo, Co., Ltd.). To calculate the strength of electrical responses, data was transferred from the tape to a dual beam memory oscilloscope VC-10 (Nihon Koden Kogyo Co., Ltd.), and loaded into a computer (Pc-8801 mkII, Nihon Electric co., Ltd.), by which the strength was automatically estimated.

Table 1. Effects of DMSP and related compounds on the striking behavior of goldfish (37)

Exp.	Compound*	Striking Frequency (No.)	Exp.	Compound*	Striking Frequency (No.)
I	Control	19	IV	Control	164
	DMSP	138		DMT	453
	Methyl(3-methylthio)propanoate	82		DMSP	575
	2-Mercapto-acetic acid	87		Dimethyl-4-buthylothetin	373
	3-Methylthio-propanoic acid	81		Dimethyl-5-pentylothetin	359
	3-Mercapto-propanoic acid	82			
II	Control	44	V	Control	124
	DMSP	314		DMT	386
	3-Methylthio-propanal	109		DMSP	503
	3-Methylthio-propylamine	101		Diethyl-3-propiothetin	340
	3-Methylthio-propanol	129		Methyl ethyl-thetin	323
				Dimethyl 2-methyl-thetin	330
III	Control	186	VI	Control	9
	DMT	553		DMSP	143
	Diethyl thetin	419		Dimethyl sulfide	69
	Dipropyl thetin	360		Methyl cysteine	56
	Dibutyl thetin	350		Methyl methionine	16

***=The compounds were tested at 10^{-3}M.**
The abbreviations: DMT=Dimethylacetothetin; DMSP = Dimethylsulfoniopropionate

Table 2. Effects of DMSP and related compounds on striking behavior of goldfish (37)

Exp.	Compound*	Striking Frequency (No.)	Exp.	Compound*	Striking Frequency (No.)
I	Control	16	III	Control	32
	DMSP	130		DMSP	136
	Dimethyl sulfide	70		Dimethyl sulfide	64
	Diethyl sulfide	58		Dimethyl sulfoxide	113
	Dipropyl sulfide	109		Dimethyl sulfone	92
	Dibutyl sulfide	81		Dimethyl sulfite	57
II	Control	14	IV	Control	1
	DMSP	230		DMSP	71
	Dimethyl disulfide	109		Dimethyl sulfide	50
	Diethyl disulfide	77		Acrylic acid	18
	Dipropyl disulfide	163		Dimethyl sulfide + Acrylic acid	33
	Dibutyl disulfide	112			

***= The compounds were tested 10^{-3}M.**
The abbreviations: Exp. = Experiment; DMSP=Dimethylsulfoniopropionate.
Roman numerals show the separate experiments.

RESULTS

Detection of Highly Active Feeding Attractants

Striking effects elicited by various sulfur-containing compounds were examined by the Kimographyone method. Compounds tested included DMSP and related sulfonium compounds having various alkyl side chains and different methylene carbon chains, dimethylsulfide (DMS), oxidation derivatives of DMS, cyclic sulfur-containing compounds and sulfur containing-amino acids. Results (Tables 1, 2) showed that DMSP was the most effective of the test compounds. This suggested that the methyl groups were the most effective among various alkyl side chains attached to the sulfur atom and that the propyl carbon chain is the most effective among various straight carbon chains, propionates and isopropionic acid. Taurine, which is contained in fish, crustaceans and shellfish (13,43), and betaine found in crustaceans, shellfish and mollusks (4,18), have been reported as attractants for fish and crustaceans, but they were inactive in our experiments (36).

Electrophysiological Experiments with Olfactory Tract of Carp

Electrophysiological experiments with carp (7), red sea bream (7a), yellowtail (16), and rainbow trout (9,44), performed with various test compounds, have shown that glutamine was the most common compound for the olfactory sense of fish. A comparison of effects of DMSP and related sulfonium compounds, and glutamine on the olfactory tract of carp showed that DMSP activated the olfactory sense more strongly than did glutamine in carp (Table 3) (37).

Effects of DMSP on Growth, Thrust, Moving Power and Stress Resistance of Various Organisms

A DMSP-supplemented diet proved to stimulate the growth of a variety of freshwater and marine fish (Figure 1) including, goldfish (26,30,37), carp (26), rainbow trout (27), and marine fish: red sea bream (30, 38), yellowtail (38), and flounder (38). It was shown that betaine and, to lesser extent, dimethylthetin (DMT) were effective growth stimulants for a flounder, Dover sole (18), whereas DMSP was not.

Stimulation effects of DMSP were also found for growth of tadpoles (Brown and Green frog) (unpublished data), salamander (Kasumi salamander) (unpublished data), crustaceans (Striped and Kuruma prawns) (25,31,32), crab (Sawagani) (unpublished data) (35), rats (28), and chickens (unpublished data).

Experiments examining thrust power of fish which ingested various diets were carried out. Results indicated that a DMSP-supplemented diet strengthened thrust power (26, 27,30) (Figure 2) and resistance of test fish against oxygen deficiency, rising temperature (30) and ammonia water (unpublished data).

Running power in a turning wheel and hanging power on to a bar in rats (23, 29) and flapping power of the wing of chickens (unpublished data) were significantly elevated when these organisms were given DMSP supplemented water. Furthermore, in rats which ingested DMSP solution, resistance against stress-induced gastric ulcers (induced by water immersion) was strengthened more than three times compared to those without DMSP (24, 28).

Distribution of DMSP Added to Carp

Incorporation of diet-supplemented DMSP into brain, kidney, heptatopancreas, heart and dorsal muscle of carp was examined. Results indicated that diet-supplemented DMSP

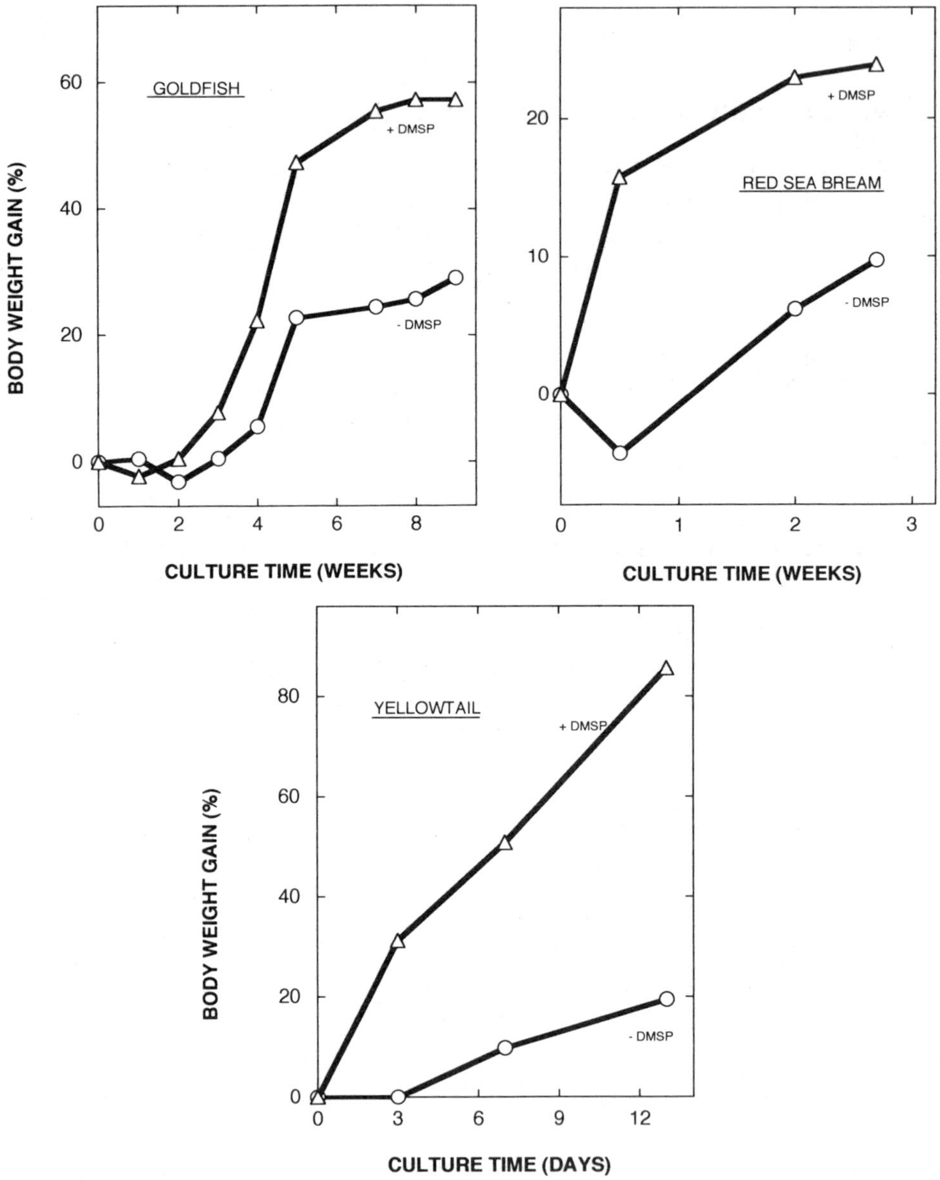

Figure 1. Effects of diet-supplemented DMSP on growth of goldfish (39), red sea bream (38) and yellowtail (38). Mean increase in body weight with time is presented. For experimental conditions, see details in the Materials and Methods section.

accumulated most rapidly, and to the highest levels, in the hepatopancreas as compared to other viscera (Figure 3). This finding suggests that DMSP may not function in the central nervous system, but rather in a detoxification organ, such as the hepatopancreas portion of the liver.

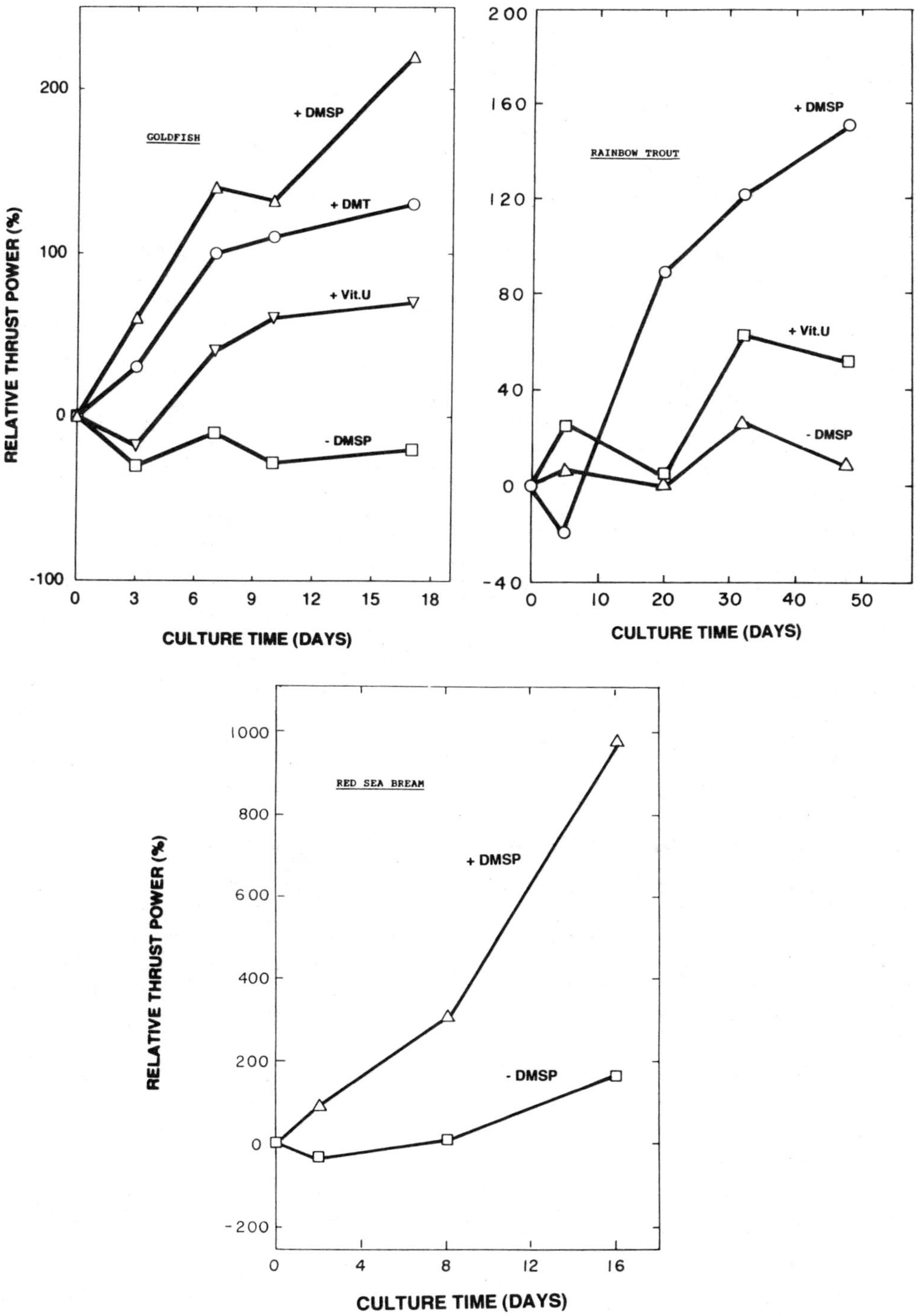

Figure 2. Effects of diet-supplemented DMSP on thrust power of goldfish (26), rainbow trout (27) and red sea bream (26). For experimental conditions, see details in the Materials and Methods section.

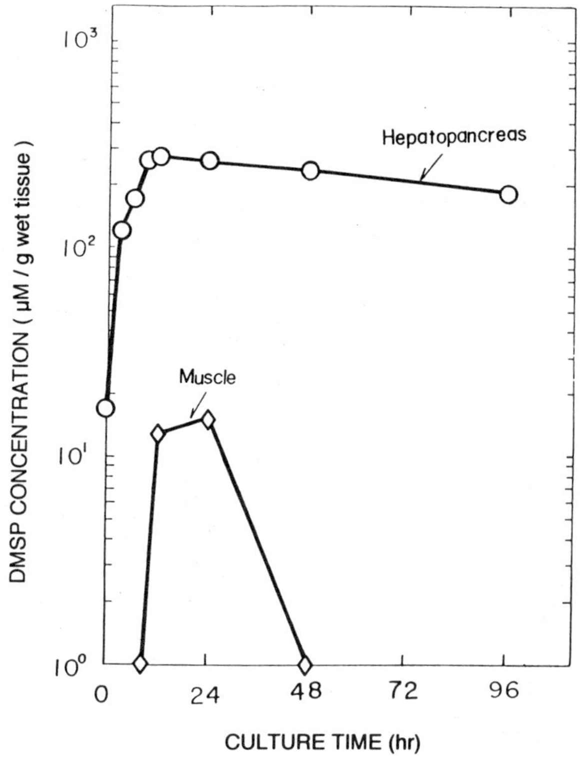

Figure 3. Incorporation of diet-supplemented DMSP into viscera of carp. Diet was prepared in the same way as in growth experiments with DMSP solution at 10^{-3}M, and initially fed to test carp (average body weight 470.4 g) for 60 min.

Effects of DMSP on Methyl Group-Transferring Enzymes

DMSP is structurally quite similar to DMT, with the latter containing an acetic acid rather than propionic acid moiety attached to the sulfur atom. Substrate specificity of DMT-homocysteine (HCYS) methyltransferase (EC. 2.1.1.3.) in acetone extracts of hepatopancreas from red sea bream was examined with DMSP and related sulfonium compounds. Among a number of compounds, DMSP proved to give the highest specific activity for the enzyme (33) (Table 4).

Specific activity of the enzyme in various viscera of carp was examined. Results showed that high activities were detected (decreasing in order) in the gall bladder, hepatopancreas, intestinal tract, spleen, ovary and kidney. No activity was found in the brain or heart (unpublished data).

Effects of Methionine on the Growth, Molt or Metamorphosis of Crustaceans or Amphibians

Effects of methionine at the concentrations of 0.02 to 0.3 mM on the growth, molt or metamorphosis of crustaceans (Striped prawn) or amphibians (tadpole of Brown frog) were examined. Results indicated that the addition of methionine to the rearing stage

Table 3. Effects of sulfur-containing compounds on the olfactory nerve of sedated carp (37)

Compound	Olfactory Response	
β-Dimethylsulfoniopropioate	37.7*	(135)**
Dimethylsulfide	26.6	(95)
Dimethylsulfoxide	29.3	(105)
Dimethylsulfone	30.5	(109)
Diprophyldisulfide	14.5	(52)
Glutatmine	27.9	(100)

***=The area of activated brain waves (μV.sec).**
****=Relative values against 100 of glutamine.**

accelerated the growth and metamorphosis of tadpoles to a lesser extent than did similar concentrations of DMSP (unpublished data).

The supplementation of methionine at 0.05 mM to the rearing water was found to promote the growth of striped prawns (Table 5). The growth stimulation by methionine at 0.05 mM was greater than observed with a similar level of DMSP. Only with DMSP levels of 1 mM was growth stimulated to the extent achieved with 0.05 mM methionine (unpublished observation).

Table 4. Dimethylthetin-homocysteine methyltransferase activities from red sea bream with various substrates (33)

Exp.*	Substrate	S.A.**	Exp.	Substrate	S.A.
I	Dimethylacetothetin	0.47	III	Dimethylacetothetin	0.34
	Diethylacetothetin	0.39		3-Methylthio-1-propanal	-***
	Dipropylacetothetin	0.67		3-Methylthio-1-propanol	-
	Dibutylacetothetin	0.62		3-Methylthio-1-propylamine	0.93
	DMSP	1.62		Methyl-3-methylthiopropanoate	1.21
				3-Methylthiopropanoic acid	1.62
				3-Mercaptopropanoic acid	0.20
				2-Mercaptoacetic acid	0.23
II	Dimethylacetothetin	0.56	IV	Dimethylacetothetin	0.53
	DMSP	1.74		2, 4, 6-Trimethyldihydro-1, 3, 5-dithiazine	0.09
	Dimethylbutylothetin	1.11			
	Dimethylpentylothetin	0.27			
	Dimethylmethylacetothetin	0.83			

***= Experiment**
****=Specific Activity (u mol/mg protein/h)**
*****=Negligible Values**

Table 5. Effect of DMSP and methionine additions on growth of striped prawn during rearing

Concentration (mM)	DMSP* (%)	Methionine* (%)
0.02	36.5	62.4
0.05	116.5	143.5
0.10	116.5	96.5
0.30	116.5	83.5

***=Mean increase in body weight is given as gain in weight after 48 days in relation to control (100%).**

The results outlined here lead to the conclusion that dietary DMSP is ingested and converted to methionine which in turn stimulates the growth of carp and crustaceans, probably by providing a source of methyl groups.

DISCUSSION

Experiments with <1 mM of the test substrates examining growth, body movement, behavior and environmental resistance showed that DMSP was a feeding attractant and promotor of growth and body movement. In addition, DMSP enhanced resistance in mammals, chickens, fish, amphibians and shellfish, acted as an anti-ulcer agent in rats, promoted metamorphosis in amphibians, and stimulated molting in crustaceans (34).

Experiments examining incorporation of DMSP into carp tissue and activity of methyltransferase enzymes suggested that diet-supplemented DMSP is first incorporated into the liver portion of the hepatopancreas. In the hepatopancreas, the methyl group of DMSP was transferred to homocysteine to form methionine via DMT-homocysteine methyltransferase, after which adenosylhomocysteine and S-adenosylmethionine were formed, respectively. This has been confirmed by experiments using various concentrations of DMSP and a mold, *Eremothecium ashbyii* (unpublished data). Thus, the methyl group of S-adenosylmethionine may be use to produce an active methylated compound, for example, a hormone. These compounds elicited various responses in the organisms tested, especially fish and crustaceans. Furthermore, the fact that DMS, a degradation product of DMSP, was not found in DMSP-incorporation experiments and that DMSP does not appear to arbitrarily decompose in fish bodies (11), suggests that DMSP plays some functional role in fish.

The methylation reaction involving DMSP appears to be enzymatic in fish (46). Therefore, the enzymatic methylation reaction of homocysteine to methionine with DMSP as methyl group donor is considered to occur rapidly, especially in fish displaying a high activity of the DMT-homocysteine methyltransferase.

The relationship of DMSP and its metabolites to the molt-stimulating hormone ecdysteroides in crustaceans and to the metamorphosis hormone tyroxine in amphibians and the growth hormone prolactin in amphibians remains quite uncertain. Also, it is not so clear if metabolic DMSP transformations are similar in different organisms. Further experiments are needed to elucidate the mechanisms of DMSP metabolism in these organisms.

ACKNOWLEDGMENTS

I am deeply grateful to Messrs H. Saito, T. Kawase, T. Kawamori, T. Oshikawa, K. Yuasa, H. Kuruma, K. Yabe and A. Ishibashi for their skillful assistance and to Dr. T. Kawai for kind gifts of several sulfonium compounds.

REFERENCES

1. Ackman, R. G., J. Hingley and A. W. May. 1967. Dimethyl-β-propiothetin and dimethylsulfide in Labrador cod. J. Fish. Res. Bd. Can., 24: 457-461.
2. Ackman, R. G. and H. J. Hingley. 1968. The occurrence and retention of dimethyl-β-propiothetin in some filter- feeding organisms. J. Fish. Res. Bd. Can. 25: 267-284.
3. Ackman, R. G., J. Hingley and T. Maskey. 1973. Dimethyl sulfide as an odor components in Nova Scotia fall mackerel. J. Fish. Res. Bd. Can., 29: 1085-1088.
4. Carr, W. E. 1973. Chemoreception in the shrimp, *Palaemonetes pugie*: the role of amino acids and betaine in elicitation of feeding response by extraction. Comp. Biochem. Physiol. 61A: 127-131.
5. Challenger, F. and M. I. Simpson. 1948. Studies on biological methylation, Part XII. A precursor of the dimethyl sulfide evolved by *Polysiphonia fastigiata*. Dimethyl-2-carboxyethyl sulfonium hydroxide and its salts. J. Chem. Soc. 1948: 1591-1597.
6. Dickson, D. M., R. G. Wyn Jones and J. Davenport. 1982. Steady state osmotic adaptation in *Ulva lactuca*. Planta 155:409-415.
7. Goh, Y. and T. Tamura. 1978. The electrical responses of the olfactory tract of amino acids in carp. Bull. Jap. Soc. Sci. Fish. 44: 341-344.
7a. Goh, Y. and T. Tamura, H. Kobayashi. 1979. Olfactory responses to amino acids in marine telosts. Biochem. Physiol. 62A: 863-868.
8. Hara, T. J. 1981. Structure-activity relationships of olfactory stimulants for fish. Pp. 48-62, In: Nippon Suisan Gakkai (Ed.), Chemical Sense of Fish and Feeding Stimulants. Koseisha Koseikaku, Tokyo.
9. Hara, T. J., Y. M. Carilina and B. R. Hobden. 1973. Comparison of the olfactory responses to amino acids in rainbow trout, brook trout and white fish. Comp. Biochem. Physiol. 45: 969-977.
10. Iida, H., K. Nakamura and T. Tokunaga. 1985. Dimethyl sulfide and dimethyl-β -propiothetin in sea algae. Bull. Jap. Soc. Sc. Fish. 51: 1145-1150.
11. Iida, H., J. Nakazoe, H. Saito and T. Tokunaga. 1986. Effect of diet on dimethyl-β-propiothetin content in fish. Bull. Jap. Soc. Sci. Fish. 52: 2155-2161.
12. Iida, H. and T. Tokunaga. 1986. Dimethyl sulfide and dimethyl-β-propiothetin in shellfish. Bull. Jap. Soc. Sc. Fish. 52: 557-563.
13. Ikeda, S., S. Kawai, M. Sakaguchi, M. Sato, Y. Makindoan, R. Yoshinaka and Y. Yamaoto. 1981. Trace elements in aquatic organisms (in Japanese). Koseisha Koseikaku, 111 pp.
14. Karsten, U., C. Wiencke and G. Kirst. 1991. The effect of salinity changes upon the physiology of eulittoral green macroalgae from Antarctica and Southern Chile. J. Exp. Bot. 42: 1533-1539.
15. Karsten, U., C. Wiencke and G. O. Kirst. 1990. The β-dimethysulphoniopropionate (DMSP) content of macroalgae from Antarctica and Southern Chile. Botanica Marina. 33: 143-146.
16. Kobayashi, H. and K. Fujiwara. 1987. Olfactory responses in the yellowtail, *Seriola quinqueradiata*. Bull. Jap. Soc. Sci. Fish. 53: 1717-1725.
17. Larher, F., J. Hamelin and G. R. Stewart. 1977. L'acide dimethylsulfonium-3- propanoique de Spartina anglica. Phytochemistry. 16: 2019-2020.
18. Mackie, A. M. 1973. The chemical basis of food detection in the lobster, *Homarus gammurus*. Mar. Biol. 21: 103-106.
19. Mackie, A. M. and A. J. Mitchell. 1982. Further studies on the chemical control of feeding behavior in the Dover sole, *Solea solea*. Comp. Biochem. Physiol. 73: 89-93.
20. Mason, T. and G. Blunden. 1989. Quaternary ammonium and tertiary sulfonium compounds of algal origin as alleviators of osmotic stress. Bot. Mar. 32: 313-316.
21. Nakajima, K. 1987. New determination method for striking behavior of fish. Bull. Koshien Univ. 15: 13-17.
22. Nakajima, K. 1988. Analysis of the biting behaviors of goldfish by the methods with a Kimographyone instrument and a series of electric equipments. Bull. Koshien Univ. 16: 1-6.

23. Nakajima, K. 1989. Effects of high concentrations of dimethyl-β- propiothetin and vitamin U on young rat. Bull. Koshien Univ. 17A: 1-8.

24. Nakajima, K. 1990. Preventive effects of dimethylthetin, dimethyl-β-priopiothetin and vitamin U on stress-induced gastric ulcer in rats. Bull. Koshien Univ. 18A: 15-22.

25. Nakajima, K. 1991. Dimethyl-β -propiothetin, a potent growth and molt stimulant for striped prawn. Bull. Koshien Univ. 20A: 7-12.

26. Nakajima, K. 1991. Effects of diet-supplemented dimethyl-β-propiothetin on growth and thrust power of goldfish, carp and red sea bream. Nippon Suisan Gakkaishi. 57: 673-679.

27. Nakajima, K. 1991. Effect of dimethyl-β-propiothetin on growth and thrust power of rainbow trout. Bull. Jap. Soc. Sci. Fish. 57: 1603

28. Nakajima, K. 1991. Dimethyl-β-propiothetin, new potent resistive agent against stress- induced gastric ulcers. J. Nutr. Sci. Vitaminol. 37: 229-238.

29. Nakajima, K. 1991. An activating agent, dimethyl-β-propiothetin, for young rat. Bull. Koshien Univ. 19A: 29-36.

30. Nakajima, K. 1992. Activation effect of a short term of dimethyl-β-propiothetin supplementation on of goldfish and rainbow trout. Bull. Jap. Soc. Sci. Fish. 58: 1453-1458.

31. Nakajima, K. 1992. Effect of various temperatures and organic acids on growth, molt and survival of striped prawn. Bull. Koshien Univ. 20A: 7-12.

32. Nakajima, K. 1993. Effect of sulfonium compound, dimethyl-β-propiothetin, on growth, molt and survival of Kuruma prawn. Bull. Koshien Univ. 21A: 11-15.

33. Nakajima, K. 1993. Dimethylthetin- and betaine-homocysteine methyltransferase activities from livers of fish and mammals. Bull. Jap. Soc. Sci. Fish., 59: 1389-1393.\

34. Nakajima, K. 1993. Combination effect of eyestalk ablation and a sulfonium compound, dimethyl-β -propiothetin, on molt of striped prawn. Bull. Koshien Univ. 21A: 3-9.

35. Nakajima, K. and J. Ito. 1993. Sulfur-containing feeding stimulants for shells. Jpn. Tokyo No. 1733003, February 17.

36. Nakajima, K., A. Uchida and Y. Ishida. 1988. Effects of hetero cyclic sulfur-containing compounds on the striking response of goldfish. Bull. Koshien Univ. 16: 7-10.

37. Nakajima, K., A. Uchida and Y. Ishida. 1989. A new feeding attractant, dimethyl-β-propiothetin, for freshwater fish. Bull. Jap. Soc. Sci. Fish. 55: 1291.

38. Nakajima, K., A. Uchida and Y. Ishida. 1989. Effects of supplemental dietary feeding attractants, dimethyl-β-propiothetin, on growth of goldfish. Bull. Jap. Soc. Sci. Fish. 56: 1151-1154.

39. Nakajima, K., A. Uchida and Y. Ishida. 1990. Effect of a feeding attractant, dimethyl- -propiothetin, on growth of a marine fish. Bull. Jap. Soc. Sci. Fish. 56: 1151-1154.

40. Pakulski, J. D. and R. P. Kiene. 1992. Foliar release of dimethylsulfoniopropionate from *Spartina alterniflora*. Mar. Ecol. Prog. Ser., 81: 277-287.

41. Reed, R. H. 1983. Measurement and osmotic significance of β-dimethylsulfoniopropionate in marine macroalgae. Mar. Biol. Lett. 4: 173-181.

42. Reed, R. H. 1983. Measurement and osmotic significance of dimethylsulfoniopropionate in marine microalgae. Mar. Biol. Ecol. 68: 169-193.

43. Sakaguchi, M. and M. Murata. 1988. Components in extracts from aquatic organisms. In: A Series of Scientific Fishery (in Japanese), Koseisha Koseikaku 72: 56-65.

44. Sato, M. and K. Ueda. 1975. Spectral analysis of olfactory response to amino acids in rainbow trout, *Salmo gairdneria*. 52A: 359-365

45. Tokunaga, T., H. Iida and K. Nakamura. 1977. Formation of dimethyl sulfide in Antarctic krill, *Euphausia superba*. Bull. Jap. Soc. Sci. Fish. 43: 1209-1217.

46. Van Diggelen, J., J. Rozema, D. M. Dickson and R. Broekman. 1986. 3-Dimethyl sulfonium propionate, proline and quaternary ammonium compounds in *Spartina anglica* in relation to sodium chloride, nitrogen and sulfur. *New Phytol., 103: 573-586.*

47. Yamauchi, K., T. Tanabe and M. Kinoshita. 1979. Trimethylsulfonium hydroxide, a new methylating agent. J. Org. Chem. 44: 638-639.

16

THE ROLE OF DMSP AND DMS IN THE GLOBAL SULFUR CYCLE AND CLIMATE REGULATION

G. Malin

School of Environmental Sciences
University of East Anglia
Norwich, NR4 7TJ, United Kingdom

SUMMARY

In 1972 Lovelock et al. published evidence of the ubiquity of DMS in surface seawater and proposed that marine DMS was the key compound transferring sulfur from the sea to the land via the atmosphere. At that time biochemical data were already available which suggested that DMSP could be the precursor of DMS in marine ecosystems. In the intervening years thousands of DMS and DMSP measurements have been made in coastal, shelf and open ocean waters and much more is now known about the water column processes which lead to DMS production. This paper will concentrate on the regional and global aspects of DMSP, which result from its role as the major precursor of DMS. DMS is now known to have 3 major environmental roles: it is the dominant volatile compound in the global sulfur cycle, the products of its atmospheric oxidation are acidic and therefore affect the acid-base balance of aerosols and rainwater, and aerosol particles derived from DMS are efficient cloud condensation nuclei (CCN) which have climatic significance. I will focus on the marine environment and discuss the sea-to-air exchange process, aspects of the atmospheric chemistry of DMS, the role of DMS in the sulfur cycle and its climatic significance.

ESTIMATING THE SEA-AIR FLUX OF DMS

Estimating the significance of DMS in the sulfur cycle necessitates quantification of sea-to-air fluxes, as well as anthropogenic inputs of sulfur to the atmosphere, on regional and global scales. DMS fluxes cannot be determined directly, so emissions are generally derived from field measurements of DMS concentration, temperature, salinity and wind-speed, using an appropriate transfer velocity model. The rate at which a gas is transferred to the atmosphere depends in part on the concentration difference between air and water. In turn the water concentration results from the balance between production, transformation and utilisation processes occurring in surface seawater. Air-sea fluxes are also governed by

Biological and Environmental Chemistry of DMSP and Related Sulfonium Compounds
edited by Ronald P. Kiene et al., Plenum Press, New York, 1996

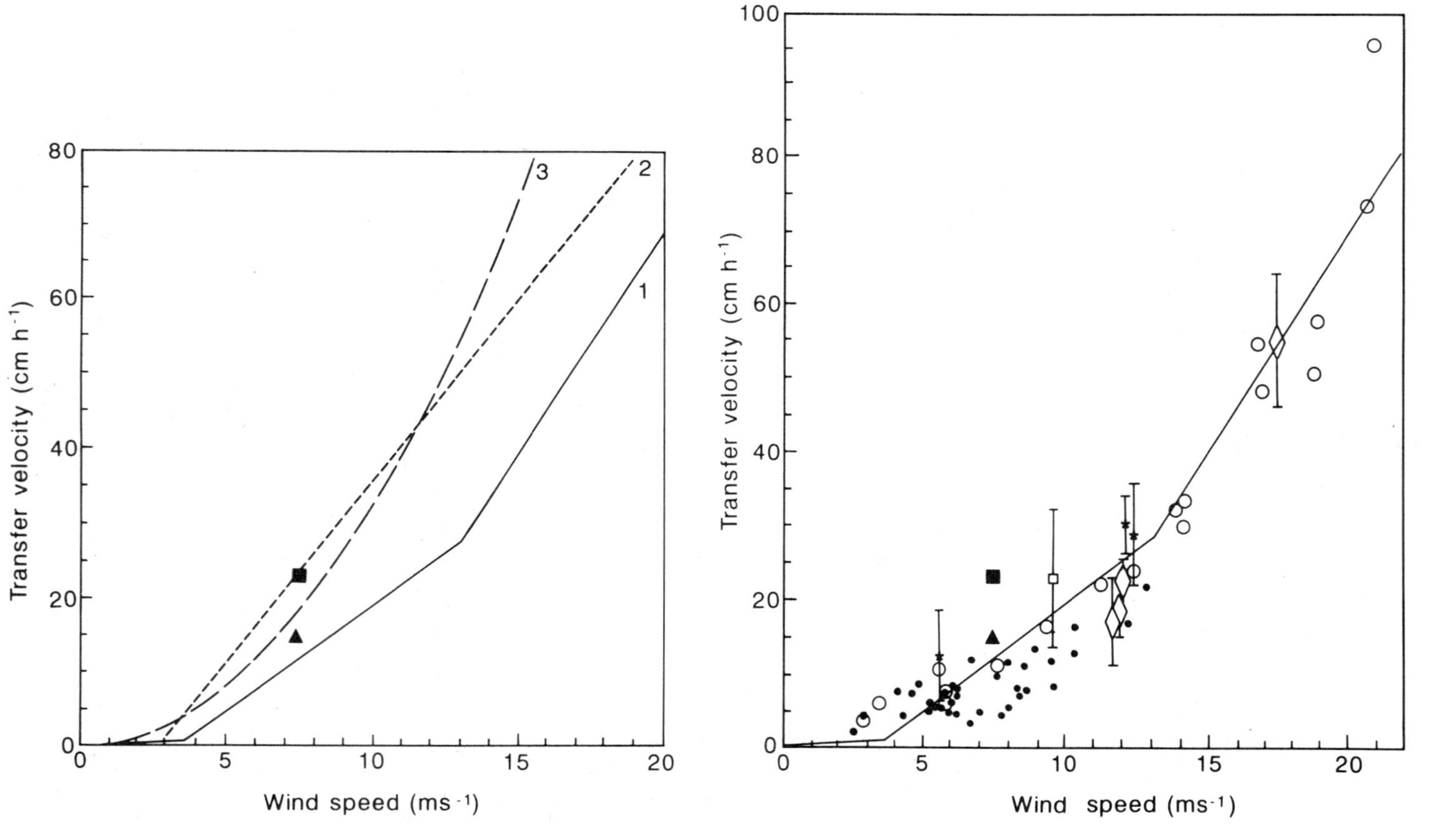

Figure 1. a) Models for air-sea transfer velocity of CO_2 as a function of windspeed at a height of 10 m, by 1. Wanninkhof (71), 2. Smethie et al. (66) as modified by Tans et al. (67), and 3. Liss and Merlivat (41). b) Data from ○ wind tunnel experiments (72), ● a lake SF_6 experiment (70), ◇ dual-tracer SF_6/ ^3He (73) and triple-tracer ❑ SF_6/spore and ★ ^3He/spore deployments (51) in the North Sea. The solid line is the Liss and Merlivat (41) relationship. In both figures, ■ is the global mean 'bomb' $^{14}CO_2$ transfer velocity (12, 71) and ▲ the global mean ^{222}Rn value (54), and all data are normalised to a Schmidt number of 600, corresponding to CO_2 at 20°C.

Table 1. Equations used to calculate DMS transfer velocity for the 3 surface condition / windspeed regimes defined by Liss and Merlivat (40)

Surface Condition	Windspeed Range u ($m\ s^{-1}$)	DMS Transfer Velocity K_{DMS} ($cm\ h^{-1}$)
Smooth	$u \leq 3.6$	$1.17\ u\ (Sc_{DMS}/600)^{-2/3}$
Rough	$3.6 < u \leq 13$	$2.85\ u - 9.65\ (Sc_{DMS}/600)^{-1/2}$
Breaking wave	$u > 13$	$5.9\ u - 49.3\ (Sc_{DMS}/600)^{-1/2}$

the Henry's Law constant and the interfacial transfer rate which depends, *inter alia*, on windspeed and for some gases, but not DMS, their chemical reactivity in the water (39). As a result of rapid dilution and oxidation of DMS in the air, atmospheric DMS concentrations are very low relative to those in seawater, and are generally insignificant for the purpose of flux calculations (see 59). Liss and Merlivat (41) summarised understanding of the processes controlling air-sea exchange of trace gases, and proposed a three-line relationship to describe transfer velocity as a function of windspeed (Fig. 1). This approach yields 3 equations describing the relationship between the water phase transfer velocity K_w ($cm\ h^{-1}$) and windspeed u ($m\ s^{-1}$) at a height of 10 metres, when the water surface is smooth, rough or has breaking waves and bubble formation (Table 1). The importance of high quality windspeed data for calculating fluxes is discussed by Turner et al. (69) and Nightingale et al. (51). Other relationships have also been proposed which are in closer agreement with the point for the globally derived bomb ^{14}C estimate of transfer velocity (67, 71 see Fig.1). It should be noted that fluxes calculated using these alternative models would give values up to 60% higher than those generated using the Liss and Merlivat approach.

In recent years considerable effort has gone into field studies designed to validate gas exchange models. The basic approach is to add 2 inert gaseous tracers, sulfur hexafluoride (SF_6) and the isotope ^3He, to a patch of seawater and to measure the change in their concentration ratio with time (described in 42, 73). The dual-tracer technique gives data for *in situ* transfer velocity over periods of tens of hours, and allows study of the effects of transient features such as storms on gas exchange. During these earlier studies a conservative nonvolatile tracer, which was chemically stable, non-toxic and easy to measure at low concentration was not available. However, in a recent triple-tracer study, we deployed *Bacillus globigii* spores in order to derive dispersion corrections for the volatile tracers (51). The inclusion of a conservative tracer is a considerable advantage in that it allows direct derivation of transfer velocity values for SF_6 and ^3He. Bacterial spores are metabolically inactive in seawater and have better detection limits than commonly used chemical tracers. They have been used for tracing sewage dispersion in coastal areas (55) and the movement of microbes in groundwater systems (32), but this represented their first use in open seawater and in the context of sea-air gas exchange.

The results of the dual and triple tracer studies support the wind speed dependence of transfer velocity proposed by Liss and Merlivat (41), thus reducing the uncertainty inherent in calculating sea-air exchange of biogenic trace gases (Fig. 1). The discrepancy with the bomb ^{14}C transfer velocity estimate is still to be resolved. However, the recent study by Hesshaimer et al. (26) suggested that the estimated ocean uptake of bomb ^{14}C should be reduced by 25%, which would bring the outlying ^{14}C point on Fig.1 somewhat closer into agreement with the Liss and Merlivat model. This has direct implications for studies on the global carbon cycle, in particular the global oceanic sink for anthropogenic CO_2. It is also of interest to note that free carbonic anhydrase in seawater might enhance CO_2 transfer and go towards explaining the remaining difference. This possibility is currently under investigation in our research group.

BIOGENIC GASES AND THE SEA-AIR INTERFACE

When considering the sea-air exchange of volatile compounds such as DMS a degree of uncertainty still remains since it is not certain that inert tracer gases behave in the same way as biogenic trace gases. In this respect the microlayer, which is defined as the top few microns of the sea surface (38), may be a particular problem. The concentration difference term used in flux calculations is invariably the concentration in a water sample taken from below the surface with a conventional water bottle, or via a pumped seawater supply taken from between 1 and 5 metres depth. Sampling the microlayer for trace gas analysis is technically difficult and very few reports have appeared in the literature. Nguyen et al. (52) found between 1.2 and 5.3 times as much DMS in microlayer samples, collected with a Garret screen, compared to samples collected below the interface. If such levels of enrichment were common DMS emissions would be considerable underestimates. However, other studies have revealed no consistent enrichment of DMS in microlayer samples (3), or concluded that simpler sampling devices, such as the Garret screen, perturbed the biota resulting in significant increases in DMS concentrations particularly during phytoplankton blooms (68).

There is clear evidence that bacteria strongly influence the production, transformation and utilisation of organic sulfur compounds in the marine biogeochemical cycle of DMS (see Taylor and Visscher, this volume). Microbiological studies on microlayer samples suggest that it represents a unique microbial habitat, and this is reflected in a bacterial flora which can be considerably different to that observed in the bulk water (45). Investigations designed to determine whether microorganisms at the air-sea interface influence emissions of DMS, and other biogenic trace gases, to the atmosphere would be worthwhile. In the future such studies might incorporate the use of immunochemical or molecular probes directed against enzymes involved in e.g. demethylation of DMSP or DMS oxidation.

It is also interesting to note that high molecular weight surfactants, including compounds produced by marine phytoplankton, have been shown in laboratory experiments to reduce gas exchange at the air-sea interface by up to 50% (21, 22). It is likely that such compounds are also produced in phytoplankton blooms.

ATMOSPHERIC CHEMISTRY OF DMS

Following emission to the atmosphere DMS undergoes a series of oxidation reactions which lead to the formation of a number of products including methanesulfonic acid (CH_3SO_3H, MSA) and sulfur dioxide (SO_2), with subsequent formation of acidic sulfate aerosol (56). Other products include dimethyl sulfoxide (CH_3SOCH_3, DMSO) and dimethyl sulfone ($CH_3SO_2CH_3$, $DMSO_2$). The atmospheric DMS oxidation mechanism proposed by Koga and Tanaka is shown in Fig. 2 (34). There is general agreement that DMS oxidation is initiated by reaction with hydroxyl radicals during the day and nitrate radicals at night. The daytime reaction of DMS with OH is the major process in unpolluted air, and at the global scale it essentially controls atmospheric DMS concentration. The reaction with NO_3 is less important although it prevents DMS accumulation at night and hence reduces the maximum-to-minimum concentration ratio (34). There are 2 major reaction pathways; hydroxyl or nitrate radicals either abstract a hydrogen atom leading to formation of SO_2, or the reaction goes via an OH^- addition pathway, in which case MSA is the dominant product (Fig. 2). As NO_x concentration increases the proportion of MSA seems to be enhanced (79, 80), and MSA is a much more significant product under certain environmental conditions, such as in Antarctica where low temperature favours the OH^- addition pathway (9, 29). Considerable

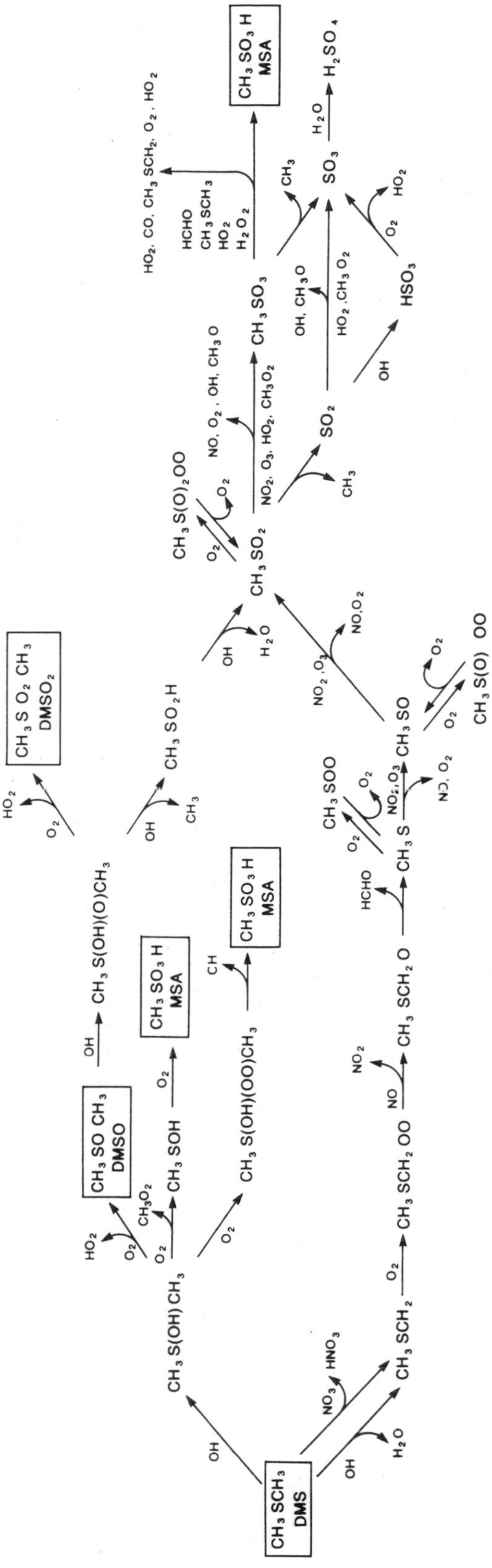

Figure 2. Atmospheric oxidation reacticn pathways of DMS from Koga and Tanaka (34).

uncertainty remains regarding the relative yields of the products, the reactions controlling the SO_2/MSA branching ratio and whether laboratory results are applicable in the field.

It has been suggested that DMS could react with IO radicals derived from photolysis of methyl iodide resulting in quantitative DMSO formation (6), but more recent studies have shown that IO concentrations are low and the reaction too slow to be of significance for the atmospheric chemistry of DMS (5, 18). However, the rate coefficient for the reaction of DMS with BrO suggests that it could be a sink for DMS in the Arctic atmosphere in spring when BrO concentrations are high (5).

The Global Sulfur Cycle

Early estimates of the quantity of sulfur emitted from the oceans were in the range 1 to 9 Tmol per annum and were based on the amount of additional sulfur required to balance the global budget (19, 23). The first estimate based on DMS field data was reported by Liss and Slater (40) who derived a flux of 0.23 Tmol per annum from the Lovelock et al. data set (44). Subsequently, estimates of approximately 1Tmol per annum were calculated using larger DMS data sets from the Atlantic and Pacific Oceans (1, 2, 4), and taking into account the strong seasonality of biogenic DMS production resulted in a flux of 0.5 Tmol per annum (7). A question mark remains over the value for DMS diffusivity used for flux calculations. Until recently the calculated value proposed by Wilke and Chang (77) and updated by Hayduk and Laudie (24) was generally used. However, Saltzman and coworkers have now reported DMS diffusivity values measured in the laboratory which are considerably lower than the calculated values (58, 60). The updated diffusivity figures serve to slightly reduce the magnitude of the biogenic flux.

During the past century anthropogenic sulfur emissions have seriously perturbed the natural biogeochemical sulfur cycle. These inputs dominate globally (Table 2), representing over 90% of the total input in the northern hemisphere (15). However, global sulfur budgets do not convey the significance of natural sulfur fluxes at the hemisphere or regional level. For example DMS emissions can account for most of the non-sea salt sulfate in the atmosphere over vast regions of the remote open oceans (62). Bates et al. (8) addressed this problem by reassessing sulfur emissions, taking into account the seasonality of natural production and refining estimates on a regional basis. Their data are summarised in Table 2. Terrestrial emissions were rather minor, but estimates were based on enclosure technique measurements which have high levels of uncertainty, and it was thought that an improved data base could increase the predicted emissions as much as 2-fold. Biogenic emissions (marine + land) accounted for 8% and 45% of the total sulfur flux in the northern and southern hemispheres respectively, and 15% on a global basis. However, natural emissions were a much more significant fraction for specific

Table 2. Sulphur emissions from natural and anthropogenic sources in units of Gmoles S per annum. Biogenic is equivalent to marine + land and natural is marine + land + volcanic sources, as a percentage of the overall total (from 8)

	Marine	Land	Volcanic	Biomass burning	Anthro-pogenic	% biogenic	% natural
Northern hemisphere	200	7	210	38	2200	8	16
Southern hemisphere	280	4	83	31	240	45	58
Global	480	11	293	69	2440	15	24

areas e.g. 30-100% for the tropical latitudes of the northern hemisphere between the equator and 20°N.

MSA AND STABLE SULFUR ISOTOPE RATIOS

Emissions of DMS, volcanic sulfur and anthropogenic SO_2 all lead to the formation of sulfate aerosol, but only DMS oxidation yields MSA. Thus this compound is regarded as an unambiguous indicator of DMS emissions. Quantification of MSA in aerosol and rainwater samples gives information regarding the influence of biogenic sulfur on adjacent landmasses. For instance, for 2 relatively remote European sampling stations a clear seasonal cycle in MSA concentration was observed (48). Whung et al. (74) investigated MSA levels in a shallow Greenland ice core covering the last 200 years. The observed trend contrasted with that for non-sea-salt sulfate, showed no correlation with sea surface temperature for the last 100 years and there was no evidence for increased yield in MSA with increased atmospheric NO_x concentrations. MSA concentrations increased from the mid 1700's to 1900 (3.01 to 4.1 ppb), and then steadily decreased to 2.34 ppb more recently. They concluded that 15-40% of the sulfur in the newer ice was of biogenic origin.

Analysis of ice cores can also provide evidence for biogenic sulfur emissions over much longer time frames. Antarctic ice cores covering 160,000 years before the present time, show elevated concentrations of MSA and sulfate during the last ice age (36, 37, 57). This may have resulted from enhanced oceanic productivity (61), or changes in the phytoplankton assemblage favouring DMS-producing species. However, changes in atmospheric circulation and snow accumulation patterns cannot be ruled out (57).

As discussed previously, the SO_2/MSA branching ratio varies in a way that is not adequately understood at present. An alternative approach is to assess the impact of volatile biogenic sulfur relative to anthropogenic sulfur by stable sulfur isotope analysis. This technique relies on different sources having characteristic isotopic signatures. Existing information shows that the isotopic signature of DMS-derived sulfur is about +20‰ (13, 53), whereas for anthropogenic sources such as power plants the value is 0-5‰ (50). Analysis of stable sulfur isotopes for samples collected at Mace Head in Eire, and at Ny Alesund, Spitsbergen show that over an annual cycle, 15 to 20% of the sampled aerosol sulfur was derived from a marine biogenic source, and the level increased to approximately 30% in spring and summer (48). The biogenic sulfur in these samples almost certainly originated from phytoplankton populations in the northeast Atlantic, an example being the large blooms of *Emiliania huxleyi* which are annual features in this area (46).

GLOBAL CHANGE AND DMS EMISSIONS

The major control on the quantity of biogenic sulfur emitted from the oceans to the atmosphere lies at the level of the marine water column production, consumption and transformation of DMS. Thus environmental changes which affect the marine biogeochemical cycle of DMS may also influence the pool of DMS available for sea-to-air exchange. There is evidence of eutrophication in coastal areas in many parts of the world. Increased levels of nitrate and phosphate relative to silicate can alter the structure of phytoplankton communities, such that flagellates, which generally produce higher cellular DMSP levels, are favoured above diatoms which commonly contain lower levels of DMSP (31, 43). For the North Sea there is also evidence for increase in total algal biomass and the duration of bloom events with elevated nutrient concentrations (for discussion see 35).

Depletion of stratospheric ozone has led to global scale increases in UV-B radiation, which can also alter species composition and decrease overall primary productivity (78). Interestingly, in the context of this paper, some of the most significant DMS producing types of phytoplankton such as dinoflagellates and *Phaeocystis pouchetii* seem better able to cope with increased UV-B levels (14, 30, 47). However, a recent study on diatom populations in fjords in the Vestfold Hills in Antarctica, revealed no changes in species composition over a 20 years period of increasing exposure to UV-B radiation (49). The authors anticipated that their data would be relevant to other coastal regions of Antarctica where the period of ice cover and timing of the phytoplankton bloom might reduce the detrimental effects of UV-B. In addition, Herndl et al. (25) demonstrated that UV-B could suppress bacterially mediated cycling of organic matter by up to 40% in the upper 5 m of the water column in nearshore waters and down to 10 m in oligotrophic areas. Since it is clear that bacterial transformation and utilisation of DMS are important pathways in the biogeochemical DMS cycle (see Taylor and Visscher, this volume), this could be a significant factor affecting DMS emissions.

Anthropogenic sulfur emissions are declining in many parts of western Europe and the U.S.A. due to a combination of fuel change, increases in efficiency, economic recession and legislative control policies. However, dramatic increases in energy use are anticipated for much of the developing world in particular the tropics, India and China (11). Interestingly, it has been noted that changes in the spatial pattern of SO_2 emissions, away from Europe and the U.S.A. to one centred on eastern Asia, could significantly alter the pattern of climate change (76). Against this background it will be necessary to improve and update current estimates of natural sulfur emissions on both regional and global scales.

CLIMATIC SIGNIFICANCE OF DMS

As previously discussed DMS is oxidised in the atmosphere to form sulfate aerosol. These particles can alter the global radiation budget directly by absorbing and scattering incoming radiation, and indirectly since they are the major source of cloud condensation nuclei (CCN) in areas remote from land influence. In a brief communication Shaw (64) highlighted the connection between biologically produced aerosols and climate, and Charlson et al (16) further proposed that the link between DMS, sulfate aerosol, CCN, cloud albedo and radiation balance could represent a biological climate regulating mechanism (Fig. 3). Concerning the magnitude of this effect; it has been calculated that a 30% change in CCN over the oceans which influenced the formation of marine stratiform clouds, would alter global heat balance by ± 1 Wm^{-2}. This level is highly significant when compared to the forcing of $+1.2$ Wm^{-2} that results from increased CO_2 (15, 16)

The hypothesis provoked considerable comment and the debate continues. Schwartz (63) contested that anthropogenic SO_2 levels in the northern hemisphere do not appear to affect temperature, rate of warming or cloud albedo, and hence it was unlikely that marine DMS emissions would play such a role. In contrast, Wigley (75) found that hemispheric temperature data *did* suggest a slower warming rate north of the equator. Further supporting evidence comes from data acquired by satellite, which were used to demonstrate that variability in low level cloud albedo in the central north Atlantic was related to chlorophyll concentration and sea surface temperature (20). They also found that anthropogenic emissions from the U.S.A. increased albedo to the immediate east of the landmass. Increased cloud reflectance has also been documented in the vicinity of shiptracks, and it is thought that this largely results from sulfur emitted with exhaust gases from ship stacks (17). Important considerations are that anthropogenic sulfur is emitted from point sources and largely deposited over land where there is no shortage of CCN, whereas DMS production occurs throughout aquatic marine environments where it represents a widely distributed

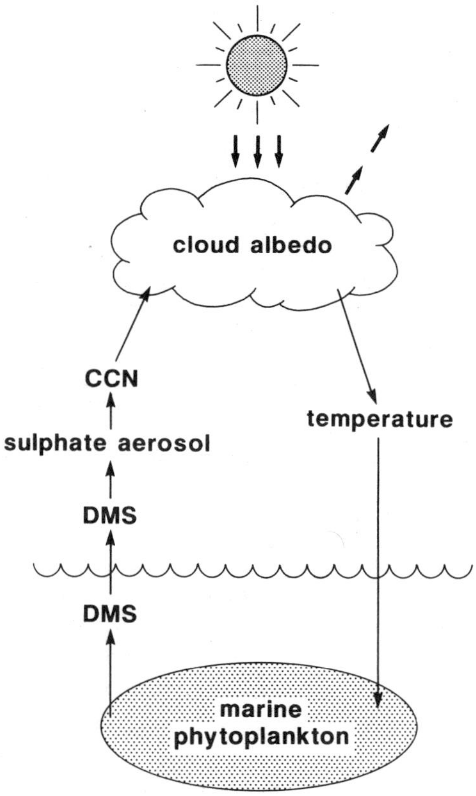

Figure 3. Schematic depicting the DMS-cloud-climate hypothesis. The central idea is that increased temperature would lead to increased DMS emissions from seawater, sulphate aerosol, CCN and cloud albedo. Increased albedo would then increase the amount of solar radiation reflected back into the upper atmosphere away from the earth. This would serve to counteract the initial temperature increase.

source of CCN. DMS-derived CCN are of particular relevance for the remote oceans where there are few alternative CCN sources. Furthermore, cloud albedo is greatest in the shallow stratiform clouds which tend to be common over the oceans and therefore contribute significantly to albedo on the global scale (65).

The central assumption in the DMS-climate hypothesis is that increased temperature would lead to increased DMS emissions. Testing this in the laboratory, in a way which is relevant to the natural environment, is not a trivial task. In open seawater any temperature change would be rather slow, occurring over periods of time representing thousands of generations at the individual phytoplankton cell level. These relatively gradual temperature changes might also lead to some species being at a slight competitive advantage compared to others which could result in changes in the species assemblage, and thus in a shift in the ability of the microalgal population as a whole to produce DMSP/DMS. Furthermore, temperature changes would affect the whole microbial community in seawater and would not necessarily result in a straightforward shift in DMSP/DMS production. Altered temperature could also lead to changes in ocean circulation and/or upwelling thus altering nutrient availability. Evidence from ice cores is, at face value, contrary to the DMS-climate hypothesis, since it suggests increased DMS production in the cooler glacial ocean (36, 37, 57). Although ocean productivity was higher during past glacials this appears to have been due to greater abundances of diatoms (61), which now appear to be relatively poor DMSP producers. However, Holligan (28) warned that ice core records should be interpreted with great caution since they may largely reflect local changes in sulfur deposition.

Nevertheless, despite the equivocal evidence for the feedback process there are several lines of evidence which support a strong relationship between DMS emissions and cloud albedo:

- Most of the particles in clean marine air are composed of non-sea-salt- sulfate or MSA derived from the atmospheric oxidation of DMS (62).
- These sub-micron particles are efficient CCN (62).
- In remote environments there is a clear relationship between atmospheric DMS concentrations and CCN abundance (27).
- CCN lead to formation of cloud drops and reflectivity of marine stratiform clouds increases with increased drop concentration.
- Coherence between CCN concentration and cloudiness has been documented using satellite data, which strongly suggest that DMS emissions can influence cloud radiative transfer properties (10).

Despite the strong link between DMS emissions and cloudiness, the feedback part of the DMS-climate hypothesis remains an unresolved problem. As stated recently by Charlson (15) it is necessary to have a much greater understanding of the microbial ecology which results in emission of DMS to the atmosphere. Available data suggest that the amount of DMS emitted to the atmosphere is a very small fraction of the marine pool of DMS (e.g. 33). Anything that significantly changes the magnitude of this DMS flux has the potential to alter climate.

CONCLUSIONS

A great deal has been learnt about DMS in the decades since it was first proposed as the volatile compound transferring sulfur from marine aquatic environments to the land by way of the atmosphere. It is clear that the marine DMS cycle has both regional and global roles, but many challenges remain for the years ahead. Research at the cellular and molecular level should lead to a greater understanding of the links between environmental factors, water column processes and DMS emissions. Further studies on the sea-to-air gas exchange process should reduce uncertainties inherent in current flux estimations. Whilst work on the detailed atmospheric chemistry of DMS and the ice core and paleogeological record should help in resolving its climatic role. Information which will allow us to predict how the system will respond to perturbation is of prime importance.

ACKNOWLEDGMENTS

I wish to thank my colleagues Peter Liss, Sue Turner, Phil Nightingale, Angela Hatton, Wendy Broadgate and Nicola McArdle for their ongoing help, advice and encouragement. Additionally, special thanks are due to Gary Creissen for his support and help and Philip Judge for preparing figures. Figure 2 appears by permission of Kluwer Academic Publishers. The research was supported by grants from the NERC and the EEC.

REFERENCES

1. Andreae, M. O. 1986. The ocean as a source of atmospheric sulfur compounds, p. 331-362. *In* P. BuatMenard (ed.), The Role of AirSea Exchange in Geochemical Cycling. Reidel, Dordrecht.
2. Andreae, M. O. 1990. Ocean-atmosphere interactions in the global biogeochemical sulfur cycle. Mar. Chem. 30: 1-29.
3. Andreae, M. O. , W.R. Barnard and J.M. Ammons. 1983. The biological production of dimethylsulfide in the ocean and its role in the global atmospheric sulfur budget. Env. Biogeochem. Ecol. Bull. 35: 167-177.
4. Andreae, M.O. and H. Raemdonck. 1983. Dimethylsulfide in the surface ocean and the marine atmosphere: a global view. Science 221: 744-747.
5. Barnes, I. 1993. Overview and atmospheric significance of the results from laboratory kinetic studies performed within the CEC project "OCEANO-NOX", p. 223-237. *In*: G. Restelli and G. Angeletti (eds.), Dimethylsulphide: Oceans, Atmosphere and Climate. Kluwer Academic Publishers, Dordrecht.
6. Barnes, I., K.H. Becker, P. Carlier. and G. Mouvier. 1987. FTIR study of the $S/NO_2/I_2/N_2$ photolysis system: the reaction of IO radicals with DMS. Int. J. Chem. Kinet. 19: 489-501.
7. Bates, T. S., J.D. Cline, R.H. Gammon and S.R. KellyHanson. 1987. Regional and seasonal variations in the flux of oceanic dimethylsulphide to the atmosphere. Journal of Geophysical Research 92: 2930-2938.
8. Bates, T. S., B.K. Lamb, A. Guenther, J. Dignon and R.E. Stoiber. 1992. Sulfur emissions to the atmosphere from natural sources. J. Atmos. Chem. 14:315-337.
9. Berresheim, H. 1987. Biogenic sulphur emissions from the sub-arctic and Antarctic oceans. J. Geophys. Res. 92:13245-13262.
10. Boers, R., G.P. Ayers and J.L. Gras. 1994. Coherence between seasonal variation in satellite-derived cloud optical depth and boundary layer CCN concentrations at a mid-latitude southern hemisphere station. Tellus 46B: 123-131.
11. Brimblecombe, P. 1992. History of atmospheric acidity, p. 267-304. *In* M. Radojevic and R.M. Harrison (eds.), Atmospheric Acidity - Sources, Consequences and Abatement. Elsevier, Essex.
12. Broecker, W.S. and T.H. Peng. 1974. Gas exchange rates between air and sea. Tellus 26: 21-35.
13. Calhoun, J.A. and T.S. Bates. 1989. Sulfur isotope ratio tracers of non-sea-salt sulfate in the remote atmosphere, p. 367-379. *In* E.S. Saltzman and W.J. Cooper (eds.), Biogenic Sulfur in the Environment. ACS Symposium Series 393, Washington.
14. Carreto, J.I., M.O. Carignan, G. Daleo and S.G. deMarco. 1990. Occurrence of mycosporine-like amino acids in the red-tide dinoflagellate *Alexandrium excavatum*: UV-photoprotective compounds? J. Plankton Res. 12: 909-921.
15. Charlson, R. J. 1993. Gas-to-particle conversion and CCN production, p. 275-286. *In* G. Restelli and G. Angeletti (eds.), Dimethylsulphide: Oceans, Atmosphere and Climate. Kluwer, Dordrecht.
16. Charlson, R. J., J.E. Lovelock, M.O. Andreae and S.G. Warren. 1987. Oceanic phytoplankton, atmospheric sulphur, cloud albedo and climate. Nature 326: 655-661.
17. Coakley, J.A., R.L. Bernstein and P.A. Durkee. 1987. Effect of ship-stack effluents on cloud reflectivity. Science 237: 1020-1022.
18. Daykin, E.P. and P.H. Wine. 1990. Rate of reaction of IO radicals with dimethylsulfide. J.Geophys.Res. 95: 18547-18553.
19. Eriksson, E. 1963. The yearly circulation of sulfur in nature. J. Geophys. Res. 68: 4001-4008.
20. Falkowski, P. G., Y. Kim, Z. Kolber, C. Wilson, C. Wirick and R. Cess. 1992. Natural versus anthropogenic factors affecting low-level cloud albedo over the North Atlantic. Science 256: 1311-1313.
21. Frew, D. C., J.C. Goldman, M.R. Dennett and A.S. Johnson. 1990. Impact of phytoplankton-generated surfactants on air-sea gas exchange. J. Geophys. Res. 95: 3337-3352.
22. Goldman, J. C., M.R. Dennett and N.M. Frew. 1988. Surfactant effects on airsea exchange under turbulent conditions. DeepSea Res. 35: 1953-1970.
23. Granat, L., H. Rodhe and R.U. Hallberg. 1976. The global sulfur cycle, p. 89-134. *In* B.H. Svensson and R. Soderlund (eds.), Nitrogen, Phosphorus and Sulfur-Global Cycles. Ecol.Bull.22, Stockholm.
24. Hayduk, W. and H. Laudie. 1974. Prediction of diffusion coefficients for nonelectrolytes in dilute aqueous solutions. A. I. Ch. E. Journal 20: 611-615.
25. Herndl, G.J., G. Mhller-Niklas and J. Frick. 1993. Major role of ultraviolet-B in controlling bacterioplankton growth in the surface layer of the ocean. Nature 361: 717-719.
26. Hesshaimer, V., M. Heimann and I. Levin. 1994. Radiocarbon evidence for a smaller oceanic carbon dioxide sink than previously believed. Nature 370: 201-203.
27. Hobbs, P.V., D.A. Hegg and R.J. Ferek. 1993. Recent field studies of sulfur gases, particles and clouds in clean marine air and their significance with respect to the DMS-cloud-climate hypothesis, p. 345-353.

In G. Restelli and G. Angeletti (eds.), Dimethylsulphide: Oceans, Atmosphere and Climate. Kluwer, Dordrecht..

28. Holligan, P. M. 1992. Do marine phytoplankton influence global climate? p. 487-501. *In* P. G. Falkowski and A. D.Woodhead (eds.), Primary Productivity and Biogeochemical Cycles in the Sea. Plenum Press, New York.
29. Hynes, A.J., P.H. Wine and D.H. Semmes. 1986. Kinetics and mechanism of OH reactions with organic sulphides. J.Phys.Chem. 90: 4148-4156.
30. Karentz, D., E. Cleaver and D.L. Mitchell. 1991. Cell survival characteristics and molecular responses of Antarctic phytoplankton to ultraviolet B radiation. J. Phycol. 27, 326-341.
31. Keller, M.D., W.K. Bellows and R.R.L. Guillard. 1989. Dimethyl sulfide production in marine phytoplankton. p. 183-200. *In* E.S. Saltzman and W.J. Cooper (eds.) Biogenic Sulfur in the Environment. American Chemical Society, Washington D.C.
32. Keswick, B.H., D-S. Wang, and C.P. Gerba. 1982. The use of microorganisms as groundwater tracers: a review. Ground Water 20: 142-149.
33. Kiene, R. P. and T. S. Bates. 1990. Biological removal of dimethyl sulphide from seawater. Nature 345, 702-705.
34. Koga, S. and H. Tanaka. 1993. Numerical study of the oxidation process of dimethylsulfide in the marine atmosphere. J. Atmos. Chem. 17, 201-228.
35. Lancelot, C., G. Billen, A. Sournia, T. Weisse, F. Colijn, M.J.W. Veldhuis, A. Davies and P.Wassmann. 1987. *Phaeocystis* blooms and nutrient enrichment in the continental zones of the North Sea. Ambio 16: 38-46.
36. Legrand, M., R.J. Delmas and R.J. Charlson. 1988. Climate forcing implications from Vostok ice-core sulphate data. Nature 334: 418-420.
37. Legrand, M., C. Feinet-Saigne, E.S. Saltzman, C. Germain, N.I. Barkov and V.N. Petrov. 1991. Ice-core record of oceanic emissions of dimethylsulphide during the last climate cycle. Nature 350, 144-6.
38. Liss, P. S. 1975. Chemistry of the Sea Surface Microlayer, p. 193-243. *In* J. P. Riley and G. Skirrow (eds.), Chemical Oceanography Volume 2. Academic Press, New York.
39. Liss, P.S. 1983. Gas transfer: experiments and geochemical imlications, p.241-298. *In* P.S. Liss and W.G.N. Slinn (eds.), Air-Sea Exchange of gases and Particles. Reidel, Dordrecht.
40. Liss, P.S. and P.G. Slater. 1974. Flux of gases across the air-sea interface. Nature 247: 181-184.
41. Liss, P. S. and L. Merlivat. 1986. Airsea gas exchange rates: introduction and synthesis, p.113-127. *In* P. BuatMenard (ed.), The role of AirSea Exchange in Geochemical cycling. Reidel, Dordrecht.
42. Liss, P.S., A.J. Watson, M.I. Liddicoat, G. Malin, P.D. Nightingale, S.M. Turner and R.C. Upstill-Goddard. 1993. Trace gases and air-sea exchanges. Phil. Trans. Roy. Soc. Lond. A. 343: 531-541.
43. Liss, P.S., G. Malin, S.M. Turner and P.M. Holligan. 1994. Dimethyl sulphide and Phaeocystis: a review. J. Mar. Syst. 5: 41-53.
44. Lovelock, J. E., R.J. Maggs and R.A. Rasmussen. 1972. Atmospheric sulphur and the natural sulphur cycle. Nature 237, 452.
45. Maki, J.S. 1993. The air-water interface as an extreme environment, p. 409-439. *In* T.E. Ford (ed). Aquatic Microbiology an Ecological Approach. Blackwell, Boston.
46. Malin, G., S. Turner, P. Liss, P. Holligan and D. Harbour. 1993. Dimethyl sulphide and dimethylsulphoniopropionate in the Northeast Atlantic during the summer coccolithophore bloom. Deep-Sea Res. 40: 1487-1508.
47. Marchant, H. J., A.T. Davidson and G.J. Kelly. 1991. UV-B protecting compounds in the marine alga *Phaeocystis pouchetii* from Antarctica. Mar. Biol. 109, 391-395.
48. McArdle, N. C. and P.S. Liss. 1995. Isotopes and Atmospheric sulphur. Atmos. Environ. 29: 2553-2556.
49. McMinn, A., H. Heijnis and D. Hodgson. 1994. Minimal effects of UVB radiation on Antarctic diatoms over the past 20 years. Nature 370: 547-549.
50. Newman, L. and J. Forrest. 1991. Sulphur isotope measurements relevant to power plant emissions in the northeastern United States, p.331-343. *In* H.R. Krouse and V.A. Grinenko (eds.), Stable Isotopes: Natural and Anthropogenic Sulphur in the Environment. Scope 43. Wiley, Chichester.
51. Nightingale, P.D., C. Law, G. Malin, A.J. Watson, P.S. Liss, R.C. Upstill-Goddard and M.I. Liddicoat. (in prep.) Air-sea gas exchange measurements at sea using dual and triple tracer releases.
52. Nguyen, B. C., A. Gaudry, B. Bonsang and G. Lambert. 1978. Reevaluation of the role of dimethyl sulphide in the sulphur budget. Nature 275, 637-639.
53. Nriagu, J.O., R.D. Coker and L.A. Barrie. 1991. Origin of sulphur in Canadian Arctic haze from isotope measurements. Nature 349: 142-145.

54. Peng, T.H., W.S. Broecker, G.G. Mathieu, Y.H. Li and A.E. Bainbridge. 1979. Radon evasion rates in the Atlantic and Pacific oceans as determined during the GEOSECS program. J. Geophys. Res. 84: 2471-2486.
55. Pike, E.B., A.W.J. Bufton and D.J. Gould. 1969. The use of *Serratia indica* and *Bacillus subtilis* var. *niger* spores for tracing sewage dispersion in the sea. J. Appl. Bacteriol. 32: 206-216.
56. Plane, J. M. C. 1989. Gas-phase atmospheric oxidation of biogenic sulphur compounds: a review, p. 404-423. *In* E.S. Saltzman and W.J. Cooper (eds.), Biogenic Sulfur in the Environment. ACS Symposium Series 393, Washington D.C.
57. Saigne, C. and M. Legrand. 1987. Measurement of methanesulphonic acid in Antarctic ice. Nature 330: 240-242.
58. Saltzman, E.S., K. Holmen and D.J. Cooper. 1988. Measurement of the piston velocity of dimethylsulfide: implications for its air-sea exchange. Eos 69: 1073.
59. Saltzman, E.S. and D.J. Cooper. 1989. Dimethyl sulfide and hydrogen sulfide in marine air, p. 330-351. *In* E.S. Saltzman and W.J. Cooper (eds.), Biogenic Sulfur in the Environment. ACS Symposium Series 393, Washington.
60. Saltzman, E.S., D.B. King, K. Holmen and C. Leck. 1993. Experimental determination of the diffusion coefficient of dimethylsulfide in water. J. Geophys. Res. 98: 16481-16486.
61. Sarnthein, M., K. Winn and R. Zahn. 1987. Paeleoproductivity of oceanic upwelling and the effect on anthropogenic CO_2 and climatic change, p. 311-317. *In* W. H. Berger and L. D. Labeyrie (eds.), Abrupt Climate Change. Reidel, Dordrecht.
62. Savoie, D. L. and J.M. Prospero. 1989. Comparison of oceanic and continental sources of non-sea-salt sulphate over the pacific Ocean. Nature 339 685-687.
63. Schwartz, S.E. 1988. Are global cloud albedo and climate controlled by marine phytoplankton? Nature 336: 441-445.
64. Shaw, G.E. 1987. Aerosols as climate regulators: a climate-biosphere linkage? Atmos. Environ. 21: 985-986.
65. Slingo, A. 1989. Sensitivity of the earth's radiation budget to changes in low clouds. Nature 343, 49-51.
66. Smethie, W.M., T.T. Takahashi, D.W. Chipman and J.R. Ledwell. 1985. Gas exchange and CO_2 flux in the tropical Atlantic Ocean determined from ^{222}Rn and pCO_2 measurements. J. Geophys. Res. 90: 7005-7022.
67. Tans, P.P., I.Y. Fung and T. Takahashi. 1990. Observational constraints on the global atmospheric CO_2 budget. Science 247: 1431-1438.
68. Turner, S. M. and P.S. Liss. 1989. A cryogenic technique for sampling the sea surface microlayer for trace gases. Chemical Oceanography 10, 379-392.
69. Turner, S. M., G. Malin, P. Nightingale and P.S. Liss. (submitted). Seasonal variation of dimethyl sulphide in the North Sea and an assessment of fluxes to the atmosphere. Marine Chemistry.
70. Upstill-Goddard, R.C., A.J. Watson, P.S. Liss and M.I. Liddicoat. 1990. Gas transfer velocities in lakes measured with SF_6. Tellus 42B: 364-377.
71. Wanninkhof, R. 1992. Relationship between gas exchange and wind speed over the ocean. J. Geophys. Res. 97:7373-7382.
72. Wanninkhof, R. and L.F. Bliven. 1991. Relationship between gas exchange, wind speed, and radar backscatter in a large wind-wave tank. J. Geophys. Res. 96: 2785-2796.
73. Watson, A.J., R.C. Upstill-Goddard and P.S. Liss. 1991. Air-sea gas exchange in rough and stormy seas measured by a dual-tracer technique. Nature 349: 145-147.
74. Whung, P.-Y., E.S. Saltzman, M.J. Spencer, P.A. Mayewski and N. Grundestrup. 1994. Two-hundred-year record of biogenic sulfur in a south Greenland ice core (20D). J. Geophys. Res. 99: 1147-1156.
75. Wigley, T.M.L. 1989. Possible climate change due to SO_2 derived cloud condensation nuclei. Nature 339: 365-367.
76. Wigley, T.M.L. 1994. Outlook becoming hazier. Nature 369: 709-710.
77. Wilke, C.R. and P. Chang. 1955. Correlation of diffusion coefficients in dilute solutions. A. I. Ch. E. Journal 1: 264-270.
78. Worrest, R.C., H. van Dyke and B.E. Thomson. 1978. Impact of enhanced simulated solar ultraviolet radiation upon a marine community. Photochemistry and Photobiology 27: 471-478.
79. Yin, F., D. Grosjean and J.H. Seinfeld. 1990a. Photoxidation of dimethyl sulfide and dimethyldisulphide. I: Mechanism development. J.Atmos. Chem. 11: 309-64.
80. Yin, F., D. Grosjean, R.C. Flagan and J.H. Seinfeld. 1990b. Photoxidation of dimethyl sulfide and dimethyldisulphide. II: Mechanism evolution. J.Atmos. Chem. 11: 365-399.

17

ORIGIN AND IMPORTANCE OF PICOPLANKTONIC DMSP

M. Corn,[1] S. Belviso,[1] F. Partensky,[2] N. Simon,[2] and U. Christaki[1]

[1] Centre des Faibles Radioactivités
Laboratoire Mixte CNRSCEA
91198 Gif-sur-Yvette Cedex, France
[2] Station Biologique
CNRS et Université Paris 6, B.P. 74
29682 Roscoff Cedex, France

SUMMARY

The importance of picoplanktonic particulate DMSP (DMSPp GF/F-2 μm) was investigated using size fractionation during a time-series experiment in the Mediterranean Sea and in three areas of the subtropical Atlantic Ocean. Picoplanktonic DMSPp accounted for up to 25% of depth-integrated total DMSPp (GF/F-200 μm) in oligotrophic waters. In order to estimate the relative contribution of picophytoplankton to DMSPp in the GF/F-2 μm fraction, we measured the DMSPp content of representative strains of the three main groups of picophytoplankton, including *Prochlorococcus (*prochlorophytes), *Synechococcus* (cyanobacteria) and picoeukaryotes belonging to four classes (Prasinophyceae, Pelagophyceae, Chlorophyceae and Prymnesiophyceae). The average cellular DMSPp concentrations (mean ± 1SD) were $2.8.10^{-4} \pm 1.8.10^{-4}$, $5.1.10^{-3} \pm 7.8.10^{-3}$ and 46.5 ± 73.7 fg/cell for *Prochlorococcus* sp., *Synechococcus* sp. and the picoeukaryotes, respectively. The DMSP content was highly variable among taxonomic groups of picoeukaryotes even when normalized per unit biovolume. The prymnesiophytes produced the most DMSP (e.g. clone CCMP625, 196 mmol.liter cell volume^{-1}) and the chlorophyte the least (0.74 mmol.liter cell volume^{-1}). Using these average DMSP contents per cell and flow cytometric counts of natural populations of picophytoplankton, it is strongly suggested that the picoeukaryotes were the main DMSP producers in the GF/F-2 μm size class. The combined contribution of the two prokaryotes was negligible (less than 1%), even in oligotrophic waters.

INTRODUCTION

Dimethylsulfide (DMS) is an important gas which is involved in the sulfur cycle. It arises mainly from the enzymatic cleavage of dimethylsulfoniopropionate (DMSP) which is produced

Biological and Environmental Chemistry of DMSP and Related Sulfonium Compounds
edited by Ronald P. Kiene et al., Plenum Press, New York, 1996

by marine algae. In an extensive survey of marine phytoplankton, Keller and coworkers (11) demonstrated that DMSP was mainly produced by members of the Dinophyceae and Prymnesiophyceae, although some other chlorophyll ccontaining algae, including species of Chrysophyceae and Bacillariophyceae (i.e. diatoms) could also contain significant amounts of DMSP. While many species of nano and microphytoplankton have been screened for DMSP production (11,12), fewer species of picophytoplankton, i.e. photosynthetic organisms passing through 2 μm pore filters, have been studied. Flow cytometric analyses have revealed that this class includes both eukaryotes, belonging to most algal classes (20), and prokaryotes mainly belonging to two genera, *Synechococcus* (cyanobacteria) and the recently discovered *Prochlorococcus* (prochlorophytes) (4, 6, 7, 14, 17, 18). The latter occur at very large concentrations in the temperate and tropical open waters of Atlantic and Pacific oceans as well as in the Mediterranean sea (7, 10, 14, 18, 21). Their contribution to the integrated photosynthetic biomass of warm oligotrophic oceans may reach up to 58% (19). Thus, due to its wide distribution and its photosynthetic activity, *Prochlorococcus* most probably plays a very significant role in the global carbon cycle. Picoeukaryotes are also significant components of the picophytoplanktonic standing stock (20-46% in terms of carbon) in both the mesotrophic and oligotrophic parts of tropical oceans. On the other hand, *Synechococcus* is most significant in mesotrophic areas (4, 19).

In this study, we investigated the contribution of picoplanktonic DMSPp (GF/F-2 μm size fraction) to total planktonic DMSPp (GF/F200 μm size fraction) in two oceanographic areas (Mediterranean Sea and subtropical Atlantic). We also estimated the relative contribution of different picophytoplankton groups to the DMSP in the GF/F-2 μm size fraction in the latter area by converting flow cytometric cell counts of *Prochlorococcus*, *Synechococcus* and picoeukaryotes into cell DMSPp using conversion factors obtained by measuring the average DMSPp per cell of representative cultured species.

MATERIALS AND METHODS

Field Samples

Seawater samples were collected at a mesotrophic site (43°25′N 7°51′E) in the central Ligurian Sea off Villefranche-sur-Mer (France). This site of the northeastern Mediterranean Sea was selected by the JGOFS-France DYFAMED Program and was occupied every month from March 1993 to November 1994. Samples were also collected in the subtropical northeastern Atlantic Ocean (Table 1), at the three sites (EU, MESO and OLIGO) selected by the JGOFS-France EUMELI program, in September-October 1991 (EUMELI 3; 2) and in May-June 1992 (EUMELI 4). All water samples were taken using 12 Niskin bottles fixed to a CTD system.

Table 1. Position and date of hydrocasts during cruises EUMELI 3 (Sept. 1991; 2) and EUMELI 4 (June 1992) in the subtropical Atlantic

SITE	DATE	POSITION	NAME
OLIGO	23 sept 91	20°55N 31°05W	BSN 11
OLIGO	23 june 92	21°02N 31°10W	GOF 49
MESO	17 june 92	18°29N 21°07W	GOF 30
EU	13 june 92	20°30N 18°30W	CTD 192

Picophytoplanktonic Cultures

The origin and mean cell volume of algal strains used in this study are summarized in Table 3. The pigment and size characteristics of picoeukaryotic strains are described elsewhere (20). Although the equivalent diameters of the unidentified prymnesiophyte, CCMP625, and of *Imantonia* are slightly larger than 2 μm (20), these organisms were included in this study on picoplankton because a significant fraction (ca. 10-40%) of these flagellates may pass through 2 μm pore Nuclepore filters (unpublished data). All eukaryotic strains were grown in K medium (13); *Synechococcus* in f/2 medium (9); and *Prochlorococcus* in a modified K medium (6) with K/10 trace metals plus 10 nM $NiCl_2$, 10 nM H_2SeO_3, 10 μM glycerophosphate, 50 μM NH_4Cl and 50 μM urea (Keller, unpublished). All cultures were maintained at 19°C under continuous low blue light (14.5 μmol quanta m^{-2} s^{-1}) provided by Daylight fluorescent tubes (Sylvania) wrapped with a "moonlight blue" Lee filter (Panavision). This low blue light simulated the light available in the lower euphotic zone in oligotrophic waters. All measurements were made in duplicate during the exponential phase of cell growth, unless otherwise specified.

Cell Enumeration

Cells were counted using two types of flow cytometers. A FACScan™(Becton Dickinson, San José, CA) flow cytometer was used aboard the ship to count live picophytoplanktonic populations during the EUMELI 3 cruise. An EPICS 541 flow cytometer (Coulter, Hialeah, Fla.) was used to analyze both the fixed samples from the EUMELI 4 cruise and live laboratory cultures of picophytoplankton. Optical configurations and set up of these systems are described in (19) and (21). With both flow cytometers, cells were counted on biparametric histograms representing right angle light scatter vs. chlorophyll red fluorescence (see e.g. 21 for illustration).

Pigment Analyses

During the EUMELI cruises, chlorophyll *a* (data obtained courtesy of J. Neveux) was measured by filtering water through 47 mm Whatman GF/F filters, extracting the filters in 90% acetone and analyzing the extracts by fluorometry (16). In the Mediterranean Sea, measurements of chlorophyll *a* and divinylchlorophyll *a* (data obtained courtesy of J.C. Marty and H. Claustre), (cumulatively termed total chlorophyll *a*), were done by reverse-phase HPLC using a modification of the method of Williams and Claustre (22), as described in (2).

Size Fractionation

The picoplankton size class was defined as that which passed through a 2.0 Nuclepore filter by gravity filtration but was retained on a Whatman GF/F filter (nominal pore size of 0.7 μm). This size class is designated as GF/F-2.0 μm.

DMSP Analysis

The DMSP content of picoplankton (GF/F-2 μm) was indirectly measured by subtracting the total dissolved DMSP + DMS of a sample filtered through a GF/F filter from the total DMSP + DMS content of an 2.0μm unfiltered sample following the method described in (2). Seawater was collected from the surface to depths of either 110 m (Mediterranean Sea) or 150 m (Atlantic) and 60 ml sub-samples were used for the DMSP analyses. These aliquots were filtered by gravity through 2 μm pore-size Nuclepore membranes (47 mm diameter) for the total DMSP + DMS sample and through Whatman GF/F glassfiber filters to produce the dissolved DMSP+DMS pool. After treatment of the samples with cold alkali, DMS analyses were performed by GC/FPD (3). Five aliquots were used to estimate the overall precision of the method with natural samples. Precision for DMSPp samples in the GF/F-2μm size class was 20-40%.

For the screening of picophytoplankton isolates, a new method was developed to increase sensitivity in order to quantify DMSPp levels in microorganisms containing small amounts of this

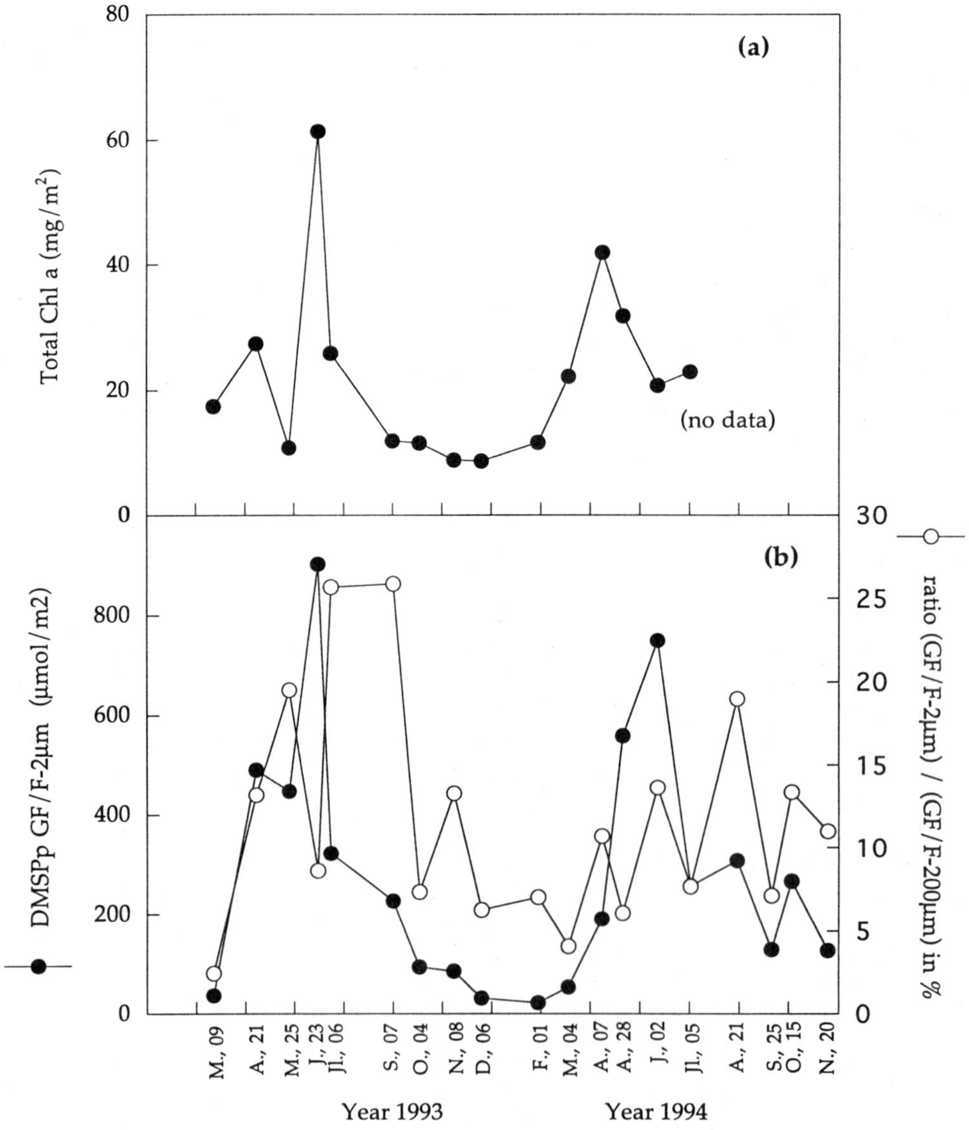

Figure 1. Seasonal variations of the depth-integrated concentrations of DMSPp in the size range GF/F-2 μm, the chlorophyll a and the ratio (DMSPp GF/F-2 μm)/(DMSPp GF/F-200 μm), in the Mediterranean Sea in 1993 and 1994 (Month in abbreviation, day).

compound. Culture samples (8 ml), either unfiltered or filtered on Whatman GF/F filters, were transferred into teflon-stoppered glassware and treated with cold alkali. After 12 h storage at room temperature, they were purged with high-grade helium. DMS was trapped cryogenically on the walls of Teflon tubing immersed in liquid nitrogen, and analyzed by gas chromatography with a Varian chromatograph equipped with a pulsed flame photometric detector (PFPD). PFPD is based on a flame source and combustible gas mixture rate that cannot sustain a continuous flame operation (5). With the PFPD, the sensitivity for sulfur is higher (detection limit of 1 pg S/sec), and the selectivity against hydrocarbons molecules is large (10^3 to 10^6 S/C), and the gas consumption is smaller than with a conventional FPD. The detection limit was 60 pg of DMS. Measurements were performed in duplicate and the agreement was generally better than 15%.

RESULTS AND DISCUSSION

Analysis of DMSPp in the Field

Mediterranean Sea. Nineteen vertical profiles were taken from March 1993 to November 1994 in the deep (> 2000 m) open waters of the central Ligurian Sea. From November to March, the seawater column was well mixed, with temperatures ca. 14°C and detectable nitrate concentrations, with a maximum of 4 μM in February (M.D. Pizay, pers. comm.). In summer, the water column became stratified, with a 10-20 m deep mixed layer and surface temperatures reaching 25°C. The upper layer was oligotrophic (nitrate was undetectable at the μM level) with a subsurface chlorophyll *a* maximum at 20-40 m (J.C. Marty, pers. comm.). The integrated concentration of chlorophyll *a* in the top 50 m, i.e. the layer where nearly all the DMSP was found, displayed interannual variability (Fig. 1a). In 1993, the Chl *a* standing stock peaked in June with a subsurface bloom of prymnesiophytes, whereas in 1994, the biomass peaked in April during a bloom of diatoms (J.C. Marty, pers. comm.). The integrated total picoplanktonic DMSPp standing stock showed a similar pattern in 1993 and 1994, with maximum levels of about 800 $\mu mol.m^{-2}$ in June and minima during winter (Fig 1b). The contribution of DMSPp in the picoplanktonic size range (GF/F-2 μm) to the total DMSPp pool (GF/F-200 μm) averaged 10% and ranged between 35% in March and 17-25% in July-September. There was a temporal displacement between the peak of total DMSPp and the maximum in the ratio of picoplanktonic to total DMSPp in both years (Fig. 1). The relative contribution of picoplanktonic DMSPp was highest during summer, i.e. when oceanographic conditions were the most oligotrophic.

Atlantic Ocean. In June 1992 (EUMELI 4), levels of particulate DMSP in the size ranges GF/F-2 μm and GF/F-200 μm were investigated in three areas of the subtropical northeastern Atlantic Ocean (Tables 1 and 2). A previously published result from the EUMELI 3 cruise (2) has been included in the data set to demonstrate that there was little seasonal variation at the OLIGO site. The depth integrated standing stock of DMSPp in the size range GF/F-2 μm was at least two times in oligotrophic waters than in eutrophic and mesotrophic waters (Table 2), although the total DMSPp (GF/F-200 μm) showed an inverse variation. This is consistent with an increase of the relative contribution of the biomass of picophytoplankton to total chlorophyll biomass along the nutrient gradient from the EU and OLIGO sites (19). The contribution of the picoplanktonic DMSPp pool to the total DMSPp pool was 12-22% in oligotrophic waters and 23% in eutrophic and mesotrophic waters.

Quantification of DMSPp in Picophytoplanktonic Species. Our filter fractionation studies in two different oceanic areas (Mediterranean Sea and Atlantic Ocean) have shown

Table 2. Depth-integrated concentrations of picoplanltonic (GF/F-2 µm) and total (GF/F-200µm) DMSPp fractions and ratio of these two fractions at three sites of the subtropical Atlantic. Intracellular DMSP was computed using cell counts for each group multiplied by the average DMSP cell content determined in cultures for the three picoplanktonic groups: Prochlorococcus = 2.8 10^{-4} fg /cell, Synechococcus = 5.1 10^{-3} fg/cell and picoeukaryotes = 46.5 fg / cell

Site-Date	Integrated depth (m)	DMSPp GF/F-2µm pool (µmol/m^2)	DMSPp GF/F-200µm pool (µmol/m^2)	(DMSPp GF/F-2µm) / (DMSPp GF/F-200µm) (%)
OLIGO-Sept 91	0-160	459	2114	21.7
OLIGO-June 92	0-140	283	2328	12.2
MESO-June 92	0-75	118	6877	1.7
EU-June 92	0-70	122	3883	3.2

that DMSPp in the size range GF/F-2 μm may account for 2-25% of the total DMSPp reservoir. However, the relative contribution of the different picophytoplankton populations to the total picophytoplanktonic DMSPp, as well as any contribution by heterotrophs or picodetritus was not resolved. To estimate these fractions, we determined the intracellular DMSP content of representative species in culture in an attempt to obtain factors to convert cell numbers (obtained by flow cytometry) from natural populations into DMSPp values for these species.

With the sensitive PFPD system used, DMSPp (> 10^4 fg cell^{1}) was detected in all the isolates screened except a clone of *Synechococcus,* EUM11, and an unidentified prasinophyte EUM16B (Table 3). The mean intracellular DMSP concentrations (±1 SD) in *Prochlorococcus*, *Synechococcus* and the eukaryote strains were 2.8 10^{-4} ± 1.8 10^{-4} (n=9), 5.1 .10^{-3} ± 7.8 10^{-3} (n=3) and 46.5 ± 73.7 fg cell^{1} (n=11) respectively.

The *Synechococcus* strains used in this study belonged to two pigment groups. Clones EUM11 and MAX42, which both possess phycoerythrin with high phycourobilin (PUB) and low phycoerythrobilin (PEB) content, were representative of the open ocean, while DC2, which possesses the converse pigment signature, was more typical of coastal waters (see e.g. 17). DMSPp was detected in clone MAX42 but not in EUM11 (Table 3). The coastal strain, DC2, was previously screened by Keller et al. (11), but with a less sensitive detection system, and they did not detect any DMSP. DMSP production does not appear to be related to the pigment signature in this genus.

Among the autotrophic picoeukaryotes, DMSP production varied up to 1,700-fold on a per cell basis (the non-DMSP producing EUM16B strain excluded). Variations in the DMSP levels per unit cell volume, which eliminates variability due to size differences among species, were also large (up to 265-fold). Three of the picoeukaryotic clones used in this study were previously screened by Keller and coworkers (11). The measured DMSP contents of *Imantonia rotunda* (11) and *Imantonia* sp. (this study) are similar. In contrast, Keller et al. (11) detected about 10 times more DMSP in *Micromonas* than we did. While we detected 0.74 mmol DMSP·liter cell volume^{-1} in *Nannochloris,* Keller et al. did not detect any DMSP in this clone. The prymnesiophytes (CCMP625 and *Imantonia*) and the prasinophytes produced the largest amounts of DMSP, while the chlorophytes produced the lowest (Table 3). *Pelagomonas*, the only representative of the newly described Pelagophyceae, previously classified as chrysophytes (1), had a moderate DMSP content. These observations support the findings of Keller et al. (11, 12), with prymnesiophytes being major producers, prasinophytes and pelagophytes (ex chrysophytes) moderate producers, and chlorophytes being non- or very small producers of DMSP.

Besides taxonomic position, which is probably the main factor of variation in DMSP production between different phytoplankton communities, intraspecific variations can also occur. Matrai and Keller (15) reported that for cultures of *Prorocentrum minimum* and *Amphidinium carterae* DMSPp varied according to the growth stage. We have confirmed that such variations can occur since DMSPp content varied by a factor of 4 in cultures of the strains EUM8 and CCMP625 sampled one week apart (Table 3).

It is possible that contaminating bacteria in xenic cultures (such as ours) may also contribute to the measured amounts of DMSP. Diaz et al. (8) showed that a marine bacterial strain could accumulate DMSP. Thus, the occurrence of contaminating bacteria may have somehow biased the determination of DMSPp in our cultures of *Prochlorococcus* and *Synechococcus*, which were both low DMSP producers. Bacteria able to pass through GF/F filters could also cause an overestimate of the dissolved DMSP pool. Since dissolved DMSP was subtracted from the total DMSP pool to calculate DMSPp, any interference of this type would be minimal. To further check possibility that bacteria in cultures might contribute to DMSPp, poststationary cultures of *Prochlorococcus,* grown at high light, were used. These conditions were deleterious to *Prochlorococcus* cells (which died), but not to contaminating

Table 3. Origin, mean cell volume and concentration of intracellular DMSP (per cell and per liter of biovolume) of the picophytoplankton surveyed. They were in exponential phase of growth except for CCMP 1192 and one of the CCMP625 samples which were in stationary phase. Clones CCMP 1426, CCMP1378 and MED4 originate from the same isolate. Average volumes of analyzed picoeukaryotes and prokaryotes species were determined conductrimetrically in a previous study (20) and by A.Morel (pers.comm.), repectively. na = information not available, <dl = <detection limit (10^{-4} fg / cell)

Family	Species	Strain	Area	Mean cell vl.	Intracellular DMSP (DMSPp)	
				μm^3	fg/cell	mmol/l
Prochlorophyceae	*Prochlorococcus*	TATL1	Tropical Atlantic	0.1	1.6E-04	0.03
	Prochlorococcus	TATL2	Tropical Atlantic	0.1	1.6E-04	0.03
	Prochlorococcus	TATL4	Tropical Atlantic	0.1	1.9E-04	0.03
	Prochlorococcus	NATL1	N.W. Atlantic	0.1	1.1E-04	0.02
	Prochlorococcus	PAC1A[a]	Tropical Pacific	0.1	2.9E-04	0.05
	Prochlorococcus	CCMP 1426[b] -	Mediterranean Sea			
		MED 4[c] -	Mediterranean Sea	0.087	2.1 to 4.0	0.03 to 0.06
		CCMP 1378[b]	Mediterranean Sea		E-04	
	Prochlorococcus	SS120[c]	Sargasso Sea	0.1	7.1E-04	0.12
Cyanophyceae	*Synechococcus*	EUM11	Tropical Atlantic	0.52	<dl	<dl
	Synechococcus	DC2[b]	Atlantic	0.65	1E-3	0.03
	Synechococcus	MAX42	Sargasso Sea	0.32	14E-3	0.7
Prasinophyceae	*Pycnococcus* sp.	CCMP 1192[b]	Oceanus 83 sta II	5.47	19.9	58
	Pycnococcus provasolii	CCMP 1203[b]	Old WHOI sta II	9.03	32.0	57
	Unidentified	EUM 16B	Tropical Atlantic	1.49	<dl	<dl
	Micromonas pusilla	CCMP 490[b]	Woods Hole. MA	1.87	2.1	17.9
	Bathycoccus prasinos	type strain	Gulf of Naples	2.04	4.5	35.6
Pelagophyceae	*Pelagomonas* sp.	EUM 8	Tropical Atlantic	4.8	4.6 & 9.3	15.4 & 31.4[e]
Chlorophyceae	*Nannochloris* sp.	CCMP 515[b]	Cape Bolinas (Philippines)	2.48	0.12	0.74
Prymnesiophyceae	Unidentified	CCMP 625[b]	na	15.07	184 & 56	196 & 60[e]
	Imantonia sp.	PCC 18561[d]	Northeast Atlantic	21.48	199	149

[a] gift of Dr. L. Campbell
[b] strains obtained from the Center For the Culture of Marine Phytoplankton, Bigelow Laboratory
[c] gift of Drs. P. Chisholm and L. Moore
[d] strains obtained from the Pasteur Culture collection, Paris, France
[e] the two values are for cultures sampled 1 week apart

Table 4. Depth-integrated cell concentrations and calculated DMSP concentrations contributed by the three main populations of picoplankton (*Prochlorococcus, Synechococcus* and picoeukaryotes), measured concentrations of picoplanktonic (GF/F-2 μm) DMSPp fraction, and ratio of total calculated picophytoplanktonic DMSP to the measured GF / F-2 μm DMSPp fraction, at three sites of the subtropical Atlantic. Intracellular DMSP was computed using the actual cell counts multiplied by the average DMSP cell content determined in cultures for the three picoplankters: *Prochlorococcus* = 2.8 10^{-4} fg / cell, *Synechococcus* = 5.1 10^{-3} fg / cell and picoeukaryoted = 46.5 fg / cell

Site-Date	Integrated depth	*Prochlorococcus* sp.		*Synechococcus* sp.		Eukaryote strains		DMSPp GF/F-2μm pool	total picopl. DMSP/ (DMSPp GF/F-2μm)
	(m)	10^{11}cell/m^2	μmol DMSP/m^2	10^{11}cell/m^2	μmol DMSP/m^2	10^{11}cell/m^2	μmol DMSP/m^2	(μmol/m^2)	(%)
OLIGO-Sept 91	0-160	235.0	0.11	2.37	0.02	1.6	120	459	26
OLIGO-June 92	0-140	192.0	0.09	3.18	0.03	1.36	102	283	36
MESO-June 92	0-75	12.9	0.006	137.3	1.12	6.58	493	118	419
EU-June 92	0-70	0.6	0.0003	0.52	0.004	0.66	49	122	40

bacteria, which remained at high concentrations. These cultures had undetectable DMSP levels. Thus, there is no evidence that contaminating bacteria contribute to the DMSP pool in cultures.

Contribution of Autotrophic Picoplankton to Total Picoplankton. The relative contribution of autotrophic prokaryotes and eukaryotes to the DMSPp pool in the GF/F-2 μm size fraction in the tropical Atlantic Ocean was estimated using the average DMSP content per cell (measured in cultures) and flow cytometric cell counts of picophytoplankton. The integrated values of estimated picophytoplanktonic DMSP and their relative contribution to measured picoplanktonic DMSP are shown in Table 4. Both in June 1991 and September 1992, *Prochlorococcus* accounted for about half of the photosynthetic biomass at the OLIGO site (19). In contrast, *Synechococcus* numerically dominated at the MESO site, occupied during the EUMELI 4 cruise (June, 1992) (19). In both cases however, the DMSPp estimated for these prokaryotic groups was very low, either when compared to that contributed by the picoeukaryotes, or to the total measured picoplanktonic DMSPp (Table 4). The contribution of prokaryotes remained low even when using the highest DMSP contents measured in cultures of *Prochlorococcus* or *Synechococccus* as conversion factors (e.g. *Synechococcus* accounted for a maximum 2.5% of the picoplanktonic DMSP at the MESO site in June 1992).

The estimation of picoeukaryotic DMSP was more difficult because of the large variability in DMSP content observed between the different taxonomic groups. Thus, a precise estimate would require accurate determination of the species distribution within each sample, a goal which is presently unattainable by flow cytometric methods. Variability in species composition occurs both horizontally and vertically. With these caveats in mind, we applied an average DMSPp value, derived from our cultures, to estimate the picoeukaryotic contribution at the different EUMELI sites. We found that the contribution of autotrophic picoeukaryotes to the DMSP pool in the GF/F to 2 μm size fraction ranged from 26 to 36% at both the OLIGO and EU sites (Table 4). This leaves the majority of the DMSP in this size fraction unaccounted for and suggests that a significant fraction of DMSPp at these stations may be due to other sources, such as heterotrophs and/or picodetritus. At the MESO site however, the simulated picoeukaryotic DMSP was greatly overestimated, possibly because the dominant species at this station were low DMSPproducers, such as chlorophytes.

ACKNOWLEDGMENTS

The authors wish to thank the captains and crew of the research vessels L'ATALANTE and TETHYS II. This work was supported by CNRS, CEA, INSU, IFREMER and EEC under contract EV5VCT930326. This is CFR contribution number 1737.

REFERENCES

1. Andersen, R. A., Saunders, G. W., Pasking, M. P. and J. P. Sexton. 1993. Ultrastructure and 18S rRNA gene sequence for *Pelagomonas calceolata* gen. and sp. nov. and the description of a new algal class, the Pelagophyceae classis nov. J. Phycol. 29: 701-715.
2. Belviso, S., Buat—Ménard, P., Putaud, J.P., Nguyen, B.C., Claustre, H. and J. Neveux. 1993. Size distribution of dimethylsulfoniopropionate in areas of the tropical northeastern Atlantic Ocean and the Mediterranean Sea. Mar. Chem. 44: 55-71.
3. Belviso, S., Kim, S.K., Rassoulzadegan, F., Krajka, B., Nguyen, B.C., Mihalopoulos, N. and P. Buat-Ménard. 1990. Production of dimethylsulfonium propionate (DMSP) and dimethylsulfide (DMS) by a microbial food web. Limnol. Oceanogr. 35: 1810-1821.

4. Campbell, L. and D. Vaulot. 1993. Photosynthetic picoplankton community structure in the subtropical North Pacific Ocean near Hawaii (station ALOHA). Deep Sea Res. 40: 2043-2060.
5. Cheskis, S., Atar, E. and A. Amirav. 1993. Pulsed-flame photometer: a novel chromatography detector. Anal. Chem. 85: 539-555.
6. Chisholm, S.W., Frankel, S.L., Goericke, R., Olson, R.J., Palenik, B., Waterbury, J., West-Johnsrud, L. and E.R. Zettler. 1992. *Prochlorococcus marinus* nov. gen. nov. sp.: an oxyphototrophic marine procaryote containing divinyl chlorophyll a and b. Arch. Microbiol. 157: 297-300.
7. Chisholm, S.W., Olson, R.J., Zettler, E.R., Waterbury, J., Goericke, R. and N. Welschmeyer. 1988. A novel freeliving prochlorophyte occurs at high cell concentrations in the oceanic euphotic zone. Nature 334: 340-343.
8. Diaz, M.R., Visscher, P.T. and B.F. Taylor. 1992. Metabolism of dimethylsulfoniopropionate and glycine betaine by a marine bacterium. FEMS Microbiol. Lett. 96: 61-66.
9. Guillard, R.R.L. and J.H. Ryther. 1962. Studies of marine planktonic diatoms. I. *Cyclotella nana* Hustedt and *Denotula confervacea* (Cleve) Gran. Can. J. Microbiol. 8: 229-239.
10. Ishizaka, J., Kiyosawa, H., Ishida, K., Ishikawa, K. and M. Takahashi. 1994. Meridional distribution and carbon biomass of autotrophic picoplankton in the Central North Pacific Ocean during Late Northern Summer 1990. Deep- Sea Res. 41: 1745-1766.
11. Keller, M.D., Bellows, W.K. and R.R.L. Guillard. 1989. Dimethylsulfide production in marine phytoplankton, p. 167-182. *In* E. Saltzman and W. Cooper (eds), Biogenic Sulfur in the Environment. Am. Chem. Soc., Washington.
12. Keller, M.D. 1989. Dimethyl sulfide production and marine phytoplankton: the importance of species composition and cell size. Biol. Oceanogr. 6: 375-382.
13. Keller, M.D.,Selvin, R.C. and R.R.L. Guillard. 1987. Media for the culture of oceanic ultraphytoplankton. J. Phycol. 23: 633-638.
14. Li, W.K.W., Dickie, P.M., Irwin, B.D. and A. M. Wood. 1992. Biomass of bacteria, cyanobacteria, prochlorophytes and photosynthetic eucaryotes in the Sargasso Sea. Deep Sea Res. 39: 501-519.
15. Matrai, P.A. and M.D. Keller. 1994. Total organic sulfur and dimethylsulfoniopropionate in marine phytoplankton: intracellular variations. Mar. Biol. 119: 61-68.
16. Neveux, J. and F. Lantoine. 1993. Spectrofluorometric assay of chlorophylls and phaeopigments using the least squares approximation technique. Deep Sea Res. 40: 1747-1765.
17. Olson, R. J., Chisholm, S. W., Zettler, E. R. and E. V. Armbrust. 1988. Analysis of *Synechococcus* pigment types in the sea using single and dual beam flow cytometry. Deep-Sea Res. 35: 425-440.
18. Olson, R.J., Chisholm, S.W., Zettler, E.R., Altabet, M. and J. Dusenberry. 1990. Spatial and temporal distributions of prochlorophytes picoplankton in the North Atlantic Ocean. Deep Sea Res. 37: 1033-1051.
19. Partensky, F., Blanchot, J., Lantoine, F., Neveux, J. and D. Marie. Vertical structure of picoplankton at different trophic sites of the subtropical Northeastern Atlantic Ocean. Deep- Sea Res, in press.
20. Simon, N., Barlow, R.G., Marie, D., Partensky, F. and D. Vaulot. 1994. Characterization of oceanic photosynthetic picoeukaryotes by flow cytometry. J. Phycol. 30: 922-935.
21. Vaulot, D., Partensky, F., Neveux, J., Mantoura, R.F.C. and C. Llewellyn. 1990. Winter presence of prochlorophytes in surface waters of the northwestern Mediterranean Sea. Limnol. Oceanogr. 35: 1156-1164.
22. Williams, R. and H. Claustre. 1991. Photosynthe tic pigments as biomarkers of phytoplankton populations and processes involved in the transformation of particulate organic matter at the BIOTRANS site (47°N, 20°W). Deep Sea Res. 38: 347-355.

18

STUDIES OF DMS+DMSP AND ITS RELATIONSHIP WITH THE PHYTOPLANKTONIC POPULATIONS IN THE EASTERN MEDITERRANEAN SEA

C. Vassilakos,[1] L. Ignatiades,[1] R. Kwint,[2] and C. Kozanoglou[1]

[1] National Research Centre for Physical Sciences "Demokritos"
153 10 Aghia Paraskevi Attikis, P.O. Box
60228, Athens, Greece
[2] Laboratory For Applied Marine Res. -TNO-IMW
P.O. Box 57
1780 AB Den Helder
The Netherlands

INTRODUCTION

The most important volatile sulfur compound in sea water has been found to be dimethylsulfide [DMS] which is mainly produced from ß-dimethylsulfoniopropionate [DMSP] by an enzymatic decomposition (21,24). DMSP is contained in marine phytoplankton possibly as a regulator of cellular osmotic pressure and it is assumed to be excreted directly from the cells as a product of algal metabolism (23). DMS constitutes about 90% of the biogenic sulfur emissions from the ocean to the atmosphere and may play an important role in climate regulation.

The main purpose of the project described herein was to measure the concentrations of DMS+DMSP in relation to certain phytoplanktonic variables (primary production, species composition, and chlorophyll *a*) as well as environmental variables (temperature, salinity, nutrients), since there are significant gaps in the literature about these relationships (11). This is the first report of field measurements of DMS+DMSP in the Eastern Mediterranean Sea.

MATERIALS AND METHODS

The experimental work (6 cruises) was carried out at 3 stations (Fig 1) in Saronicos Gulf, Aegean Sea, during the period December 1994-February 1995. Water samples were collected from the surface microlayer (upper 1 mm) as described in Ignatiades (6) and from

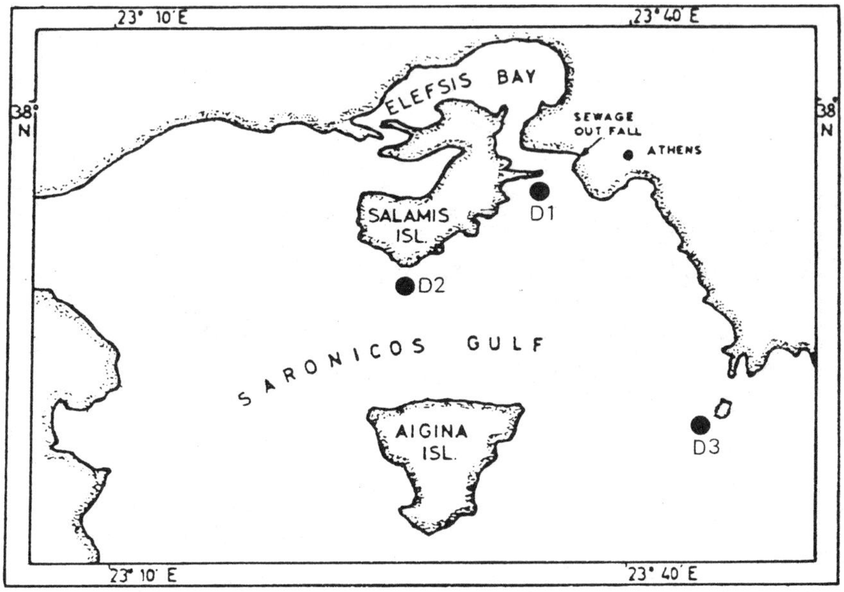

Figure 1. Location of sampling stations.

the bulk water (1, 10, 20, 30, 40 m depth) with a van Dorn water bottle. Temperature and salinity were recorded for each sampling depth.

All samples were filtered on board using glass fiber filters (Sartorious, SM13400) and analysed for chlorophyll *a*, phosphates, nitrates, nitrites, and silicates (20) and ammonia (15). Lugol's iodine was used to preserve subsamples for species identification and enumeration in an inverted microscope.

Photosynthetic productivity was measured by the ^{14}C -technique (19). The samples from the surface microlayer and the different depths of bulk water were dispensed to 125 ml polycarbonate bottles (two light and one dark for each depth) and each bottle was injected with 4 μCi $NaH^{14}CO_3$. The bottles with the subsurface samples were incubated *in situ* at the collection depths (1,10,20,30,40 m) whereas the bottles with the surface microlayer samples were placed in a specially designed apparatus (7), allowing incubation at the surface microlayer.

Samples for DMS+DMSP were also collected from each sampling depth, dispensed to 9 ml vials (8 ml sea water plus 1 ml 10 N NaOH) and sent to Dr. Kwint for analysis. The analytical procedure was as follows: The DMS+DMSP water samples were poured into a glass purge vessel equipped with a glass fritt and purged with high grade helium at 45 $ml.min^{-1}$ for 10 min. The purge gas, containing the volatile compounds, was dried using a Nafion permeation drier (Dupont, model MD-125 P/F). Nitrogen was used as a drying gas (100 $ml.\ min^{-1}$). The dried helium was led through a cold trap, consisting of a straight glass tube containing 200 mg Tenax-ta 60/80 (Chromopack). This cold trap was placed horizontally over a Dewar flask filled with liquid nitrogen. Cooling was achieved with aluminium strips, placed over the glass tube, into the nitrogen in order to achieve a temperature of -120°C. The temperature was checked periodically. After purging, the collection tube was closed at both ends with Swagelock caps (stainless steel, fitted with teflon ferrules) and stored in liquid nitrogen until analysis (adapted from Lindqvist (16)). Storage tests with calibration gas show that samples can be stored in this way for at least 8 weeks without change.

Table 1. Physical and chemical parameters recorded at the three stations during the six sampling periods

Date	St.	DEPTH (m)	TEMP. ^0C	SAL ppt	P-PO_4^{3-} μM	N-NH_3 μM	N-NO_3^- μM	N-NO_2^- μM	Si-SiO_2 μM
5/12/1994	D1	S	18.2	36.0	0.83	3.08	0.8	0.27	4.79
		1	18.1	36.0	0.72	2.57	0.74	0.23	4.62
		10	18.1	36.2	0.72	2.66	0.69	0.29	1.18
		20	18.1	36.2	0.57	2.03	0.52	0.27	3.24
		30	18.1	36.5	0.54	2.01	0.54	0.26	2.81
		40	18.1	37.2	0.48	1.07	0.56	0.23	3.10
		MEAN	18.1	36.4	0.64	2.24	0.64	0.26	3.29
6/12/1994	D2	S	18.2	37.1	0.41	1.76	0.72	0.27	3.62
		1	18.2	37.2	0.29	1.41	0.30	0.23	3.26
		10	18.2	37.7	0.26	0.81	0.41	0.21	2.58
		20	18.2	37.3	0.05	0.36	0.32	0.20	1.69
		30	18.2	37.1	0.05	0.16	0.30	0.20	2.14
		40	18.2	37.2	0.19	0.43	0.26	0.18	2.08
		MEAN	18.2	37.3	0.21	0.82	0.38	0.22	2.56
24/1/1995	D1	S	14.7	38.1	0.14	0.87	0.61	0.16	6.04
		1	14.7	38.0	0.15	0.49	0.56	0.07	3.24
		10	14.7	38.3	0.15	0.48	0.38	0.08	3.80
		20	14.8	38.2	0.14	0.40	0.25	0.14	3.20
		30	14.8	38.2	0.19	0.45	0.31	0.14	2.22
		40	14.8	38.8	0.17	0.45	0.45	0.14	2.10
		MEAN	14.7	38.3	0.16	0.52	0.42	0.12	3.43
25/1/1995	D3	S	15.2	39.0	0.15	0.41	0.21	0.07	2.86
		1	15.2	38.9	0.07	0.14	0.15	0.07	2.68
		10	15.2	38.8	0.10	0.05	0.06	0.04	2.56
		20	15.1	38.8	0.05	0.09	0.07	0.03	2.31
		30	15.1	38.8	0.08	0.09	0.06	0.01	2.35
		40	15.1	38.8	0.09	0.08	0.06	0.11	1.92
		MEAN	15.2	38.9	0.09	0.14	0.10	0.05	2.45
8/2/1995	D3	S	14.8	38.1	0.11	0.15	0.31	0.05	7.15
		1	14.8	38.1	0.09	0.12	0.08	0.04	4.9
		10	14.8	38.2	0.17	0.11	0.07	0.06	6.13
		20	14.7	38.1	0.09	0.10	0.08	0.09	4.87
		30	14.7	38.1	0.06	0.12	0.09	0.08	3.96
		40	14.8	38.1	0.05	0.13	0.10	0.06	4.62
		MEAN	14.8	38.1	0.07	0.12	0.12	0.06	5.27
9/2/1995	D1	S	14.8	38.7	0.17	0.37	0.60	0.14	5.80
		1	14.8	38.7	0.19	0.33	0.33	0.13	5.46
		10	14.8	38.7	0.15	0.09	0.48	0.10	5.28
		20	14.8	38.4	0.19	0.07	0.29	0.09	4.06
		30	14.8	38.6	0.11	0.11	0.30	0.11	4.69
		40	14.8	38.7	0.14	0.31	0.24	0.13	3.24
		MEAN	14.8	38.6	0.16	0.21	0.37	0.12	4.76

The samples were analysed according to Lindqvist (16) on a VARIAN 3700 gas chromatograph equipped with a capillary linear plot column and a photo-ionization detector (PID) of 10.2 eV and with nitrogen as a carrier gas (column flow 10 ml.min^{-1}). The detection limit for DMS was 1.5 pmol; calibration was performed using DMS permeation tubes in a dynamic dilution system. The coefficient of variation (CV) for the DMS analyses for independent analyses was no larger than 5% (13,14).

Results and Discussion

Stations D1 and D2 were located in the western part of the Saronicos Gulf and are affected by sewage (after its primary treatment) whereas station D3 was located in the eastern Gulf quite far from the sewage outfall (Fig 1). The depth of the water column at each station was 40 m.

Table 1 shows the values of certain physical and chemical parameters in seawater sampled at the experimental stations. Temperature and salinity values did not display a significant difference from surface to bottom indicating the strong mixing conditions that prevailed for December to February in the Gulf. In December the seawater was rather warm (mean temp. 18.1 -18.2 °C) but temperature dropped in January (mean: 14.7-15.6 °C) and February (14.7-14.8 °C). Salinity was lower in December (mean: 36.35-37.2 ppt) but increased in January (mean: 38.2-38.8 ppt) and February (mean: 38.1-38.6 ppt).

The highest nutrient values (Table 1) were recorded in December at station D1 (mean: P-PO_4^{3-} : 0.64 μM; N-NH_3 : 2.32 μM; N-NO_2^-: 0.64 μM; N-NO_2^-: 0.08 μM) and these levels characterised this station as eutrophic (Ignatiades et al., 1992). However, trophic conditions at station D1 changed in January and February since the nutrient concentrations decreased by 2-3 times corresponding to mesotrophic levels and approaching the values of the mesotrophic station D2. Station D3 was characterised as oligotrophic since the mean nutrient concentrations in January and February ranged as follows: P-PO_4^{3-} : 0.09-0.07 μM; N-NH_3 : 0.14-0.12 μM; N-NO_3^- : 0.10-0.12 μM; N-NO_2^-: 0.05-0.06 μM . Silicate concentrations were lower in the December and January samples (mean range: 2.45-3.34 μM) and higher in the February (mean range: 4.76-5.27 μM) samples. Nutrient values at the surface microlayer were in most cases higher than at 1 m depth.

The graphical presentations of the vertical distribution of chlorophyll *a,* primary production and DMS+DMSP for the three stations are given in Figs 2a, 2b and 2c. Chlorophyll *a* concentrations showed a gradient among sampling periods and stations being higher at stations D1 (mean over depth: 1.34 mg/m^3) and D2 (mean over depth: 1.43 mg/m^3) in December and lower in January and February (station D1: mean over depth 0.42, 0.20 mg/m^3 respectively; station D3: means over depth 0.2, 0.11 mg/m^3 respectively). The vertical distribution of chlorophyll *a* was higher at the surface microlayer in most cases but irregular in the bulk water.

Primary production was also higher in December (mean: 2.27 and 1.59 mgC/m^3.h) at stations D1 and D2 respectively) but declined in January and February with means of 0.50-1.03 mgC/m^3.h (station D1) and 0.21-0.91 mgC/m^3·h (station D2). The vertical profiles of primary production showed that in all cases maximum values were recorded at 1 m depth and these data are in agreement with previous observations in this area (7), which showed reduced rates of carbon fixation in the surface microlayer.

The levels of DMS+DMSP in seawater showed a well defined trend from the eutrophic to oligotrophic conditions. Thus, the mean concentration at station D1 was 11.30 nM in December, 7.2 nM in January and 10.86 nM in February, at station D2 12.67 nM (January) whereas at station D3 the mean values ranged from 4.79 nM (January) to 8.84 nM (February). The vertical distribution of this parameter did not show a clearly defined pattern.

Table 2. Quantitative and qualitative data of phytoplanktonic communities collected at the three stations during the six sampling periods (D = Diatoms; DF = Dinoflagellates; C = Coccolithophores; S = Silicoflagellates; O = Others)

Date/Station	Total cells	Taxa					Dominant species
	(cells/l)	D %	DF %	C %	S %	O %	
5/12/1994 D1	95000	65	1	3	1	30	*Phaeocystis poucheti* *Chaetoceros curvisetus*
6/12/1994 D2	98000	48	1	3	1	49	*Phaeocystis poucheti* *Chaetoceros curvisetus*
24/1/1995 D1	25000	69	12	12	1	7	*Chaetoceros curvisetus* *Pontosphaera steueri*
25/1/1995 D3	4000	36	25	34	2	5	*Emiliania huxleyi* *Nitzschia delicatissima*
8/2/1995 D3	4100	58	8	30	2	4	*Leptocylindrus danicus* *Bacteriastrum delicatulum*
9/2/1995 D1	20100	64	2	18	1	21	*Chaetoceros curvisetus* *Phaeocystis poucheti*

Table 2 presents the taxonomic groups of phytoplankton recorded at each station and sampling depth. The total mean number of cells was higher at station D1 (range: $2.0x10^4$ - $9.5x10^4$ cells/l) and station D2 ($9.8x10^4$cells/l) whereas at station D2 it was considerably lower (range: $4.0x10^3$- $4.1x10^3$ cells/l). Diatoms were the dominant taxa (36-68% of the total population) at all stations and depths. "Others" (unclassified blue greens, flagellates, *Phaeocystis poucheti*) were very important (30-49% of the total population) at stations D1 and D2 in December. The coccolithophores exhibited significant growth at station D3 (29-33% of the total population) in January and February.

It has been well documented that DMS production in sea water results mostly from phytoplankton activity (18,23). However, there is a disagreement in the literature concerning the relationships among DMS concentration and certain phytoplanktonic variables. Some investigators (9,17) reported a good correlation between DMS and chlorophyll *a* concentrations in sea water whereas others (18,22) found no relationship between these parameters. Proportionality of DMS production to cell biomass in unialgal cultures has been reported by Ackman *et al.*, (1) and Vairavamurthy *et al.*, (23).

In the present investigation, the Spearman's rank correlation analysis showed statistically significant relationships between DMS+DMSP and a) chlorophyll *a* ($r = 0.62$; $p=0.0001$), b) primary production ($r =0.41$; $p= 0.01$), c) total cell concentration ($r=0.66$; $p=0.0001$) and d) temperature ($r=0.51$; $p=0.001$). These relationships suggest that DMS+DMSP concentrations result mostly from phytoplankton activity as reported by Iverson *et al.*, (9), but temperature may also play a significant role as reported by Kiene and Service (12). The significant relationship between primary production and DMS+DMSP concentration has been also demonstrated by Andreae and Barnard (2).

The taxonomic composition of phytoplankton may also play an important role in DMS production in sea water. Thus, Turner *et al.*, (22) found no correlation between total chlorophyll *a* and DMS concentration in their samples but they did find significant relationships of DMS and the chlorophyll *a* content of certain taxa (coccolithophores, some flagellates). Barnard *et al.*, (4) attributed the high concentrations of DMS in the Bering Sea to high abundances of the prymnesiophyte *Phaeocystis pouchetii* and Malin *et al.*, (17) found that the coccolithophore species *Emiliania huxleyi* was a significant DMS/DMSP producer in the N.E. Atlantic Ocean.

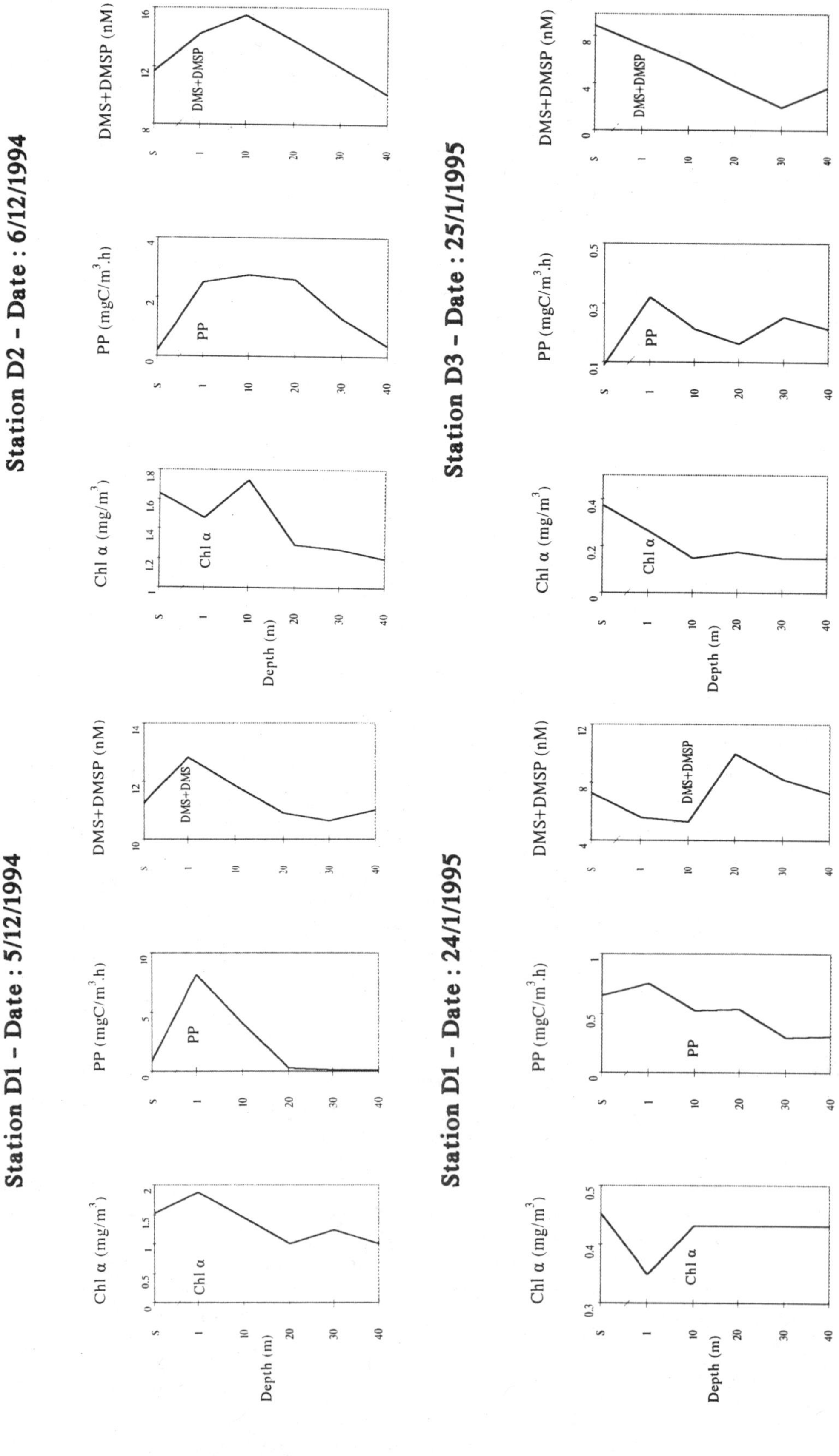
Station D2 - Date : 6/12/1994
Chl α (mg/m^3)
PP (mgC/m^3.h)
DMS+DMSP (nM)
Chl α
PP
DMS+DMSP
Depth (m)
Station D3 - Date : 25/1/1995
Chl α (mg/m^3)
PP (mgC/m^3.h)
DMS+DMSP (nM)
Chl α
PP
DMS+DMSP
Depth (m)
Station D1 - Date : 5/12/1994
Chl α (mg/m^3)
PP (mgC/m^3.h)
DMS+DMSP (nM)
Chl α
PP
DMS+DMS
Depth (m)
Station D1 - Date : 24/1/1995
Chl α (mg/m^3)
PP (mgC/m^3.h)
DMS+DMSP (nM)
Chl α
PP
DMS+DMSP
Depth (m)

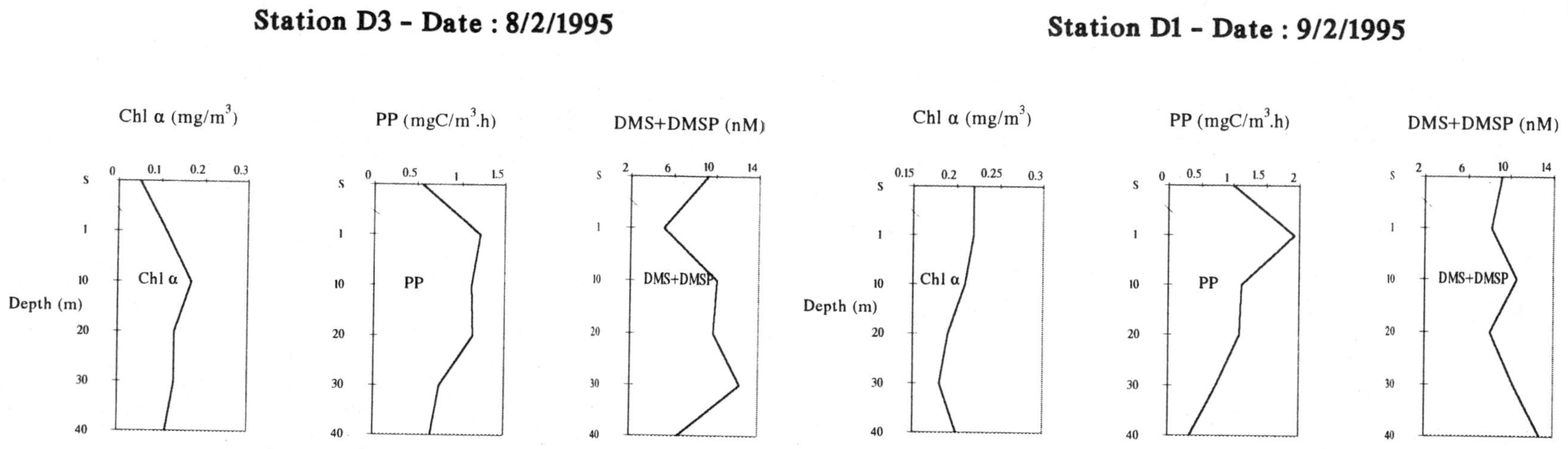

Figure 2. Vertical distribution of chlorophyll *a* (mg/m3), primary production (PP, mgC/m^{3}h) and DMS+DMSP (nM) recorded during the six cruises.

The taxonomic analysis of the present investigation has shown the presence (as dominant) of certain species reported as DMS producers (*Phaeocystis poucheti*, *Emiliania huxleyi*) in some samples (Table 2) but their contribution to the total DMS+DMSP production in these samples cannot be evaluated, since, they coexisted with a strongly developed diatom community which may also play a role in the DMS+DMSP production (18). It is obvious that the species composition as a factor controlling DMS production has not been fully examined in so far it will be one of the major tasks of our investigations in the future.

ACKNOWLEDGMENTS

This work was supported by European Community Contract No EV5V-CT93-0326 (DG 12 SOLS).

REFERENCES

1. Ackman, R.G., C.S. Tocher, and J. McLachlan, 1966. Occurrence of dimethyl-ß-propiothetin in marine phytoplankton. J. Fish. Res. Bd. Can., 23:357-364.
2. Andreae, M.O. and W.R. Barnard, 1984. The marine chemistry of dimethylsulfide. Mar. Chem. 14: 267-279.
3. Belviso, S., P. Buat-Menard, J.P. Putaud, B.C. Nguyen, H. Claustre and J. Neveux, 1993. Size distribution of dimethylsulphoniopropionate (DMSP) in areas of the tropical northeastern Atlantic Ocean and the Mediterranean Sea. Mar. Chem., 44:55-71.
4. Barnard, W.R., M. O. Andreae, and R.L. Iverson, 1984. Dimethylsulphide and *Phaeocystis poucheti* in the southeastern Bering Sea. Cont. Shelf Res., 3: 103-113.
5. Holligan, P.M., S.M. Turner and P.S. Liss, 1987. Measurement of dimethylsulphide in frontal regions. Continental Shelf Res., 7: 213-224.
6. Ignatiades, l., 1987. A simple apparatus for sampling sea surface microlayers. Mar. Ecol. Prog. Ser., 39, 207-208.
7. Ignatiades, L., 1990. Photosynthetic capacity at the surface microlayer during the mixing period. J. Plankton Res., 12: 851-860.
8. Ignatiades, L., M. Karydis and P. Vounatsou, 1992. A possible method for evaluating oligotrophy and eutrophication based on nutrient concentration scales. Mar. Poll. Bull. 24: 238-243.
9. Iverson, R.L., F.L. Nearhoof and M.O. Andreae, 1989. Production of dimethylsulfonium propionate and dimethylsulfide in estuarine and coastal waters. Limnol. Oceanogr. 34: 53-67.
10. Kiene, R.P., 1993. Measurement of dimethylsulphide (DMS) and dimethylsulphoniopropionate (DMSP) in seawater and estimation of DMS turnover rates. In: Handbook of Methods in Aquatic Microbial Ecology, Lewis Publ., 601-610 pp.
11. Kiene, R.P., 1992. Dynamics of dimethyl sulfide and dimethylsulfoniopropionate in oceanic water samples. Mar. Chem., 37: 29-52.
12. Kiene, R.P. and S.K. Service, 1991. Decomposition of dissolved DMSP and DMS in estuarine waters: dependence on temperature and substrate concentration. Mar. Ecol. Prog. Ser., 76:1-11.
13. Kwint, R.L.J., and K.J.M. Kramer, 1995. DMS production by plankton communities. Mar. Ecol.-Prog. Ser., 121: 227-237.
14. Kwint, R.L.J., K.J.M. Kramer, A.C. Baart, and H.L.M. Verhagen, 1992. The production of DMS by a plankton community: A mesocosm experiment. In Dimethylsulphide:Oceans, Atmosphere and Climate. Restelli, G. and G. Angeletti (eds) Int. Symp. Proc. Belgirate. Kluwer Acad. Publ. Dordrecht, 53-62 pp.
15. Liddicoat, M.I., J.S. Tibbitis, and I.E. Buttler, 1976. The determination of ammonia in natural waters. Water Res., 10: 567-568.
16. Lindqvist, F., 1989. Sulphur-specific detection in air by photoionization in a multiple detector gas chromatographic system. J. High Res. Chrom. 12:628-631.
17. Malin, G., S. Turner, P. Liss, P. Holligan and D. Harbour, 1993. Dimethylsulphide and dimethylsulphoniopropionate in the Northeast Atlantic during the summer coccolithophore bloom. Deep-Sea Res., 40:1487-1508.

18. Nguyen, B.C., S. Belviso, and N. Michalopoulos, 1988. Dimethylsulphide production during natural phytoplanktonic blooms. Mar. Chem. 24:133-141.
19. Steemann-Nielsen, E., 1952. The use of radioactive carbon ^{14}C for measuring organic production in the sea. J. Cons. Int. Explor. Mer, 18: 117-140.
20. Strickland, T.R. and T.R. Parsons, 1968. A practical handbook of seawater analysis. Fish. Res. Board Can., 167: 1-311.
21. Turner, S.M., and P.S. Liss, 1985. Measurements of various sulphur gases in a coastal marine environment. J. Atmos. Chem., 2:223-232.
22. Turner, S., G. Mallin and P. Liss, 1988. The seasonal variation of dimethyl sulfide and dimethylsulfoniopropionate concentrations in nearshore waters. Limnol. Oceanogr., 33: 364-375.
23. Vairavamurthy, A., M.O. Andreae, and R.L. Iverson, 1985. Biosynthesis of dimethylsulfide and dimethylpopiothetin by *Hymenomonas carterae in relation to sulfur source and salinity variations. Limnol. Oceanogr. 30: 59-70.*
24. Wakeham, S.G., B.L. Howes and J.W.H. Dacey, 1984. Dimethyl sulfide in a stratified coastal salt pond. Nature, 310:770-772.

19

ACRYLATE AND DIMETHYLSULFONIOPROPIONATE (DMSP) CONCENTRATIONS DURING AN ANTARCTIC PHYTOPLANKTON BLOOM

New Sources of Reduced Sulfur Compounds

John A. E. Gibson,[1] Kerrie M. Swadling,[2] and Harry R. Burton[1]

[1] Australian Antarctic Division
Channel Highway, Kingston
Tasmania 7050, Australia
[2] Department of Zoology
University of Tasmania
GPO Box 252C, Hobart
Tasmania 7001, Australia

SUMMARY

Very high concentrations of dimethylsulfide (DMS) and dimethylsulfoniopropionate (DMSP) have been regularly reported from around the Antarctic coastline, and have been associated with blooms of the prymnesiophyte alga *Phaeocystis* sp. affin. *antarctica*. In this study, acrylate and DMSP concentrations were measured at an inshore marine site during the 1994-5 summer Antarctic phytoplankton bloom. The concentration of DMSP was determined by measuring the difference between the amount of acrylate before and after hydrolysis with NaOH. The concentration of acrylate rose from below the detection limit (ca. 10 nM) in early November to over 1.20 μM by 7 December. Total DMSP also reached a maximum of 2.47 μM on this date. Concentrations then dropped to below the detection limit by the end of January. Correlation matrix analysis of acrylate and DMSP concentrations and various parameters revealed significant positive correlations with cell counts of *Cryptomonas sp.*, the dominant phytoplankton species during December, and dinoflagellates. *Phaeocystis* sp. affin. *antarctica* was present during the study in low numbers and probably contributed little to the DMSP pool. This is the first study which suggests that cryptomonads and dinoflagellates may be important producers of DMSP in the Antarctic environment.

Biological and Environmental Chemistry of DMSP and Related Sulfonium Compounds
edited by Ronald P. Kiene et al., Plenum Press, New York, 1996

INTRODUCTION

A number of recent studies have suggested that the Antarctic region is very important in the global sulfur cycle. Very high dimethylsulfide (DMS), acrylate and dimethylsulfoniopropionate (DMSP) concentrations have been recorded in inshore waters close to Australia's Davis scientific base, East Antarctica (8, 12, 13, 18, 25), and on the opposite side of the continent in the Weddell Sea (11). Anecdotal evidence (4, P. Nichols and H. Marchant, personal communication) suggests that high concentrations of these compounds are widespread near the Antarctic continent. The peaks in DMS, acrylate and DMSP concentrations are very seasonal (12, 13), occurring for only relatively short periods during intense summer phytoplankton blooms. In most of these studies, it has been concluded that the high levels of the compounds were produced by the Prymnesiophyte alga *Phaeocystis* sp. affin. *antarctica* (this species was previously included in *P. pouchetii* (19)). Similar conclusions have also been reached in studies undertaken in the northern hemisphere (2, 17, 24), where other species of *Phaeocystis* were present. In Antarctic studies undertaken during periods other than the peak phytoplankton bloom, much lower DMS and DMSP concentrations have been recorded (3, 9).

Phaeocystis sp. affin. *antarctica* is often the dominant phytoplankton species in early summer blooms near the Antarctic continent (6, 7, 22). The species is colonial, producing spherical colonies, up to 5 mm across, containing thousands of individual cells embedded in an organic matrix. It is often associated with extremely high biomass during intense Antarctic blooms (6). However, its distribution is quite patchy (H. Marchant, personal communication), and there is no certainty that it will appear at a particular site in every year.

Very few measurements of the concentration of acrylate (the form of acrylic acid dominant at the pH of seawater) have been made in the marine system. Sieburth reported the occurrence of the compound during *Phaeocystis* blooms in the Antarctic (23), and also reported its antibacterial properties. A recent study by Yang et al. (25), carried out in essentially the same area considered in the present study, found high concentrations of acrylate over the summer period, presumably due to the breakdown of DMSP.

As the breakdown of DMSP by base hydrolysis yields equimolar amounts of acrylate and DMS, acrylate released upon base treatment may be used as a proxy for DMSP concentrations in much the same way as DMS routinely is. Because the breakdown of DMSP appears to be the only major source of acrylate in the marine environment, acrylate in the water column reflects DMSP broken down *in situ*. The compound will be lost from the system by bacterial activity (acrylate appears to have antibacterial properties only at high concentration and low pH (5,23)), but is unlikely to be lost by ventilation to the atmosphere. Therefore, any concentrations observed will be the balance of production and loss terms.

In this paper, we report DMSP and acrylate concentrations at a coastal Antarctic site at which *P.* sp. affin. *antarctica* was only a minor component of the summer phytoplankton bloom. The paper highlights the role that other phytoplankton species may play in the Antarctic sulfur cycle.

METHODS

Water samples were collected at a site approximately 1 km offshore from Australia's Davis Base (68° 35′ S, 78° 00′ E) (Figure 1) on 11 occasions over the period from early November 1994 to the end of January 1995. The water depth at this site was 23 m. The sample site was covered by a layer of ice approximately 1.6 m thick from the start of the study until ice breakout, which occurred as a result of a combination of high winds and tidal

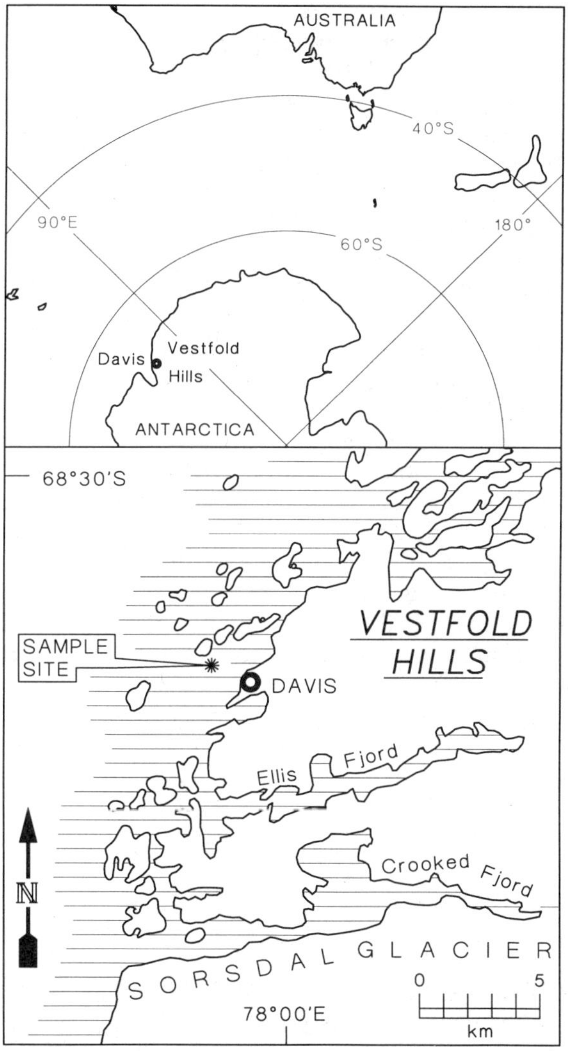

Figure 1. Map of the Davis region showing the sample site.

action on 13 January 1995. Prior to ice breakout, samples were collected from depths of 2 and 10 m using a polycarbonate Kammerer bottle deployed through a hole drilled in the sea ice. Extra samples were obtained from 5, 15 and 20 m on 7 December 1994. After ice breakout, samples were collected from 0 and 10 m from a small boat.

Single water samples for analysis of acrylate and DMSP were transferred to acid-washed polyethylene bottles, and were stored at -20°C until analysis was performed. Analyses were undertaken within 2 to 4 weeks of sampling. Reanalysis of some individual samples after different periods of storage within this time range revealed that delays in analysis had little effect on concentrations. Samples for nutrient analyses were also stored in acid-washed polyethylene bottles at -20°C. Samples for the enumeration and identification of phytoplankton were preserved with Lugol's Iodine, and were stored in glass bottles at room temperature.

Acrylate and DMSP were determined in the following way. Samples for the measurement of dissolved acrylate (A) were filtered gently through a 0.2 μm syringe filter. To determine total DMSP (DMSP(T)+A, 0.25 mL of 5 M NaOH was added to 10 mL of an unfiltered seawater sample. This resulted in the conversion of DMSP to acrylate and DMS. The basified samples were stored in sealed vials at 4°C overnight, reacidified with 100 μL concentrated HCl to redissolve the precipitated calcium carbonate and filtered as above. The total concentration of DMSP (DMSP(T)) which included both particulate and dissolved fractions was calculated by subtracting dissolved acrylate concentration from the acrylate measured in the DMSP(T)+A samples.

Acrylate was measured using a Dionex HPLC system. Separation of acrylate from other short chain organic acids was achieved using a Waters μBondapak C-18 column employing 0.1% H_3PO_4 as the eluent at a flow rate of 1.0 mL min^{-1}. Acrylate was detected with a Waters 410 UV detector at 210 nm. The injection volume was 250 μL. Peaks attributable to other short chain fatty acid anions (e.g. formate and acetate) were also observed, but the acrylate peak (retention time = 9.1 min) was well separated from those of the other acid anions. Peak areas were measured using a Maclab integration package. Standard solutions were prepared from pure acrylic acid (Aldrich) in artificial seawater. The detection limit for acrylate was approximately 10 nM. Analysis of each sample was repeated and the mean calculated. Precision for repeat analyses was generally better than 3%.

Nutrients (NO_3^-, PO_4^{3-} and H_4SiO_4) and chlorophyll *a* (Chl *a*) concentrations were determined by standard wet chemical methods (21). Water temperature and conductivity were measured using a submersible data logger (Platypus Engineering, Hobart, Tasmania) and salinity calculated (10). Subsamples of preserved water samples collected for phytoplankton identification and enumeration were allowed to sediment first in measuring cylinders and then in counting chambers. At least 15 fields of view containing at least 300 cells were counted using a Leitz Laborlux inverted microscope at a magnification of 400x.

RESULTS

Seasonal distributions of acrylate and DMSP(T) concentrations are shown in Figure 2. During November, the concentrations were relatively low, with only DMSP(T) being detected early in the month. Acrylate was not detected until 7 December,when the concentration at 2 m increased to 1.21 μM. DMSP(T) also increased dramatically to 2.47 μM. Concentrations decreased during the rest of December, before rising briefly in early January. By the end of January, both acrylate and DMSP(T) were present at levels which were below the detection limit of the method used. Before the breakout of the ice (when stratification precluded water mixing), concentrations of acrylate and DMSP(T) were lower at 10 m than directly under the ice at 2 m, but were comparable later in the study when the water column was mixed (Fig. 2B).

Chl *a*, a measure of phytoplankton biomass, was low during November, rose sharply on 7 December and dropped again by 14 December (Figure 3). The concentration of Chl *a* rose again in late December and early January, before tapering off late in the study.

The major groups of phytoplankton present in the water samples were *Phaeocystis* sp. affin. *antarctica*, *Cryptomonas sp.,* diatoms and dinoflagellates (Figure 4). *P.* sp. affin. *antarctica* was observed throughout the study, but only at low numbers in comparison to previous studies (6, 13, 22, 26). Cell counts for this species ranged from close to zero early and late in the study to ca. 6 x 10^5 cells L^{-1} on 7 December (Fig. 4C). Most cells were motile flagellates, and very few colonies were observed.

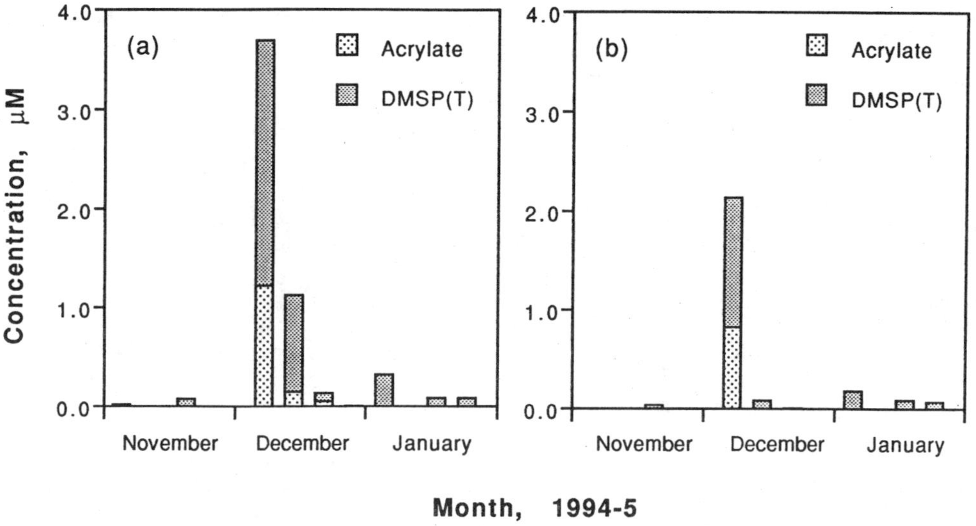

Figure 2. Concentrations of acrylate and DMSP(T) (µM) at (a) 2m and (b) 10 m. Both acrylate and DMSP(T) were beneath the limit of detection on 28 November and 30 January. The total height of the columns is equivalent to the concentration of DMSP(T)+A.

Cryptomonas sp. was the dominant phytoplankter during the initial bloom in early December (Fig. 4A). Cell numbers were approximately 2 x 10^5 cells L^{-1} early in the study, but increased to 1 x 10^8 cells L^{-1} directly beneath the ice on 7 December. The population decreased after this date to approximately 5 x 10^6 cells L^{-1} late in the month before rising again to ca. 2 x 10^7 cells L^{-1} in early January. Cell numbers dropped for the rest of the month to about 5 x 10^5 cells L^{-1}.

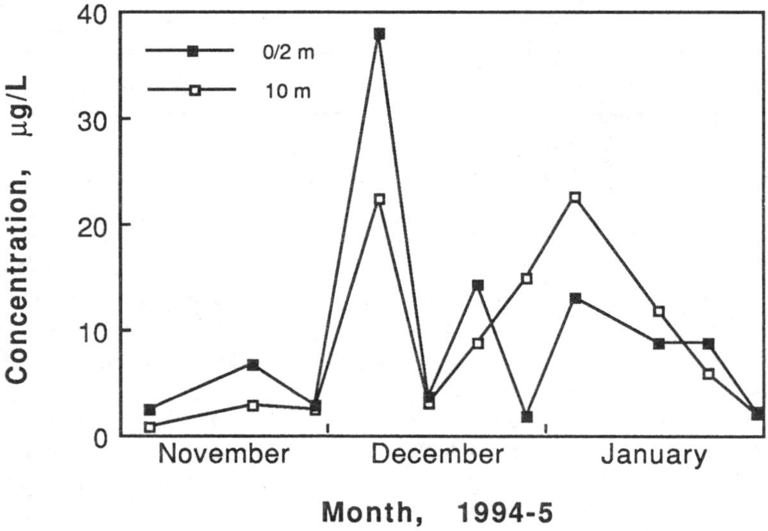

Figure 3. Concentrations of chlorophyll *a* (µg L^{-1}) at the sampling station near Davis, Antarctica.

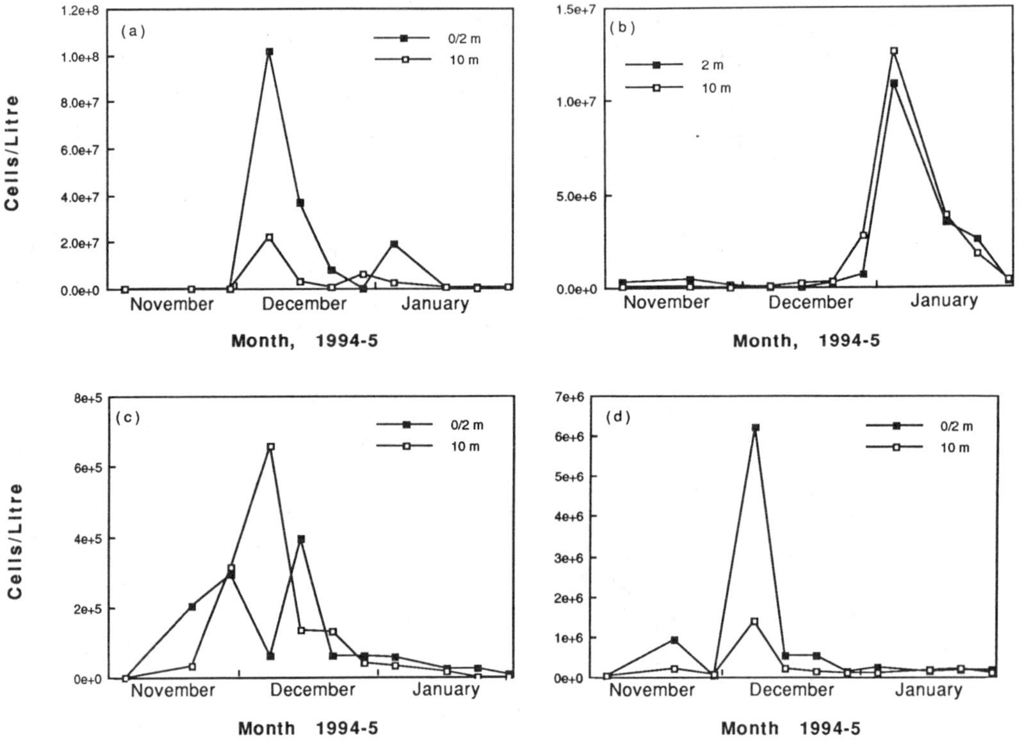

Figure 4. Phytoplankton cell counts: (a) *Cryptomonas sp.*, (b) diatoms, (c)*Phaeocystis sp.* affin. antarctica and (d) dinoflagellates. Note the different scales in each plot.

Diatom numbers were low at the start of the study, and did not begin to increase significantly until the end of December (Fig. 4B). The maximum population was recorded on 4 January, when cell numbers reached 1.26 x 10^7 cells L^{-1} at 10 m. The species of diatoms present varied during the study. In early November, *Entomoneis kjellmanii* (Cleve) Thomas was the most abundant taxon. This species is a member of the sea-ice community that develops in spring. In late December, *Nitzschia* and *Fragilariopsis* species were the dominant taxa, but were replaced by *Thalassiosira dichotomica* (Kozl.) Fryx. and Hasle after ice breakout.

Dinoflagellates were present in significant numbers throughout the study (Fig. 4D). In November, cell numbers were in the range 2 x 10^4 - 1 x 10^6 cells L^{-1}, with a small heterotrophic *Gymnodinium* species (10 x 15 μm) the most common taxon. This species increased in numbers to 6 x 10^6 cells L^{-1} beneath the ice on 7 December before largely disappearing. Other species, including *Gyrodinium* sp., *Dinophysis* sp., *Protoperidinium* spp. and *Amphidinium* sp., became more common at the end of December, and cell numbers remained in the range 1 - 5 x 10^5 cells L^{-1} for the rest of the study.

In order to gain an insight into which parameters or species were possibly involved with the production of DMSP and acrylate, a correlation matrix was calculated to investigate relationships between DMSP(T)+A, DMSP(T) and acrylate concentrations, and a variety of physical and biological parameters, including phytoplankton cell counts. The relevant parts of the matrix are shown in Table 1. The strongest correlations occurred between the

Table 1. Linear correlation (r) table of acrylate (A), DMSP(T) and DMSP(T)+A, phytoplankton counts and chemical and physical parameters. All data was usedfor the correlations

	Acrylate	DMSP(T)	DMSP(T)+A
Cryptomonas	0.904*	0.944*	0.940*
Diatoms	-0.217	-0.144	-0.172
Phaeocystis	0.551*	0.526*	0.541*
Dinoflagellates	0.922*	0.875*	0.902*
Chlorophyll *a*	0.798*	0.762*	0.784*
Salinity	0.261	0.178	0.211
Temperature	-0.319	-0.289	-0.302
Phosphate	0.121	0.004	0.044
Silicate	0.269	0.224	0.241
Nitrate	-0.166	-0.220	-0.203

* = $P \leq 0.01$

concentrations of the compounds and cell counts of *Cryptomonas sp.* and total dinoflagellates. Weaker correlation was observed with Chl *a* and still lower correlation with cell counts of diatoms, *P.* sp. affin. *antarctica,* or with any of the physical or chemical parameters. The correlation with Chl *a* should be viewed with some caution, however, as it is strongly biased by the data from one day when the concentrations of Chl *a* and DMSP(T)+A were highest (7 December: r = 0.973 - 0.998, n = 5). Exclusive of this date, the correlation was much weaker (r = 0.045 - 0.089, n = 20). The correlation coefficient between *Cryptomonas sp.* and total dinoflagellate cell counts was 0.939.

DISCUSSION

The data for acrylate in this paper are among the first to be reported for the marine environment. Concentrations observed in this study were of the same order of magnitude as those of acrylate and DMS found previously at essentially the same site (12, 13, 25). DMSP(T) has not been previously measured during an Antarctic bloom, but the ratio of acrylate:DMSP(T)+A appears to be higher (ca. 0.35) than the equivalent ratio of DMS:(DMSP+DMS) found in studies of European coastal waters (17). This ratio may, in part, reflect the efficiency of the microbial community in converting DMSP to DMS and acrylate.

The correlation of acrylate, DMSP(T) and DMSP(T)+A, to Chl *a* throughout the study, was not particularly strong, except on 7 December. This was related in part to the composition of the phytoplankton community. Diatoms predominated in January and they are not large producers of DMSP on a per cell basis (14). On 7 December, the date of highest acrylate (A), DMSP(T) and DMSP(T)+A concentrations, there was a much stronger correlation, suggesting that the species contributing most of the Chl *a* at this time were involved in the production of DMSP. The ratio of acrylate to Chl *a* on this date was 3.28×10^{-5} mol (mg Chl a)$^{-1}$ (n = 5), which was 2 - 10 times greater than the values of the equivalent DMS:Chl a ratio found in studies of phytoplankton blooms in European coastal and shelf waters (17). The ratio DMSP(T)+A:Chl *a*, 8.47×10^{-5} mol (mg Chl *a*)$^{-1}$, was approximately twice the ratio found for DMSP:Chl *a* in the European studies. Obviously these ratios do not reflect total acrylate and DMSP production, as they do not take into account consumption or degradation of the compounds. The conclusion to be drawn from these results is that the phytoplankton community producing the high acrylate and DMSP(T) concentrations were more significant producers of these substances, per unit biomass (Chl *a*), than those in the European populations, or, conversely, that biological degradation was less efficient.

Even though there was a statistically significant correlation ($P \leq 0.01$) between cell counts of *Phaeocystis* sp. affin. *antarctica* and DMSP(T)+A, DMSP(T) and acrylate, it is unlikely that *Phaeocystis* sp. affin. *antarctica* was the major producer of DMSP during this study. The coefficients of correlation, circa 0.5, of *P.* sp. affin *antarctica* cell counts with DMSP(T)+A, DMSP(T) and acrylate were considerably less than for *Cryptomomonas sp.* (0.9) and dinoflagellates (0.9). Cell counts of *P.* sp. affin. *antarctica* were much lower than in previous studies at the same site (13, 25), and, if this species were the major producer, the per cell production of DMSP (ca. 2500 fmol cell^{-1}) would be approximately 2 orders of magnitude higher than found previously at this site (13, 25). The *P.* sp. affin. *antarctica* present undoubtedly contributed some DMSP to the total, but it must be concluded that this was a relatively minor fraction of total DMSP.

The correlation between cell counts of *Cryptomonas sp. and DMSP(T)+A suggests that this species was a major producer of DMSP. Cryptomonas sp.*, a member of the class Cryptophyceae, is reported regularly from throughout the Antarctic region and has been has been an important member of the phytoplankton community of the Davis area during summers over the period 1992-5 (J. Gibson, unpublished results). It was not recorded during earlier studies in the 1980s (6, 22).

Assuming the DMSP(T) and acrylate were totally produced by this species, the apparent per cell production of acrylate and DMSP+A by *Cryptomonas sp.* was 12 fmol cell^{-1} and 34 fmol cell^{-1} respectively. The acrylate quota is of the same order of magnitude as previously reported for DMS produced by *Phaeocystis* sp. in polar waters (13 - 87 fmol cell^{-1}) (2, 13, 7, 25). *Cryptomonas sp. has a slightly larger cell size (10-15* μm) than *P.* sp. affin. *antarctica*, but any conclusions regarding the amount of DMSP on a per unit carbon basis should also take into account the large quantity of mucilage produced by the latter species.

Keller et al. (14) found that only one member of the class Cryptophyceae, out of eight tested, produced significant amounts of DMSP. This species was an unidentified *Cryptomonas* isolated from an oceanic area. If the high concentrations of DMSP and acrylate in this study were produced by *Cryptomonas sp.*, this species could be added to the list of Cryptophyceae that produce relatively large amounts of DMSP.

The correlation matrix (Table 1) also demonstrates a strong correlation between total dinoflagellate numbers and acrylate and DMSP(T)+A. If dinoflagellates were the major producer (*Cryptomonas sp.* producing little if any DMSP), the amount of acrylate and DMSP(T)+A would be approximately 190 and 500 fmol cell^{-1} respectively, which are far higher than previously recorded for *Phaeocystis* sp (17). Several genera of dinoflagellates (e.g. *Amphidinium, Prorocentrum*) were found by Keller (14) to produce considerable DMSP (> 500 fmol·cell^{-1}), whereas others (e.g. *Ceratium, Gyrodinium*) produced little, if any. In this study, the dinoflagellate counts were not separated into different genera or species. Considerable variation occurred during the study in both species and size of the dominant type within the class (ranging from small (less than 15 μm) *Gymnodinium* species early in the study to much larger *Protoperidinium* sp. and *Gyrodinium lachrymosum* in January). Since DMSP production does not appear to be consistent across different dinoflagellate genera, it is unclear how to interpret the strong correlation. It cannot be concluded that dinoflagellates as a class were not the major producers of DMSP during the study period.

Epifluorescence microscopy revealed that the small, unidentified, *Gymnodinium* species which was the dominant dinoflagellate species on 7 December, did not contain chlorophyll, and was thus heterotrophic. Only one heterotrophic dinoflagellate, *Crypthecodinium cohnii*, has been previously demonstrated to produce DMSP (14, and references therein). If *Gymnodinium sp.* were responsible for the bulk of the production of the DMSP and acrylate measured on this date, the correlation between these compounds, the cell counts of *Cryptomonas sp.*, and the concentration of Chl *a*, must have been fortuitous. The coefficient of correlation between dinoflagellates and

Cryptomonas sp. on this date was 0.962 (n = 5). It is also possible that this heterotrophic dinoflagellate could accumulate DMSP from its prey.

The data support the contention that diatoms, on a per cell or per unit biomass level, are not significant producers of DMSP in the oceans near the Antarctic continent. At the time of maximum DMSP concentrations, diatoms were nearly absent, although the smaller increase in DMSP during early January, the period of the main diatom bloom, may be attributable to production by diatoms or *Cryptomonas sp.* and dinoflagellates, which also increased in numbers at this time. Diatoms do produce some DMSP (16), and the diatoms present would have undoubtedly added to the DMSP pool.

The identity of phytoplankton species that were responsible for the production of the majority of the DMSP measured in this study cannot be identified with complete certainty from the data obtained. Though it appears likely that *Cryptomonas sp.* and/or dinoflagellates were the taxa responsible, it remains possible that other, possibly fragile or small species that were not efficiently sedimented or identified in the settling/microscopy process produced the bulk of the DMSP, or at least contributed significantly.

None of the other parameters listed in Table 1 show strong positive or negative correlations to DMSP. Perhaps most importantly, there was no correlation between DMSP and the *in situ* concentration of nitrate. It has been suggested that increased DMSP production by phytoplankton is a response to a N-limited environment (1, 24). This interaction was not evident in the sea ice community studied here.

The enhanced production of DMSP by Antarctic phytoplankton could be related to the life histories of the species. *Cryptomonas sp.*, Phaeocystis sp. affin. *antarctica* and numerous dinoflagellate species have been reported to occur in the sea-ice (19), where they probably survive during winter. It appears that the high salinitiy and low temperature condition in the ice during winter are common experiences for many species. Kirst et al. (15) reported very high DMSP concentrations in sea ice, and attributed this to the production of the compound by ice algae. Synthesis of high internal concentrations of DMSP, to combat osmotic or cold stress could be a characteristic shared by many types of phytoplankton that exist in this environment.

The conclusion to be drawn from this study is that, in the Antarctic marine environment, other species of phytoplankton apart from *P.* sp. affin. *antarctica*, are capable of producing significant amounts of DMSP. Thus, if *P.* sp. affin. *antarctica* is absent from a particular phytoplankton bloom, considerable amounts of DMSP, and therefore DMS, can still be produced by the other organisms comprising the bloom. A key factor in the production of high concentrations of DMSP in the Antarctic region appears to be the existence of the sea-ice. A greater understanding of the behaviour of algal cells in this environment, as well as their response (in terms of DMSP production) to release from the melting ice, will be required to understand the role of biota in the Antarctic sulfur cycle more completely.

ACKNOWLEDGMENTS

We wish to thank the 1994 wintering expeditioners at Davis Station, Antarctica, especially Dr H. Cooley and T. Newton, for assistance in the field, as well as the Australian Antarctic Division, who provided logistic support. Andrew Davidson provided useful discussion. Three reviewers provided suggestions which substantially improved the manuscript.

REFERENCES

1. Andreae, M.O. 1986. The ocean as a source of atmospheric sulfur compounds, p 331-362. *In* P. Buat-Menard (ed), The Role of Air-Sea Exchange in Geochemical Cycling. Reidel, Dordrecht.

2. Barnard, W.R., M.O. Andreae and R.L. Iverson. 1984. Dimethylsulfide and *Phaeocystis pouchetii* in southeastern Bering Sea. Continental Shelf Research, *3*: 103-113.
3. Berresheim, H. 1987. Biogenic sulfur emissions from the Subantarctic and Antarctic Oceans. Journal of Geophysical Research, *92*: 2930-2938.
4. Bolter, M. and R. Dawson. 1982. Heterotrophic utilization of biochemical compounds in Antarctic waters. Netherlands Journal of Sea Research, *16*: 315-332.
5. Davidson, A.T. and H.J. Marchant. 1987. Binding of manganese by Antarctic *Phaeocystis pouchetii* and the role of bacteria in its release. Marine Biology, *95*:481-487.
6. Davidson, A.T. and H.J. Marchant. 1992. Protist abundance an carbon concentration during a *Phaeocystis*-dominated bloom at an Antarctic coastal site. Polar Biology, *12*: 387-395.
7. Davidson, A.T. and H.J. Marchant. 1992. The biology and ecology of *Phaeocystis*. Progress in Phycological Research, *8*: 1-40.
8. Deprez, P.P., P.D. Franzmann and H.R. Burton. 1986. Determination of reduced sulfur gases in Antarctic lakes and seawater by gas chromatography after solid absorbent preconcentration. Journal of Chromatography, *362*: 9-21.
9. DiTiullio, G.R. and W.O. Smith. 1993. Dimethyl sulfide concentrations near the Antarctic Peninsula: November 1992. Antarctic Journal of the United States, *28*: 130-132.
10. Fofonoff, N.P and R.C. Millard Jr. 1983. Algorithms for the computation of fundamental properties of seawater. UNESCO Technical Papers in Marine Sciences *44*. 53 pp.
11. Fogelqvist, E. 1991. Dimethylsulphide (DMS) in the Weddell Sea surface and bottom water. Marine Chemistry, *35*: 169-177.
12. Gibson, J.A.E., R.C. Garrick, H.R. Burton and A.R. McTaggart. 1988. Dimethylsulfide concentrations close to the Antarctic Continent. Geomicrobiology Journal, *6*: 179-185.
13. Gibson, J.A.E., R.C. Garrick, H.R. Burton and A.R. McTaggart. 1990. Dimethylsulfide and the alga *Phaeocystis pouchetii* in Antarctic coast waters. Marine Biology, *104*: 339-346.
14. Keller, M.D., W.K. Bellows and R.R.L. Guillard. 1989. Dimethyl sulphide production in marine phytoplankton, p183-200. *In* E.S. Saltzmann and W.J. Cooper (ed), Biogenic Sulfur in the Environment. American Chemical Society, Washington, D.C.
15. Kirst, G.O., C. Thiel, H. Wolff, J. Nothnagel, M. Wazek and R. Ulmke. 1991. Dimethysulfoniopropionate (DMSP) in ice-algae and its possible biological role. Marine Chemistry, *35*: 381-388.
16. Levasseur, M., M. Gosselin and S. Michaud. 1994. A new source of dimethylsulfide (DMS) for the arctic atmosphere: ice diatoms. Marine Biology, *121*:381-387.
17. Liss, P.S., G. Malin, S.M. Turner and P.M. Holligan. 1994. Dimethyl sulphide and *Phaeocystis*: A review. Journal of Marine Systems, *5*: 41-53.
18. McTaggart, A.R. and H.R. Burton. 1992. Dimethyl sulfide concentrations in the surface waters of the Australasian antarctic and subantarctic oceans during an austral summer. Journal of Geophysical Research, *97*: 14407-14412.
19. Medlin, L.K., M. Lange and M.E.M. Baumann. 1994. Genetic differentiation among three colony-forming species of *Phaeocystis*: further evidence for the phylogeny of the Prymnesiophyta. Phycologia, *33*:199-212.
20. Palmisano, A.C. and D.L. Garrison. 1993. Microorganisms in Antarctic sea ice, p167-218. *In* E.I. Friedmann (ed), Antarctic Microbiology. Wiley-Liss, New York.
21. Parsons, T.R., Y. Maita and C.M. Lalli. 1984. The Manual of Chemical and Biological Methods of Seawater Analysis. Pergamon Press, Oxford.
22. Perrin, R., P. Lu and H.J. Marchant. 1987. Seasonal variation in marine phytoplankton and ice algae at a shallow Antarctic coastal site. Hydrobiologia, *146*: 33-46.
23. Sieburth, J. M. 1960. Acrylic acid, an "antibiotic" principal in *Phaeocystis* blooms in Antarctic waters. Science, *132*:676-677.
24. Turner, S.M., G. Malin, P.S. Liss, D.S. Harbour and P.M. Holligan. 1988. The seasonal variation of dimethyl sulphide and dimethylsulphoniopropionate concentrations in nearshore waters. Limnology and Oceanography, *33*: 364-375.
25. Yang, H., A.T. Davidson and H.R. Burton. 1994. Measurement of acrylic acid and dimethyl sulphide in Antarctic coastal water during a summer bloom of *Phaeocystis pouchetii*. Proceedings of the NIPR Symposium on Polar Biology, 7: 43-52.

PARTICULATE DIMETHYLSULFONIOPROPIONATE REMOVAL AND DIMETHYLSULFIDE PRODUCTION BY ZOOPLANKTON IN THE SOUTHERN OCEAN

Kendra L. Daly[1] and Giacomo R. DiTullio[2]

[1] Department of Ecology and Evolutionary Biology
University of Tennessee
Knoxville, Tennessee 37996
[2] University of Charleston
Grice Marine Biology Laboratory
Charleston, South Carolina 29412

SUMMARY

The influence of Antarctic krill, *Euphausia superba,* on particulate dimethlysulfoniopropionate (DMSP(p)) and dimethylsulfide (DMS) concentrations in surface waters of the Southern Ocean was investigated by shipboard experiments during austral spring near the Antarctic Peninsula. Chlorophyll concentrations were low in the water column, but substantially higher in sea ice due to the high biomass of ice algae, predominantly pennate diatoms. A comparison of DMSP(p) concentrations and algal accessory pigments indicated that DMSP(p) was associated primarily with diatoms (fucoxanthin) and to a minor extent with *Phaeocystis* spp. (19′-hexanoyloxyfucoxanthin) in sea ice algae. Maximum DMSP(p) and fucoxanthin concentrations also occurred in the 100-200 µm size fraction. We interpret this to mean that high biomass of diatoms in sea ice contributes significantly to DMSP(p) pools in the Antarctic.

Juvenile krill were the dominant biomass component of the zooplankton community and were often observed grazing along the edge of ice floes. In experiments, krill grazing on phytoplankton or ice algae produced 3 to 16 times the amount of DMS, respectively, relative to that in control bottles without krill. On average, juvenile krill produced 0.64 nmol DMS krill^{-1} h^{-1}. Control bottles containing autotrophic and microheterotrophic organisms released only 0.03 nmol DMS L^{-1} h^{-1} and the net change in DMS due to microbial processes was negligible. Hence, krill appear to be very efficent in converting DMSP(p) to DMS. Krill also egested 1.06 nmol DMSP(p) (mg fecal pellet)$^{-1}$ h^{-1}. Results from sediment traps, however, indicate that little DMSP(p) flux to deep water occurred even though particle fluxes were dominated by fecal pellets. DMSP(p) in fecal pellets may rapidly degrade to DMSP(d) and diffuse from pellets into the surrounding water. Our study indicates that mesozooplankton grazing activity plays a significant role in the release of DMSP and DMS from phytoplankton and ice algae. Because krill feed on ice algae during seasons when phytoplankton production is low, DMS production from grazing occurs during all seasons in the Southern Ocean.

INTRODUCTION

Oceanic emissions of dimethylsulfide (DMS) to the atmosphere may influence global climate (10), yet little is known about the biogeochemical and physical processes which control DMS concentration in seawater. DMS is the major volatile sulfur compound in the open ocean and concentrations within the euphotic zone vary spatially and temporally (2), with higher concentrations generally occurring in regions of higher biological productivity (20). Some of the highest concentrations of DMS in seawater are reported from the Southern Ocean (21, 24, 44). Recent studies suggest that the structure of marine pelagic food webs is the primary determinant governing differences in DMS concentrations between oceanic regions (5, 35, 37).

Marine phytoplankton are the major source of DMS, which is produced by enzymatic cleavage of the precursor β-dimethylsulfoniopropionate, an intracellular compound. High DMSP(p) concentrations per unit cell volume are found in some taxonomic groups, such as dinoflagellates and prymnesiophytes (e.g., coccolithophores and *Phaeocystis* spp.), while diatoms have variable concentrations, and other groups, such as chlorophytes, have less significant concentrations (33). Particulate and dissolved fractions of dimethylsulfoniopropionate are released from phytoplankton during cell lysis; consequently, large blooms of DMSP(p)-producers that typically occur in high latitudes (e.g., *Emiliania huxleyi, Phaeocystis* spp.) may result in significant emissions of DMS to the atmosphere in these regions (40, 43, 24). Recently, ice algae also were shown to produce significant amounts of DMSP(p) in both Antarctic and Arctic regions (36, 38). In addition to cell lysis, zooplankton grazing is likely to be an important mechanism leading to the formation of DMS through the mechanical disruption of cells and digestive processes (12, 37); however, little is known about the influence of grazing in the natural environment. Recent field and laboratory studies have indicated that microzooplankton grazing plays an important role in producing DMS as well (5, 52).

Biological processes also affect the removal of DMSP(p), DMSP(d), and DMS from surface water. Sinking phytoplankton and fecal pellets from zooplankton may carry DMSP(p) out of the surface layer, but recent reports (43, 4) do not support this hypothesis. Bacteria consume DMSP(d) and may produce DMS or alternative products (34). Conversely, significant rates of bacterial consumption of DMS were reported for the eastern, tropical Pacific Ocean (35); hence, ocean regions with an active microbial loop may recycle DMS too quickly to sustain a high net flux to the atmosphere. Physical and chemical processes that act as a sink for DMS in seawater include exchange with the atmosphere (3), photochemical oxidation (8), and demethylation (34). The relative importance of these sources and sinks, and the factors controlling the production of DMS in different regions, remain poorly known.

High DMS concentrations associated with phytoplankton blooms have been reported from coastal regions in the Antarctic during spring and summer (16, 24). During winter, most of the ocean is covered by ice and primary production is low. Ice algae, however, are abundant in ice floes during all seasons and may have high concentrations of DMSP(p) (36, 38).

Because the Antarctic krill, *Euphausia superba*, is a dominant member of the Antarctic food web and feeds on ice algae as well as phytoplankton (14, 15), we hypothesized that grazing by krill would be a significant mechanism of DMS production in the Southern Ocean. We present here the results of shipboard experiments designed to examine the role of krill in the consumption of algal DMSP(p) and the subsequent release of DMS in Antarctic coastal waters during spring.

MATERIALS AND METHODS

Several areas in the vicinity of the Antarctic Peninsula were sampled from the RV *Polar Duke* from 2-23 November 1992 (Fig. 1). Environmental conditions were typical for spring and much of the near-shore region was still covered by sea ice. In coastal bays, vertical profiles of temperature and salinity indicated extensive vertical mixing, with weak to no stratification (A. Amos, pers. comm.), and primary production was low (245 mg C m^{-2} d^{-1}; W. O. Smith, pers. comm.). Two small bays, Charlotte Bay and Paradise Harbour, were sites of intensive sampling for zooplankton and dimethylsulfide (DMS) and (DMSP(p) concen-

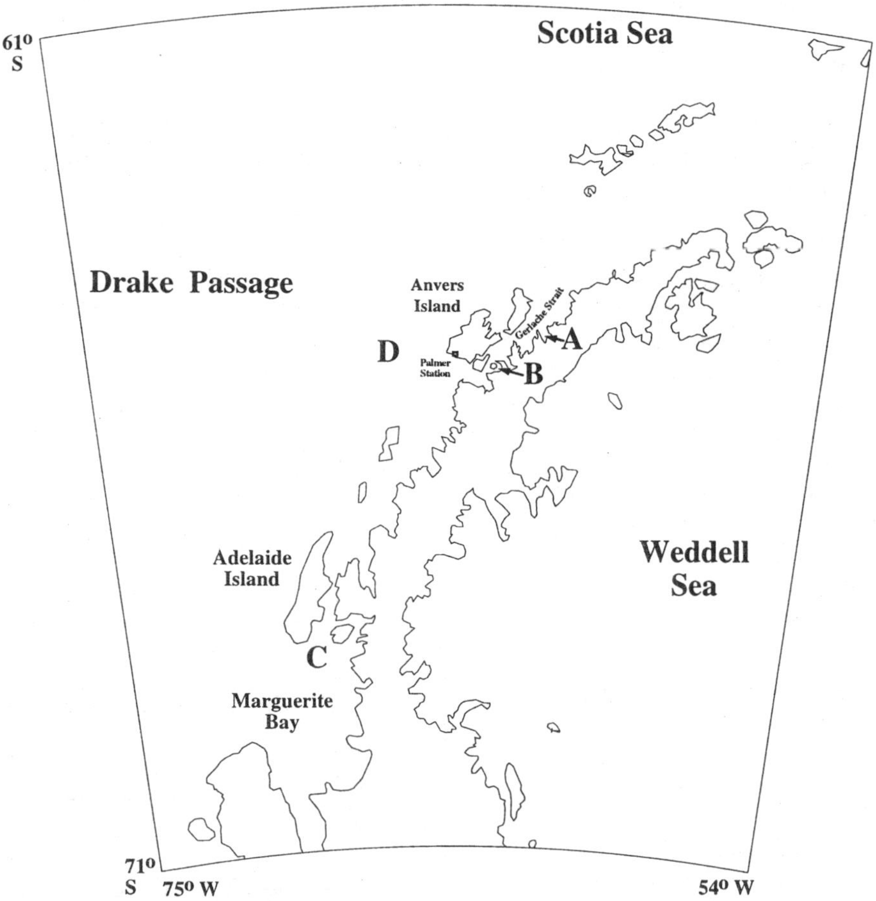

Figure 1. Location of sampling sites near the Antarctic Peninsula, (A) Charlotte Bay, (B) Paradise Harbour, (C) northern end of Maruerite Bay, and (D) Drake Passage.

trations in phytoplankton and ice algae. In addition, samples of ice algae were collected from ice floes at the northern end of Marguerite Bay, and water-column samples were collected in open water in the Drake Passage offshore from Anvers Island.

Phytoplankton pigment samples were collected from the water column with 10-L Niskin bottles and filtered onto GF/F filters, then frozen in liquid nitrogen (N_2) until analyzed. Bottom-layer assemblages of sea ice biota were collected from small pieces of sea ice and from the undersurface of large floes immediately after being overturned by the ship. Ice showing brownish coloration was slowly melted in 5:1 filtered (0.7 µm) seawater to sample at ca. 0°C in order to prevent algal cells from rupturing. Aliquots (385 - 560 ml) of the melted sea ice were filtered onto GF/F filters and frozen in liquid N_2 for subsequent pigment analyses by high-performance liquid chromatography (HPLC) using a Beckman system and a Kratos fluorescence detector. Pigment absorption was monitored at 430 nm. In the laboratory, filters were homogenized and pigments extracted in 90% acetone for 2 h at -10°C. Samples were microcentrifuged, then injected onto a Alltech Spherisorb ODS-2 (3 µm, 150 mm) reverse phase C-18 column and eluted with a gradient similar to that reported by Wright et al. (53). Pigment standards were obtained from Sigma Chemical Co. (chl a, b, and beta carotene) or were purified from phytoplankton cultures by collecting known peaks as they eluted off the column. Standards for pigment degradation products were made from the parent pigments following methods in Jeffrey and Hallegraeff (31) and Holm-Hansen et al. (27).

In addition, ice algae from Marguerite Bay were size fractionated (0.70 - 20, 20 - 100, 100-200, and > 200 µm) in order to assess DMSP(p), HPLC pigments, and particulate organic carbon (POC) and nitrogen (PON) concentrations associated with each size fraction. Ice was thawed as above, and the sample filtered by gravity through stacked Nitex screens in descending mesh size, and then through a Whatman GF/F glass fiber filter under low vacuum. Particulates retained on the screens were backwashed onto additional GF/F filters for DMSP and HPLC analyses. POC and PON samples were washed onto combusted GF/F filters. All filters were frozen in liquid N_2 until analyzed. In the laboratory, filters for carbon and nitrogen were dried at 60°C and analyzed on a Carlo-Erba Model EA1108 elemental analyzer using acetanilide as a standard.

Zooplankton were collected by vertical tows with a 1-m Ring net (333 µm mesh) and by oblique tows with a 1-m Issacs-Kidd Midwater trawl (IKMT) (6 mm mesh on top of net and 333 µm on bottom). *Euphausia superba* were collected from ice floes with a hand-held dip net from a deployed zodiac and krill swarms were detected by a 50 kHz Simrad echosounder. *E. superba* were sorted immediately after collection for live experiments and the remaining sample was preserved in buffered 4% formalin.

Some freshly collected krill and all individuals from experiments were identified to life history stage (39) and measured for length to the nearest mm, from the base of the eye to the tip of the telson. Then they were rinsed very briefly in distilled water, blotted dry, and frozen for dry weight analyses. In the laboratory, individual krill were placed in preweighed aluminum boats, dried at 60°C for at least 24 h, then weighed on a Mettler AC100 balance. Rate measurements were not normalized to krill weight because all krill were juveniles and the difference in weights among individuals was relatively minor.

Freshly collected krill from the water column and ice floes were assessed for relative gut fullness by the gut fluorescence method (13). Individual krill were homogenized by grinding in 10 ml of 90% acetone, then pigments were passively extracted in 90% acetone in the dark for 24 h at -20°C. Krill extracts were centrifuged and fluorescence in krill filtrates was measured on a Turner Designs fluorometer before and after acidification. Pigment content was calculated by summing chlorophyll *a* and phaeopigment fluorescence.

Fecal pellet production rates were obtained by placing one to four individuals in 2-L polycarbonate containers filled with either unfiltered seawater from the chlorophyll maximum or ice algae in filtered seawater. The containers, which held a suspended inner plastic cylinder having a 1-mm mesh bottom to prevent the ingestion of fecal material, were submerged in a darkened aquarium at ambient seawater temperature (ca. -1°C). After three to four hours, the contents of the container were examined under a dissecting scope and the fecal pellets were individually siphoned off in order to separate them from phytoplankton. Pellets were placed on separate preweighed, combusted Whatman GF/F glass fiber filters and rinsed with a small amount of deionized, distilled water to remove salts. The excess water was removed by hand pump under low vacuum and the filters were frozen in liquid N_2. In the laboratory, the pellet filters were dried, then weighed on a Sartorius Pro II microbalance. Fecal pellet production rates (mg dry weight of pellets krill^{-1} h^{-1}) were determined from the difference in weight of filters before and after the addition of pellets and drying.

Additional experiments using the same experimental containers were performed to evaluate the pigment content of food and fecal pellets. The same procedures were followed as above except that subsamples of food were filtered onto GF/F filters prior to the experiment, then after three to four hours fecal pellets were siphoned onto GF/F filters. Filters were frozen in liquid N_2 until HPLC analyses.

DMS production by krill grazing was determined in two experiments. In the first, surface water was collected in open water in Charlotte Bay and transferred into a 20-L carboy. Initial samples were taken for phytoplankton pigment analyses (2 L) and DMS concentration (300 ml). The remaining water was siphoned into a control bottle (phytoplankton + microzooplankton) and three 4-L experimental bottles along with three juvenile krill. A filtered seawater (0.7 μm) control also was sampled before and after the incubation in order to determine the level of microbial activity on DMS cycling, and a bottle of filtered seawater containing three krill. All labware was HCl cleaned. Controls were held in stoppered, glass bottles and experimental bottles were narrow-mouthed polycarbonate bottles sealed with parafilm. All bottles were incubated with zero headspace for 48 h at ca. 0°C in a darkened aquarium with no agitation beyond the movement of the ship. Initial and final pigment samples were filtered, then frozen in liquid N_2. DMS production by krill was calculated as the increase in DMS in experimental bottles minus that in the microplankton control.

The second DMS experiment was a time-series experiment conducted in Paradise Harbour. Sea ice biota (ice algae + microzooplankton) were collected from the undersurface of ice floes and slowly melted in cold, filtered seawater. Juvenile krill were prefed on this diet for 24 h for digestive acclimation. Initial concentrations of pigments, DMS, and DMSP(p) were determined for filtered seawater and ice algae. Samples for pigments and DMS were processed as described above. DMSP(p) samples were filtered onto GF/F filters and frozen in liquid N_2 for subsequent analyses. The controls were (1) filtered seawater, (2) filtered seawater and krill that had been starved for 10 days, and (3) sea ice biota. For the experiment, three juvenile krill were incubated in each of eight 6-L polycarbonate bottles containing ice algae in a darkened aquarium at ca. 0°C. One control (sea ice biota) and two experimental bottles were processed after 3, 6, 12, and 22.5 h, and the remaining control bottles were processed at the end of the experiment for pigments, DMS, and DMSP(p). Fecal pellets were pipetted onto pre-weighed GF/F filters, frozen in liquid N_2, freeze-dried, then reweighed. DMS samples were refrigerated and run on shipboard within a few hours of collection, while DMSP(p) samples were frozen for six months before being analyzed.

DMSP(p) may degrade when stored in liquid N_2 (M. Keller, pers. comm.), but samples collected in the Antarctic have showed little change between those analyzed at sea and those frozen until analyzed (18) and our DMSP(p): chlorophyll ratios for ice algae are similar to the mean ratio (3.2 ± 2.8) obtained from samples recently collected in the Ross Sea and preserved in methanol (J. Dacey, pers. comm.). Furthermore, our algal DMSP(p) concentrations are very similar to other values reported for the Antarctic in spring (24, 36).

DMSP(p) concentrations in pellets, however, were corrected for a 40% (± 10%) loss that resulted from lyophilization, based on replicate analyses. Amaral et al. (1) found a similar mean loss (36 ± 27%) of total sulfur for lyophilized sediments. Due to the uncertainties introduced by methodology, our DMSP(p) concentrations should be considered as lower limits.

DMS and DMSP(p) measurements were made following methods described in DiTullio and Smith (18). Briefly, DMS was analyzed using a liquid N_2 cryogenic purge and trap system. Sulfur gas was detected by a Hewlett Packard 5890 Series II Gas Chromatograph (GC) equipped with a flame photometric detector (FPD) and an HP 3396 Series II Integrator. Precision for replicate standards was < 4% and for replicate samples was < 10%. DMSP(p) filters were hydrolyzed in 2.8 ml 6N NaOH in the dark for 24 h at room temperature to convert DMSP(p) to DMS. An aliquot was injected into a stripping chamber containing sparged Burdick and Jackson water and bubbled for 20 min, then analyzed as for DMS.

Results

Coastal waters along the Antarctic Peninsula in the vicinity of the Palmer Achipelago (Fig. 1) remained partially covered with sea ice during November 1992. Phytoplankton biomass, based on chlorophyll concentrations, was very low in both Charlotte Bay and Paradise Harbour (Table 1). Chlorophyll concentrations associated with bottom assemblages of sea ice biota, however, suggested that substantial biomass had accumulated on the undersurface of ice floes. Microscopical examination of live phytoplankton samples at sea indicated that the water-column community was dominated by nanoflagellates, primarily cryptomonads, concomitant with low abundances of two species of centric diatoms (D. Bird, pers. comm.). In contrast, ice algae samples were dominated by a small pennate diatom, *Navicula glaciei*, at all sites, but several other species of diatoms and heterotrophic dinoflagellates also were present (D. Garrison, pers. comm.).

Algal pigments, used as a qualitiative chemotaxonomic tool to indicate the major classes of autotrophic organisms present in the water column and in sea ice, supported

Table 1. Pigment characteristics and DMSP(p) concentrations for phytoplankton and ice algae and pigments in fecal pellets from *Euphausia superba* for different sites along the Antarctic Peninsula*

	Chl ($mg L^{-1}$)	% Fuco	% Hex	DMSP(p) (nM)	DMSP(p) Chl
Phytoplankton					
Charlotte Bay	0.06	60	40	1.6	8.33
Paradise Harbour	0.17	80	20	1.3	4.56
Ice Algae					
Charlotte Bay	33.96	97	3	nd	nd
Paradise Harbour	23.58	99	1	46.7	1.98
Marguerite Bay	13.85	98	2	12.2	0.88
Fecal Pellets	(μg)				
Charlotte Bay	1.21	95	5	nd	nd
Paradise Harbour	3.04	91	9	nd	nd
Marguerite Bay	1.58	96	4	nd	nd

*Pigments are median concentrations in the euphotic zone, Fuco and Hex expressed as percent of total accessory pigments, pigments in pellets from krill feeding on ice algae. *Chl* chlorophyll *a*, Fuco fucoxanthin, *Hex* 19'-hexanoyloxyfucoxanthin, *nd* no data microscopic evidence. Fucoxanthin was the dominant accessory pigment in both phytoplankton and ice algae, followed by 19′-hexanoyloxyfucoxanthin (Hex) (Table 1). Although the relative proportion of accessory pigments in cells may vary, we assumed that fucoxanthin was an indicator for diatoms because the concentrations of other accessory pigments (i.e., 19′-butanoyloxyfucoxanthin for chrysophytes and peridinin for dinoflagellates) that would indicate the presence of additional fucoxanthin-containing groups were < 1% of total accessory pigments in samples. Hex is the dominant xanthophyll and an indicator pigment for Antarctic prymnesiophytes, such as *Phaeocystis* spp. (9). Fucoxanthin also occurs in *Phaeocystis* spp., but was found to be a small percentage (3%) of the total carotenoid content in Antarctic cultures (50). When Hex pigments were present, we assumed that *Phaeocystis* spp. occurred as nanoflagellates since no colonial forms were observed during this study.

DMSP(p) concentrations in ice algae samples were higher than those in phytoplankton samples of identical volume, yet when normalized to chlorophyll, they were lower, owing to the high chlorophyll concentrations of ice algae (Table 1). This serves to illustrate the point that because cellular concentrations of chlorophyll and DMSP vary as a function of physiology and environmental conditions, the intercomparison of DMSP(p):Chl ratios is problematic, particularly between different communities, and should be used with caution. Nevertheless, in the absence of cell numbers or other measures of biomass, we include the ratios here for comparison with other studies.

The size fractionation of Marguerite Bay sea ice biota revealed that the largest proportion of chlorophyll *a* and particulate organic carbon (POC) and nitrogen (PON) was associated with the smallest size of ice algal cells (0.70 - 20 μm) (Table 2). In contrast, the 100 - 200 μm size fraction contained the highest concentration of DMSP(p) and was second highest in terms of carbon and nitrogen content. Hence, chlorophyll as a proxy for plant biomass was not a good indicator of DMSP(p) in ice algae. Fucoxanthin was the dominant accessory pigment in all size fractions, suggesting that diatoms were the primary source of DMSP(p) in ice algae. The presence of 19′ Hex in the smallest size fraction suggests that *Phaeocystis* flagellates were present but were not a significant contributor to DMSP(p) inventories. The relatively high C:N ratios also indicate that microheterotrophs, as well as detritus, may have been present in sea ice biota. The summed size fractions, or total, gives a composite depiction of the characteristics of ice algae from this location.

Euphausia superba dominated the biomass of mesozooplankton in the study area as the abundance of other zooplankton grazers, such as copepods, amphipods, and salps, was low. Few acoustic targets were detected in open water of the coastal bays, Gerlache Strait, or in Drake Passage off Anvers Island. Instead, small swarms were detected in the upper 20 m of the water column, primarily in embayments near

Table 2. Size fractionation of Marguerite Bay ice algae for DMSP(p), HPLC pigments, and particulate matter. Fuco and Hex expressed as percent of total accessory pigments. *Chl* chlorophyll *a*, *Fuco* fucoxanthin, *Hex* 19′-hexanoyloxyfucoxanthin, *POC* particulate organic carbon, *PON* particulate organic nitrogen

Size Fraction (mm)	DMSP(p) (nM)	DMSP(p) Chl	Chl (μg L^{-1})	% Fuco	% Hex	POC (μg L^{-1})	PON (μg L^{-1})	C:N
0.7 - 20	1.83	0.19	9.43	97	3	746.2	120.3	6
20 - 100	0.82	0.33	2.50	98	2	183.0	8.3	22
100 - 200	8.58	6.00	1.43	99	1	230.4	13.6	17
200	1.01	2.10	0.48	99	1	139.1	4.8	29
Total	12.24	0.88	13.85	98	2	1298.7	147.1	8.83

ice floes. IKMT hauls (tows < 30 min) in Charlotte Bay caught hundreds of juvenile krill near ice floes, particularly in areas of acoustically detected swarms, while none were caught in open water at the mouth of the bay. Krill also were visually observed on the undersurface of ice floes overturned by the ship and were grazing along the edge of ice floes in many areas. All *E. superba* collected in the water column or from ice floes were juveniles, with a mean ($\bar{x}\pm$ 1SD) length and dry weight of 27.5 mm (± 3.1 mm, n = 146) and 24.18 mg (± 13.54 mg, n = 36), respectively.

Grazing activity by juvenile krill showed striking differences within Charlotte Bay. For example, the mean chlorophyll concentration in guts of individuals collected from net tows in the water column was 15.1 ± 5.7 ng pigment ($\bar{x}$ ± 1SD, n=60), while the gut content of individuals collected from ice floes was 21 times higher (322.6 ± 212.8 ng pigment, n = 18). Gut fluorescence also indicated that krill were feeding during both day and night and no vertical migration was evident from acoustic signals or net collections. Pigments in fecal pellets from krill feeding on ice algae (Table 1) suggested that juveniles fed primarily on diatoms but ingested *Phaeocystis* flagellates as well. Fecal pellet production was variable among individuals that fed on phytoplankton and ice algae with no discernible trend; therefore, these data were pooled. On average, the dry weight of pellets produced by krill was 0.261 mg $krill^{-1}$ h^{-1} (0.061 - 0.499) (geometric mean (95% CI), n = 17). The geometric mean is appropriate for rate data that do not have a normal distribution (54). The size of fecal pellets egested by juvenile krill also was similar for both habitats; about 140 μm in diameter and from 1 to 4 mm in length.

The experiments designed to test DMS production demonstrated that krill produce substantial amounts of DMS as a by-product of grazing. In the first experiment (Fig. 2), krill ingested phytoplankton that was collected from the water surface in an ice-free section of Charlotte Bay. Athough fucoxanthin (69%) was the dominant accessory pigment in the experimental water, Hex (31%) was a substantial component as well, indicating that *Phaeocystis* was present. If microzooplankton, such as small protozoans, were abundant in the surface water, then the result of their grazing activity should be accounted for by the seawater control. Although the initial chlorophyll concentration was relatively low (0.31 μg L^{-1}), grazing processes in two of the botttles generated a greater than threefold increase in DMS relative to that in the initial and final control treatments after a 48 h incubation. Two of the krill in the third bottle

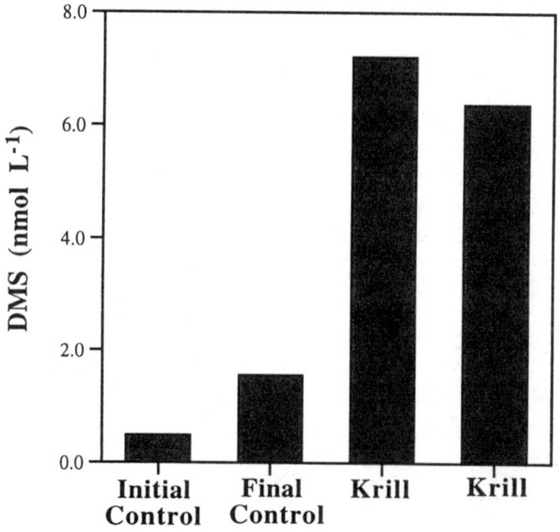

Figure 2. DMS production from *Euphausia superba* feeding on phytoplankton. DMS concentrations in initial seawater sample containing phytoplankton and, after a 48 h incubation, in a phytoplankton control and two experimental bottles containing phytoplankton and three juvenile krill. Phytoplankton collected from surface water in an ice-free section of Charlotte Bay.

were in poor condition at the end of the experiment, therefore, the results from this bottle were not included here. On average, krill produced 0.14 nmol DMS krill^{-1} h^{-1}. There was virtually no change in DMS concentration in the filtered seawater control, only a 0.02 nmol L^{-1} h^{-1} increase in the microplankton community (phytoplankton + microheterotrophs) control, and a 0.03 nmol L^{-1} h^{-1} increase in DMS due to non-feeding krill in filtered seawater.

In the second experiment (Fig. 3), the production of DMS from krill grazing on ice algae collected in Paradise Harbour was examined over time. The initial chlorophyll concentration was 14.71 μg L^{-1} and the relative concentration of fucoxanthin (99% of accessory pigments) was high, indicating that diatoms dominated the biomass of food. Net changes in DMS concentrations in two of the controls, filtered seawater and filtered seawater containing starved krill, were negligible, and the sea ice biota (ice algae + microheterotrophs) control only increased 0.73 nmol L^{-1} in 22.5 h (i.e., 0.03 nmol L^{-1} h^{-1}). Hence, we assume that the increase in DMS in experimental bottles was due to grazing activity by krill. There was no evidence of any other kind of contribution by krill or of microbial or microzooplankton activity influencing net DMS concentrations over the time period of the experiment. Little difference in DMS concentration in experimental bottles was observed for the first 3 h, but at all subsequent time points, DMS concentrations were 2 to 16 times higher in krill treatments than in the sea ice biota controls.

There was not, however, a continuous increase in DMS concentration over time (Fig. 3). DMS concentrations at 6, 12, and 22.5 h were significantly higher ($p < 0.05$) than the concentration at 3 h, but not significantly different from each other. The maximum rate of DMS production at 6 h was 3.73 nmol DMS krill^{-1} h^{-1} and the average rate for the experiment was 0.82 nmol krill^{-1} h^{-1} (0.38 - 1.39) (geometric mean (95% CI)). Chlorophyll concentrations were variable and not significantly different ($p = 0.565$) between time points. Although chlorophyll appeared to be positively correlated with DMS, there was no significant linear association between DMS and any of the phytoplankton pigments. Furthermore, if there had been a functional relationship between the removal of phytoplankton pigments and the subsequent release of DMS by krill, then a negative correlation between these variables would be expected.

The change in DMSP(p) concentration in ice algae and fecal pellets in experimental bottles also was monitored over time (Fig. 4). DMSP(p) associated with ice algae was not

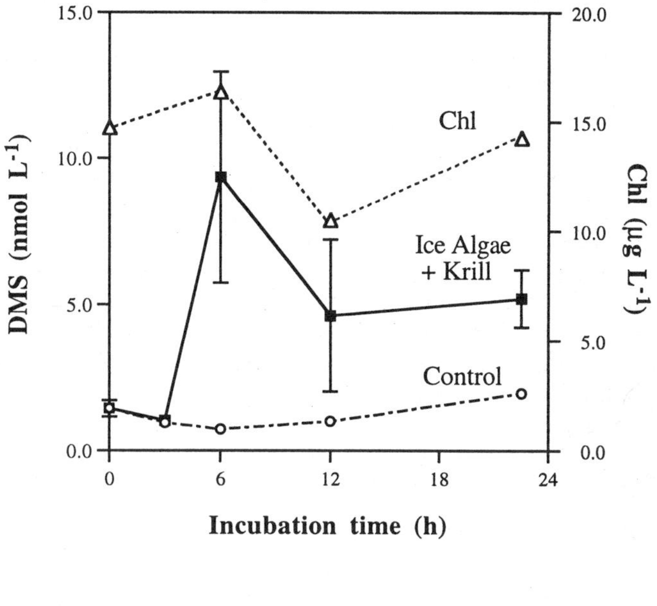

Figure 3. DMS production from *Euphausia superba* feeding on ice algae. Results of time course incubations showing (1) chlorophyll *a* concentrations remaining in experimental bottles, (2) DMS concentrations as the mean (± 1 SE) of duplicate experimental treatments containg ice algae and three krill, and (3) DMS concentrations in ice algae controls. Ice algae collected in Paradise Harbour. *SE* standard error.

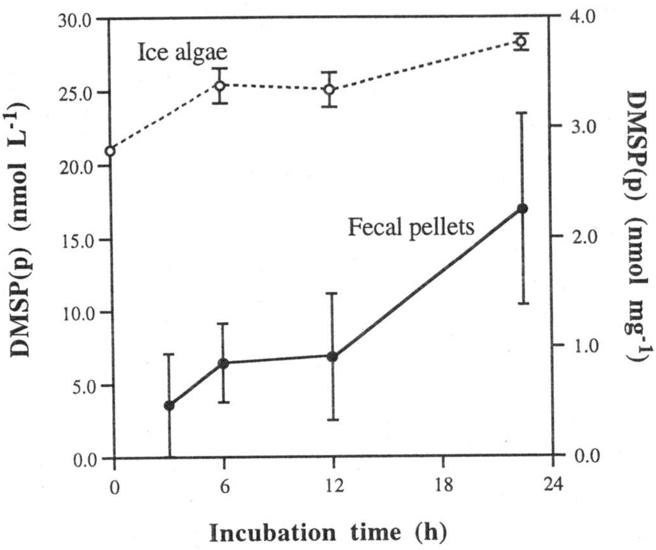

Figure 4. DMSP(p) concentrations in ice algae and in fecal pellets produced by juvenile *Euphausia superba* by the end of each incubation period. DMSP(p) concentrations are the means (±1SE) of duplicate experimental treatments containing ice algae and three krill; ice algae scale on left, fecal pellet scale on right. DMSP(p) concentration in fecal pellets normalized to dry weight of pellets produced at each time point.

significantly different ($p = 0.308$) between time points. Considering DMSP(p) only in ice algae and not in pellets, DMSP(p): DMS ratios decreased from an initial value of 17.1 to 2.7 at 6 h and then rose to ca. 5 for the duration of the experiment. The total amount of fecal pellets increased with time, indicating that krill continued to feed and egest pellets throughout the experiment. Although DMSP(p) concentrations normalized to pellet weight appeared to increase between 12 and 22.5 h, none of the concentrations were significantly different ($p = 0.409$). In other words, krill efficiency in degrading DMSP(p) did not change during the experiment. Nevertheless, DMSP(p) concentration in fecal pellets was positively correlated with fecal pellet weight (Pearson correlation coefficient: $r = 0.98$, $n = 8$); hence, the amount of DMSP(p) increased with the increased number of pellets over time. On average, krill egested 1.06 nmol DMSP(p) per mg dry weight of fecal pellet (0.78 - 1.39) (geometric mean (95% CI), $n = 8$; $\bar{x} \pm 1SD = 1.23 \pm 0.99$; median = 0.86). The rate of krill DMSP(p) egestion in fecal pellets combined with the rate of fecal pellet production is equivalent to 0.28 nmol DMSP(p) krill^{-1} h^{-1} or 6.72 nmol DMSP(p) krill^{-1} d^{-1}.

DISCUSSION

During early spring in the Antarctic, the seasonal pack ice contains assemblages with dense algal concentrations which may have a higher rate of primary production than communities in the underlying seawater (23, 36). Although ice algal production was not measured during our study, biomass accumulations were more substantial in sea ice than in the water column, thereby providing a concentrated source of food for grazers. HPLC-pigment analyses and microscopic identifications indicate that diatoms were an important component of the phytoplankton and ice algae communities during spring and, therefore, a likely source of dimethlysulfoniopropionate (DMSP(p)). Diatoms generally contain less DMSP(p) relative to other groups, such as *Phaeocystis* spp. or dinoflagellates (33), but Levasseur et al. (38) observed significant amounts of DMSP(p) associated with high biomass of ice diatoms in the Arctic. Kirst et al. (36) also noted that diatoms were the dominant component in their Antarctic sea ice samples, but attributed DMSP(p) concentrations to the much lower abundances of dinoflagellates and *Phaeocystis* colonies present. Even though our evidence suggests that diatoms were the primary source of DMSP(p) during our study, *Phaeocystis* is a persistent and often dominant member of Antarctic communities and contributes substantially to DMSP(p) and dimethylsulfide (DMS) inventories elsewhere (18, 24).

DMS concentrations in the water column during this study were spatially variable and ranged from 0.05 to 10 nM (17). The lowest concentrations occurred in Drake Passage ($\bar{x} \pm 1SD = 1 \pm 1.1$ nM, n=31) offshore from the Antarctic Peninsula, while average concentrations were similar in Charlotte Bay (3.5 ± 3.4 nM, n = 74) and Paradise Harbour (2.4 ± 2.2 nM, n = 55). DMS concentrations also varied spatially and temporally within the bays. For example, on 11-12 November open-water surface concentrations in Charlotte Bay were relatively low (< 1 nM), but concentrations were higher (4 - 6 nM) near ice floes where juvenile krill were observed feeding along the edge of ice floes. Surface DMS concentrations also fluctuated between successive days, from < 1 nM to > 10 nM. A variety of factors may have contributed to the large variability in DMS concentrations, including an advective physical regime, low primary productivity in the water column, sporadic release of ice algae from melting sea ice, low microbial activity, and spatially variable grazing activity. Nevertheless, our DMS concentrations were similar to those observed by Gibson et al. (24) near the Australian Antarctic research station in Prydz Bay during November. Water-column DMSP(p) concentrations (0.21 - 9.40 nM) during this study (17) also occurred within the same range as those in Prydz Bay in November (24), the Weddell Sea in November (36), as well as concentrations in the Ross Sea during summer (18). Furthermore, our DMSP(p) concentrations in ice algae are comparable to concentrations in brine samples reported by Kirst et al. (36), but an order of magnitude lower than their concentrations in 'brown ice'.

The zooplankton composition observed during this study is similar to that found in previous studies of this region during spring. The abundance and biomass of net samples was dominated by the juvenile Antarctic krill, *Euphausia superba*, whose distribution was strongly related to ice cover. Both acoustic and net samples indicated that few krill occurred in open water where phytoplankton concentrations were low. Instead, krill were concentrated near pack ice and were observed to be feeding along the edges of ice floes. These observations suggest that krill were feeding primarily on sea ice biota during spring in this region. Siegel (48) also found juvenile krill predominantly on the shelf of the Antarctic Peninsula during spring, while adults were found offshore.

Physiological rates of krill also are similar to reported values. For example, molting and growth rates (33 d and 0.049 ± 0.041 mm d^{-1} (n=7), respectively; Daly, unpubl.) of these individuals are typical for juvenile krill (29) and gut fluorescence measurements for juveniles feeding on ice algae are similar to the mean gut fullness of adults in summer (47). Our geometric mean fecal pellet production rate, normalized to wet weight, also is within the range reported by Clarke et al. (11) for summer. Hence, krill behavior during this study was representative of populations in this region as well as elsewhere in the Southern Ocean.

Our experiments clearly demonstrate that grazing activity by krill produces significant amounts of DMS . The maximum amount of DMS produced was by krill feeding on ice algae, which was probably a result of grazing on higher concentrations of food and thus higher DMSP(p) quotas. Krill feeding rates for ice algae in filtered seawater are not comparable to the rate at which they acquire ice algae from sea ice (which are unknown), but krill also may feed on sinking ice algae released by melting pack ice. Juvenile krill produced up to 16 times more DMS than ice algae alone, for a maximum rate of 3.73 nmol DMS $krill^{-1}$ h^{-1}. For both experiments, grazing krill produced on average 0.64 nmol DMS $krill^{-1}$ h^{-1} (0.34 - 1.01) (geometric mean (95% CI); $\bar{x} \pm 1SD = 0.88 \pm 1.23$; median = 0.30) or 15.36 nmol DMS $krill^{-1}$ d^{-1}. The mean rate is slightly higher than the rate reported for two species of copepods feeding on high concentrations of a dinoflagellate in a laboratory study (12). It is likely, however, that krill production rates of DMS are underestimated because container effects strongly influence krill. For example, Price et al. (46) found significantly higher and less variable feeding rates for krill in 50-L tubs than in 5-L bottles.

At this time, we cannot reconcile the lack of mass balance between DMSP(p) and DMS in the second experiment (Figs. 3 and 4). Despite the trends in chlorophyll, DMS, and DMSP(p) in food and pellets, none of the mean values after 3 h were significantly different due to the large variability among the small number of replicates. Pigment and DMSP(p) concentrations were not correlated and did not appear to decrease as a result of grazing. It is unlikely that ice algal biomass or DMSP(p) increased during the experiment because bottles were incubated in the dark and growth rates of ice algae, like polar phytoplankton, typically are low (ca. 0.5 doublings d^{-1}; 23). Although the initial food supply was well mixed before filling the experimental bottles, the dominant diatom, *Navicula glaciei*, tended to aggregate or clump together and some aggregates sank to the bottom of bottles during the experiment. Furthermore, since the amount of fecal pellets increased with time, it is clear that krill continued to feed during the entire incubation but possibly at highly variable rates. We suspect that variability in krill feeding rates and artifacts of sub-sampling (aggregated food) led to the apparent differences between time points. The changes in DMS levels also are puzzling. Based on gut turnover time (Daly, unpubl.), only DMS concentrations in the first three hours would have been influenced by the gut contents of prefed krill. Although pellets were egested during the first three hours (Fig. 4), DMS didn't increase until between 3 and 6 h. Moreover, DMS may not have decreased after 6 h, but simply reflect levels of variable grazing, or digested DMSP(p) may have been channeled to the DMSP(d) pool rather than to DMS. We also cannot discount the possibility that some DMS was lost from containers. Clearly, variability in degradation of DMSP(p) by gut processes and production of DMSP(d) or alternative products should be considered in future experiments.

Despite the variability, our measured rates of krill DMS production contrast with results reported by Berresheim (6) who suggested that live krill were not a significant source of DMS. Moreover, our experiments showed that both starved krill and krill transferred to filtered seawater after feeding produced negligible amounts of DMS; hence, it appears that live krill only produce DMS through the transformation of DMSP(p) by digestive processes. Berresheim (6) also found relatively high concentrations of DMS in seawater containing dead krill, presumably from the decay of body tissue containing DMS and DMSP(p) (49). It is unlikely, however, that decaying krill would be a significant source of DMS, because they should sink out of the water column before they could substantially decompose. On the other hand, predators, such as penguins, seals, fish, and whales, also may release DMS after ingesting krill.

Although bacteria and microzooplankton significantly influence DMS cycles in other marine systems (5, 35), these components of the food web may not be as important in the Southern Ocean, especially during spring. Microbial abundances were low ($9 - 36 \times 10^4$ cells ml^{-1}; D. Bird, pers. comm.) during this study and may remain low even when chlorophyll concentrations are relatively high during summer (32). Also, DMS concentrations in filtered seawater controls were not significantly different from initial concentrations after 22 to 48 h incubations, indicating little net accumulation from microbial activity. Although microheterotroph abundance was not assessed during this study, a cursory examination of live and preserved samples did not indicate substantial numbers and *in situ* concentrations were represented in the microplankton community treatments in experiments. Because the net increase in DMS due to release by phytoplankton, ice algae, and/or grazing microheterotrophs was very small, microheterotroph grazing was assumed to be a minor source for DMS compared to krill grazing. Microheterotrophs, however, can be an important component in Antarctic food webs, both in the water column and in sea ice (22, 45); therefore, their role in DMS production deserves further study.

The sedimentation of krill fecal pellets may be a major pathway for DMSP(p) transport to deep water since krill fecal pellets have been reported to be a dominant component of the particle flux in sediment traps, particularly near the Antarctic Peninsula (51). Although krill egested a significant amount of DMSP(p) (6.67 nmol DMSP(p) krill^{-1} d^{-1}) in fecal pellets per day, it was only about 43% of the average amount of DMS (15.36 nmol DMS krill^{-1} d^{-1}) produced per day by the same individuals in the second experiment. Hence, krill appear to be very efficient at converting DMSP(p) to dissolved DMSP (DMSP(d)) and DMS. In addition, results from sediment traps suggest that the DMSP(p) flux was low. Drifting sediment trap arrays, with collectors at 60, 80, and 100 m, were deployed for ca. 24 h in Charlotte Bay and Paradise Harbour during our study. Despite the fact that particle fluxes were dominated by zooplankton fecal pellets about the size produced by krill (D. Karl and A. Diercks, pers. comm.), DMSP(p) fluxes in both bays were less than 1% of the integrated DMSP(p) (19). It appears that little DMSP(p) was removed from surface waters by sinking particles. Leaching solutes from fecal pellets would act to increase the contribution by grazers to the DMSP(d) and DMS pools. Consequently, the importance of DMSP(p) removal from surface waters via zooplankton fecal pellet flux remains uncertain.

A variety of factors will influence the seasonal production of DMS by grazers in the Southern Ocean, including the abundance and distribution of grazers, the abundance and distribution of autotrophic cells available as food, the DMSP concentrations associated with different sizes and types of cells, and selective feeding by grazers. *E. superba* is a dominant component of the herbivore community and remains in the surface layer feeding throughout the year. During spring and summer, other grazers, such as the copepods, *Calanoides acutus* and *Calanus propinqus*, and the salp, *Salpa thomsoni*, may be very abundant in some regions (7, 28). In these seasons, *E. superba* may aggregate along the ice edge and in coastal regions where large phytoplankton blooms form, and are believed to influence the density and species composition of phytoplankton over large areas (26, 25). Krill tend to be more efficient at ingesting particles > 20 μm (30), primarily diatoms, but they also are known to feed on nanoplankton, such as *Phaeocystis* flagellates (41). It is not known whether they feed on the colonial form of *Phaeocystis*. From autumn to spring when much of the Southern Ocean is covered by sea ice and phytoplankton concentrations are low, krill also feed on sea ice biota (14, 15). It is unlikely that krill feed selectively when they scrape off material from the undersurface of ice floes (42), and the amount of ice algae that they ingest relative to heterotrophic protozoans and detritus is unknown. Because pack ice covers about 20 million km^2 of sea surface, ice algae is a vast food resource available for much of the year.

In conclusion, regional food web composition and dynamics influenced the production, transformation, and removal of particulate dimethlysulfoniopropionate (DMSP(p)) and dimethylsulfide (DMS) in surface waters adjacent to the Antarctic Peninsula. The results of our study indicate that high biomass of diatoms in sea ice contained substantial amounts of DMSP(p) and that DMS release by primary producers alone or by microzooplankton grazing was relatively small. Furthermore, vertical transport of phytoplankton and ice algae and microbial consumption, both important loss terms in other regions, did not appear to be important during this study. Strong tidal currents and wind mixing did occur in the study area, therefore, horizontal transport may have been a major determinant of DMSP(p) and DMS distributions, but the influence of physical factors is unknown. Most importantly, our results suggest that zooplankton grazing plays a significant role in the removal of DMSP(p) and the production of DMS from the ingestion of both phytoplankton and ice algae. Because krill feed on ice algae during seasons when phytoplankton concentrations are low, DMS release from grazing must occur during all seasons in the Southern Ocean.

ACKNOWLEDGMENTS

We thank W. O. Smith, Jr. and D. Karl for the opportunity to join their research cruise to the Antarctic Peninsula, D. Bird for identifying phytoplankton at sea, and D. Garrison for identifying ice algae. M. Pascual-Dunlap, J. D. Goodlaxson, A. Schauer, and J. Ustach provided assistance in the field, M. Garrett and S. Polk provided assistance in the laboratory, and two anonymous reviewers provided helpful comments. We also thank the crew of the R/V *Polar Duke* for their aid in the collection of krill and sea ice. This research was supported by a National Science Foundation grant OPP 91-16872 to W. O. Smith, Jr.

REFERENCES

1. Amaral, J. A., R. H. Hesslein, J. W. M. Rudd and D. E. Fox. 1989. Loss of total sulfur and changes in sulfur isotopic ratios due to drying of lacustrine sediments. Limnol. Oceanogr. *34*: 1351-1358.
2. Andreae, M. O. 1992. The global biogeochemical sulfur cycle: A review, p. 87-128. *In* B. Moore III, D. Schimel (eds), Trace Gases and the Biosphere. UCAR/Office for Interdisciplinary Earth Studies, Boulder, Colorado.
3. Bates, T. S., J. D. Cline, R. H. Gammon and S. R. Kelly-Hanson. 1987. Regional and seasonal variations in the flux of oceanic dimethylsulfide to the atmosphere. J. Geophys. Res. *92*: 2930-2938.
4. Bates, T. S., R. P. Kiene, G. V. Wolfe, P. A. Matrai, F. P. Chavez, K. R. Buck, B. W. Blomquist and R. L. Cuhel. 1994. The cycling of sulfur in surface seawater of the northeast Pacific. J. Geophys. Res. *99*, C4: 7835-7843.
5. Belviso, S., S.-K. Kim, F. Rassoulzadegan, B. Krajika, B. C. Nguyen, N. Mihalopoulos and P. Buat-Menard. 1990. Production of dimethylsulfonium proprionate (DMSP) and dimethlysulfide (DMS) by a microbial food web. Limnol. Oceanogr. *35*: 1810-1821.
6. Berresheim, H. 1987. Biogenic sulfur emissions from the subantarctic and Antarctic oceans. J. Geophys. Res. *92*, No. D11: 13,245-13,262.
7. Boysen-Ennen, E. and U. Piatkowski. 1988. Meso- and macrozooplankton communities in the Weddell Sea, Antarctica. Polar Biol. *9*: 17-35.
8. Brimblecombe, P. and D. Shooter. 1986. Photo-oxidation of dimethlysulfide in aqueous solution. Mar. Chem. *19*: 343-355.
9. Buma, A. G. J., N. Bano, M. J. W. Veldhuis and G. W. Kraay. 1991. Comparison of the pigmentation of two strains of the prymnesiophyte *Phaeocystis* sp. Netherland J. Sea Res. *27*: 173-182.
10. Charlson, R. J., J. E. Lovelock, M. O. Andreae and S. G. Warren. 1987. Oceanic phytoplankton, atmospheric sulfur, cloud albedo, and climate. Nature *326*: 655-661.
11. Clarke, A., L. B. Quetin and R. M. Ross. 1988. Laboratory and field estimates of the rate of faecal pellet production by Antarctic krill, *Euphausia superba*. Mar. Biol. *98*: 557-563.
12. Dacey, J. W. H. and S. G. Wakeham. 1986. Oceanic dimethylsulfide: Production during zooplankton grazing on phytoplankton. Science *233*: 1314-1316.
13. Dagg, M. J. and E. Walser, Jr. 1987. Ingestion, gut passage, and egestion by the copepod *Neocalanus plumchrus* in the laboratory and in the subarctic Pacific Ocean. Limnol. Oceanogr. *32*: 178-188.
14. Daly, K. L. 1990. Overwintering development, growth, and feeding of larval *Euphausia superba* in the Antarctic marginal ice zone. Limnol. Oceanogr. *35*: 1564-1576.
15. Daly, K. L. and M. C. Macaulay. 1991. Influence of physical and biological mesoscale dynamics on the seasonal distribution and behavior of *Euphausia superba* in the antarctic marginal ice zone. Mar. Ecol. Progr. Ser. *79*: 37-66.
16. Deprez, P. P., P. D. Franzmann and H. R. Burton. 1986. Determination of reduced sulfur gases in Antarctic lakes and seawater by gas chromatography after solid adsorbent preconcentration. J. Chromatogr. *362*: 9-21.
17. DiTullio, G. R. and W. O. Smith, Jr. 1995. Relationship between dimethylsulfide and phytoplankton pigment concentrations in the Ross Sea, Antarctica. Deep-Sea Res. 42: 873-892.
18. DiTullio, G. R. and W. O. Smith, Jr. 1993. Dimethyl sulfide concentrations near the Antarctic Peninsula: November 1992. Antarctic J. U.S. 28: 130-133.
19. DiTullio, G. R. and W. O. Smith, Jr. Studies on dimethylsulfide in Antarctic coastal waters. In: Battaglia, B., J. Valencia and D. W. H. Walton (eds.), Proceedings from SCAR VI Biology Symposium, Cambridge University Press, Cambridge. (in press)

20. Erickson III, D. J., S. J. Ghan and J. E. Penner. 1990. Global ocean-to-atmosphere dimethyl sulfide flux. J. Geophys. Res. *95* (D6): 7543-7552.
21. Fogelqvist, E. 1991. Dimethylsulfide (DMS) in the Weddell Sea surface and bottom water. Mar. Chem. *35*: 169-177.
22. Garrison, D. L. and K. R. Buck. 1989. Protozooplankton in the Weddell Sea, Antarctica: Abundance and distribution in the ice-edge zone. Polar Biol. *9*: 341-351.
23. Garrison, D. L. and K. R. Buck. 1991. Surface-layer sea ice assemblages in Antarctic pack ice during the austral spring: environmental conditions, primary production and community structure. Mar. Ecol. Prog. Ser. *75*: 161-172.
24. Gibson, J. A. E., R. C. Garrick, H. R. Burton and A. R. McTaggart. 1990. Dimethylsulfide and the alga *Phaeocystis pouchetti* in antarctic coastal waters. Mar. Biol. *104*: 339-346.
25. Graneli, E., W. Graneli, M. M. Rabbani, N. Daugbjerg, G. Fransz, J. Cuzin-Roudy and V. A. Alder. 1993. The influence of copepod and krill grazing on the species composition of phytoplankton communities from the Scotia-Weddell sea. An experimental approach. Polar Biol. *13*: 201-213.
26. Holm-Hansen, O. and M. Huntley. 1984. Feeding requirements of krill in relation to food sources. J. Crust. Biol. *4,* Spec. No. 1: 156-173.
27. Holm-Hansen, O., C. J. Lorenzen, R. W. Holmes and J. D. H. Strickland. 1965. Fluorometric determination of chlorophyll. J. Cons. Perm. Int. Explor. Mer. *30*: 3-15.
28. Huntley, M. E., P. F. Sykes and V. Marin. 1989. Biometry and trophodynamics of *Salpa thompsoni* Foxton (Tunicata: Thaliacea) near the Antarctic Peninsula in austral summer, 1983-1984. Polar Biol. *10*: 59-70.
29. Ikeda, T. and P. G. Thomas. 1987. Moulting interval and growth of juvenile Antarctic krill (*Euphausia superba*) fed different concentrations of the diatom *Phaeodactylum tricornutum* in the laboratory. Polar Biol. *7*: 339-343.
30. Ishii, H. 1986. Feeding behavior of the Antarctic krill, *Euphausia superba* Dana II. Effects of food condition on particle selectivity. Memoirs National Inst. Polar Res., Special Issue *44*: 96-106.
31. Jeffrey, S. W. and G. M. Hallegraeff. 1987. Chlorophyllase distribution in ten classes of phytoplankton: a problem for chlorophyll analysis. Mar. Ecol. Progr. Ser. *35*: 293-304.
32. Karl, D. M., O. Holm-Hansen, G. T. Taylor, G. Tien and D. F. Bird. 1991. Microbial biomass and productivity in the western Bransfield Strait, Antarctica during the 1986-87 austral summer. Deep-Sea Res. *38*: 1029-1055.
33. Keller, M.D., W. K. Bellows and R. R. L. Guillard. 1989. Dimethyl sulfide production in marine phytoplankton, p. 167-200. *In* E. S. Saltzman and W. J. Cooper (eds), Biogenic Sulfur in the Environment. ACS Symposium Series *393*.
34. Kiene, R. P. 1992. Dynamics of dimethyl sulfide and dimethylsulfonioproprionate in oceanic water samples. Mar. Chem. *37*: 29-52.
35. Kiene, R. P. and T. S. Bates. 1990. Biological removal of dimethyl sulfide from sea water. Nature *345*: 702-705.
36. Kirst, G.O., C. Thiel, H. Wolff, J. Nothnagel, M. Wanzek and R. Ulmke. 1991. Dimethlysulfonioproprionate (DMSP) in ice algae and its possible biological role. Mar. Chem. *35*: 381-388.
37. Leck, C., U. Larsson, L. E. Bågander, S. Johansson and S. Hajdu. 1990. Dimethyl sulfide in the Baltic Sea: Annual variability in relation to biological activity. J. Geophys. Res. *95* (C3): 3353-3363.
38. Levasseur, M., M. Gosselin and S. Michaud. 1994. A new source of dimethyl sulfide for the Arctic atmosphere: ice diatoms. Mar. Biol. *121*: 381-387.
39. Makarov, R. R. and C. J. Denys. 1981. Stages of sexual maturity of *Euphausia superba* Dana. BIOMASS Handbook Ser. *11*: 1-11.
40. Malin, G., S. Turner, P. Liss, P. Holligan and D. Harbour. 1993. Dimethylsulfide and dimethylsulphoniopropionate in the Northeast Atlantic during the summer coccolithophore bloom. Deep-Sea Res. *40*: 1487-1508.
41. Marchant, H. J. and G. V. Nash. 1986. Electron microscopy of gut contents and faeces of *Euphausia superba* Dana. Mem. Natl Inst. Polar Res., Spec. Issue, *40*: 167-177.
42. Marschall, H.-P. 1988. The overwintering strategy of Antarctic krill under the pack-ice of the Weddell Sea. Polar Biol. *9*: 129-135.
43. Matrai, P. A. and M. D. Keller. 1993. Dimethylsulfide in a large-scale coccolithophore bloom in the Gulf of Maine. Contin. Shelf Res. *13*: 831-843.
44. McTaggart, A. R. and H. Burton. 1992. Dimethyl sulfide concentrations in the surface waters of the Australian Antarctic and subantarctic oceans during an austral summer. J. Geophys. Res. *97* (C9): 14,407-14,412.
45. Palmisano, A. C. and D. L. Garrison. 1993. Microorganisms in Antarctic sea ice, p. 167-218. *In* E. I. Friedman (ed.), Antarctic Microbiology. Wiley-Liss, New York.

46. Price, H. J., K. R. Boyd and C. M. Boyd. 1988. Omnivorous feeding behavior of the Antarctic krill *Euphausia superba*. Mar. Biol. *97*: 67-77.
47. Priddle, J., J. Watkins, D. Morris, C. Ricketts and F. Buchholz. 1990. Variation of feeding by krill in swarms. J. Plankton Res. *12*: 1189-1205.
48. Siegel, V. 1989. Winter and spring distribution and status of the krill stock in Antarctic Peninsula waters. Arch. FischWiss. *39*: 45-72.
49. Tokunaga, T., Iida, H. and K. Nakamura. 1977. Formation of dimethyl sulfide in Antarctic krill, *Euphausia superba*. Bull. Jap. Soc. Sci. Fish. *43*: 1209-1217.
50. Vaulot, D., J.-L. Birrien, D. Marie, R. Casotti, M. J. W. Veldhuis, G. W. Kraay and M.-J. Chrétiennot-Dinet. 1994. Morphology, ploidy, pigment composition, and genome size of cultured strains of *Phaeocystis* (Prymnesiophyceae). J. Phycol. *30*: 1022-1035.
51. Wefer, G., G. Fischer, D. Fuetterer and R. Gersonde. 1988. Seasonal particle flux in the Bransfield Strait, Antarctica. Deep-Sea Res. *35*: 891-898.
52. Wolfe, G. V., E. B. Sherr and B. F. Sherr. 1994. Release and consumption of DMSP from *Emiliania huxleyi* duirng grazing by *Oxyrrhis marina*. Mar. Ecol. Prog. Ser. *111*: 111-119.
53. Wright, S. W., S. W. Jeffrey, R. F. C. Mantoura, C. A. Llewellyn, T. Bjørnland, D. Repeta and N. Welschmeyer. 1991. Improved HPLC method for the analysis of chlorophylls and carotenoids from marine phytoplankton. Mar. Ecol. Prog. Ser. *77*: 183-196.
54. Zar, J. H. 1984. Biostatistical analysis. Prentice Hall, Englewood Cliffs, N. J.

21

COPEPODS AND DMSP

Rik L. J. Kwint,[1] Xabier Irigoien,[2] and Kees J. M. Kramer[3]

[1] University of Groningen and Laboratory for Applied Marine Research
Institute of Environmental Sciences-TNO
P.O. Box 57, 1780 AB Den Helder, The Netherlands
[2] Institut de Ciències del Mar CSIC
P. Joan de Borbó, s/n
08039 Barcelona, Spain
[3] Mermayde
P.O. Box 109, 1860 AC Bergen
The Netherlands

SUMMARY

The fate of DMSP produced by phytoplankton cultures incubated with and without copepods of the species *Eurytemora affinis* was monitored to investigate the influence of zooplankton grazing activity on the release of dissolved DMSP and DMS. Processes related to the flux of DMSP through (sloppy) feeding, gut passage, fecal pellet formation and fecal pellet stability were studied. It appeared that with the phytoplankton species used (*Phaeodactylum tricornutum* and *Thalassiosira weissflogii*), no increase in DMS formation was detected during grazing but the release of dissolved DMSP increased slightly. Depending on the phytoplankton species, minor or no loss in DMSP was observed during passage trough the copepod guts. Almost all DMSP ingested leaves the copepods in the form of fecal pellets, a possible sink for DMSP. Based on the experimental results the relative importance of the zooplankton related removal of DMSP was quantified and found to be minor with respect to DMS and DMSP production by phytoplankton. The main effect of the zooplankton grazing activity was the repackaging of DMSP present in phytoplankton cells into fecal pellets.

INTRODUCTION

DMSP (ß-dimethylsulfoniopropionate), a compound suspected to be involved in the osmoregulation of certain algal species (10) is the main precursor of dimethylsulfide (DMS) in the marine environment. The conversion of DMSP may be influenced by algal senescence, bacterial activity, phytoplanktonic enzymes and zooplankton grazing (2, 6, 15, 16, 22, 31).

Some of the DMS may be ventilated into the atmosphere, where it can contribute to the formation of cloud-condensation-nuclei, which influence cloud albedo and climate (21).

Of the various routes and processes involved in the transformation of DMSP relatively little is known about the role of zooplankton (3, 15, 18, 19). Dacey and Wakeham (6) found that about one third of the *Gymnodinium nelsoni* and *Prorocentrum micans*-derived DMSP ingested by the copepod species *Labidocera aestiva* or by *Centropages hamatus* was released as DMS into the water column of cultures. Leck et al. (19) found, when following a natural phytoplankton bloom in the Baltic Sea, that copepod numbers were correlated to DMS concentrations, without specifying cause-effect relations or quantifying the role of zooplankton. Belviso et al. (2, 4) studied the role of micro/nano- zooplankton using different mesh sizes and sieving procedures; they concluded that significant amounts of DMSP could be temporarily stored by microzooplanktonic predators. The same group used sediment traps in the Mediterranean and found that differences between day and night occurred in DMSP concentrations in the deposited material (3). They attributed this to the diurnal rhythm in the vertical migration of the zooplankton. Wolfe et al. (31) found that the ciliate *Oxyrrhis marina* metabolized up to 70% of the DMSP ingested without DMS production, while the rest was released as dissolved DMSP.

In a mesocosm study we investigated the importance of zooplankton in DMS formation and found little difference in mesocosms with and without zooplankton > 50 μm (16, 18). Because mesocosm experiments shorten the time of a phytoplankton bloom it may have been possible that the growth of the copepod zooplankton may have been too slow to reveal interactions between phytoplankton and copepods. Some additional evidence has been found in field surveys which suggest that zooplankton grazing activity may be an important factor in the production of DMS (19, 20, 29), but these studies never quantified the role of zooplankton in the DMSP dynamics.

Despite these discussions, most processes related to the flux of DMSP, notably sloppy feeding, ingestion of phytoplankton, the passage through the gut, the production of fecal pellets and the stability of the fecal pellets with respect to DMSP have not been studied in detail.

By analyzing all compartments for the DMSP-related sulfur species: dissolved and particulate DMSP (DMSP(d), DMSP(p)), DMSP inside the copepods (DMSP(z)), DMSP contained in fecal pellets (DMSP(f)), and the DMS in the water column, the final aim is to obtain a complete budget of DMSP through the zooplankton linked part of the system. In this paper we describe a series of experiments performed under laboratory conditions that aim at the description of the routes in a quantitative and qualitative sense.

After an investigation of the individual variation between copepods, including the effect of sex, incubation experiments will be described where the kinetics of uptake and excretion (gut passage time), fecal pellet production and DMSP stability in the fecal pellets were monitored.

MATERIAL AND METHODS

Phyto- and Zooplankton Cultures

Stock cultures of copepods *Eurytemora affinis* were grown in natural filtered seawater (volume about 80 l) on a suspension of the diatom *Phaeodactylum tricornutum* and at 15°C, a salinity of 28.5 ‰ and a light cycle of 14 h light, 10 h dark. Only individuals of the size class > 300 μm (adults only) obtained through sieving were used for the experiments.

Distinction between males and females was made by hand-picking under a microscope. For the investigation of the individual variation, adult copepods were sampled directly from the culture and analyzed individually for their DMSP content.

Phytoplankton were cultured in 10 l vessels under continuous light on standard F2 medium at the same temperature and salinity as the zooplankton. Weekly, the phytoplankton culture was harvested and concentrated by centrifugation. The concentrate was stored at 4°C. From this concentrate defined volumes ($cells.ml^{-1}$) could be added to the different incubation experiments.

Incubation Experiments

At the start of the experiments, batches of 5 individual copepods were manually selected and incubated overnight in filtered artificial seawater to allow them to defecate. These starved copepods were then transferred to 100 ml dark bottles, that were filled with a suspension of the algae *Phaeodactylum tricornutum* ($60\text{-}1000 \times 10^3$ $cells.ml^{-1}$). For each determination a separate bottle was prepared. The bottles were closed without a headspace to prevent the ventilation of DMS into the atmosphere. Incubation bottles without zooplankton served as controls. The concentrations of DMS and DMSP were followed for 4 hours. About every 30 minutes grazed and non-grazed incubation bottles were harvested and water samples were taken to be analyzed for phytoplankton cell numbers (Elzone particle counter), DMSP(p), DMSP(d), and DMS. The 5 copepods from each sample were analyzed for DMSP content (DMSP(z)). At several points duplicate analyses were performed using duplicate incubation bottles.

Gut Clearing Experiment

The gut content of well fed copepods was monitored over time after they were transferred to filtered seawater. For this experiment, a large volume (50 l) from the copepod culture was sieved over a 300 μm mesh sieve and the copepods remaining on the sieve were incubated for two hours on a suspension of *P. tricornutum* (20×10^6 $cells.ml^{-1}$) to ensure maximum feeding. After this feeding, the copepods were sieved again (180 μm) to separate them from the phytoplankton. They were rinsed with 0.45 μm filtered aged seawater and transferred into a clean bucket containing 0.45 μm filtered seawater to allow them to defecate. Sub-samples of copepods were taken by means of a 180 μm mesh sieve every 5 to 10 minutes during 1 hour. The sieve was folded into aluminum foil and submerged into liquid N_2 to preserve the copepods and the DMSP in their guts. For each analysis 20 males and 20 females were manually collected and analyzed for DMSP(z).

This experiment is a classic experiment to evaluate the ingestion rate and gut passage time in zooplankton, which is normally performed using chlorophyll-*a* as tracer (23). As DMSP is also a component linked to the algae the experiment can also be performed using DMSP as the tracer compound instead of chlorophyll-*a* (17). The ingestion rate (I) and gut passage time (GPT) can then be calculated from the decrease in gut content (G) according to: $G_t=G_0\ e^{-gt}$ and $GPT = 1/g$, where g is the gut clearance rate constant in min^{-1}, G_t is the gut content at each time step in $pmol\ DMSP.ind^{-1}$. The ingestion rate is calculated as:

$$I = G_0\ 60/GPT \text{ (in pmol DMSP.ind}^{-1}\ h^{-1}\text{) (23).}$$

Fecal Pellet Experiment

A large number of copepods were concentrated into a 2 l glass beaker and incubated for 2 h on a dense suspension of *P. tricornutum* at 15°C. The copepods, the fecal pellets produced and the phytoplankton were then separated by sieving over a 180 μm mesh and a 25 μm mesh respectively. The fecal pellets were carefully and briefly rinsed with a little filtered seawater and transferred into 2 l of 0.45 μm filtered aged seawater. The temporal changes in the amount of DMSP(f) in the pellets was monitored as well as the changes in DMSP(d) and DMS concentrations. Occasionally subsamples of fecal pellets were preserved with Lugol's solution (10%) and counted using a microscope (mag. 40 ×) and a 1 ml counting chamber.

Total Budget Experiment

This experiment had a similar setup as the first incubation experiment, only this time the amount of DMSP(p) in the phytoplankton was adjusted to be approximately at the same level as the amount of DMSP(z) in the copepods added. The effects of two different diatom species was studied. For each determination, independent replicate bottles were used. Batches of 20 copepods per bottle were incubated in triplicate on a suspension of the diatom *P. tricornutum* (20×10^3 cells.ml^{-1}) and on a suspension of the diatom *Thalassiosira weissflogii* (5×10^3 cells.ml^{-1}), while samples were taken at t = 0 and t = 24 h. For each diatom species duplicate ungrazed samples served as controls. Analyses were made for DMSP(p,d,z,f) and DMS.

DMS and DMSP Analysis

To determine the DMS and the DMSP(d) concentrations, a 25 ml water sample was filtered gently over a GF/C filter into a purge vessel and purged with helium for 10 minutes (40 ml.min^{-1}). DMS was cryogenically trapped on Tenax-ta (Chrompack) after drying through a Nafion drier (Dupont). From the purged filtrate, a 10 ml subsample was taken and transferred to a crimp-cap vial. After 1 ml NaOH (10 M) was added, the vial was immediately closed with a teflon lined septum. After overnight hydrolysis at 4 °C, the DMS formed was purged from the vial with helium through hypodermic syringes and trapped cryogenically. The DMS formed by hydrolysis with cold NaOH is believed to represent only DMSP (14, 30).

For DMSP(p) a 5 ml unfiltered sample was put into a crimp cap vial and treated in the same way, DMSP(p) concentrations were calculated by subtracting the DMS and DMSP(d) concentration from this total sample.

For the DMSP(z) and DMSP(f), the zooplankton or pellets respectively were collected from the batch experiments, rinsed briefly with a little artificial seawater and treated the same way as the DMSP(p) samples. In the fecal pellet experiment we analyzed a 0.45 μm filtered water sample for DMS (25 ml) and DMSP(d) (10 ml). For the determination of the DMSP(f) concentration a total water sample was analyzed (10 ml). The DMSP(f) concentration was then calculated by subtracting the DMS and DMSP(d) concentrations from this total sample.

DMS was analyzed on a Varian 3700 gas chromatograph equipped with a capillary linear plot column and a photo-ionization detector (PID, 10.2 eV). Hydrogen was used as the carrier gas. Calibration was performed using DMSP-HCl salt standards, that were hydrolyzed to DMS and treated the same way as the other samples. The detection limit for DMS and DMSP was 1.5 pmol. For a detailed description of the analysis of DMSP and DMS, see Kwint & Kramer (16).

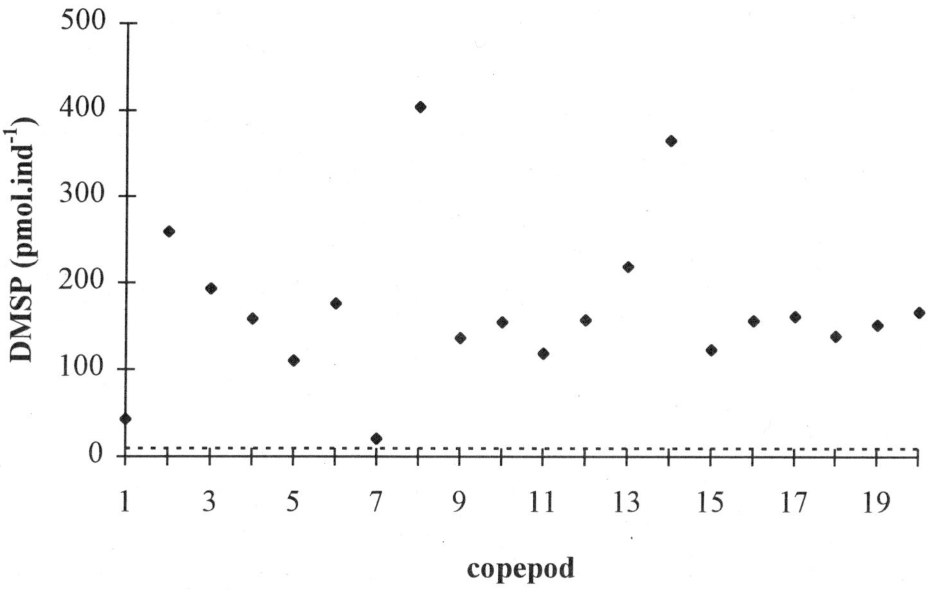

Figure 1. Individual variation of the DMSP gut content in adult copepods >300 μm (*Eurytemora affinis*) from the culture. The dashed line represents the DMSP content after starvation for 24 h.

RESULTS AND DISCUSSION

Individual Variation of DMSP(Z)

The results of the experiment to asses the individual variation of DMSP gut content between adult copepods taken directly from the culture becomes clear from Figure 1. The dashed line indicates the mean DMSP(z) content in copepods that were starved for at least 24 h. There appears to be a common concentration level (per copepod) of about 160 pmol.ind^{-1}. However, several data points are well below or well above this level. Although differences between males and females occur and differences between experimental conditions/batches (see also Figure 4), it is expected that the differences between individuals observed from one experiment are related to the characteristics of the individuals. From visual inspection of copepods it can be seen that individuals may contain 0, 1, 2 or 3 fecal pellets in the gut. This explains the 'outliers' in Figure 1. For this reason, in later experiments we used either starved, or well fed copepods that were incubated (2 h) in phytoplankton suspensions of much higher density than the original culture.

Four Hour Incubation Experiment

The incubation experiments where the concentrations of DMSP(p,d,z) and DMS were monitored over time in systems with and without zooplankton are presented in Figures 2 and 3. Figure 2 shows the change in DMSP concentrations expressed as nM in the particulate (phytoplankton only) fraction (2a), in the zooplankton (2b) and dissolved fraction (2c) as well as the changes in DMS concentration (2b). The DMSP(p) concentration in this experiment was about 3 orders of

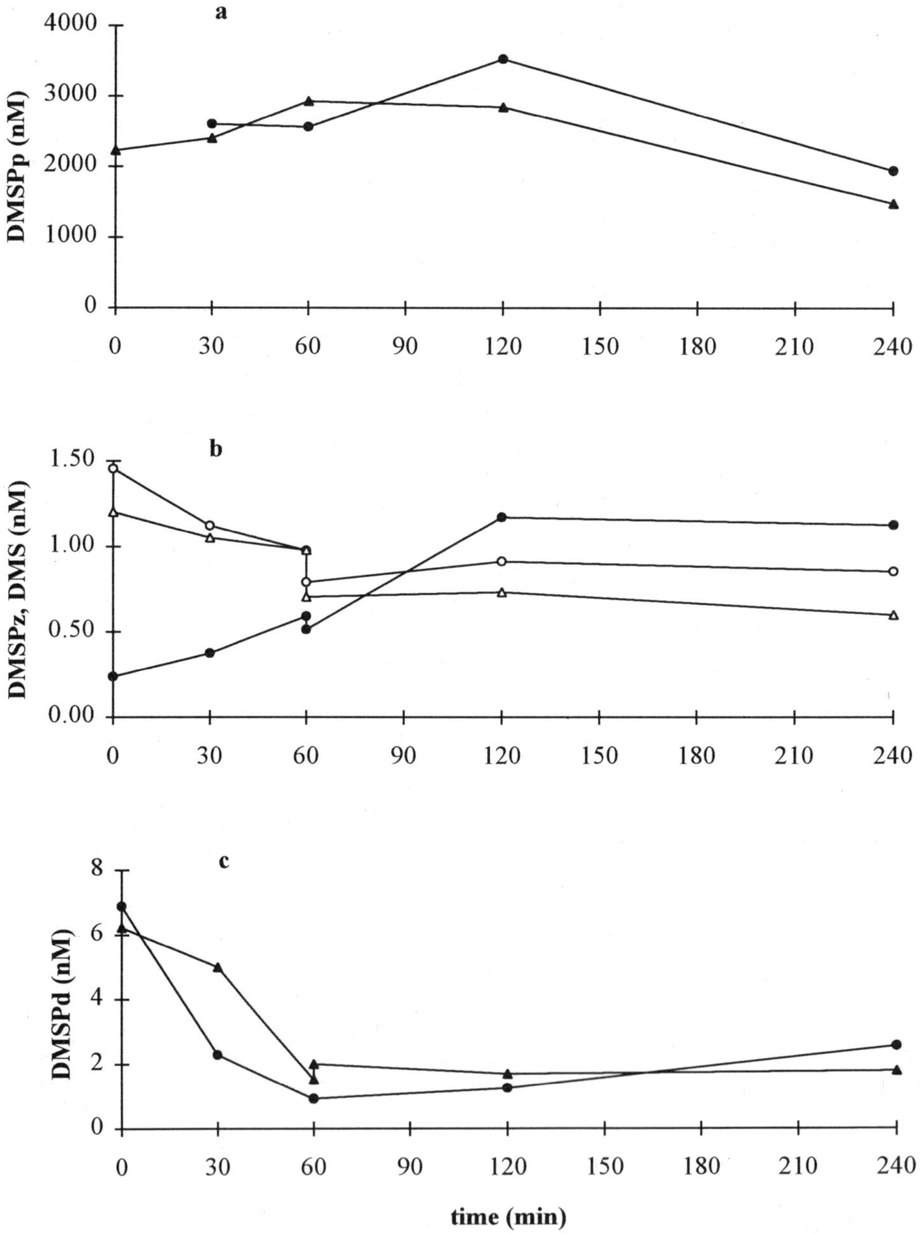

Figure 2. Changes in DMSP (closed symbols) and DMS (open symbols) concentrations in suspensions of the diatom *Phaeodactylum tricornutum* (1000×10^3 cells.ml^{-1}) incubated in 100 ml glass bottles without copepods (triangles) and with 5 copepods (circles) of the species *Eurytemora affinis*. 2a: particulate DMSP (DMSP(p)). 2b: DMS and DMSP in the zooplankton gut (DMSP(z)). 2c: dissolved DMSP (DMSP(d)).

magnitude above the DMS, the DMSP(d) and DMSP(z) concentration. The change of DMSP(p) in the systems with and without zooplankton is more or less parallel.

In the copepods there was an increase in the DMSP(z) gut content in the first two hours of about 0.5 $nM.h^{-1}$ (calculated as 10 $pmol.ind^{-1}.h^{-1}$ for 5 individuals in 100 ml medium), after which the gut content remains stable at about 1.2 nM (24 $pmol.ind^{-1}$), indicating that the copepods were completely saturated and were possibly excreting DMSP

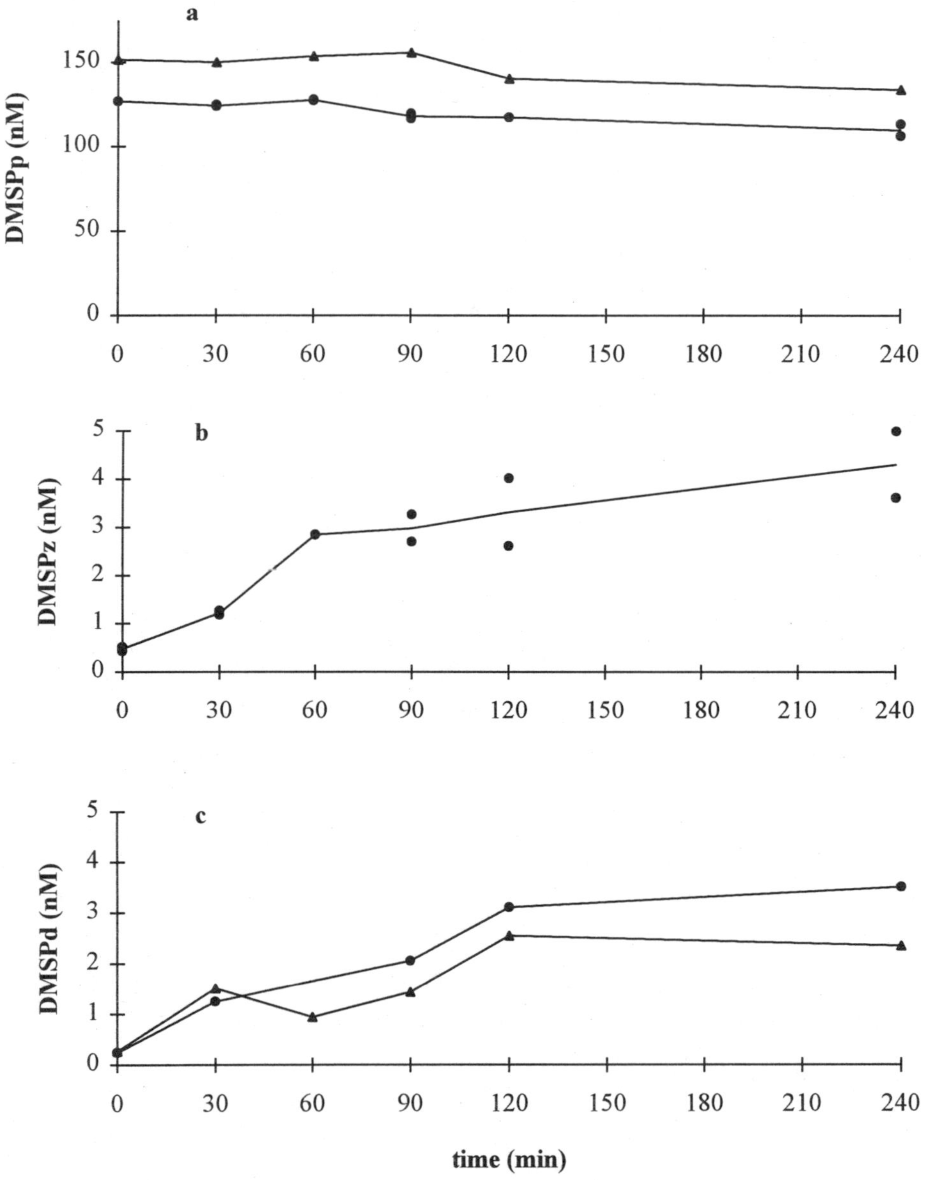

Figure 3. Changes in DMSP concentration in suspensions of the diatom *Phaeodactylum tricornutum* (60×10^3 $cells.ml^{-1}$) incubated in 100 ml glass bottles without copepods (triangles) and with 5 copepods (circles) of the species *Eurytemora affinis*. 2a: particulate DMSP (DMSP(p)). 2b: DMSP in the zooplankton gut (DMSP(z)). 2c: dissolved DMSP (DMSP(d)), DMS could not be detected (detection limit 0.15 nM).

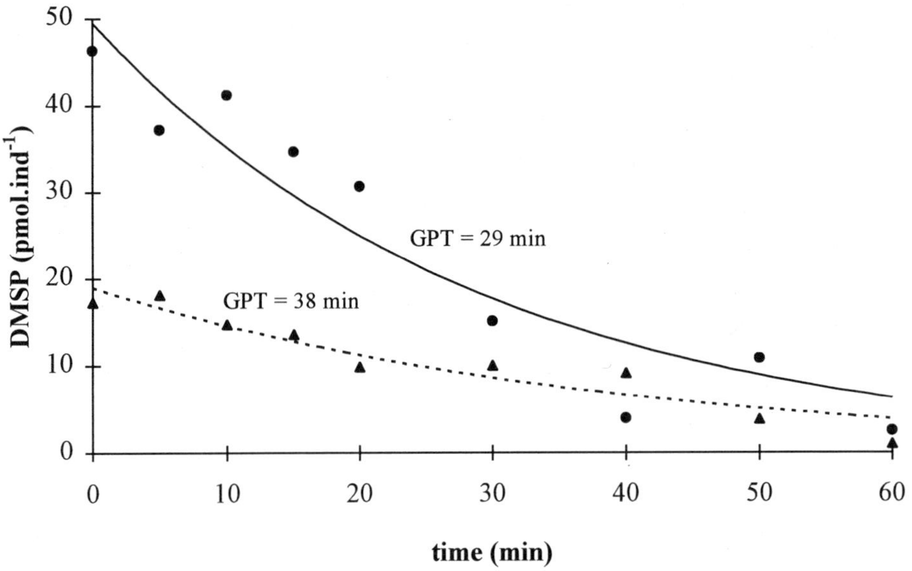

Figure 4. Decrease of DMSP gut content in fully satisfied *Eurytemora affinis* transferred to filtered seawater, to allow them to defecate. Circles: adult females. Triangles: adult males. GPT = gut passage time from mouth to rear end.

at the same rate as ingesting it. The reason for using nM.h^{-1} instead of pmol.ind^{-1}.h^{-1} was to facilitate budget calculations.

The DMS concentration initially decreased from about 1.3 nM to about 0.8 nM, after which it remained more or less stable in both systems. The DMSP(d) concentration also decreased initially in both systems, but in the systems with copepods there was an increase of about 0.5 nM.h^{-1} between the 60 and 240 minutes, while in the ungrazed systems the concentration remained constant. This pattern seemed to match the changes in DMSP(z).

As the DMSP(p) concentration in the previous experiment appeared to be very high compared to the other fractions, the experiment was repeated with a less dense and more realistic suspension of *P. tricornutum* (Figure 3 a-c). Again the changes in DMSP(p) in both systems appeared to be rather parallel. The DMSP gut content in the copepods increased initially with a rate of about 2 nM.h^{-1} (40 pmol.ind^{-1} h^{-1}) and leveled off after about 1.5 h. We were unable to detect any DMS in these samples (detection limit 0.15 nM). The DMSP(d) concentration started in this experiment at a low level and increased with time in both systems. In the grazed systems, however, the DMSP(d) concentration showed an increase of 0.5 nM.h^{-1} in the last 2 h, compared to a decrease of 0.2 nM.h^{-1} in the ungrazed systems.

Residence Time of DMSP in Copepods

In order to assess the residence time of the DMSP in the copepods (thus the DMSP(z)), the results of the gut clearance experiment are displayed in Figure 4. A large difference in the initial DMSP gut content between male (about 20 pmol DMSP .ind^{-1}) and female (about 50 pmol DMSP.ind^{-1}) copepods could be observed. The gut passage time for both sexes was in the order of 0.5 hour. According to Dam and Peterson (8) a normal value at this incubation temperature for natural conditions would be 26 min. The ingestion rate for

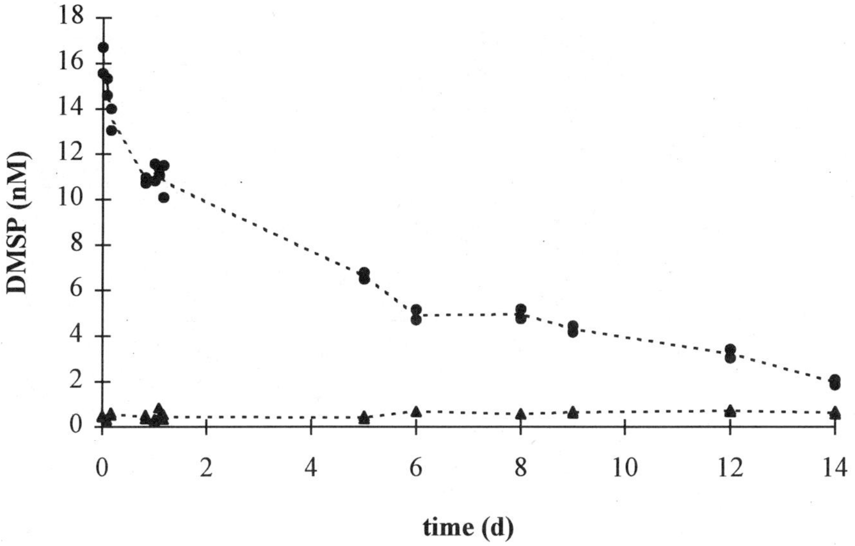

Figure 5. Decrease of the DMSP concentration in fresh fecal pellets (3000 pellets.l^{-1}) placed in 2 l 0.45 μm filtered seawater. Circles: DMSP in pellets (DMSP(f)). Triangles: dissolved DMSP (DMSP(d)).

females was about 100 pmol DMSP.ind^{-1} h^{-1} and for males about 30 pmol DMSP.ind^{-1} h^{-1}. The large differences between males and females can be explained by the difference in size and the fact that the females of *Eurytemora* carry their eggs with them, which requires more energy for swimming and hence a higher feeding rate.

The effect of zooplankton upon DMSP(p) appears to be that, via DMSP(z), the plankton cells are packaged into fecal pellets (DMSP(f)). The release of DMSP(d) and the stability of DMSP(f) was followed in a separate experiment (Figure 5).

DMSP Stability in Fecal Pellets

The amount of DMSP(f) in the pellets appeared to decrease about 30% during the first day; after 5 days about 70% had disappeared and after 2 weeks only 10% of the initial amount of DMSP(f) was left (Figure 5). Surprisingly, no increase in the DMSP(d) concentration could be detected (mean concentration about 0.5 ± 0.15 nM). The DMS concentration remained below the detection limit (0.15 nM). Apparently, the DMSP in the fecal pellets was metabolized by bacteria present in the pellets to other sulfur compounds (e.g. methyl-mercaptopropionate (MMPA), mercaptopropionate (MPA) or DMSO) that could not be detected by the analysis method used, or the DMS formed was immediately assimilated to CO_2 and particulate material (bacteria cells) (12, 15, 31). As the headspace of the incubation bottle was relatively small compared to the water volume, it is unlikely that all DMS escaped to the atmosphere without being detected in the water column. Visual inspection of the fecal pellets showed that their number did not change and that they remained intact for at least 8 days.

Zooplankton Related DMSP Budget

The results of the complete budget over a 24 h experiment become clear from Figure 6. The column t=0 shows respectively the amounts of DMSP(p), DMSP(d) and DMSP(z)

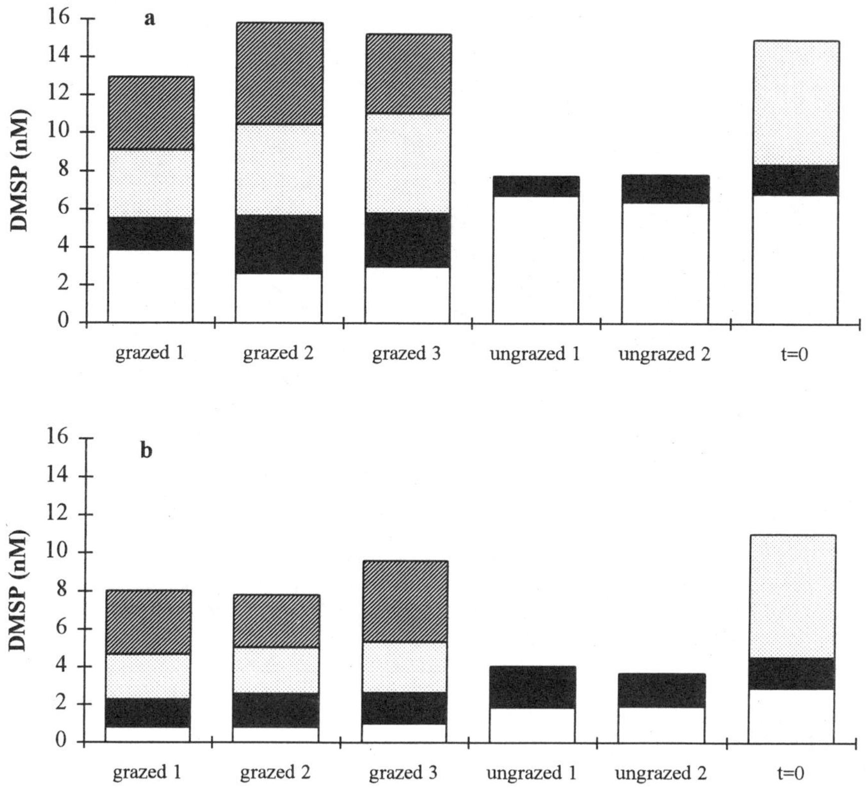

Figure 6. DMSP concentration in 100 ml suspensions of the diatom species *Phaeodactylum tricornutum* (20×10^3 cells.ml^{-1}) (Panel A) and *Thalassiosira weissflogii* (5×10^3 cells.ml^{-1}) (Panel B) incubated for 24h with 20 copepods (grazed 1-3) or without copepods (ungrazed 1-2) of the species *Eurytemora affinis*. The column t=0 represent the initial concentrations. White column: DMSP(p). Black column: DMSP(d). Grey column: DMSP(z). Shaded column: DMSP(f).

(no pellets at t=0) each expressed in nM in order to make budget calculations easy. The other columns give the results for the amount of DMSP in all fractions after 24 h. Obviously the ungrazed systems contain no DMSP(z) or DMSP(f) fractions. The data presented are the result of triplicate (grazed) or duplicate (ungrazed) independent analyses.

It is clear, that the total amount of DMSP(p) + DMSP(d) in the ungrazed systems shows only a minor change, if any. There is a small shift from DMSP(p) to DMSP(d) in the experiment with *T. weissflogii*.

In the grazed systems there is a distinct replacement of DMSP(p) to the fecal pellets, while there is also a decrease of DMSP(z). In the experiment with *P. tricornutum* the total amount of DMSP in all the fractions after 24 h is remarkably similar to the total amount of DMSP at t=0, on average 98% of the DMSP added at t=0 was accounted for at the end of the experiment. In the *T. weissflogii* experiment there appeared to be a loss of about 20% of the DMSP to sulfur compounds that could not be detected. Grazing rates calculated from the disappearance of cells (measured by the Elzone particle counter) and calculated from the disappearance of DMSP from the particulate fraction appeared to be the same (about 2 ml.ind^{-1} d^{-1}) for the *P. tricornutum* experiment and about 2.5 ml.ind^{-1} d^{-1} for the *T. weissflogii* experiment). No DMS formation could be detected in this experiment.

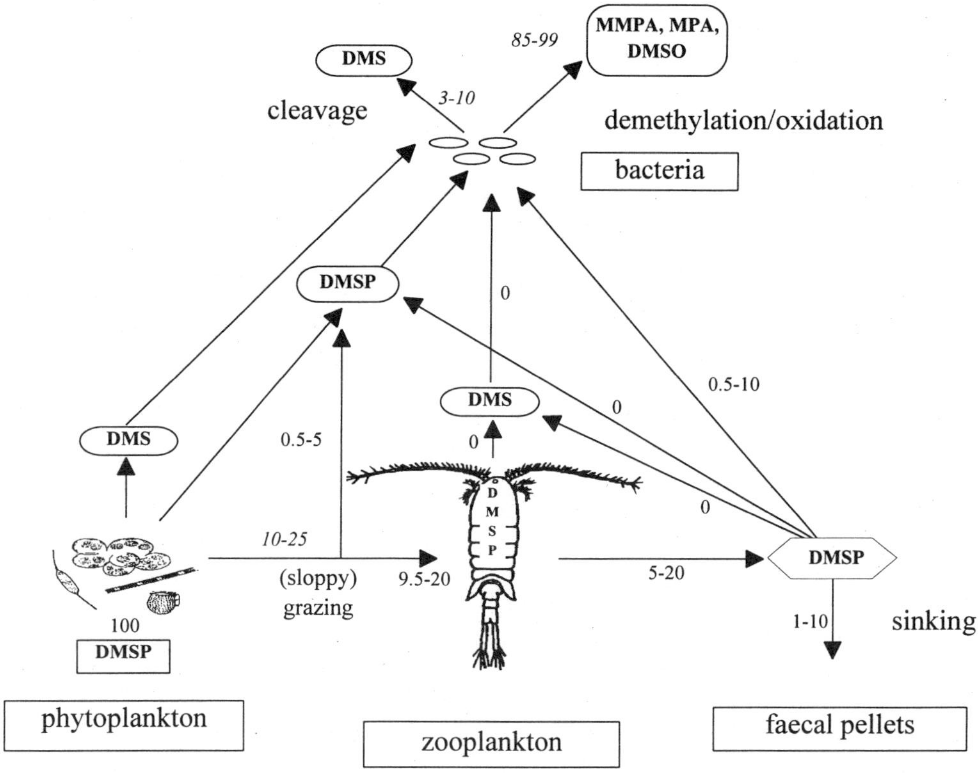

Figure 7. Possible pathways by which copepod grazing activity may influence the release of DMSP and DMS. Numbers represent percentages of primary or DMSP production by phytoplankton. Italics represent numbers derived from literature, other numbers were calculated from this study. For further explanation see text.

This total budget experiment allows us to construct a scheme where the different pathways are quantified in a relative sense (Figure 7). Phytoplankton is shown as the only producer of DMSP and this production is set at 100%. According to the literature (1, 28) 10 to 25% of the daily primary production in Dutch coastal waters is consumed by copepods in summer and autumn. In our experiments this grazing led to a small release of 0.5-5% DMSP(d) (based on the results of Figure 6). We attribute this to sloppy feeding, as the results presented in Figures 5 and 6 show that no net DMSP(d) was released by the fecal pellets.

In our experimental system no net DMS was formed and the major part of the ingested DMSP was excreted unmetabolized (when 25% of the production was ingested, about 20% would pass the copepods undigested), some loss in the gut is sometimes observed, however (see Figure 6b). The most important role of copepod grazing on phytoplankton appears to be the repackaging of phytoplanktonic DMSP into fecal pellets. Sinking rates of the fecal pellets are in the order of 20-200 $m.d^{-1}$ depending on the size and the copepod species used and the food quality (5). These high sinking velocities mean that a significant amount (up to 10%) of the DMSP daily production can disappear from the surface waters in this way, and thus at least temporarily forms a sink of DMS(P). According to Small et al. 1979 (25), sinking rates for fecal pellets from small copepods can be described with $\log S = 0.374 \log V - 0.416$ (with S, sinking rate and V, Volume). The fecal pellets produced in our experiment had an average volume of 1.5×10^5 μm^3 which results in an average sinking rate of 33 $m.d^{-1}$. Deposition is not the only process affecting the fecal pellets; zooplankton feeding on fecal pellets may influence the amount of DMSP disappearing from the surface waters (13).

No release of DMSP(d) or DMS from the fecal pellets was found in our experiments. Due to the rinsing of the pellets before measurement, loosely associated DMSP(d) and DMS could have been lost from the pellets during this procedure. One may question whether sulfur removed by this gentle treatment should be considered part of the pellet. A loss of DMSP from the pellets during the experiment could be observed (Figure 5), which must be attributed to bacterial consumption within the pellets involving pathways to compounds that were not detected.

How do these findings compare with other (field) observations? According to Belviso et al. (3), the length of the deposition path of fecal pellets in the water column affects the amount of DMSP in fecal pellet material which reaches sediment traps held at 200 m depth. This supports the idea that the pellets may be a sink for DMS(P). Daly & DiTullio (7) found low concentrations of DMSP(p) in sediment traps in deep water, suggesting that break down of the DMSP in the fecal pellets in the water column is a process that should not be neglected (see also Figure 5).

Our findings that no net DMS is formed either from the grazing activity or from the pellets appears to be in contrast with the results of Dacey & Wakeham (6) and Daly & DiTullio (7) who found that the grazing activity of copepods and juvenile krill respectively, caused a remarkable increase of the DMS production compared to ungrazed control systems. The latter study was, however, performed in the presence of *Phaeocystis* sp., which is known to be able to cleave extracellular DMSP(d) very rapidly into DMS and acrylate due to the presence of extracellular DMSP-lyase (26, 27). Apparently, the diatom species we used in our experiments do not possess this lyase activity. Of the species *Gymnodinium nelsoni* and *Prorocentrum micans* (6) it is not known whether they have this enzyme. Another explanation is the possibility of DMS formation by bacteria associated with the plankton species used. We should also not forget that the selection of species may play an important role in the outcome of the results. Different zooplankton species (and development stages!) tend to select for favored food sources (9, 11), with different DMSP contents and lyase activity. In addition the shape/size and density (thus sinking rate) of the fecal pellets changes with the zooplankton species. The physiological state of a phytoplankton bloom may affect the deposition rates as well (13). Estep et al. (9) and Hansen & van Boekel (11) showed that the grazing activity of copepod species on *Phaeocystis* may depend on the physiological state of the algae and that the copepod *Temora longicornis* is able to switch from phytoplankton to heterotrophic nanoflagellates as a food source during a *Phaeocystis* dominated bloom.

This all means that in addition to our findings the various pathways should be investigated for a number of different species combinations, before they can be generally applied.

CONCLUSIONS

The experiments with the copepod *Eurytemora affinis* grazing on the diatoms *Phaeodactylum tricornutum* and *Thallassiosira weissflogii* suggest that most of the DMSP ingested is packaged into fecal pellets. The pellets are subject to DMSP degradation ($t_{1/2}$ of several days). This allows part of the fecal pellets to be deposited beyond the mixing zone, thus forming a potential sink for DMS(P) from the surface water.

No net formation of DMS (from the DMSP(z) or DMSP(f)) could be observed. Only a small fraction was released as DMSP(d) due to sloppy feeding.

ACKNOWLEDGMENTS

We would like to thank the following people without whom this work would not have been possible: Robbert Jak for fruitful discussions on the grazing experiments. And last, but certainly not least, RLJK thanks Marlien van Lunen, for being my best friend, support and mother of our son Reinier Kwint (22nd May 1995). This work was funded by the Commission of the European Communities, Environment Programme (project no: 93-0326).

REFERENCES

1. Baars, M.A. and Fransz, H.G. 1984. Grazing pressure of copepods on the phytoplankton stock of the central North Sea. Neth. J. Sea Res. 18(12): 120-142.
2. Belviso, S. Kim, SK. Rassoulzadegan, F. Krajka, B. Nguyen, B.C. Mihalopoulos, N. and Buat–Menard, P. 1990. Production of dimethylsulfonium propionate (DMSP) and Dimethylsulfide (DMS) by a microbial food web. Limnol. Oceanogr. 35(8): 1810-1821.
3. Belviso, S. Corn, M. and Buat–Menard, P. 1992. Assessment of the role of zooplankton in the cycling of DMSP and DMS in the water column During EUMELI4 (FranceJGOFS). Dimethylsulphide: Oceans, Atmosphere and Climate. Restelli, G. and Angeletti, G. (eds). Int. Symp. Proc. Belgirate. Kluwer Acad. Publ. Dordrecht, p. 15-19.
4. Belviso, B. Buat–Menard, P. Putaud J.P. Nguyen, B.C. Claustre, H. and Neveux, J. 1993. Size distribution of dimethylsulfoniopropionate (DMSP) in areas of the tropical northeastern Atlantic Ocean and the Mediteranean Sea. Mar. Chem. 44: 55-71.
5. Butler, M. and Dam, H.G. 1994. Production rates and characteristics of fecal pellets of the copepod *Acartia tonsa* under simulated phytoplankton bloom conditions: implications for vertical fluxes. Mar. Ecol. Prog. Ser. 114: 81-91.
6. Dacey, J.W.H. and Wakeham, S.G. 1986. Oceanic dimethylsulfide: production during zooplankton grazing on phytoplankton. Science 233: 1314-1316.
7. Daly, K., and G. R. DiTullio. 1996. Particulate dimethylsulfoniopropionate removal and dimethylsulfide production by zooplankton in the southern ocean. *In* R. P. Kiene, P. T. Visscher, M. D. Keller, and G. O. Kirst (ed.), Biological and environmental chemistry of DMSP and related sulfonium compounds. Plenum, New York, pp. 223-238.
8. Dam, G.H. and Peterson, W.T. 1988. The effect of temperature on the gut clearance rate constant of planktonic copepods. J. Exp. Mar. Biol. Ecol. 123: 1-14.
9. Estep, K.W. Nejstgaard, J.Ch. Skjoldal, H.R. and Rey, F. 1990. Predation by copepods upon natural populations of *Phaeocystis pouchettii* as a function of the physiological state of the prey. Mar. Ecol. Prog. Ser. 67: 235-249.
10. Gröne, T. and Kirst, G.O. 1992. The effect of nitrogen deficiency, methionine and inhibitors of methionine metabolism on the DMSP contents of *Tetraselmis subcordiformis* (Stein). Mar. Biol. 112: 497-503.
11. Hansen, F.C. and Van Boekel, W.H.M. 1991. Grazing pressure of the calanoid copepod *Temora longicornis* on a *Phaeocystis* dominated spring bloom in a Dutch tidal inlet. Mar. Ecol. Prog. Ser. 78: 123-129.
12. Hatton, A.D., Malin, G. and Liss P.S. 1994. Determination of dimethylsulfoxide in aqueous solution by an enzymelinked method. Anal. Chem. 66: 4093-4096.
13. Jackson, G.A. 1993. Flux feeding as a mechanism for zooplankton grazing and its implications for vertical particulate flux. Limnol. Oceanogr. 38(6): 1328-1331.
14. Kiene, R.P. and Gerard, G. 1994. Determination of trace levels of dimethylsulfoxide (DMSO) in seawater and rainwater. Mar. Chem. 47: 1-12.
15. Kiene, R.P. and Service, S.K. 1991. Decomposition of dissolved DMSP and DMS in estuarine waters: dependence on temperature and substrate concentration. Mar. Ecol. Prog. Ser. 76: 1-11.
16. Kwint, R.L.J. and Kramer, K.J.M. 1995. Dimethylsulphide production by plankton communities. Mar. Ecol. Prog. Ser. 121: 227-237.
17. Kwint, R.L.J. and Kramer, K.J.M. 1995. A new sensitive method for the determination of zooplankton grazing activity. Submitted to J. Plankt. Res.
18. Kwint, R.L.J. Kramer, K.J.M., Baart, A.C. and Verhagen, H.L.M. 1992. The production of DMS by a plankton community: A mesocosm experiment. Dimethylsulphide: Oceans, Atmosphere and Climate. Restelli, G. and Angeletti, G. (eds). Int. Symp. Proc. Belgirate. Kluwer Acad. Publ. Dordrecht, p. 53-62.

19. Leck, C., Larsson, U., Bågander, L.E., Johansson, S. and Hajdu, S. 1990. Dimethyl sulfide in the Baltic Sea: Annual variability in relation to biological activity. J. Geophys. Res. 95(C3): 3353-3363.
20. Levasseur, M., Keller, M.D., Bonneau, E., D'Amours, D. and Bellows, W.K. 1994. Oceanographic basis of a DMSrelated Atlantic cod (*Gadus morhua*) fishery problem: Blackberry Feed. Can. J. Fish. Aquat. Sci. 51: 881-889.
21. Malin, G., Turner, S.M. and Liss, P.S. 1992. Sulphur: the plankton/climate connection. J. Phycol. 28: 590-597.
22. Nguyen, B.C., Belviso, S., Mihalopoulos, N. and Gostan, J., Nival, P. 1988. Dimethyl sulfide production during natural phytoplanktonic blooms. Mar. Chem. 24(2): 133-142.
23. Peterson, W., Painting, S. and Barlow, R. 1990. Feeding rates of *Calanoides carinatus*: a comparison of five methods including evaluation of the gut fluorescence method. Mar. Ecol. Prog. Ser. 63: 85-92.
24. Quist, P., Kwint, R.L.J., Hansen, T.A., Dijkhuizen, L. and Kramer, K.J.M. 1995. Turnover of dimethylsulphoniopropionate and dimethylsulphide in the marine environment: The role of bacteria . submitted to Mar. Ecol. Prog Ser.
25. Small, L.F., Fowler, S.W. and Ünlü, M.Y. 1979. Sinking rates of natural copepod fecal pellets. Mar. Biol. 51: 233-241.
26. Stefels, J., Dijkhuizen, L. and Gieskes, W.W.C. 1995. DMSPlyase activity in a spring phytoplankton bloom off the Dutch coast, related to *Phaeocystis* sp. abundance. Mar. Ecol. Prog. Ser. 123: 235-243
27. Stefels, J. and Van Boekel, W.H.M. 1993. Production of DMS from dissolved DMSP in axenic cultures of the marine phytoplankton species *Phaeocystis* sp. Mar. Ecol. Prog. Ser. 97: 11-18.
28. Tackx, M.L.M., Bakker, C. and Van Rijswijk, P. 1990. Zooplankton grazing pressure in the Oosterschelde (the Netherlands). Neth. J. Sea Res. 25(3): 405-415.
29. Tokunaga T., Iida H. and Nakamura K. 1977. Formation of Diethyl Sulfide in Antarctic krill *Euphausia superba*. Bull. Jpn. Soc. Sci. Fish. 43(10): 1209-1218.
30. White, R.H. 1982. Analysis of dimethylsulfonium compounds in marine algae. J. Mar. Res. 40(2): 529-536
31. Wolfe, G.V. and Kiene, R.P. 1993. Radioisotope and chemical inhibitor measurements of dimethyl sulfide consumption rates and kinetics in estuarine waters. Mar. Ecol. Prog. Ser. 99: 261-269
32. Wolfe, G.V., Sherr, E.B. & Sherr, B.F. 1994. Release and consumption of DMSP from *Emiliania huxleyi* during grazing by *Oxyrrhis marina*. Mar. Ecol. Prog. Ser. 111: 111-119.

22

OSMOREGULATION IN BACTERIA AND TRANSPORT OF ONIUM COMPOUNDS

D. Le Rudulier, J. -A. Pocard, E. Boncompagni, and M. C. Poggi

Laboratoire de Biologie Végétale et Microbiologie
URA CNRS 1114, Université de Nice - Sophia Antipolis
Parc Valrose, 06108 NICE Cedex 2, France

SUMMARY

Bacteria are able to adapt to environmental changes and generally respond to increases in the osmotic pressure of their surroundings by elevating the intracellular concentrations of osmoprotective compounds. Besides potassium, the most prevalent cellular cation, the preferred solute is glycine betaine which is accumulated either by uptake from the environment or via synthesis from choline. In *Escherichia coli*, osmoregulatory uptake of glycine betaine is mediated by two transport systems, the constitutive low-affinity ProP system ($K_m = 40\ \mu M$), and the osmotically inducible ProU high-affinity system ($K_m = 1\ \mu M$). ProP is a single polypeptide and the energy for substrate transport is provided by the proton motive force. ProU is a periplasmic-binding-protein-dependent system encoded by three structural genes, *proV*, *proW* and *proX*.

Dimethylsulfoniopropionate (DMSP) also behaves as a powerful osmoprotectant in various bacteria. Its ability to restore growth of *E. coli* in media of high osmolarity is dependent on the ProP and ProU systems. The osmoprotective effect of DMSP is abolished in strains lacking *proP* and *proX,* although DMSP has no detectable affinity for ProX, the periplasmic glycine betaine-binding protein. Dimethylsulfonioacetate (DMSA = dimethylthetin) is equally as effective as glycine betaine in stimulating the growth of *E. coli* in medium of high salt concentration. In *Rhizobium meliloti* DMSA is an efficient competitor for ^{14}C-glycine betaine uptake but does not bind to the glycine betaine-binding protein.

INTRODUCTION

In order to proliferate in an environment subjected to fluctuations in osmolarity, bacterial cells must maintain a positive turgor, i.e. a cytoplasmic osmotic pressure higher than that of the extracellular environment. The mechanisms responsible for cellular adaptation to osmotic stress (osmoregulation) have been elucidated mainly during the past ten years, especially in bacteria (6, 7, 28). It is now clear that organisms as diverse as algae, fungi,

Biological and Environmental Chemistry of DMSP and Related Sulfonium Compounds
edited by Ronald P. Kiene et al., Plenum Press, New York, 1996

higher plants, and animals have evolved similar strategies to cope with osmotic stress. Indeed, only a few solutes, termed compatible solutes, are accumulated under conditions of osmotic stress. The main compatible solutes found in bacteria include potassium ions, a few amino acids (proline, glutamate), several sugars and polyols (sucrose, trehalose, glucosylglycerol), the most widely adopted glycine betaine, and other N- or S-methylated amino acid derivatives. The accumulation of osmolytes in bacteria relies on transport of osmoprotectants from the culture medium and/or *de novo* synthesis. The mechanisms of osmoregulation have been studied intensively by biochemical and molecular approaches, mainly in the enteric bacteria *E. coli* and *Salmonella typhimurium* (30). In this review article, we report on the main properties of betaine uptake systems and on the regulation of osmotica by genetic mechanisms. We also consider the role of DMSP as an osmoprotectant, and some characteristics of its uptake.

THE PRIMARY ROLE OF POTASSIUM

The most prevalent cellular cation in the cytoplasm of bacteria is potassium. Its intracellular concentration in *E. coli* is nearly proportional to the osmolarity of the growth medium, ranging from about 100 mM to nearly 1 M. The transport systems which are active in potassium accumulation have been studied in detail by Epstein and coworkers (14). Potassium uptake is dependent on the activity of the osmotically-inducible Kdp system and several constitutive Trk systems. Kdp is an ATP-driven system with a high affinity for K^+ (K_m = 1.5 μM). Three structural genes, *kdpABC*, are adjacent to a regulatory element, *kdpED*, which ensures induction of the Kdp system whenever the external K^+ concentration is too low for the Trk system to operate. A recent genetic analysis of K^+ constitutive uptake suggested the presence of three different systems: a system capable of low rate transport, TrkD, dependent solely on the product of the *trkD* gene, and two systems capable of high rate transport, TrkG and TrKH, formerly lumped together as the single TrkA system (11). Besides these uptake systems, there are at least three efflux systems encoded by the *kefABC* genes (12).

GLYCINE BETAINE AND RELATED COMPOUNDS AS MAJOR COMPATIBLE SOLUTES

Besides the PutP protein which is a Na^+/proline symporter with a high affinity for proline (K_m = 1 μM) and a rather low V_{max} (4.6 nmol≠mg protein^{-1}≠min^{-1}), *E. coli* and *S. typhimurium* possess two other independent proline transport systems, ProP and ProU. The *putP*, *proP* and *proU* operons map at 23, 93 and 58 min on the chromosome (43). The PutP system is required for the transport of proline as a sole nitrogen, carbon, or energy source. Since its activity is not stimulated in cells grown at high osmolarity, this system is not important for the transport of proline as an osmoprotectant. ProP and ProU were originally discovered by virtue of their ability to transport toxic proline analogs and were subsequently shown to transport glycine betaine.

Accumulation of Glycine Betaine Is Mediated by ProP and ProU. ProP has a very poor affinity for proline (K_m = 300 μM; V_{max}= 32 nmo≠mg protein^{-1}≠min^{-1}) and would not seem to contribute significantly to proline uptake in cells in which PutP is functional. In fact, ProP plays a major role in glycine betaine uptake and mutations in *proP* reduce the ability of this betaine to serve as an osmoprotectant (3). Studies of *S. typhimurium* strains lacking

PutP and ProP uptake systems have suggested the existence of a third proline permease, the ProU system, that functions only in media of elevated osmolarity (5). However, this system has a low affinity for proline (K_m = 200 μM), and Cairney et al. (2) have demonstrated that proline uptake via ProU is negligible. Rather, these authors have shown that *proU* encodes a high-affinity (K_m = 1.3 μM; V_{max} = 12.5 nmol≠mg protein^{-1}≠min^{-1}) transport system for glycine betaine. Both systems, ProP and ProU are controlled at two levels. Increased uptake of glycine betaine at high osmolarity results largely from *de novo* synthesis of the ProP and ProU transport components, since the transcription of *proP-lacZ* and *proU-lacZ* fusions is enhanced approximately 3- and 100-fold, respectively, when the fusion strains are grown at high osmolarity or subjected to a sudden osmotic upshock (2, 3, 16). In addition, the activity of the transport systems themselves is increased by the osmotic pressure of the medium. This stimulation of transport activity, independent of *de novo* protein synthesis, might be caused by conformational changes in transport proteins. As a consequence of the activity of both ProP and ProU, glycine betaine is accumulated at very high concentrations in *E. coli* cells grown at high osmolarity (800 mM or more) while its intracellular concentration is always a very low (10 - 15 mM) in cells grown in minimal medium (36). Similar results have been obtained in *Rhizobium meliloti*: intracellular glycine betaine concentration is only 18 mM in cells grown in low-salt medium and reaches 650 mM in the presence of 0.5 M NaCl (25).

In a nitrogen- and carbon-free medium, glycine betaine does not support the growth of *E. coli* either in low-salt or in high-salt medium. This molecule is not catabolized (36). Recent data suggest that *E. coli* and *S. typhimurium* possess efflux systems for glycine betaine, separate from ProP and ProU, that permit a rapid depletion of the intracellular pool of the osmoprotectant (21). This system is responsive to changes in the osmotic pressure of the medium: a sudden osmotic downshock induces a rapid efflux of glycine betaine. However, this loss may not be a glycine betaine-specific phenomenon since osmotic downshock with serine-loaded cells provokes similar loss of the accumulated amino acid. The molecular basis of the efflux system and its mode of control remain to be elucidated.

By contrast, in *R. meliloti*, glycine betaine can be used as a building block for cellular components, as well as an osmoprotectant. However, the NaCl concentration of the growth medium regulates the levels of enzymatic activities responsible for glycine betaine catabolism in a way that allows its accumulation when the cells are osmotically stressed (25, 39). This catabolic repression prevents a futile cycle of uptake and degradation and ensures that glycine betaine is preserved to function as an osmoprotectant.

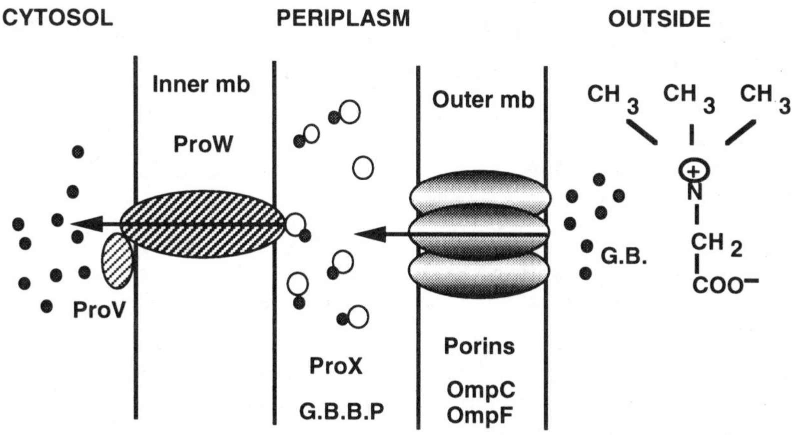

Figure 1. Structural organization of the binding protein dependent ProU transport system in *E. coli*.

Structure of ProP and ProU and Genetic Organization. Wood and coworkers (8) have used functional complementation of multiple proline/betaine transport defects to isolate the *proP* gene from *E. coli*. The nucleotide sequence was determined. The ProP transporter is an integral cytoplasmic membrane protein of 42 kDa. Comparison of protein sequence data reveals that ProP shows significant homology to the citrate/proton symporters from *E. coli* and *Klebsiella pneumoniae,* and to an α-ketoglutarate transporter from *E. coli*. Members of this transporter superfamily are predicted to have cytoplasmic amino and carboxyl termini and to cross the cell membrane 12 times. However, in contrast to α-ketoglutarate transporters, ProP includes an extended central hydrophilic loop, and an extended hydrophilic carboxyl terminal domain of about 50 amino acids that is likely to participate in the formation of an α-helical bundle (8).

Recently, it has been shown that *proP* utilizes two different promoter sites: P1 is responsible for strong basal expression and response to osmotic signals, while P2 is involved in *rpoS*-dependent stationary phase induction (34). Thus, *proP* can be added to the growing list of dual-signal genes responsive to both osmotic and stationary-phase signals.

The ProU transporter is a multicomponent binding-protein-dependent transport system (Fig. 1). The *proU* locus contains three genes organized in a single operon and designated *proV, proW,* and *proX* (9 ; Fig. 2). The sequence of 4,362 nucleotides encompassing the *proU* operon of *E. coli* has been determined (17).

ProV, a peripheral membrane protein found on the cytoplasmic surface of the membrane, is predicted to be a 400-amino-acid-long polypeptide, relatively hydrophilic (Mr 44,162). Interestingly, it is devoid of any tryptophan residues, and it seems that the C-terminal residues of the native protein are not essential for its function as a component of the glycine betaine transporter. The inferred amino acid sequence of ProV shows two regions of significant similarity to HisP, a component of the L-histidine transporter of *S. typhimurium*

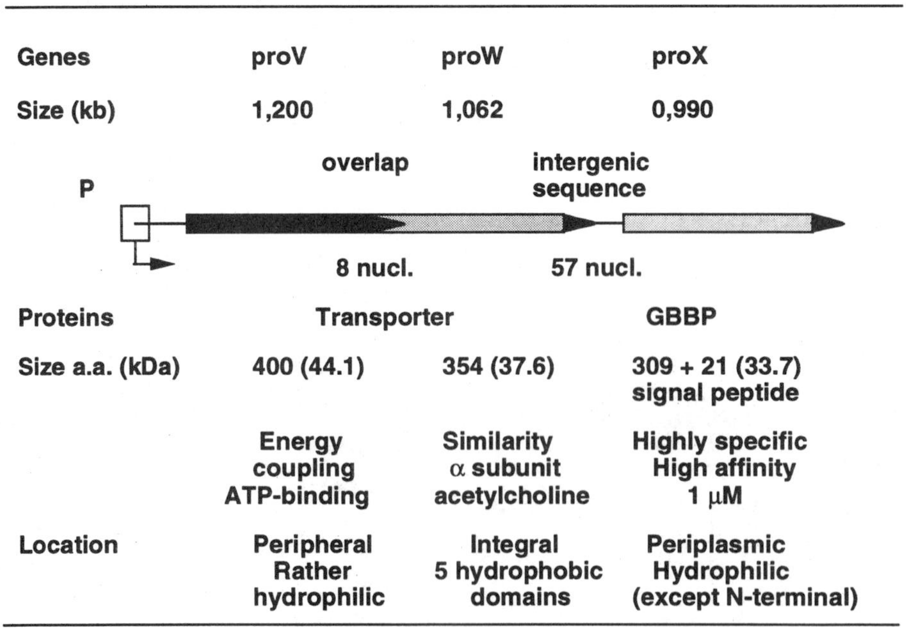

Figure 2. Genetic organization of the *proU* locus of *E. coli*.

(17). ProV also shows extensive homologies to the ATP-binding domains of other binding-protein-dependent transport systems.

ProW is a hydrophobic polypeptide of 354 amino acids (Mr 37,619). There is an 8-nucleotide overlap between the end of *proV* and the start of *proW*, suggesting that there may be translational coupling in the expression of the two genes. ProW has the features of an integral membrane protein with several hydrophobic stretches capable of spanning the membrane. An interesting similarity between a region of ProW and the α-subunit of the acetylcholine receptor protein is observed. This similarity might relate to the fact that the choline moiety of acetylcholine, which binds the α-subunit, is structurally similar to glycine betaine.

The predicted ProX polypeptide (330 amino acids), is hydrophobic at its N-terminal end, and hydrophilic thereafter. The sequence of the first 21 amino acids of ProX shows the characteristics typical of a leader peptide which is expected to be present in a periplasmic protein. The calculated Mr for the mature protein is 33,729. This protein was purified from *E. coli* and was shown to bind glycine betaine with a high affinity, $K_D = 1\ \mu M$ (1). The purified protein has no detectable affinity for proline, but proline betaine competes with glycine betaine for binding (18). Periplasmic glycine betaine-binding proteins have also been detected in the soil microorganisms *R. meliloti* and *Azospirillum brasilense* (27, 28). As observed in *E. coli*, high osmolarity does not increase the affinity of the glycine betaine-binding protein from *R. meliloti* for its substrate (33, 38).

A number of structural analogs of proline and glycine betaine such as proline betaine, γ-butyrobetaine, homobetaine, L-pipecolate betaine, taurine, N,N-dimethylglycine, pipecolic acid, and ectoine also have osmoprotecting effects in *E. coli* (26, 15, 32, 20). Both the ProP permease and the ProU system exhibit a broad substrate affinity for most of these osmoprotectants (15, 30). With the exception of proline betaine, none of these osmoprotectants is able to bind to ProX.

Regulation of the ProU Locus. Several studies with *lacZ* and *phoA* gene fusions (2, 6) have revealed that the transcription of *proU* is negligible at low osmolarity but is induced several hundred-fold at high osmolarity. This induction is detected within a few minutes after osmotic upshock, and *proU* is expressed at a high level as long as elevated osmolarity is maintained. Various facets of *proU* regulation have been covered in recent articles (6, 30, 7), and only essential features are briefly presented here. It has been proposed that the intracellular concentration of potassium/ glutamate functions as a "second messenger" which selectively stimulates the transcription of genes such as *proU* that enhances survival under osmotic stress (37). An alternative model suggests that *proU* transcription is regulated primarily through changes in the DNA supercoiling of the *proU* promoter region (19). Both mechanisms are not mutually exclusive and may contribute in a synergistic manner to mediate osmotic regulation of *proU* transcription (30). The level of *proU* expression is also altered by mutations in the *hns* gene (previously designated *osmZ*) which maps at 27.4 min on the *E. coli* chromosome (19, 30). *hns* is the structural gene for a non-specific DNA-binding protein H-NS (or H1) that is tightly associated with chromosomal DNA. Sequences located upstream and downstream from the *proU* promoter are targets for H-NS binding (for a recent review see ref. 30). The model of direct action of H-NS on *proU* expression is supported by a recent report showing that H-NS can act as a specific repressor of *proU* transcription in a purified *in vitro* system (42). In addition, deletion analysis of chromosomal sequences 5′ of *proU* have revealed an "upstream activating region", UAR, extending approximately 200 bp 5′ to the -35 box. This UAR sequence is required for maximal expression of the promoter under conditions of both low and high osmolarity (29). Furthermore, genetic experiments have demonstrated the presence of extended transcriptional "silencer" sequences at the 5′

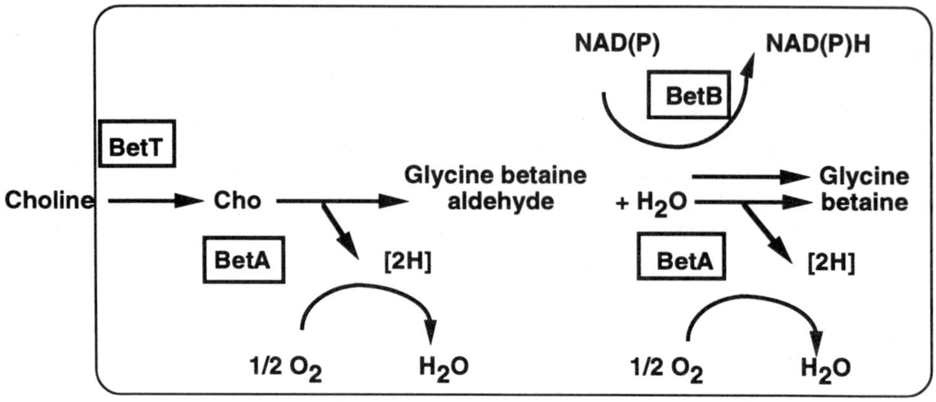

Figure 3. The osmoregulatory choline-glycine betaine pathway in *E. coli*.

end of the *proV* genes from both *E. coli* and *S. typhimurium* that keep *proU* expression repressed in medium of low osmolarity (35). Despite these advances, there are still a large number of questions that remain unanswered about *proU* regulation.

Genes Encoding the Osmoregulatory Choline-Glycine Betaine Pathway (Bet Genes). *E. coli* can synthesize glycine betaine from exogenously supplied choline (40) under conditions of osmotic stress. The *bet* genes which govern this pathway cluster at about 7.5 min on the chromosomal map of *E. coli*, i. e. counterclockwise to *lac* (41). *S. typhinurium* lacks corresponding genes and therefore is unable to synthesize glycine betaine from choline. The *bet* genes of *E. coli* are expressed only under aerobic conditions. They are strongly induced by osmotic stress, but the presence of both choline (1 mM) and high salt is required for full expression (23). Using *E. coli* strains carrying *bet-lacZ* transcriptional fusions, Eshoo

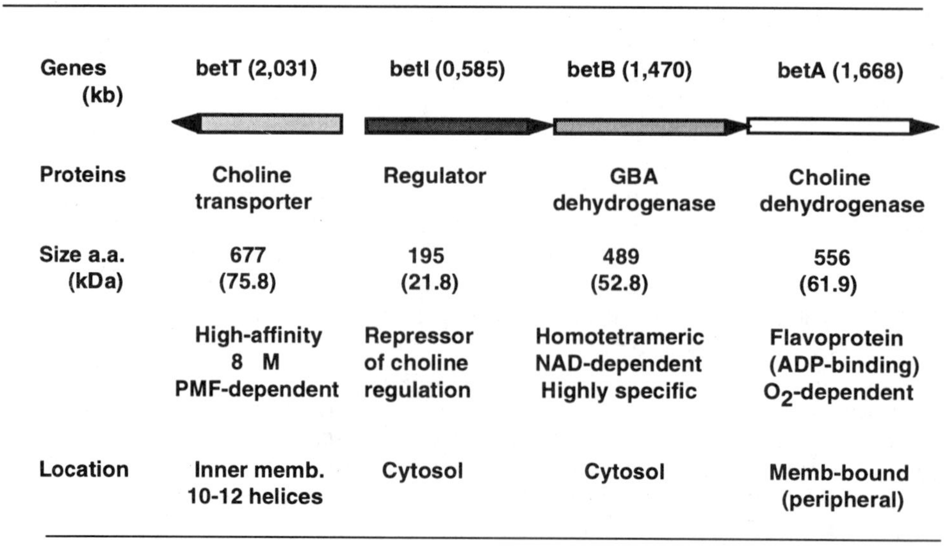

Figure 4. Genetic organization of the *bet* locus of *E. coli*.

(13) has shown that osmotic stress elicited a 7- to 10-fold increase in transcription; these genes are also regulated by temperature, oxygen, glycine betaine, and choline.

The synthesis of glycine betaine from choline entails two oxidation steps (40), with glycine betaine aldehyde as the intermediate (Fig. 3). The first enzyme is an oxygen-dependent choline dehydrogenase which has both choline and glycine betaine aldehyde dehydrogenase activities. This enzyme is membrane-bound and independent of soluble cofactors. Its activity is coupled to electron transport and a terminal electron acceptor such as 0_2 is needed; under anaerobic conditions, glycine betaine cannot be produced. In addition, *E. coli* also possesses a specific NAD-dependent glycine betaine aldehyde dehydrogenase.

The sequence of 6,493 nucleotides encompassing the *bet* genes of *E. coli* has been determined (22). Four open reading frames were identified (Fig. 4). The *betA* gene (1,668 nucleotides) encodes the choline dehydrogenase, a 556-amino acid protein of 61.9 kDa. BetA is a flavoprotein, presenting a typical binding site for the ADP moiety of FAD. The rather hydrophilic amino acid profile of BetA suggests that it is a peripheral membrane-bound protein. The *betB* gene (1,470 nucleotides) encodes a 489 amino acid soluble protein (glycine betaine aldehyde dehydrogenase) of 52.8 kDa. The sequence demonstrates extensive amino acid similarity to the specific glycine betaine aldehyde dehydrogenase from spinach chloroplasts, as well as unspecific aldehyde dehydrogenases from other eukaryotes. The *betBA* region contains an additional gene, *betI* (585 nucleotides), which encodes a polypeptide of 21.8 kDa (195 amino acids) with homology to several repressor proteins. BetI is probably involved in the choline regulation of *bet* expression. Besides the *betIBA* genes, *betT* (2,031 nucleotides) is located upstream of this operon and transcribed divergently (Fig. 4). BetT corresponds to a proton motive force driven, high-affinity transport system for choline (K_m = 8 μM). An interesting feature of BetT is the presence of a long hydrophilic C-terminal region of nearly 200 amino acid residues (22).

R. meliloti also converts choline to glycine betaine (39), and we have recently cloned and sequenced the *betAB* genes in this species (unpublished results).

DMSP AS AN OSMOPROTECTANT IN BACTERIA

Biological Activity. DMSP, the major precursor of dimethyl sulfide (DMS), is well-known as the principal sulfonium compound found in marine environments. There are remarkable structural similarities between DMSP and glycine betaine. Therefore it is interesting to determine whether this sulfur-containing analog can substitute for glycine betaine as an osmoprotectant in bacteria.

Table 1. Increased salt resistance of *E. coli* strain 31 in the presence of sulfonium compounds (Values are the highest NaCl concentration (M) in which growth occurred)

	Osmoprotectants		
Concentrations	GB	DMSP	Dimethylthetin
0	0.7	0.7	0.7
1 μM	0.9	0.7	0.9
10 μM	1.0	0.8	1.0
100 μM	1.0	0.8	1.0
1 mM	1.0	0.8	1.0

The data shown are from Chambers *et al.*, 1987.

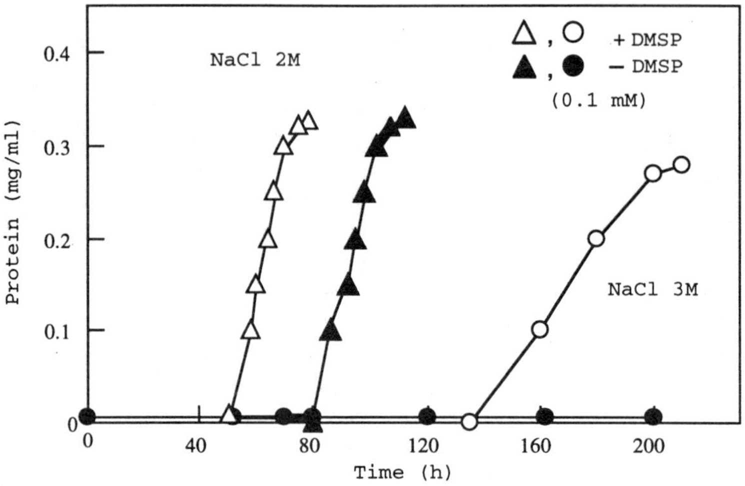

Figure 5. Effect of DMSP on the growth of strain MD 14-50, a bacterium isolated from a colony of the cyanobacterium *Trichodesmium*. The data shown are from Diaz M.R. *et al.*, 1992.

Chambers et al. (4) reported that the growth of *E. coli* in minimal medium containing high NaCl concentrations is enhanced by the addition of increasing quantities of DMSP (Table 1). More specifically, the highest NaCl concentration in which growth occurs is increased when 10 mM or more of DMSP is added to the medium. Lower concentrations have no effect. 3-methylsulfonio 2-methylproprionate has a similar effect whereas dimethyl sulfonioacetate (dimethylthetin) shows a higher activity. In fact, dimethylthetin is equally as effective as glycine betaine in promoting growth of *E. coli* in high-salt medium. Moreover, several betaines and tertiary sulfonium compounds found in marine algae have been tested using *K. pneumoniae* in a salt stress alleviation assay to quantify their respective growth-promoting activities (31). A pronounced activity is observed for glycine betaine, proline betaine, choline (26), and also for DMSP (31). If the ability of glycine betaine to alleviate growth inhibition is estimated at 100%, the relative activity of DMSP is slightly higher (110%). A recent study using an unspecified bacterium (strain MD 14-50) isolated from a colony of the cyanobacterium *Trichodesmium* indicates that strain MD 14-50 can grow over a wide range of salinities (0.2 to 2 M NaCl) in the absence of organic osmolytes, and clearly demonstrates that the addition of 0.1 mM DMSP to the growth medium is essential for growth at 3M NaCl (10). At 2M NaCl, the presence of DMSP reduces the lag phase during growth on glucose (Fig. 5). Thus, DMSP clearly functions as an osmoprotectant in various bacteria including marine bacteria. DMSP also plays a role as a carbon and energy source in bacteria. Strain MD 14-50, for example, grows well on DMSP alone (10), and rapidly oxidizes the substrate at low NaCl concentration (0.2 M), producing DMS and acrylate through the activity of a DMSP lyase. However, when strain MD 14-50 is grown in the presence of 2 M NaCl, only about 30% of the exogenous DMSP (0.1 mM) is catabolized during the lag phase. Interestingly, DMSP accumulation occurs when the cells enter the stationary phase. Catabolism of DMSP also occurs in other marine bacteria such as *Pseudomonas doudorofii* and strain LFR, a Gram-negative bacterium isolated from the Sargasso Sea (24).

Very little is know about the possible accumulation of other sulfonium compounds in bacteria. However, dimethylthetin which stimulates growth of *E. coli* under osmotic stress is removed from the medium and essentially recovered from the bacterial cell pellets (4).

Table 2. Effect of DMSP on growth rate of *E. Coli* (Doubling times, h)

	Minimal medium + 0.7 M NaCl		
Strains (genotype)	Control	+ GB 1 mM	+ DMSP 1 mM
MC 4100 (WT)	12.5	2.5	3.6
GM 50 (*proU*)	12.5	2.7	3.8
BK 32 (D*proP*)	16.5	2.5	10
MKH 13 (*proU*, D*proP*)	n.g.	n.g	n.g

n.g. = no growth. The data shown are from Gouesbet G. *et al.*, 1994.

Transport of DMSP. From the previous experiments it is clear that DMSP is taken up from the medium. Hence, questions concerning the specificity of transport system(s) in bacteria deserve investigation. During the last five years, several authors have suggested that ProP and ProU are the only uptake systems for osmoprotective molecules in *E. coli.* The question of whether ProP and/or ProU are involved in DMSP uptake was recently addressed by Gouesbet et al. (15). In a comparative study, the authors investigated the uptake of DMSP and the uptake of various betaines, such as proline betaine, β-alanine betaine, γ-butyrobetaine, and trigonelline by different *E. coli* mutants. In wild type cells (strain MC4100), DMSP (1 mM), is a powerful osmoprotectant, although its stimulatory effect on growth at 0.7 M NaCl is less than that observed in the presence of glycine betaine (Table 2). In a mutant strain lacking the ProU system (strain GM50) the beneficial effect of DMSP is approximately the same as in the wild type strain: doubling times of 3.8 and 3.6 h, respectively, compared to 12.5 h in the absence of DMSP. However, in strain BK32 which lacks ProP, DMSP is far less efficient at improving growth: doubling time of 10 h. In a double mutant strain MKH13 (*proU, ΔproP*) neither DMSP nor glycine betaine is able to promote growth in the presence of 0.7 M NaCl. Thus, it is clear that the ability of DMSP to restore growth of *E. coli* in media of high osmolarity depends essentially on the ProP system, but the ProU system is also involved. The authors determined the relative affinity of both systems for DMSP by measuring ^{14}C-glycine betaine and ^{14}C-proline uptake in the presence of cold DMSP. Their conclusion is that DMSP shows a higher affinity for ProP than for ProU. In addition, the binding protein encoded by *proX* is essential for osmoprotection by DMSP through ProU, although DMSP does not seem to have any detectable affinity for ProX. In *R. meliloti* we have also observed a strong ^{14}C-glycine betaine uptake inhibition by dimethylthetin without measuring any binding between the glycine betaine binding protein and this substrate. In order to clearly understand the role of this protein, further investigations using radioactive DMSP and dimethylthetin are needed.

CONCLUSIONS

During the last decades, studies of dimethylsulfide (DMS) from marine organisms (bacteria, phytoplankton, macroalgae, higher plants) and their environments (salt marsh sediments, microbial and algal mats) led to the demonstration that DMSP is the precursor of DMS. Despite the fact that research focusing on DMSP represent a very active area, we still know much less about the physiological function(s) of DMSP and the osmotic control of its accumulation than what we know about glycine betaine. In order to assess the importance of DMSP in osmoregulation, it will be important to understand: (1) the intimate mechanisms of uptake and the genetic regulation of the transport system(s) in model organisms such as *E. coli* as a model and also in marine bacteria, (2) the different pathways for degradation including enzymatic studies and gene characterization, (3) the intracellular localization and pathways of synthesis in higher plants. In the longer term, genes encoding these pathways could be cloned and transferred into plant species which are naturally unable to accumulate this compound. Further analyses of such transgenic plants will allow testing of whether DMSP accumulation can play a role in osmoadaptation. Because of the linkages between DMSP and the better known betaine system such experimental approaches should be productive.

ACKNOWLEDGMENTS

This work was funded by the Centre National de la Recherche Scientifique(URA 1114) and, in part, by the European Communities' BIOTECH Programme, as part of the Project of Technological Priority 1993-1996.

REFERENCES

1. Barron, A, J.U. Jung and M. Villarejo. 1987. Purification and characterization of a glycine betaine binding protein from *Escherichia coli*. J. Biol. Chem. *262*: 11841-11846.
2. Cairney, J., I.R. Booth and C.F. Higgins. 1985. Osmoregulation of gene expression in *Salmonella typhimurium*: *proU* encodes an osmotically induced betaine transport system. J.Bacteriol. *164*: 1224-1232.
3. Cairney, J., I.R. Booth and C.F. Higgins. 1985. *Salmonella typhimurium proP* gene encodes a transport system for the osmoprotectant betaine. J. Bacteriol. *164*: 1218-1223.
4. Chambers, S.T. C. M. Kunin, D. Miller and A. Hamada. 1987. Dimethylthetin can substitute for glycine betaine as an osmoprotectant molecule for *Escherichia coli*. J. Bacteriol. *169* : 4845-4847.
5. Csonka, L.N. 1983. A third L-proline permease in *Salmonella typhimurium* which functions in media of elevated osmotic strength. J. Bacteriol. *151* : 1433-1443.
6. Csonka, L.*N*. 1989. Physiological and genetic responses of bacteria to osmotic stress. Microbiol. Rev. *53*: 121-147.
7. Csonka, L.N. and A.D. Hanson. 1991. Prokaryotic osmoregulation: genetics and physiology. Annu. Rev. Microbiol. *45*: 569-606.
8. Culham, D.E., B. Lasby, A.G. Marangoni, J.L. Milner, B.A. Steer, R.W. van Nues and J.M. Wood. 1993. Isolation and sequencing of *Escherichia coli* gene *proP* reveals unusual structural features of the osmoregulatory proline / betaine transporter, ProP. J. Mol. Biol. *229*: 268-276.
9. Dattananda,C.S. and J. Gowrishankar. 1989. Osmoregulation in *Escherichia coli*: complementation analysis and gene-protein relationships in the *proU* locus. J. Bacteriol. *171*: 1915-1922.
10. Diaz M.R., P.T. Visscher and B.F. Taylor. 1992. Metabolism of dimethylsulfoniopropionate and glycine betaine by a marine bacterium. FEMS Microbiol. Lett. *96* : 61-66.
11. Dosch, D., G.L. Helmer, S.H. Sutton, F.F. Salvacion and W. Epstein. 1991. Genetic analysis of potassium transport loci in *Escherichia coli* : evidence for three constitutive systems mediating uptake of potassium. J. Bacteriol. *173*:687-696.
12. Douglas, R.M., J.A. Roberts, A.W. Munro, G.Y. Ritchie, A.J. Lamb and I.R. Booth. 1991. The distribution of homologues of the *Escherichia coli* KefC K^+-efflux system in other bacterial species. J. Gen. Microbiol. *137*: 1999-2005.
13. Eshoo, M. 1988. *lac* fusion analysis of the *bet* genes of *Escherichia coli*: regulation by osmolarity, temperature, oxygen, choline and glycine betaine. J. Bacteriol. *170*: 5208-5215.
14. Esptein, W. 1986. Osmoregulation by potassium transport in *Escherichia coli*. FEMS Microbiol. Rev. *39*: 73-78.
15. Gouesbet, G., M. Jebbar, R. Talibart, T. Bernard and C. Blanco. 1994. Pipecolic acid is an osmoprotectant for *Escherichia coli* taken up by the general osmoporters ProU and ProP. Microbiology, *140*: 2415-2422.
16. Gowrishankar, J. 1986. *proP*-mediated proline transport also plays a role in *Escherichia coli* osmoregulation. J. Bacteriol. *166*: 331-333.
17. Gowrishankar, J. 1989. Nucleotide sequence of the osmoregulatory *proU* operon of *Escherichia coli*. J. Bacteriol. *171*: 1923-1931.

18. Haardt, M., B. Kempf, E. Faatz and E. Bremer. 1995. The osmoprotectant proline betaine is a major substrate for the binding-protein-dependent transport system ProU of *Escherichia coli* K-12. Mol. Gen. Genet. *246*: 783-786.
19. Higgins, C.F., C.J. Dorman, D.A. Stirling, L. Waddell, I.R. Booth, G. May and E. Bremer. 1988. A physiological role for DNA supercoiling in the osmotic regulation of gene expression in *Salmonella typhimurium* and *Escherichia coli*. Cell. *52*: 569-584.
20. Jebbar, M., R. Talibart, K. Gloux, T. Bernard and C. Blanco. 1992. Osmoprotection of *Escherichia coli* by ectoine : uptake and accumulation characteristics. J. Bacteriol. *174*: 5027-5035.
21. Koo, S.P., C.F. Higgins and I.R. Booth. 1991. Regulation of compatible solute accumulation in *Salmonella typhimurium*. Evidence for a glycine betaine efflux system. J. Gen. Microbiol. *137*: 2617-2625.
22. Lamark T., I. Kaasen, M.W. Eshoo, P. Falkenberg, J. Mc Dougall and A.R. Strøm. 1991. DNA sequence and analysis of the *bet* genes encoding the osmoregulatory choline-glycine betaine pathway of *Escherichia coli*. Mol. Microbiol. *5* : 1049-1064.
23. Landfald, B. and A.R. Strøm. 1986. Choline-glycine betaine pathway confers a high level of osmotic tolerance in *Escherichia coli*. J. Bacteriol. *165* : 849-855.
24. Ledyard, K.M., E.F.De Long and J.W.H. Dacey. 1993. Characterization of a DMSP-degrading bacterial isolate from the Sargasso sea. Arch. Microbiol. 160 : 312-318.
25. Le Rudulier, D. and T. Bernard. 1986. Salt tolerance in *Rhizobium*: a possible role for betaines. FEMS Microbiol. Rev. *39*: 67-72.
26. Le Rudulier, D., T.Bernard, G. Goas and J. Hamelin. 1984. Osmoregulation in *Klebsiella pneumoniae* : enhancement of anaerobic growth and nitrogen fixation under stress by proline betaine, γ-bytyrobetaine and other related compounds. Can. J. Microbiol. *30*: 299-305.
27. Le Rudulier, D., K. Gloux and N. Riou. 1991. Identification of an osmotically induced periplasmic glycine betaine-binding protein from *Rhizobium meliloti*. Biochim. Biophys. Acta *1061*: 197-205.
28. Le Rudulier, D., A.R. Strøm, A.M. Dandekar, L.T. Smith and R.C. Valentine. 1984. Molecular biology of osmoregulation. Science, *224*: 1064-1068.
29. Lucht, J. M. and E. Bremer. 1991. Characterization of mutations affecting the osmoregulated *proU* promoter of *Escherichia coli* and identification of 5′ sequences required for high-level expression. J. Bacteriol. *173* : 801-809.
30. Lucht, J.M. and E. Bremer. 1994. Adaptation of *Escherichia coli* to high osmolarity environments : osmoregulation of the high-affinity glycine betaine transport system ProU. FEMS Microbiol. Rev. *14* : 3-20.
31. Mason, T.G. and G. Blunden. 1989. Quaternary ammonium and tertiary sulphonium compounds of algal origin as alleviators of osmotic stress. Bot. Mar. *32* : 313-316.
32. McLaggan, D. and W. Epstein. 1991. *Escherichia coli* accumulates the eukaryotic osmolyte taurine at high osmolarity. FEMS Microbiol. Lett. *81*: 209-214.
33. May, G., E. Faatz, M. Villarejo and E. Bremer. 1986. Binding protein-dependent transport of glycine betaine and its osmotic regulation in *Escherichia coli* K-12. Mol. Gen. Genet. *205*: 225-233.
34. Mellies, J., A. Wise and M. Villarejo. 1995. Two different *Escherichia coli proP* promoters respond to osmotic and growth phase signals. J. Bacteriol. *177*: 144-151.
35. Overdier, D.G. and L.N. Csonka. 1992. A transcriptional silencer downstream of the promoter in the osmotically controlled proU operon of *Salmonella typhimurium*. Proc. Natl. Acad-Sci. USA *89* : 3140-3144.
36. Perroud, B. and D. Le Rudulier. 1985. Glycine betaine transport in *Escherichia coli*: osmotic modulation. J. Bacteriol. *161*: 393-401.
37. Prince, W.S. and M.R. Villarejo. 1990.Osmotic control of *proU* transcription is mediated through direct action of potassium glutamate on the transcription complex. J. Biol. Chem. *265*: 17673 -17679.
38. Riou, N., M.C. Poggi and D. Le Rudulier. 1991. Characterization of an osmoregulated periplasmic glycine betaine-binding protein in *Azospirillum brasilense* sp. 7. Biochimie, *73*: 1187-1193.
39. Smith, L.T., J.A. Pocard, T. Bernard and D. Le Rudulier. 1988. Osmotic control of glycine betaine biosynthesis and degradation in *Rhizobium meliloti*. J. Bacteriol. *170*: 3142-3149.
40. Strøm, A.R., D. Le Rudulier, M.W. Jakowec, R.C. Bunnell and R.C. Valentine. 1983. Osmoregulatory (Osm) genes and osmoprotective compounds, p. 39-59. *In* T. Kosuge, C.P. Meredith and A. Hollaender (eds), Genetic Engineering of Plants, an Agricultural Perspective. Plenum Press, New York.
41. Styrvold, O.B., P. Falkenberg, B. Landfald, M.W. Eshoo, T. Bjørnsen and A.R. Strøm. 1986. Selection, mapping, and characterization of osmoregulatory mutants of *Escherichia coli* blocked in the choline-glycine betaine pathway. J. Bacteriol. *165*: 856-863.
42. Ueguchi, C. and T. Mizuno. 1993. The *Escherichia coli* nucleoid protein H-NS functions directly as a transcription repressor. EMBO J. *12*: 1039-1046.
43. Wood, J.M. 1988. Proline porters effect the utilization of proline as nutrient or osmoprotectant for bacteria. J. Membrane Biol. *106*: 183-202.

23

METABOLIC PATHWAYS INVOLVED IN DMSP DEGRADATION

Barrie F. Taylor[1] and Pieter T. Visscher[2]

[1] Division of Marine and Atmospheric Chemistry
Rosenstiel School of Marine and Atmospheric Science
University of Miami, Miami, Florida 33149
[2] Department of Marine Science
University of Connecticut
Groton, Connecticut 06340

SUMMARY

Dimethylsulfoniopropionate (DMSP) is initially biodegraded by cleavage into dimethyl sulfide (DMS) and acrylate or by demethylation to 3-methylmercaptopropionate (MMPA). Demethylation of MMPA produces 3-mercaptopropionate (MPA) which is catabolized with the elimination of H_2S to leave acrylate. MMPA is also metabolized with the formation of methanethiol by unknown mechanisms. DMSP lyases which catalyze the cleavage of DMSP into DMS and acrylate, occur in a variety of organisms; aerobic and anaerobic bacteria, phytoplankton, macroalgae and possibly higher plants. Biochemical properties reveal the occurrence of more than one DMSP lyase probably because of the different physiological functions of the enzyme. Demethylations of DMSP to MMPA and thence to MPA are performed by aerobic and anaerobic bacteria. In anaerobes the first demethylation step was documented for a species of *Desulfobacterium* and the second step from MMPA to MPA was established with species of *Methanosarcina*. MPA degradation has been observed only with anoxygenic phototrophic bacteria (*Rhodopseudomonas* sp. strain BB1, *Thiocapsa roseopersicina*) and occurs with H_2S elimination to leave acrylate. DMS and methanethiol are degraded by a variety of aerobes and anaerobes. Strict aerobes may use monooxygenases to oxidize the methyl groups whereas facultative and strict anaerobes probably employ transmethylases and a C_1-folate system of oxidation. Methanogens probably funnel methyl groups from DMS, methanethiol and MMPA via specific methyl transferases to methyl coenzyme M reductase.

INTRODUCTION

The function of DMSP as an osmolyte in marine plants ensures its importance as a major carbon source for the growth of a variety of marine bacteria. The initial events involved in the metabolic pathways for DMSP degradation are summarised in Fig. 1. Cleavage of DMSP gives dimethylsulfide (DMS) and acrylate whereas demethylation produces 3-methylmercaptopropionate (MMPA). MMPA is metabolized with the formation of methanethiol or by demethylation to 3-mercaptopropionate (MPA), which is degraded with the elimination H_2S to produce acrylate. The metabolic pathways for DMSP degradation therefore encompass those for DMS, methanethiol and acrylate.

INITIAL TRANSFORMATIONS OF DMSP

The cleavage of DMSP:

$$(CH_3)_2S^+CH_2CH_2COOH \rightarrow CH_3SCH_3 + CH_2{=}CHCOOH + H^+$$

is catalyzed by DMSP lyases that occur in bacteria, eukaryotic algae and perhaps higher plants. The enzymatic activity is present in aerobic bacteria (2,5,6,8,9,24,29,30), anaerobic bacteria (48,58), phytoplankton (17,18,39,47) and possibly flowering plants (3,13). Several strains of marine bacteria, including *Pseudomonas doudoroffii* and *Shewanella* spp. produced DMS and/or methanethiol from DMSP (31).

There are indications of lyases with different biochemical properties, probably reflecting their varying physiological roles (5,7). For example, biochemical differences were observed for lyase activity in cell-free extracts of an aerobic marine bacterium (strain MD 14-50) that grows on DMSP, as compared to that in extracts of a green macroalga (*Ulva lactuca*) in which DMSP is an osmolyte (7) (Table 1). The bacterial but not the algal enzyme cleaved diethyl sulfide (DES) from diethylsulfoniopropionate (DESP). The bacterial enzyme was more sensitive to inhibition by 2-methyl DMSP

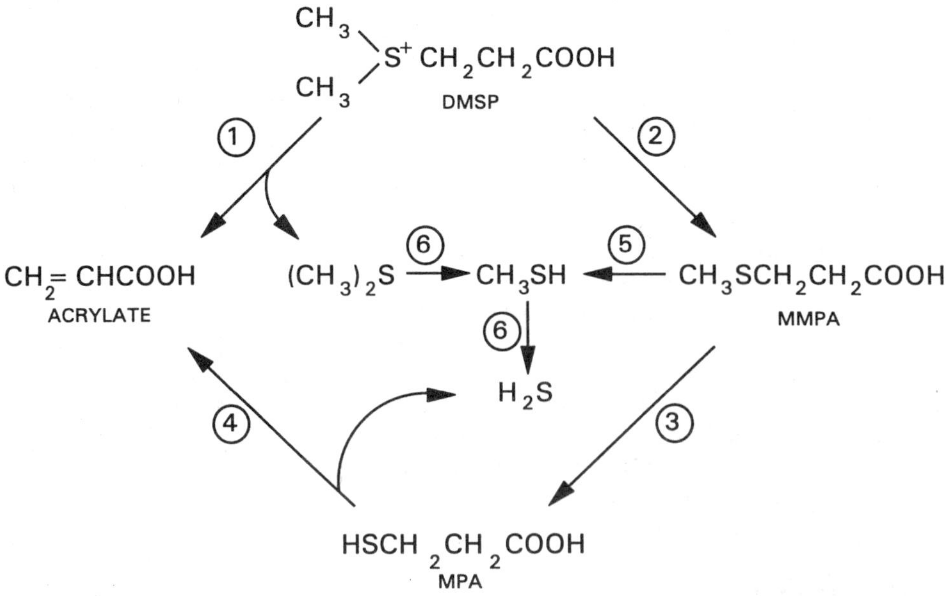

Figure 1. Initial transformations of DMSP. (1) DMSP lyase; (2) Demethylation of DMSP; (3) Demethylation of MMPA ; (4) Elimination of H_2S from MPA; (5) "Demethiolation" of MMPA; (6) Demethylations of methylated sulfides.

or DESP than the algal enzyme. An intracellular location for the bacterial enzyme appears reasonable in light of its sensitivity to 0.5M NaCl and neutral pH optimum. The algal enzyme was not inhibited by 0.4M NaCl and showed an alkaline pH optimum thereby being suited to function in seawater. The lyase activity in extracts of *Gyrodinium cohnii* required 0.4M NaCl for maximal activity and was still fully active at 3M NaCl (18). Unlike the enzyme of *U. lactuca*, the lyase of *G. cohnii* was optimally active at pH 6.0-6.5 but displayed very little activity at pH 8.

Some strict anaerobes grow on glycine betaine, an important osmolyte for a wide range of organisms (36,60), with a reductive cleavage to yield trimethylamine and acetate (14,32,34):

$$(CH_3)_3N^+CH_2COOH + 2H \rightarrow (CH_3)_3N + CH_3COOH + H^+$$

The presence of DMSP lyase in some anaerobes allows their exploitation of the carbon chain of DMSP without an expenditure of reducing power on an initial cleavage reaction. DMSP lyase activity in a bacterium was originally detected in a *Clostridium* sp. isolated from river mud (58) and more recently in a *Desulfovibrio* sp. (48). The *Clostridium* sp. grew on DMSP with the production of DMS and fermented acrylate to a mixture of propionate and acetate whereas the sulfate reducer used acrylate as an additional electron acceptor to produce only propionate.

The demethylation of DMSP to MMPA has been established for aerobic and anaerobic bacteria (41,50,53). The demethylation of DMSP to MMPA was first shown with an aerobe, strain DG-C1 (41) which was isolated from a culture of the marine coccolithophore *Emiliania huxleyi*, a DMSP producer (19). Metabolism of MMPA by strain DG-C1 was predominantly with the formation of methanethiol although traces of MPA were detected in cultures growing on MMPA. Strain BIS-6, isolated from a coastal sediment, grew aerobically on a range of methylated compounds that included DMSP, DMS, glycine betaine and methylamines (53). Strain BIS-6 quantitatively demethylated both DMSP and MMPA to MPA. A sulfate-reducing bacterium, *Desulfobacterium* strain PM4, that grew on glycine betaine with demethylation to yield dimethylglycine (14) was recently shown to grow on DMSP with the accumulation of MMPA (50). Even though acetogens such as *Eubacterium limosum* can demethylate glycine betaine (15), acetogenic bacteria that demethylate DMSP or MMPA have not been isolated or detected, even though their presence was anticipated (28).

Table 1. Biochemical Properties of DMSP Lyases [a]

	Organism	
Property	Strain MD 14-50	*Ulva lactuca*
pH optimum	7.0	8.5
Inhibition by NaCl (0.5M)	+	-
Inhibition by GB (0.5M)[b]	-	+
DESP as a substrate	+	-
Inhibition by DESP	+++	+
Inhibition by 2-methyl-DMSP	+++	+

[a] Diaz and Taylor (7).
[b] GB = glycine betaine.

DEGRADATION OF 3-METHYLMERCAPTOPROPIONATE (MMPA) AND 3-MERCAPTOPROPIONATE (MPA)

Catabolism of MMPA by aerobes or anaerobes proceeds with either demethylation to MPA or the production of methanethiol (28,41,49,53). Three isolates of *Methanosarcina* species, from marine habitats, were recently shown to use MMPA as a methanogenic substrate with the quantitative accumulation of MPA (49). Possible mechanisms for the anaerobic demethylation of MMPA by methanogens were discussed by Hansen and his colleagues (49). They concluded that transmethylation from MMPA to coenzyme M was more likely than direct use of MMPA as a substrate in the reaction catalyzed by methyl-S-coenzyme M reductase (57). This conclusion was based mainly on the limited ability of methanogens (3 out of 11 strains examined) to use MMPA in spite of the ubiquity of methyl-S-coenzyme M reductase in methanogens. The ability to use or not use MMPA, however, may reside in differences in transport systems. Methane production from MMPA also proceeds after demethiolation of MMPA to generate methanethiol which is consumed by both methanogenic and sulfate reducing bacteria (10,23,27,40). Experiments with slurries of diluted sediment, at 25°C and with 0.5 mM MMPA, indicated that demethiolation of MMPA predominated over its demethylation as the route for methane formation (49). In earlier studies of a coastal environment (28), demethylation of MMPA to MPA was favored over its demethiolation and even though environments may vary, studies are needed at environmental substrate levels to understand the relationships between the two pathways.

Aerobic demethylation of MMPA could involve a monooxygenase:

$$CH_3SCH_2CH_2COOH + O_2 + NAD(P)H + H^+ \rightarrow HCHO + HSCH_2CH_2COOH + H_2O + NAD(P)^+$$

or a corrinoid-based transmethylation with oxidation to HCOOH via the folate coenzyme system, and then oxidation catalyzed by a formate dehydrogenase. Interestingly, it appears

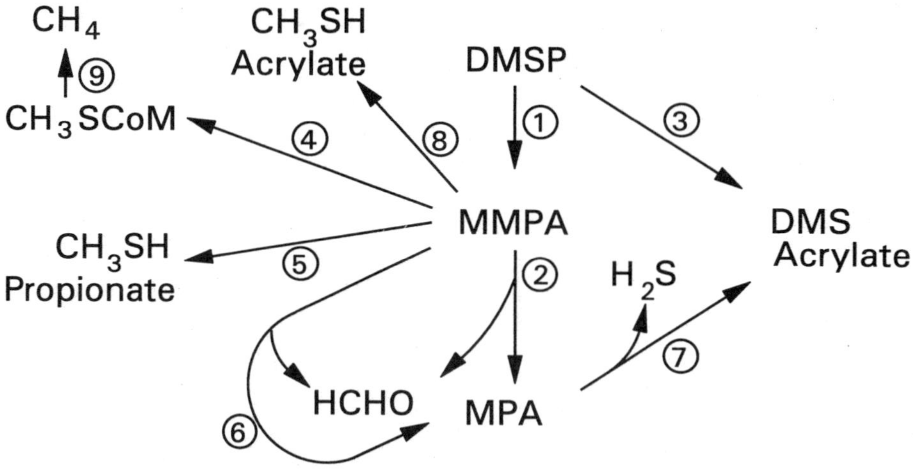

Figure 2. Proposed metabolic routes for MMPA production and degradation. Methyl transferases specific for (1) DMSP and (2) MMPA; (3) DMSP lyase; (4) methyl-S-CoM reductase; (5) MMPA reductase; (6) MMPA monooxygenase; (7) MPA desulfhydrase; (8) MMPA lyase; (9) MMPA: methyl-S-CoM transferase.

more difficult to obtain demethylating bacteria on MMPA than on DMSP (53,55). In enumeration studies of water samples and sediments in which product formation was measured from MMPA, methanethiol production always occurred more rapidly than MPA production. More detailed studies are needed to determine true differences of rates and mechanisms *in situ*.

Mechanisms for the formation of methanethiol from MMPA are unknown but there appear to be several (Fig 2). As noted above, strain DG-C1 which grows aerobically on DMSP or MMPA (41) produced methanethiol from either substrate and also from S-methyl cysteine (44). Kiene and Capone (25) previously observed methanethiol formation from S-methyl cysteine in slurries of coastal sediments and it seemed possible that this activity coincidentally resided in an enzyme of MMPA degradation, since MMPA is probably more abundant than S-methylcysteine in coastal environments. However, we were able to isolate aerobic bacteria from coastal marine habitats that grew on either MMPA or S-methyl cysteine, or both compounds (16,42). Thus MMPA and S-methyl cysteine degradation are not necessarily linked by a common enzyme. We also isolated a nitrate-respirer which produced methanethiol from MMPA either aerobically or anaerobically (16,42). MMPA degradation with methanethiol production may include ß-oxidation in which MMPA is metabolized as a fatty acid (Fig. 3). The final hydrolytic reaction of the sequence in Fig. 3 may be a purely chemical decomposition. Support for the ß-oxidation route comes from the ability of a marine aerobic isolate, that grows on MMPA with methanethiol production and also grows on butyrate, to oxidize MMPA when grown on butyrate and the oxidation of MMPA by butyrate-grown cells (43). Strain BIS-6, however, grows only poorly on butyrate and so mechanisms other than ß-oxidation undoubtedly exist for the dissimilation of MMPA with methanethiol formation (Fig. 2). Possibilities include an elimination reaction:

$$CH_3SCH_2CH_2COOH \rightarrow CH_3SH + CH_2{=}CHCOOH$$

or a reduction:

$$CH_3SCH_2CH_2COOH + 2H \rightarrow CH_3SH + CH_3CH_2COOH$$

The microbial degradation of MPA has received little attention but it was metabolized by two species of anoxygenic phototrophic bacteria that were isolated from marine habitats (54). *Rhodopseudomonas* sp. strain BB1, derived from a coastal marine sediment, and *Thiocapsa roseopersicina*, from a marine microbial mat, metabolized MPA or mercaptomalate with the elimination of H_2S and the formation of acrylate or fumarate:

$$HSCH_2CH_2COOH \rightarrow H_2S + CH_2{=}CHCOOH$$

$$HOOCCHSHCH_2COOH \rightarrow H_2S + HOOCCH{=}CHCOOH$$

Rhodopseudomonas sp. strain BB1 grew with subsequent consumption of both the sulfide and the carbon moiety whereas *T. roseopersicina* grew only photolithoautotrophically on the liberated sulfide, leaving behind the acrylate or fumarate. The desulfurylation activity was demonstrated in cell-free extracts of *Rhodospeudomonas* sp. strain BB1 and an identical or similar activity would allow the growth of nonphototrophic bacteria, either aerobes or anaerobes, on MPA and other low molecular weight thiols that occur in marine sediments (26, 33).

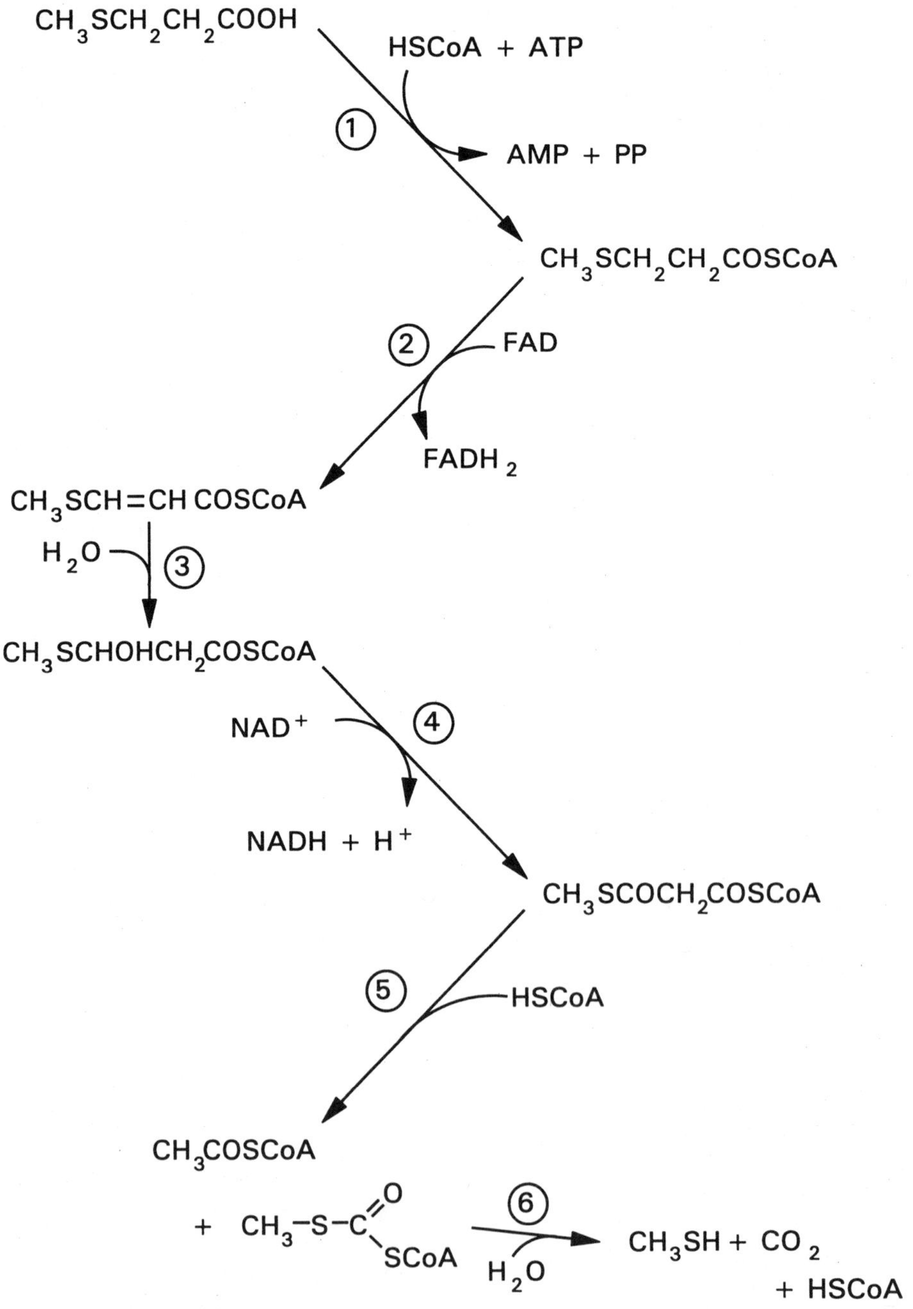

Figure 3. β-Oxidation of 3-methylmercaptopropionate. (1) Acyl-CoA synthetase; (2) acyl-CoA dehydrogenase; (3) 3-hydroxyacyl-CoA hydratase; (4) 3-hydroxyacyl-CoA dehydrogenase; (5) ß-ketothiolase; (6) Chemical/biochemical hydrolysis (?).

DEGRADATION OF DMS AND METHANETHIOL

DMS is metabolized by pathways that involve demethylation to methanethiol and then sulfide (Fig. 4), and it is also oxidized to dimethylsulfoxide (DMSO). Aerobic bacteria that metabolize methylated sulfides include thiobacilli and hyphomicrobia (Fig. 4) (22,63). With some aerobic bacteria, O_2 directly participates in the oxidation of the methyl groups and degradation is by the action of DMS monooxygenase and methanethiol oxidase:

$$CH_3SCH_3 + O_2 + NADH + H^+ \rightarrow HCHO + CH_3SH + NAD^+ + H_2O$$

$$CH_3SH + O_2 + H_2O \rightarrow HCHO + H_2S + H_2O_2$$

with further oxidation of the HCHO to HCOOH and then CO_2 and H_2O in reactions catalyzed by formaldehyde and formate dehydrogenases. H_2S is oxidized to sulfate in both thiobacilli and hyphomicrobia. A facultatively aerobic thiobacillus (*Thiobacillus* sp. strain ANS-1) grew anaerobically with denitrification (51) and in this, and other anaerobes, another mechanism for methyl group catabolism operates (Fig. 5). Presumably corrinoid-dependent methyltransferases catalyze the transfer of methyl groups from the methylated sulfides to C_1 carriers, such as folates, which mediate a stepwise oxidation to HCOOH which is subsequently oxidized to CO_2 and H_2O in a reaction catalyzed by a formate dehydrogenase. Nitrate-respirers, sulfate reducers and methanogens presumably also utilize methyl groups via methyltransferases. DMS metabolism by *Thiobacillus* sp. strain ASN-1 was inhibited by $CHCl_3$, an established inhibitor of corrinoid-dependent methyltransferases (59), but not by methyl butyl ether which blocked DMS oxidation in the strict aerobe *Thiobacillus thioparus* strain T5 (52). DMS oxidation, presumably by an oxygenase, was not inhibited by $CHCl_3$ in *T. thioparus* strain T5.

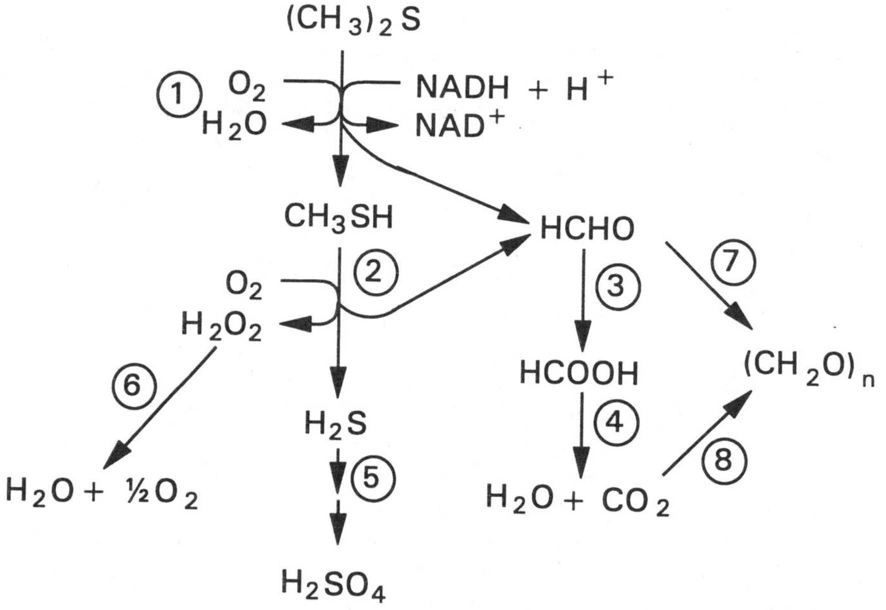

Figure 4. Catabolism of methylated sulfides by thiobacilli and hyphomicrobia. (1) Dimethyl sulfide monooxygenase; (2) methanethiol oxidase; (3) formaldehyde dehydrogenase; (4) formate dehydrogenase; (5) sulfide oxidizing enzymes; (6) catalase; (7) serine pathway for carbon assimilation; (8) Calvin cycle for CO_2 fixation.

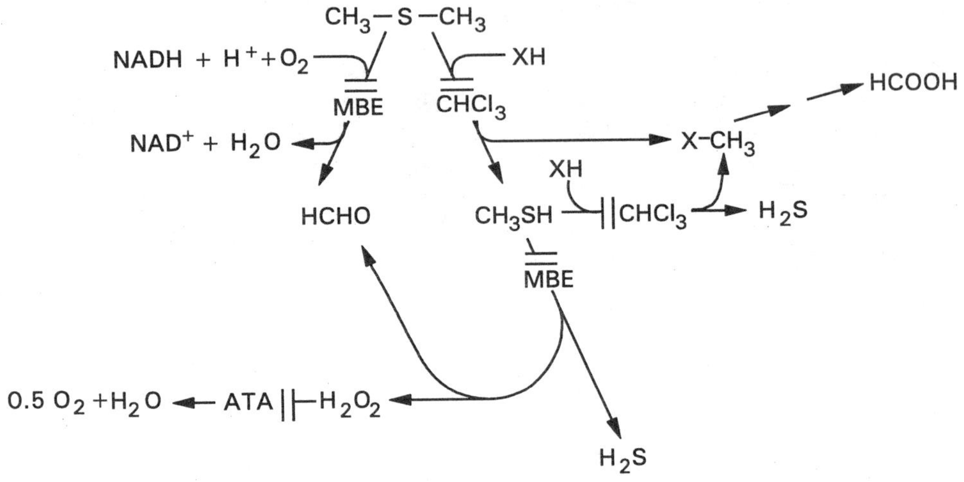

Figure 5. Oxygenase-oxidase and methyl-transferase pathways for catabolism of methylated sulfides. In methanogens the methylcobalamin transfers the methyl group to coenzyme M and the methyl-S-coenzyme is reduced to yield methane.

The photochemical oxidation of DMS in the marine atmosphere generates methane sulfonic acid, which returns to the Earth's surface in rain and snow, and is degraded by aerobic bacteria from marine and other environments (1,21,46). However, the catabolism of methane sulfonate involves a monooxygenase (21) and the anaerobic degradation of methane sulfonate has not been reported. If the hydrolysis of MSA occurs, then, as originally considered by Kelly and Baker (20), its anaerobic breakdown is possible with the production of methane or methanol:

$$CH_3SO_3H + H_2O \rightarrow CH_4 + H_2SO_4$$

$$CH_3SO_3H + H_2O \rightarrow CH_3OH + H_2SO_3$$

The hydrolytic cleavage of the C-P bond in methane phosphonate by *Pseudomonas testosteroni* produces methane (4). An analogous cleavage of MSA would require the presence of an additional carbon and energy source for the growth of an anaerobic microorganism. If biochemically possible, the hydrolysis of MSA to methanol and sulfite would benefit sulfate-reducers by providing both an oxizable substrate and electron acceptor; methanol would also be available to other anaerobes such as nitrate-reducers and methanogens (20).

DMS is biochemically oxidized to DMSO by anoxygenic phototrophs (56,61) and aerobically by a pseudomonad (62). There is as yet no evidence for its further oxidation to dimethyl sulfone $(CH_3)_2SO_2$ even though haloperoxidases, which are common in marine algae, catalyze this conversion (11,35). It appears that the link between DMS and MSA is chemical rather than biochemical although oxygenases may exist that catalyze the transformation.

CATABOLISM OF ACRYLATE

Interest was aroused in the fate of acrylate because of its observed antibacterial properties (37,38). Acrylate is, however, a good carbon source for the growth of both aerobic and anaerobic bacteria.

Acrylate is metabolized with an initial activition to acrylyl coenzyme A. Further metabolism to lactyl CoA, catalyzed for example by lactyl dehydratase, provides entry into aerobic or anaerobic metabolic pathways and reduction to propionyl CoA, catalyzed for example by propionyl CoA dehydrogenase, provides entry into a variety of fermentative routes (12,45). An interesting and novel physiological function of acrylate is its recent discovery as a respiratory electron acceptor generated from DMSP by a *Desulfovibrio* sp. that possesses DMSP lyase activity; the higher free energy yield on acrylate as opposed to sulfate might provide the organism with a competitive advantage over other sulfate reducers (48).

DISCUSSION

With respect to the varied pathways available for DMSP degradation in the marine environment it is worth noting that we have evidence for multiple pathways operating in a nonmarine environment, namely the alkaline, hypersaline Mono Lake of California (8). Cultures that cleave DMSP and use DMSP or MMPA and generate methanethiol were isolated and they functioned best at an alkaline pH (9.5).

Strain BIS-6, *T. roseopersicina* and some methanogenic Archae individually participate in multiple steps of the routes for DMSP degradation. Strain BIS-6 demethylates DMSP and MMPA and grows on DMS, *T. roseopersicina* grows phototrophically with either DMS or H_2S (liberated from MPA), and *Methanosarcina* strain MTP4 demethylates MMPA, DMS and methanethiol. Whether or not environmental conditions exist that permit the simultaneous use of DMSP and its immediate degradation products by the same organism requires investigation.

Metabolism of DMSP and glycine betaine show both similarities and differences; some organisms engage in demethylation of both compounds whilst others are more restricted. Cleavage of DMSP via a lyase enzyme may be more common in bacteria than the reductive cleavage of glycine betaine. In environments where concentrations of substrates are typically low, more versatile generalists, like strain BIS-6 (53), may have advantages over specialists. However, other factors, such as rapid adaptation to changing environmental conditions are possibly more important in determining population composition.

DMSP and its degradation products, even acrylate which has been identified as an "antibiotic", that are released by phytoplankton and macroalgae, may provide a carbon substrate that promotes the growth of epiphytic bacteria, which in turn might fix nitrogen or provide growth factors for the plant. The association of bacteria that degrade DMSP with marine phytoplankton (9,41) suggests possible interactions between DMSP producers and their associated bacterial flora. The bacteria could provide the enzymes needed for DMSP removal, rendering superfluous a DMSP lyase in the phototroph. Finally, if a ß-oxidation pathway functions for MMPA degradation in some bacteria then the possibility of the formation of poly-ß-hydroxybutyrate with methylthio-groups exists via an intermediate ($CH_3SCHOHCH_2COSCoA$) in the ß-oxidation route for MMPA catabolism. Such a polymer would be a store of carbon (especially methyl groups), energy and sulfur, and might have novel plasticity characteristics.

ACKNOWLEDGMENTS

Financial support by the National Science Foundation that allowed the preparation of this paper is gratefully acknowledged (grant OCE 9012157).

REFERENCES

1. Baker, S. C., D. P. Kelly and J. C. Murrell. 1991. Microbial degradation of methane sulfonic acid: a missing link in the biogeochemical sulfur cycle. Nature *350:* 627-628.
2. Dacey, J. W. H. and N. V. Blough. 1987. Hydroxide decomposition of dimethylsulfoniopropionate to form dimethyl sulfide. Geophys. Res. Lett. *14:* 1246-1249.
3. Dacey, J. W. H., G. M. King and S. G. Wakeham. 1987. Factors controlling emission of dimethyl sulfide from salt marshes. Nature *330:* 643-645.
4. Daughton, C. G., A. M. Cook and M. Alexander. 1979. Biodegradation of phosphate toxicants yields methane or ethane on cleavage of the C-P bond. FEMS Microbiol. Lett. *5:* 91-93.
5. de Souza, M. P., and D. C. Yoch. 1996. N-terminal amino acid sequences and comparison of DMSP lyases from *Pseudomonas doudoroffii* and *Alcaligenes* strain M3A. *In Biological and environmental chemistry of DMSP and related sulfonium compounds.* R. P. Kiene, P. T. Visscher, M. D. Keller, and G. O. Kirst (eds.), Plenum Press, New York.
6. De Souza, M. P. and D. C. Yoch. 1995. Purification and characterization of dimethylsulfoniopropionate lyase from an *Alcaligenes*-like dimethyl sulfide-producing marine isolate. Appl. Environ. Microbiol. *61:* 21-26.
7. Diaz, M. R. and B. F. Taylor. 1994. Comparison of dimethylsulfoniopropionate lyase activity in a prokaryote and a eukaryote. Annu. Gen. Meet. Amer. Soc. Microbiol. Abstract N18, p. 319.
8. Diaz, M. R. and B. F. Taylor. 1996. Metabolism of methylated osmolytes by aerobic bacteria from Mono Lake, a moderately hypersaline, alkaline environment. FEMS Microbiol. Ecol. In press.
9. Diaz, M. R., P. T. Visscher and B. F. Taylor. 1992. Metabolism of dimethylsulfoniopropionate and glycine betaine by a marine bacterium. FEMS Microbiol. Lett. *96:* 61-66.
10. Finster, K., Y. Tanimoto and F. Bak. 1992. Fermentation of methanethiol and dimethyl sulfide by a newly isolated methanogenic bacterium. Arch. Microbiol. *157:* 425-430.
11. Geigert, J., S. K. De Witt, S. L. Neidleman, G. Lee, D. J. Dalietos and M. Moreland. 1983. DMSO is a substrate for chloroperoxidase. Biochem. Biophys. Res. Commun. *116:* 82-85.
12. Gottschalk, G. 1986. Bacterial metabolism. Springer-Verlag, New York.
13. Hanson, A. D. and D. A. Gage. 1996. 3-Dimethylsulfoniopropionate biosynthesis and use by flowering plants. *In Biological and environmental chemistry of DMSP and related sulfonium compounds* R. P. Kiene, P. T. Vischer, M. D. Keller, and E. O. Kirst (eds.). Plenum Press, New York.
14. Heijthuijsen, J. H. F. G., and T. A. Hansen. 1989. Anaerobic degradation of betaine by marine *Desulfobacterium* strains. Arch. Microbiol. *152:* 393-396.
15. Heijthuijsen, J. H. F. G., and T. A. Hansen. 1990. C_1 metabolism in anaerobic non-methanogenic bacteria,pp 163-193. In: G. A. Codd, L. Dijkhuizen and F. R. Tabita (ed.), Autotrophic Microbiology and One-Carbon Metabolism. Kluwer, Boston.
16. Hunkele, G. E., P. T. Visscher and B. F. Taylor. 1996. Aerobic formation of methanethiol production from organosulfur precursors by bacteria isolated from marine environments. In preparation.
17. Ishida, Y. 1968. Physiological studies on the evolution of dimethyl sulfide from unicellular marine algae. Mem. Coll. Agric. Kyoto *94:* 47-82.
18. Kadota, H. and Y. Ishida. 1968. Effects of salt on the enzymatic production of dimethyl sulfide from *Gyrodinium cohnii.* Bull. Jap. Soc. Sci. Fish. *34:* 512-518.
19. Keller, M. D., W. K. Bellows and R. R. L. Guillard. 1989. Dimethyl sulfide production in marine phytoplankton, pp. 167-200. In E. S. Saltzman and W. C. Cooper (ed.), Biogenic Sulfur in the Environment. American Chemical Society, Washington D.C.
20. Kelly, D. P. and S. C. Baker. 1990. The organosulfur cycle: aerobic and anaerobic processes leading to turnover of C_1-sulfur compounds. FEMS Microbiol. Rev. *87:*241-246.
21. Kelly, D. P., S. C. Baker, J. Trickett, M. Davey and J. C. Murrell. 1994. Methane sulfonate utilization by a novel methylotrophic bacterium involves an unusual monooxygenase. Microbiology *140:* 1419-1426.
22. Kelly, D. P. and N. A. Smith. 1990. Organic sulfur compounds in the environment. Biogeochemistry, microbiology and ecological aspects. Adv. Microbial Ecol. *11:* 345-385.

23. Kiene, R. P. 1988. Dimethyl sulfide metabolism in salt marsh sediments. FEMS Microbiol. Ecol. 53: 71-78.
24. Kiene, R. P. 1990. Dimethyl sulfide production from dimethyl sulfoniopropionate in coastal seawater samples and bacterial cultures. Appl. Environ. Microbiol. *56:* 3292-3297.
25. Kiene, R. P. and D. G. Capone. 1988. Microbial transformations of methylated sulfur compounds in anoxic saltmarsh sediments. Microbial Ecol. *15:* 275-291.
26. Kiene, R. P., K. D. Malloy and B. F. Taylor. 1990. Sulfur-containing amino acids as precursors of thiols in anoxic coastal sediments. Appl. Environ. Microbiol. *56:* 156-161.
27. Kiene, R. P., R. S. Oremland, A. Catena, L. G. Miller and D. G. Capone. 1986. Metabolism of reduced methylated sulfur compounds in anaerobic sediments and by a pure culture of an estuarine methanogen. Appl. Environ. Microbiol. *52:* 1037-1045.
28. Kiene, R. P. and B. F. Taylor. 1988. Demethylation of dimethylsulfoniopropionate and production of thiols in anoxic marine sediments. Appl. Environ. Microbiol. *54:* 2208-2212.
29. Ledyard, K. M., and J. W. H. Dacey. 1996. Kinetics of DMSP-lyase activity in coastal seawater. *In* R. P. Kiene, P. T. Visscher, M. D. Keller, and G. O. Kirst (ed.), Biological and environmental chemistry of DMSP and related sulfonium compounds. Plenum, New York.
30. Ledyard, K. M. and J. W. H. Dacey. 1994. Dimethyl sulfide production from dimethylsulfoniopropionate by a marine bacterium. Mar. Ecol. Progr. Ser. *110:* 95-103.
31. Ledyard, K. M., E. F. DeLong and J. W. H. Dacey. 1993. Characterization of a DMSP-degrading bacterial isolate from the Sargasso Sea. Arch. Microbiol. *160:* 312-318.
32. Moller, B., R. Ossmer, B. H. Howard, G. Gottschalk and H. Hippe. 1984. *Sporomusa*, a new genus of gram-negative anaerobic bacteria including *Sporomusa sphaeroides* sp. nov. and *Sporomusa ovata* sp. nov. Arch. Microbiol. *139:* 388-396.
33. Mopper, K. and B. F. Taylor. 1986. Biogeochemical cycling of sulfur: thiols in coastal marine sediments, pp. 324-339. *In* M. Sohn (ed.), Organic Marine Geochemistry. American Chemical Society, Washington, D.C.
34. Naumann, E., H. Hippe and G. Gottschalk. 1983. Betaine: a new oxidant in the Stickland reaction and methanogenesis from betaine and L-alanine by a *Clostridium sporogenes-Methanosarcina barkeri* coculture. Appl. Environ. Microbiol. *45:* 474-483.
35. Neidleman, S. L. and J. Geigert. 1986. Biohalogenation: Principles, basic rules and applications. Wiley, New York.
36. Rhodes, D. and A. D. Hanson. 1993. Quaternary ammonium and tertiary sulfonium compounds in higher plants. Annu. Rev. Plant Physiol. Plant Mol. Biol. *44:* 357-384.
37. Sieburth, J. M. 1960. Acrylic acid, an "antibiotic" principle in *Phaeocystis* blooms in Antarctic waters. Science *132:* 676-677.
38. Sieburth, J. M. 1961. Antibiotic properties of acrylic acid, a factor in the gastrointestinal antibiosis of polar marine animals. J. Bacteriol. *82:* 72-79.
39. Stefels, J. and W. H. M. van Boekel. 1993. Production of DMS from disolved DMSP in axenic cultures of the marine phytoplankton species *Phaeocystis* sp. Mar. Ecol. Progr. Ser. *97:* 11-18.
40. Tanimoto, Y. and F. Bak. 1994. Anaerobic degradation of methylmercaptan and dimethyl sulfide by newly isolated thermophilic sulfate-reducing bacteria. Appl. Environ. Microbiol. *60:* 2450-2455.
41. Taylor, B. F. and D. C. Gilchrist. 1991. New routes for the aerobic biodegradation of dimethylsulfoniopropionate. Appl. Environ. Microbiol. *57:* 3581-3584.
42. Taylor, B. F., G. E. Hunkele and R. Baynard. 1993. Methanethiol formation from from 3-methiolpropionate and S-methylcysteine by marine bacteria. Annu. Gen. Meet. Amer. Soc. Microbiol. Abstract Q40, p. 353.
43. Taylor, B. F. Unpublished data.
44. Taylor, B. F. and D. C. Gilchrist. Unpublished data.
45. Tholozan, J. L., J. P. Touzel, E. Samain, J. P. Grivet, G. Prensier and G. Albagnac. 1992. *Clostridium neopropionicum* sp. nov., a strict anaerobic bacterium fermenting ethanol to propionate through acrylate pathway. Arch. Microbiol. *157:* 249-257.
46. Thompson, A. S., N. J. P. Owens and J. C. Murrell. 1995. Isolation and characterization of methanesulfonic acid-degrading bacteria from the marine environment. Appl. Environ. Microbiol. *61:* 2388-2393.
47. Vairavamurthy, A., M. O. Andreae and R. L. Iverson. 1985. Biosynthesis of dimethyl sulfide and dimethylsulfoniopropionate by *Hymenomonas carterae* in relation to sulfur source and salinity variations. Limnol. Oceanogr. *30:* 59-70.
48. van der Maarel, M. J. E. C., and T. A. Hansen. 1996. Anaerobic microorganisms involved in the degradation of DMS(P). *In Biological and environmental chemistry of DMSP and related sulfonium*

compounds. R. P. Kiene, P. T. Visscher, M. D. Keller, and G. O. Kirst (eds.), Plenum Press, New York. p. 351-360.

49. van der Maarel, M. J. E. C., M. Jansen and T. A. Hansen. 1995. Methanogenic conversion of 3-*S*-methylmercaptopropionate to 3-mercaptopropionate. Appl. Environ. Microbiol. *61:* 48-51.
50. van der Maarel, M. J. E. C., P. Quist, L. Dijkhuizen and T. A. Hansen. 1993. Anaerobic degradation of dimethylsulfoniopropionate to 3-*S*-methylmercaptopropionate by a marine *Desulfobacterium* strain. Arch. Microbiol. *160:* 411-412.
51. Visscher, P. T. and B. F. Taylor. 1993. Aerobic and anaerobic degradation of a range of alkyl sulfides by a denitrifying marine bacterium. Appl. Environ. Microbiol. *59:* 4083-4089.
52. Visscher, P. T. and B. F. Taylor. 1993. A new mechanism for the aerobic catabolism of dimethyl sulfide. Appl. Environ. Microbiol. *59:* 3784-3789.
53. Visscher, P. T. and B. F. Taylor. 1994. Demethylation of dimethylsulfoniopropionate to 3-mercaptopropionate by an aerobic marine bacterium. Appl. Environ. Microbiol. *60:*4617-4619.
54. Visscher, P. T. and B. F. Taylor. 1993. Organic thiols as organolithotrophic substrates for growth of phototrophic bacteria. Appl. Environ. Microbiol. *59:* 93-96.
55. Visscher, P. T. and B. F. Taylor. Unpublished data.
56. Visscher, P. T. and H. van Gemerden. 1991. Photoautotrophic growth of *Thiocapsa roseopersicina* on dimethyl sulfide. FEMS Microbiol. Lett. *81:* 247-250.
57. Wackett, L. P., J. F. Honek, T. P. Begley, V. Wallace, W. H. Orme-Johnson and C. T. Walsh. 1987. Substrate analogues as mechanistic probes of methyl-S-coenzyme M reductase. Biochemistry *26:* 6012-6018.
58. Wagner, C. and E. R. Stadtman. 1962. Bacterial fermentation of dimethyl-ß-propiothetin. Arch. Biochem. Biophys. *98:* 331-336.
59. Wood, J. M., F. S. Kennedy and R. S. Wolfe. 1968. The reaction of multihalogenated hydrocarbons with free and bound reduced vitamin B_{12}. Biochemistry *7:* 1707-1713.
60. Yancey, P. H., M. E. Clark, S. C. Hand, R. D. Bowlus and G. N. Somero. 1982. Living with water stress: evolution of osmolyte systems. Science *217:* 1214-1222.
61. Zeyer, J., P. Eicher, S. G. Wakeham and R. P. Schwarzenbach. 1987. Oxidation of dimethyl sulfide to dimethyl sulfoxide by phototrophic purple bacteria. Appl. Environ. Microbiol. *53:* 2026-2032.
62. Zhang, L., I. Kuniyoshi, M. Hirai and M. Shoda. 1991. Oxidation of dimethyl sulfide by *Pseudomonas acidovorans* DMR-11 isolated from peat biofilter. Biotechnol. Lett. *13:* 223-228.
63. de Zwart, J. M. M. and J. G. Kuenen. 1992. C_1-cycle of sulfur compounds. Biodegradation *3:* 37-59.

24

ACCUMULATION OF DISSOLVED DMSP BY MARINE BACTERIA AND ITS DEGRADATION VIA BACTERIVORY

Gordon V. Wolfe

College of Oceanic and Atmospheric Sciences
Oceanography Administration Bldg. 104
Oregon State University
Corvallis, Oregon 97331-5503

SUMMARY

Several bacterial isolates enriched from seawater using complex media were able to accumulate dimethylsulfoniopropionate (DMSP) from media into cells over several hours without degrading it. Uptake only occurred in metabolically active cells, and was repressed in some strains by the presence of additional carbon sources. Accumulation was also more rapid in osmotically-stressed cells, suggesting DMSP is used as an osmotic solute. Uptake could be blocked by inhibitors of active transport systems (2,4-dinitrophenol, azide, arsenate) and of protein synthesis (chloramphenicol). Some structural analogs such as glycine betaine and S-methyl methionine also blocked DMSP uptake, suggesting that the availability of alternate organic osmolytes may influence DMSP uptake. Stresses such as freezing, heating, or osmotic down shock resulted in partial release of DMSP back to the medium. One strain which contained a DMSP-lyase was also able to accumulate DMSP, and DMS was only produced in the absence of alternate carbon sources. Bacteria containing DMSP were prepared as prey for bacterivorous ciliates and flagellates, to examine the fate of the DMSP during grazing. In all cases, predators metabolized the DMSP in bacteria. In some cases, DMS was produced, but it is not clear if this was due to the predators or to associated bacteria in the non-axenic grazer cultures. Bacterivores may influence DMSP cycling by either modulating populations of DMSP-metabolizing bacteria, or by metabolizing DMSP accumulated by bacterial prey.

INTRODUCTION

Recent investigations have focused attention on the potential role of marine bacteria in the breakdown of phytoplankton-produced dimethylsulfoniopropionate (DMSP) to dimethyl sulfide (DMS) and other products (8, 16, 17, 24, 25, 38, 39). It is likely that bacteria

Biological and Environmental Chemistry of DMSP and Related Sulfonium Compounds
edited by Ronald P. Kiene et al., Plenum Press, New York, 1996

are responsible for a substantial fraction of the DMS produced in the ocean. Since methylotrophic bacteria may also consume DMS in surface waters (18, 42), the balance between DMS production and removal, which affects DMS concentration, is strongly dependent on bacterial metabolism.

Though not generally considered, other, indirect pathways for DMSP metabolism via bacterial uptake may exist. Much previous work has shown that many bacteria accumulate betaines and similar organic molecules from their environment for use as osmotic solutes (2, 4, 5, 13, 23, 30, 32, 33). Frequently, structurally-related compounds function equally well (1, 31), including DMSP and other sulfonium compounds (5), and further evidence suggests that DMSP added to growth media counteracts salt-stress in bacteria (8, 29). It is therefore possible that many marine bacteria may accumulate DMSP and other solutes from seawater without directly metabolizing them. However, this DMSP may still be metabolized by bacterivorous protists during grazing, as has been shown for herbivorous protists (43), thus representing an additional pathway for degradation of phytoplankton-derived DMSP. Predation may also influence the availability of bacteria capable of scavenging and metabolizing DMSP and DMS, further mediating the abundance of those compounds.

In this study, isolated marine heterotrophic bacterial strains, including one which produced DMS, were tested for their ability to accumulate DMSP from seawater. Bacteria which had accumulated DMSP were used as prey for bacterivorous ciliates and flagellates in order to examine the roles of bactervory in the cycling of this compound, and in the production of DMS.

MATERIALS AND METHODS

Media for Isolation of Bacteria

Heterotrophic marine bacteria from phytoplankton cultures and Oregon coastal seawater (9-12° C) were enriched and isolated by streaking on filtered-seawater agar (1.5%) amended with 1% peptone, glucose, and 0.5% yeast extract. Cultures were incubated in the dark at room temperature and individual colonies were restreaked to purity. Bacteria were prepared for experiments by picking colonies from the plates, inoculating into 20-200 ml of similar liquid media, and incubating 1-2 days on a shaker at 100 rpm.

Selection of DMSP-Metabolizers

Bacteria were isolated which either lysed or demethylated DMSP by taking advantage of the fact that the lysis reaction produces free acid (protons) and the demethylation reaction does not. Agar plates were prepared as follows: 1 liter of 0.2-µm filtered seawater was autoclaved with 15 g agar (Difco) and 10 mg bromthymol blue and then cooled to 60°C in a water bath. Two or 5 g (10 or 25 mM) DMSP-Br and 1.6 g Tris-OH (10 mM), were dissolved in 40 ml of cold, 0.2µm-filtered seawater and pH-adjusted as necessary to 7.6 with NaOH. This solution was sterile-filtered into the 60°C agar which was then poured immediately into sterile petri plates. Because DMSP showed thermal degradation to DMS and acrylate above 60°C, it was not autoclaved. Some plates were made with trace nutrients (0.01% glucose, peptone, and yeast extract) added as well to promote general heterotrophic growth or provide trace nutrients.

Bromthymol blue, the pH indicator dye in these plates, is blue at pH 7.5, colorless at pH 6.5, and yellow at pH 6. Typically, colonies producing acid appear green, since the yellow layer around the cells lies on top of a thicker layer of unmodified blue agar. DMSP-lysing colonies were easily detectable by this method, while other colonies (DMSP-demethylators or those growing on other trace carbon sources) appeared white or clear. DMS odor could be detected in all plates, including sterile, unstreaked controls, suggesting that some autolysis of DMSP occurred. This was a small fraction of the total and sterile plates did not shift color over several months or more; however, it did mask biological production of sulfur gases. Plates were streaked with various inocula and incubated at room temperature in the dark.

Demonstration of DMSP Uptake by Bacteria

Bacteria were either incubated 24-48 hours with 10-100μM dissolved DMSP during growth on complex substrates (peptone, glucose, yeast extract), or were centrifuged after growth, washed in filtered, autoclaved seawater (FASW), and resuspended in FASW with dissolved DMSP and incubated 4-24 hours at 100 rpm at room temperature. Typical bacterial densities were 10^7-10^8 ml^{-1}. After incubation, bacteria were again centrifuged, and a subsample of the supernatent was assayed for remaining dissolved DMSP. The cells were then washed and resuspended in DMSP-free FASW, and a subsample assayed for accumulated DMSP. Samples were placed in 10-ml crimp-top vials, 2 ml 10 N NaOH was added by pipette and the vial was quickly capped with Teflon-lined septa. The bottles were incubated along with DMSP standards at room temperature for 2-24 hours and the DMS produced was sampled in 10-25μl headspace samples taken by gastight syringe. DMS was measured by GC-FPD detection.

Bacterial uptake of DMSP was also measured by filtering cell solutions with GF/F or Millipore type HA 0.45μm filters and measuring particulate DMSP retained on the filter. Samples were assayed as for centrifuged samples. Loss of cell DMSP was examined by similar methods after heating to 60, 80, or 100 °C for 2 hours, freezing, or other treatments. The effect of salinity changes on DMSP uptake was tested by growing cells in FASW, then centrifuging and resuspending them in FASW diluted with distilled water (hypotonic) or amended with 25 g $liter^{-1}$ sodium chloride (hypertonic), along with 10-100μM dissolved DMSP. Release of DMSP due to salinity changes was measured by centrifuging DMSP-containing bacteria, then resuspending and incubating them in appropriate media. Salinity was measured by refractometer.

Inhibitor studies. Inhibitors were prepared as 100 mM solutions in deionized water, except for chloramphenicol which was dissolved in ethanol, and then added to bacterial samples (strain 1030) at 1 mM final concentrations (final volume 2 ml). Ethanol was added to the no-inhibitor control as well to account for any solvent effect. 2,4dinitrophenol was dissolved by adding NaOH dropwise. Formalin used as a positive control was added directly to cultures to a final concentration of 3.7%. After adding inhibitors, samples were pre-incubated at room temperature in the dark with shaking for 0.5 hour, then DMSP was added to 100μM and the samples incubated overnight at room temperature in the dark with shaking. 0.5-ml subsamples were removed at 0, 3, and 22.5 hours and centrifuged to remove cells. 0.4 ml of the supernatent was assayed for remaining dissolved DMSP by headspace analysis.

Ciliate and Flagellate Cultures

Protists were enriched and cultured from seawater. Cultures were maintained on bacteria growing on sterilized wheat berries, or on bacterial strains heat-killed for 1 hr at 80 °C. Cultures were grown in 0.25 or 0.5 L polycarbonate flasks in the dark. Prior to the experiments, the wheat berries were removed to allow the bacterivores to graze down their prey as much as possible. The cultures were diluted into appropriate experimental samples with FASW and allowed to incubate overnight at 15°C before prey bacteria, DMSP, or other amendments were added.

Bottle Incubations

Protists and prey were incubated in 500- or 250-ml Nalgene polycarbonate flasks, filled completely to minimize headspace. Duplicate bottles of each treatment were prepared. Sampling during the experiments typically introduced headspace volumes less than 10% of the total bottle volume over several days. Bottles were handled gently to avoid aeration and were incubated in the dark at 15 °C.

Cell Enumeration. Heterotrophic flagellate and bacterial prey cells were enumerated every 12-24 hours during incubations. 0.5-2 ml culture samples were preserved with alkaline Lugol's reagent (10µl ml^{-1}) followed by sodium tetraborate-buffered formalin (3.7% final concentration). The Lugol's was bleached by the addition of 1 drop ml^{-1} 3% sodium thiosulfate (35). Samples were then stained with acridine orange (AO) or 4′,6-diamidino-2-phenylindole (DAPI) and filtered onto black 0.2 or 0.8µm membrane filters (Poretics, Livermore CA, #11053, 11021) immediately after preservation, and counted by epifluorescence microscopy. Actively swimming ciliates were enumerated directly with a dissecting microscope (Wild M3Z, Leica, Inc.) in replicate 1-20µl drops. Bacterial viability was determined during incubations by the Live/Dead® BacLight™ viability kit (Molecular Probes, Inc., #L-7007).

Sulfur Analyses

Sulfur analyses were made by GC using a Shimadzu GC-14 chromatograph equipped with a flame photometric detector. The column packing was Chromosil 330 (Supelco, Bellefonte, PA), operated isothermally at 60 or 90°C. Helium was the carrier gas. Headspace samples (10-25µl) were collected by gastight syringe and injected onto the column (injector 200°C) and were sampled in triplicate. Stock DMSP solutions were treated as other headspace samples for standards. For DMS analyses, 1- or 2-ml samples were sparged with He, cryotrapped in liquid nitrogen, and subsequently introduced onto the GC column with heating. For DMSP, a separate 1-2 ml sample was filtered through a GF/F filter under gentle filtration, and the filter was placed in 10 N NaOH for at least 6 hours. A subsample of the NaOH was then sparged/cryotrapped for DMS produced from the alkaline hydrolysis of particulate DMSP. For dissolved DMSP, the filtrate was first sparged to remove DMS, and 1 ml was then sparged with an equal volume of 10 N NaOH and cryotrapped as DMS. Minimum detection limit was approximately 100 pg S. For additional analytical details see (43).

Chemicals. DMSP·HCl was obtained from Research Plus (Bayonne, NJ) and was prepared in concentrated solutions in water. Stocks were kept frozen until use, and after dilution into seawater the pH was checked to make sure samples were not acidified. Stocks for GC standards were further acidified with HCl to prevent bacterial growth and stored at room temperature. DMSP-HBr (> 90% purity) used for isolation media was synthesized from DMS and 3-bromo-propionic acid (Aldrich) according to the method of Kondo (20) and was verified by melting point (112-113 °C) and NMR spectroscopy. Sources for inhibitors were: glycine betaine HCl, DL-S-methyl methionine, sodium azide, sodium arsenate, L-proline, and chloramphenicol, Sigma Chemicals; N,N-dimethyl glycine HCl, Aldrich Chemicals; 2,4-dinitrophenol, Kodak Chemicals.

RESULTS

Marine Bacteria Accumulated Dissolved DMSP

Five strains of marine heterotrophic bacteria were isolated on rich media. Four were gram-negative, and one (strain 0030) was gram-positive. All strains but 4030 readily took up most of the 10-100μM dissolved DMSP from seawater (figure 1). In three strains the DMSP appeared to be stored inside cells rather than metabolized, and one accumulated DMSP with subsequent lysis to DMS (strain 0010). Some cells appeared to increase DMSP uptake at higher salinity (figure 1, strains 0030, 1010), suggesting an osmotic use, but others showed no such response (figure 1, strain 1030).

One accumulator (strain 1030) and the lyser (strain 0010) were selected for more detailed studies. Uptake times typically ranged from minutes (figure 2) to hours (figure 4a). However, bacterial numbers or biomass were not constant between experiments, so rate comparisons are not meaningful. Uptake occurred both during growth on complex substrates, or following growth when washed cells were resuspended in seawater amended with 10-100μM DMSP. Occasionally, bacteria appeared to take up DMSP more rapidly in the presence of other C compounds, but more typically strains accumulated DMSP more quickly in their absence. The additional carbon sources may have provided alternative osmolytes which competitively block uptake of DMSP (see below); this inhibition was overcome in many strains when bacteria were incubated in double-salinity seawater (data not shown). Uptake of 10-100μM dissolved DMSP resulted in internal cell concentrations of approximately 10-100 mM, based on whole-cell volumes.

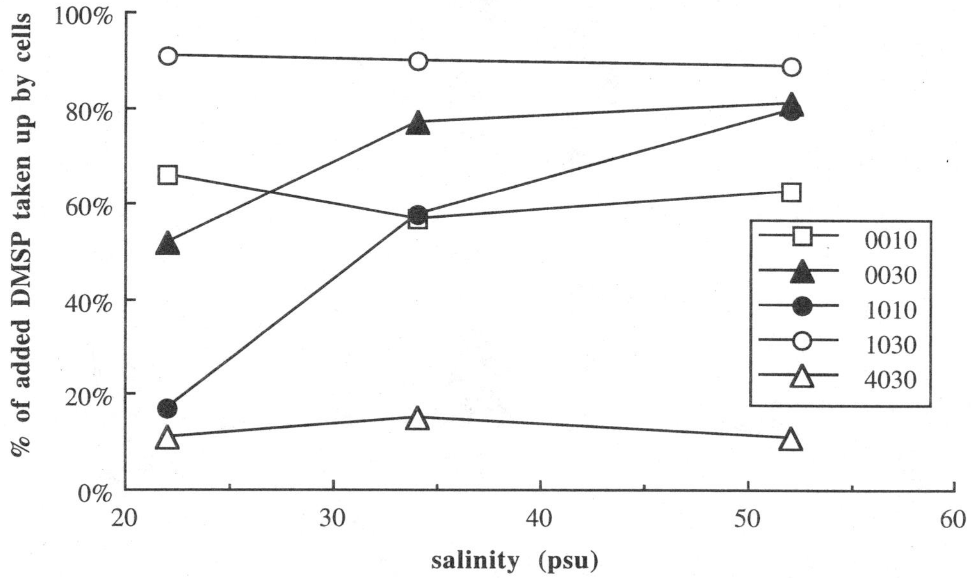

Figure 1. Example of cell uptake of dissolved DMSP. Dissolved DMSP was added to filtered, autoclaved seawater and assayed at time zero; bacterial culture (strain 0010) was added at 0.5 min (arrow).

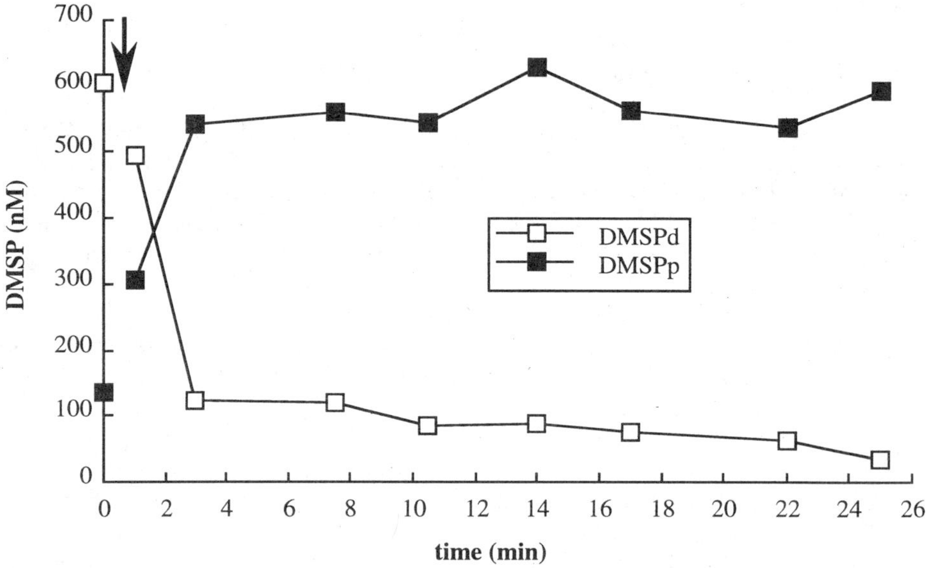

Figure 2. Effect of salinity on accumulation of dissolved DMSP from seawater by marine bacteria. Bacteria were incubated with 10μM DMSP for 16 hr in either filtered, autoclaved seawater, seawater amended with NaCl, or seawater diluted with deionized water.

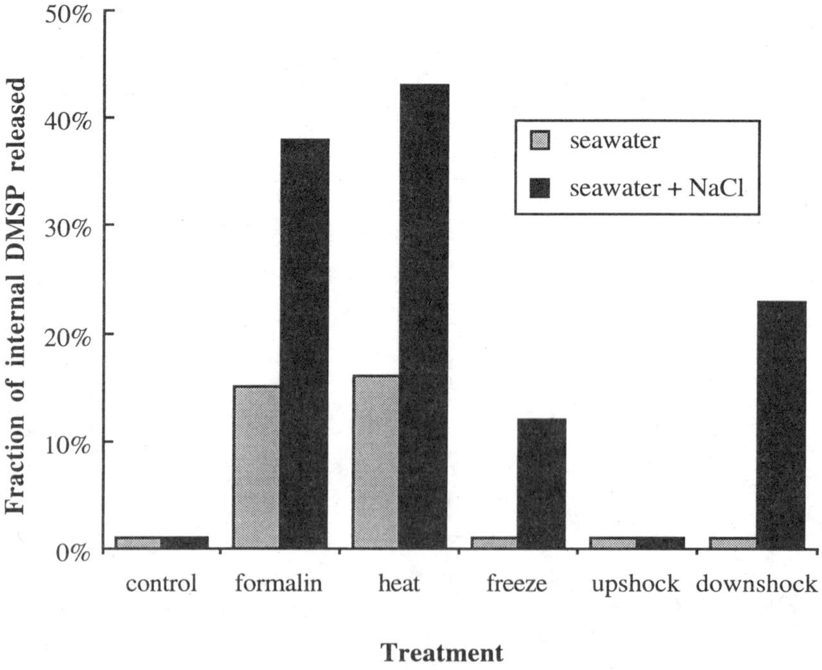

Figure 3. Release of DMSP from bacteria following shock treatments, including poisons, heating, freezing, and osmotic downshock. More DMSP was released when cells were already osmotically stressed. Data are for strain 1030.

DMSP Uptake Could Be Blocked by Uptake Inhibitors and Other Osmolytes

DMSP appeared to be taken up actively by cells. Formalin-treated bacteria did not take up DMSP from seawater and DMSP did not attach onto ion exchange resin particles such as Sephadex, so adsorption was probably not a factor. Cells heated 1-2 hrs at 80 or 100°C showed no uptake, but cells heated to only 60 °C remained active (data not shown) and were capable of growth when plated onto rich media. Pre-incubating bacteria (strain 1030) with inhibitors of transport systems or protein synthesis generally prevented uptake (table 1). Some substrate analogs effectively blocked DMSP uptake (glycine betaine, S-methyl methionine) while other organic osmolytes had reduced (dimethyl glycine) or no (L-proline) effect . Staining cells with the BacLight™ viability kit showed that the glycine betaine was not lethal to cells, so uptake was presumably blocked by competition.

DMSP could be stored inside cells for days without degradation, either in concentrated cell solutions or after cells were resuspended in seawater. In many instances, though, after resuspension and dilution of DMSP-containing cells, DMSP returned to the "dissolved" fraction over several days (see grazing results below). This may have been due to slight osmotic differences between the concentrated bacterial stock media and the seawater used for the grazing experiments, or possibly due to handling shock. Other stresses such freezing or heating cells resulted in release of DMSP to the media (figure 3). Poisons or osmotic down shock also resulted in release, while osmotic upshock did not. Greater release occurred in cells already stressed osmotically in seawater amended with NaCl (figure 3).

A DMSP-Lysing Strain Also Accumulated DMSP Depending on External Conditions

Strain 0010, able to cleave DMSP to form DMS, was able to accumulate and retain DMSP from solution as well. Uptake was greatest under conditions of osmotic stress (figure 4a) and was partially repressed when other C substrates (peptone, yeast extract, glucose) were present. Production of DMS following uptake was greatest at higher salinities as well (figure 4b), but DMS production was completely suppressed when other carbon substrates were present. These observations suggest that some DMSP-lysing bacteria may also be able to use DMSP as an osmolyte, and that cleavage to DMS and acrylate may depend on the presence of other carbon sources.

Bacterivores Were Able to Utilize DMSP Accumulated in Their Prey

When live, DMSP-containing bacteria were fed to a bacterivorous scuticociliate (*Uronema* sp.), bacterial cell numbers began to decline immediately due to grazing (figure 5b). After a 24-hr lag, ciliate numbers increased (figure 5a) and bacterial DMSP decreased (figure 6b). These results suggest that the grazers metabolized prey DMSP within 24 hr of ingestion. Ungrazed bacteria containing DMSP showed little decrease in cell number (figure 5b), but they did release DMSP to seawater (figure 6a). However, in ungrazed bacteria, total DMSP was conserved and no DMS was produced (figure 6a). A small amount of DMS was produced in the grazed cultures, probably from bacterial associates of the scuticociliates, since scuticociliate cultures streaked onto DMSP agar plates gave positive results for lysers. No axenic ciliate

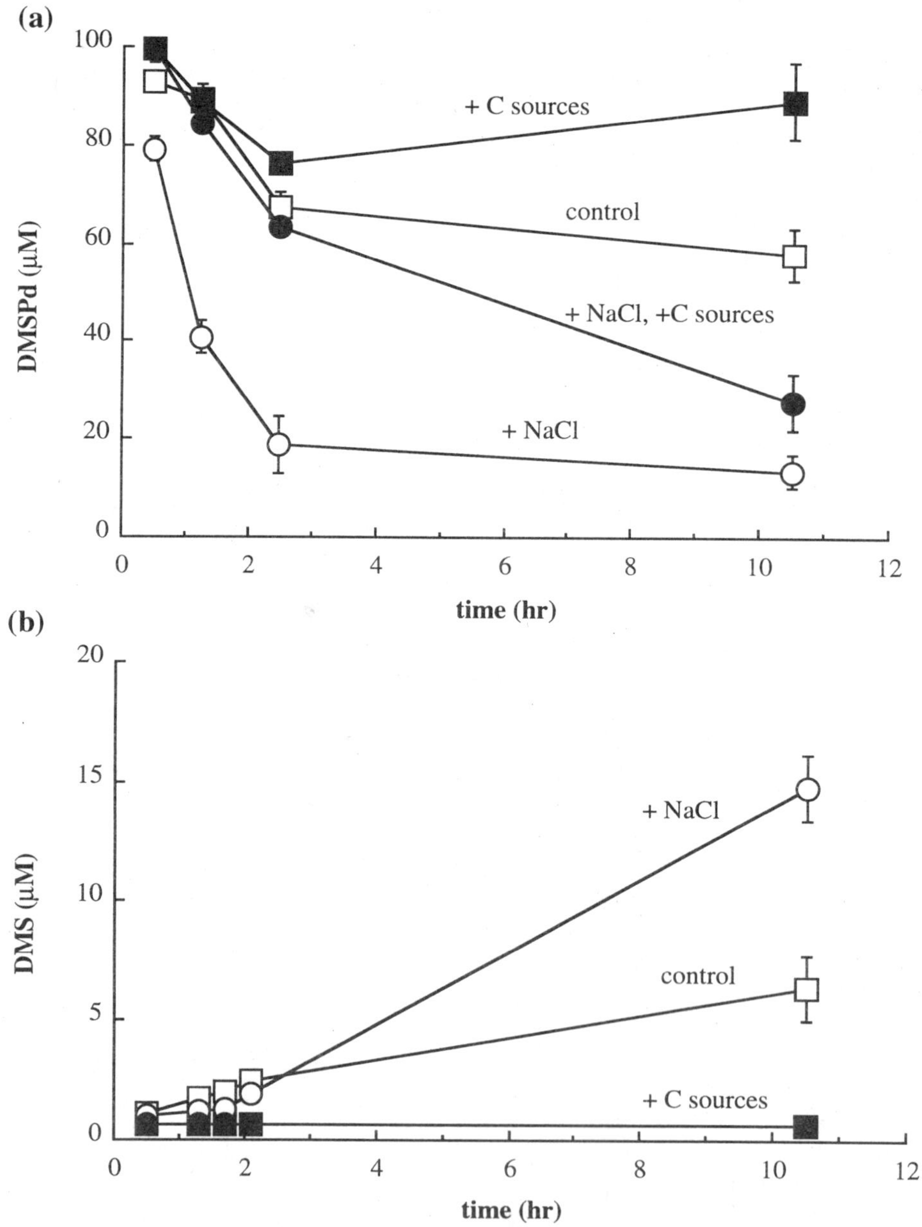

Figure 4. Accumulation of DMSP by DMSP-cleaving bacterium (strain 0010), and dependence of DMS production on external conditions. (a): uptake of dissolved DMSP by strain 0010 was hastened when grown in 56 psu seawater compared to uptake in 32 psu seawater. Addition of other C sources (peptone, glucose, yeast extract) decreased uptake partially in all cases, but at higher salinities uptake still occurred. (b) production of DMS by 0010 culture was completely suppressed by addition of other C sources and production was increased at higher salinities. Note different scales for two graphs.

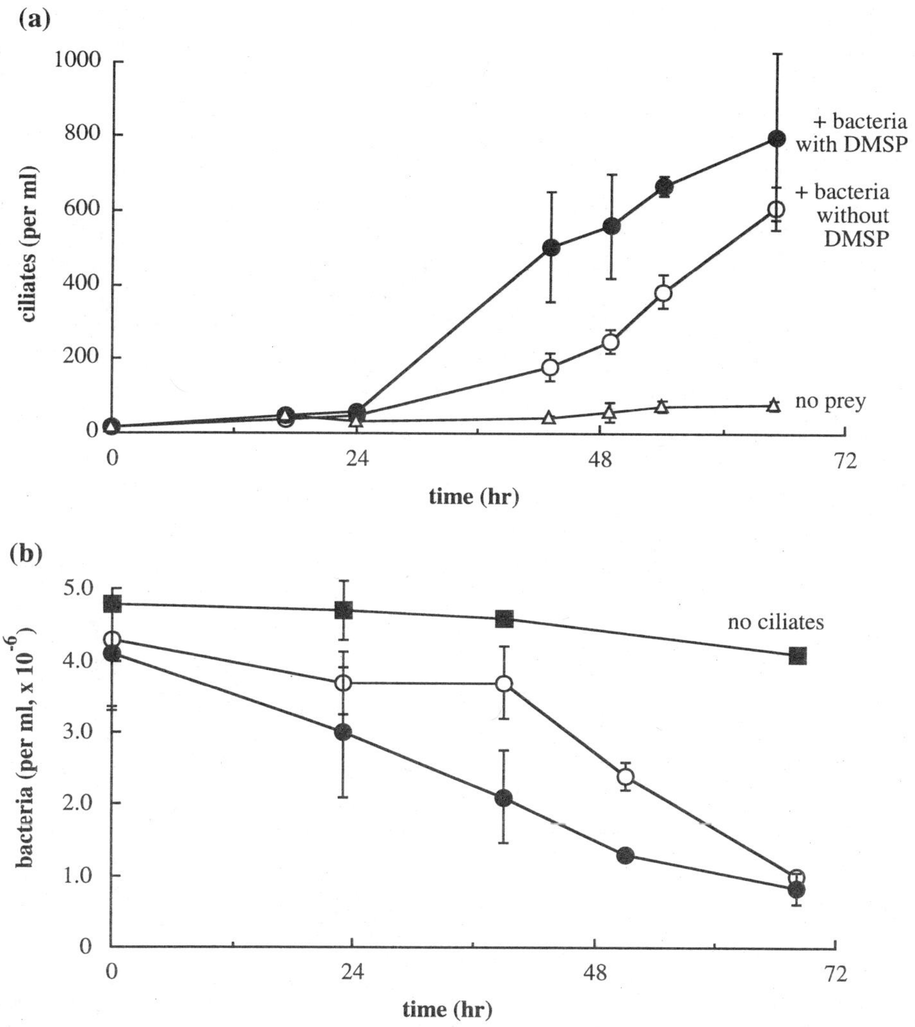

Figure 5. Grazing by scuticociliate on bacterial strain 1030 pre-incubated with DMSP (●) or without DMSP (○). (a): ciliate numbers vs. time; (Δ) are ciliates without added prey. (b): bacterial numbers vs. time; (■) is strain 1030 with DMSP ("1030+") but no ciliates.

cultures were available to test whether the DMSP lysis might in fact be due to ciliates themselves.

Similar results were also obtained in grazing experiments with a flagellate (*Cafeteria* sp., data not shown). Once again, although ungrazed bacteria released DMSP, total DMSP was conserved, as were bacterial numbers, while in the grazed samples both decreased. In this case, though, a fraction of DMSP was converted to DMS in both grazed and ungrazed samples.

With both predators, bacteria which contained up to 100 mM internal DMSP were not grazed any faster than those without, nor did grazer appearance or the increase in grazer number vary with prey DMSP, suggesting that even high internal DMSP pools in bacterial prey made little nutritional difference to predators.

DISCUSSION

Accumulation of Dissolved DMSP by Marine Bacteria

Bacteria are recognized to play a role in the cycling of phytoplankton-derived DMSP in marine systems. Most work on bacterial pathways for DMSP has focused on those which metabolize it as a carbon source, either by lysis to DMS and acrylate or by demethylation to 3-methiol propionate (8, 16, 17, 24, 25, 38-40). Bacteria which could contribute to these pathways may be significant (10%) fraction of the total bacterial population in some marine surface waters (39), and several strains have been isolated.

However, bacteria may also take up dissolved DMSP without metabolizing it, and this process has not been emphasized in studies of DMSP cycling in marine surface waters, despite the fact that much work has documented bacterial uptake of other osmotic solutes such as choline and glycine betaine (1, 4, 5, 22, 30-32). *De novo* synthesis of betaine is rare

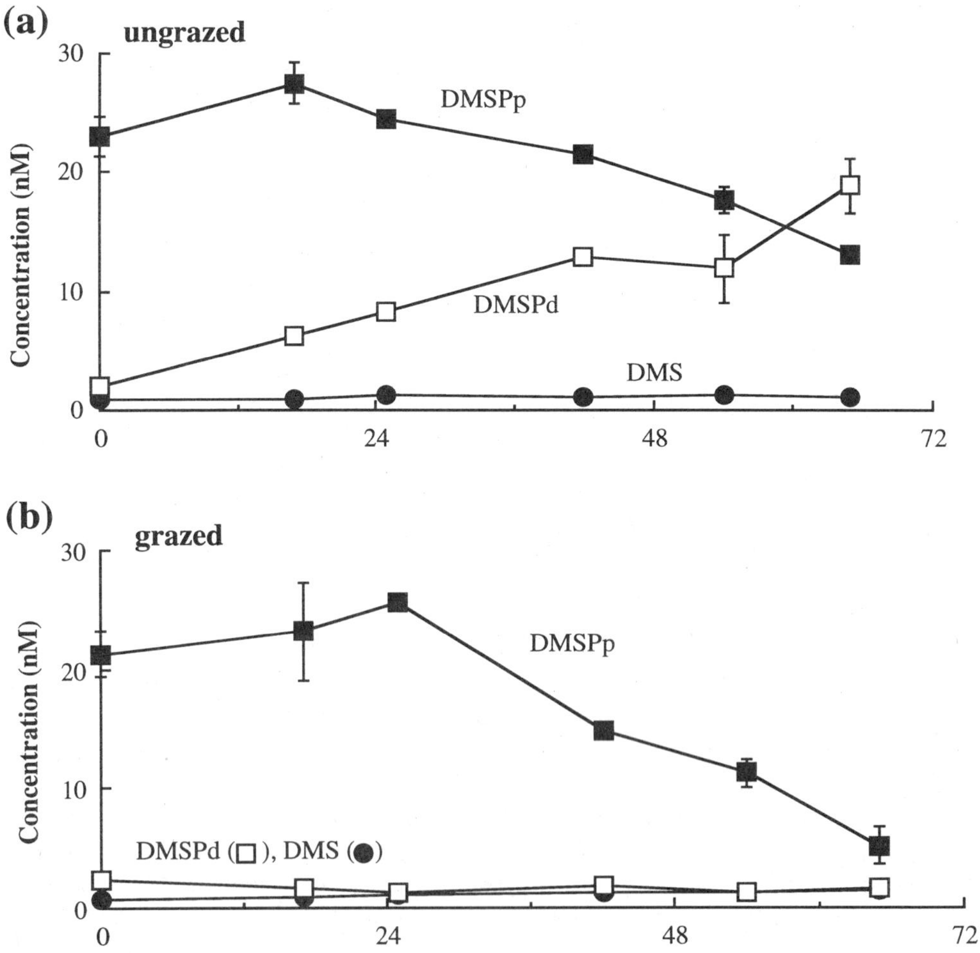

Figure 6. Sulfur pools for scuticociliate grazing on bacterial strain 1030 pre-incubated with DMSP. Symbols: (■) DMSPp; (□) DMSPd; (●): DMS. (a) Bacteria without grazers. DMSPp in the bacteria leaked out over 3 days to the dissolved pool but no DMS was formed, and there was no net loss of DMSP. (b) Bacteria with grazers. No dissolved DMSP appears but bacterial DMSP disappears as bacteria are grazed.

in eubacteria and seems limited to moderately or extremely halophilic eubacteria, especially phototrophs (41). More slightly halophilic eubacteria either accumulate betaine from their environment or produce it from externally-acquired choline (2, 12). Many eubacteria, particularly those in the Enterobacteriacae, accumulate these solutes but do not utilize them as carbon sources, while others such as *Rhizobium* can both accumulate and metabolize these solutes, depending on both salinity stress and nutrient needs (36). The non-metabolizers may also actively export these solutes to maintain osmotic balance (21, 37). In addition to the quaternary amines, enteric bacteria have been shown to accumulate DMSP and related sulfonium compounds (5, 31), and this study suggests that many marine bacteria may do so as well. Only one previous study (8) demonstrated that a marine bacterium could accumulate DMSP at high salinities, although that work focused on the ability of that strain to degrade DMSP at normal seawater salinities.

The majority of bacterial strains in this study appeared to take upμM DMSP from seawater over a range of salinities (figures 1b, 4a). Uptake rates varied among strains, and were influenced by environmental conditions such as the availability of alternate carbon substrates. In all cases, though, uptake appeared to be the result of active metabolism. Heat- or formalin-killed cells did not take up DMSP, and inhibitors of transport systems such as 2,4-dinitrophenol, azide, or arsenate prevented uptake (table 1).

DMSP was accumulated to mM internal concentrations, based on whole-cell volume calculations. However, it is possible that DMSP is stored in the cell periplasm or other sub-volume. Stresses such as freezing, heating, osmotic down shock, or poisoning all resulted in partial release of DMSP to the medium (figure 3), and frequently, dilution of concentrated cultures into seawater resulted in slow leakage of DMSP out of cells (figure 6a). This may have been due to handling stress during dilution. However, strain 1030, which was used for most of the grazing experiments, was shown by fatty acid MIDI analysis to be an enteric, corresponding most closely to *Enterobacter agglomerans* (R. Herwig, personal communication). This strain did not appear to metabolize DMSP. Since export pumps for osmotic solutes such as glycine betaine are well known for enterics which do not further catabolize it as a carbon source (21, 37), the release of DMSP by this strain may also have been an osmoregulatory effect.

Most of the strains appeared to retain DMSP without metabolism, but one DMSP-lysing strain (0010) also accumulated DMSP from the medium (figure 4a). Although the presence of other carbon sources slowed DMSP uptake, it did not prevent it. However, the production of DMS by this strain was completely blocked when other carbon sources were

Table 1. Effect of inhibitors (1 mM) on accumulation of 100:M DMSP by strain 1030

Metabolic type	Compound	% Inhibition [1]
Controls	none (- control)	0
	formalin (+ control)	100
Uptake system poisons	dinitrophenol	100
	sodium azide	100
	sodium arsenate	81
Protein synthesis inhibitor	chloramphenicol	100
Substrate analogs	Glycine betaine	99
	S-methyl methionine	86
Other osmolytes	dimethyl glycine	67
	L-proline	n.s. [2]

[1] percent reduction in regression slope of DMSP uptake over time relative to no-inhibitor control

[2] not significantly different than no-inhibitor control

available, suggesting a possible dual-role of DMSP as both an osmotic solute and a carbon source. This is similar to the use of glycine betaine by *Rhizobium* (36) which can exploit it for either purpose, depending on environmental conditions.

These observations suggest that the role of marine bacteria in the cycling of phytoplankton-derived DMSP may be greater than previously suspected, since it is not limited only to those which can utilize DMSP as a carbon source. Furthermore, although lysis of DMSP is usually quite specific to that compound (7, 31), uptake of DMSP for use as an osmotic solute may be less tightly controlled, and may be influenced by the relative availability of DMSP and other organic solutes which can compete for uptake system proteins. Such competition has been demonstrated for glycine betaine and DMSP (1, 5, 8), and other solutes present in seawater may also influence DMSP uptake. In this study, both glycine betaine and S-methyl methionine blocked DMSP uptake in strain 1030, while other solutes (L-proline, dimethyl glycine) did not. Glycine betaine has recently been used to inhibit bacterial DMSP lysis in seawater (19), and it is possible that this effect is due to competition for bacterial uptake systems rather than for the DMSP-lyase itself. The presence of yeast extract, known to be a rich source of glycine betaine (12), along with other carbon sources appeared to decrease uptake of DMSP by most strains, although this effect could be reduced when additional salt stress was imposed (data not shown).

Fate of Bacterial DMSP during Grazing by Microzooplankton

Although bacterial metabolism is probably a major sink for phytoplankton-derived DMSP, bacteria which accumulate DMSP without metabolizing it may also lead to its removal indirectly, during grazing by bacterivores. Similar to results with herbivorous flagellates (43), the passage of DMSP across trophic levels results in its degradation by bacterial grazers, including both ciliates and flagellates (figure 6). It is not yet clear whether the protists themselves are able to produce DMS from prey DMSP. Although DMS was produced in some grazing experiments, it may have been the result of other bacteria associated with the grazer cultures utilizing dissolved DMSP released from the test prey, since the grazer cultures tested positive for DMSP-lysing bacterial strains.

Because DMSP and other osmotic solutes accumulate to high internal concentrations inside prey, it is possible that they may influence the nutritional quality of prey to predators. DMSP and betaine have been shown to function as methyl donors in plants (3), flagellates (15), fish and birds (28), and mammalian cells (9, 27, 28). Whether bacterivores can utilize DMSP as a methyl donor is not clear, but at least some appear to metabolize it without significant production of DMS, consistent with this function. It is therefore possible that prey containing large concentrations of these compounds might be nutritionally advantageous for predators. However, neither ciliates nor flagellates grazing on bacteria which contained up to 100 mM internal DMSP showed any increase in feeding or growth rates compared to those grazing on bacteria containing no DMSP (data not shown). It is unlikely therefore that DMSP confers any nutritional advantage to predators, especially in natural waters, where prey are less abundant and more heterogeneous. It is possible that uptake of organic solutes such as DMSP effects other changes in bacterial physiology (e.g. size, motility) which in turn influences their desirability as prey for bacterivores.

Implications for DMSP Cycling in Marine Surface Waters

There are two separate mechanisms by which DMSP accumulates in marine microorganisms: (a) by biosynthesis, which to date has been shown only for phytoplankton and for one heterotrophic dinoflagellate (14); and (b) by uptake of dissolved DMSP for use as osmotic solutes, without degradation. A third pool of accumulated DMSP resides in meso- and macroorganisms from DMSP ingested in the diet (see for example Iida (11), Levasseur (26), and Dacey (6) and references therein). Although the latter has received notice because it can lead to odor problems in commercial fish and shellfish, it is likely insignificant compared to the pool of DMSP in microorganisms. Similar accumulation in microzooplanktonic herbivores and bacterivores may also occur, although cultures examined to date appear to metabolize the DMSP rapidly (43).

Despite the evidence shown here that the bacterial uptake mechanism may occur, it is not at all obvious that it plays an important role in DMSP cycling in natural systems. In particular, there are three questions which need to be addressed: (I) Could DMSP accumulated in bacteria be a significant fraction of total DMSP? (ii) Does accumulation occur in natural waters where bacterial concentrations and DMSP concentrations are low? (iii) Could grazing of DMSP-containing bacteria be a major loss pathway for phytoplankton-derived DMSP?

It is not clear whether a significant fraction of the "dissolved" or "particulate" DMSP measured in natural seawater might actually be DMSP stored in bacteria. Although the bacteria in this study were retained on GF/F filters, they were also extremely large cells, compared to typical marine bacteria. It is quite likely that some marine bacteria may pass through GF/F filters (nominal size retention 0.8μm). However, these probably do not contribute significantly to the "dissolved" DMSP pool. For example, if 10% of the typical 10^6 bacteria ml^{-1} contained 100 mM DMSP, and all these were spheres of diameter 0.8μm and could pass through a GF/F filter, they would contribute about 2.7 nM DMSP to the "dissolved" pool. Since "dissolved" DMSP concentrations are frequently 10 nM or greater, this seems a small contribution, especially given the generous assumptions. Furthermore, filtration with 0.2μm-pore filters and tests with dialysis membranes have shown that there really is a pool of dissolved DMSP, at least in some waters (R.P. Kiene, personal communication). Clearly, larger bacteria, as well as those attached to surfaces, will contribute to the particulate DMSP pool. But because particulate DMSP concentrations are usually greater than for dissolved DMSP, similar calculations show the bacterial contribution is again likely to be minor. However, it is still possible that in certain environments, bacterial DMSP may contribute significantly to either DMSP pool.

Does bacterial accumulation of DMSP occur at the low dissolved DMSP concentrations (typically well below 0.1 μM) which occur in most marine environments? Accumulating DMSP against enormous concentration gradients from very dilute solutions is metabolically expensive. Furthermore, other solutes may be more abundant than DMSP in seawater and may compete with DMSP for cell receptor sites. Glycine betaine blocked accumulation of DMSP by some of the bacteria in this study (table 1), and betaine appears to block metabolism of dissolved DMSP by natural microbial assemblages (19), possibly by preventing its uptake into the cells. Betaine and choline probably occur in marine waters and sediments in concentrations similar to dissolved DMSP (34). Therefore, it is possible that accumulation of DMSP by marine bacteria may not occur to the degree indicated by these bottle experiments where both DMSP and bacterial concentrations were artificially elevated.

Whether the bactivory pathway is an important sink for phytoplankton-derived DMSP in natural environments is difficult to evaluate. The great majority of marine bacteria do not appear to be metabolically active, based on selective staining techniques which measure respiratory activity (44) or DNA (45). Because DMSP uptake requires metabolically active cells, it is therefore likely that only a small fraction of marine bacteria may accumulate DMSP, even though many genera may be able to take up this compound. This seems to imply that if bacterivores utilize random selection of their prey, their chance of ingesting and degrading DMSP inside bacteria would be low. However, there is good evidence that some bacterivores do preferentially graze metabolically-active bacteria (10), raising the possibility that DMSP-containing bacteria might be preferentially selected, especially if DMSP confers any increase in size or motility to cells. It is also not yet clear whether many bacterivorous ciliates and flagellates can metabolize DMSP in their prey, or whether they may produce DMS from this DMSP.

Clearly, the experiments reported here, using high DMSP concentrations, high bacterial and bacterivore populations, and long incubation periods, present only the possibility that bacterial accumulation of DMSP and its subsequent degradation by bacterivores may be important in natural waters. Further work is needed to assess the importance of this mechanism on the removal of phytoplankton-produced DMSP in natural waters.

ACKNOWLEDGMENTS

This work was supported by NASA grant # NAGW-3737. Lynne Fessenden isolated protist cultures. Russell Herwig performed the fatty acid MIDI analysis. I thank Barry and Evelyn Sherr for discussions, and Ron Kiene for helpful comments and for sharing an unpublished manuscript.

REFERENCES

1. Abdel-Ghany, Y. S., M. A. Ihnat, D. D. Miller, C. M. Kunin and H. H. Tong. 1993. Structure-activity relationship of glycine betaine analogs on osmotolerance of enteric bacteria. J. Med. Chem. *36*: 784-489.
2. Boch, J., B. Kempf and E. Bremer. 1994. Osmoregulation in *Bacillus subtilis*: synthesis of the osmoprotectant glycine betaine from exogenously provided choline. J. Bact. *176*: 5364-5371.
3. Byerrum, R. U., C. S. Sato and C. D. Ball. 1956. Utilization of betaine as a methyl group donor in tobacco. Plant Physiol. *31*: 374-377.
4. Chambers, S. and C. M. Kunin. 1985. The osmoprotective properties of urine for bacteria: the protective effect of betaine and human urine against low pH and high concentrations of electrolytes, sugars, and urea. J. Infect. Dis. *152*: 1308-1316.
5. Chambers, S. T., C. M. Kunin, D. Miller and A. Hamada. 1987. Dimethylthetin can substitute for glycine betaine as an osmoprotectant molecule for *Escherichia coli*. J. Bact. *169*: 4845-4847.
6. Dacey, J. W. H., G. M. King and P. S. Lobel. 1994. Herbivory by reef fishes and the production of dimethylsulfide and acrylic acid. Mar. Ecol. Prog. Ser. *112*: 67-74.
7. de Souza, M. P. and D. C. Yoch. 1995. Purification and characterization of dimethylsulfoniopropionate lyase from an *Alcaligenes*-like dimethyl sulfide producing marine isolate. Appl. Environ. Microbiol. *61*: 21-26.
8. Diaz, M. R., P. T. Visscher and B. F. Taylor. 1992. Metabolism of dimethylsulfoniopropionate and glycine betaine by a marine bacterium. FEMS Micro. Lett. *96*: 61-66.
9. du Vigneaud, V., A. W. Moyer and J. P. Chandler. 1948. Dimethylthetin as a biological methyl donor. J. Biol. Chem. *174*: 477-480.
10. González, J. M., E. B. Sherr and B. F. Sherr. 1993. Differential feeding by marine flagellates on growing versus starving, and on motile versus nonmotile, bacterial prey. Mar. Ecol. Prog. Ser. *102*: 257-267.
11. Iida, H., K. Nakamura and T. Tokunaga. 1985. Dimethyl sulfide and dimethyl-β-propiothetin in shellfish. Bull. Jap. Soc. Sci. Fish. *51*: 1145-1150.
12. Imhoff, J. F. 1986. Osmoregulation and compatible solutes in eubacteria. FEMS Micro. Rev. *39*: 57-66.
13. Imhoff, J. F. and F. Rodriguez-Valera. 1984. Betaine is the main compatible solute of halophilic eubacteria. J. Bact. *160*: 478-479.
14. Ishida, Y. and H. Kadota. 1967. Isolation and identification of dimethyl-β-propiothetin from *Gyrodinium cohnii*. Agr. Biol. Chem. *31*: 765-767.
15. Ishida, Y. and H. Kadota. 1968. Participation of dimethyl-β-propiothetin in transmethylation reaction in *Gyrodinium cohnii*. Bull. Jap. Soc. Sci. Fish. *34*: 699-705.
16. Kiene, R. P. 1990. Dimethyl sulfide production from dimethylsulfoniopropionate in coastal seawater and bacterial cultures. Appl. Environ. Microbiol. *56*: 3292-3297.
17. Kiene, R. P. 1992. Dynamics of dimethyl sulfide and dimethylsulfoniopropionate in oceanic water samples. Mar. Chem. *37*: 29-52.
18. Kiene, R. P. and T. S. Bates. 1990. Biological removal of dimethyl sulfide from seawater. Nature *345*: 702-705.

19. Kiene, R. P. and G. Gerard. 1995. Evaluation of glycine betaine as an inhibitor of dissolved dimethylsulfoniopropionate degradation in marine waters. Mar. Ecol. Prog. Ser. 128: 121-131.
20. Kondo, H. and M. Ishimoto. 1987. Thetin. Meth. Enzymol. *143*: 227.
21. Koo, S.-P., C. F. Higgins and I. R. Booth. 1991. Regulation of compatible solute accumulation in *Salmonella typhimurium*: evidence for a glycine betaine efflux system. J. Gen. Micro. *137*: 2617-2625.
22. Lamark, T., O. B. Styrvold and A. R. Strom. 1992. Efflux of choline and glycine betaine from osmoregulating cells of *Escherichia coli*. FEMS Micro. Lett. *96*: 149-154.
23. Le Rudulier, D. and B. Perroud. 1983. Glycine betaine, an osmotic effector in *Klebsiella pneumoniae* and other members of the *Enterobacteriaceae*. Appl. Environ. Microbiol. *46*: 152-159.
24. Ledyard, K. M. and J. W. H. Dacey. 1994. Dimethylsulfide production from dimethylsulfoniopropionate by a marine bacterium. Mar. Ecol. Prog. Ser. *110*: 95-103.
25. Ledyard, K. M., E. F. DeLong and J. W. H. Dacey. 1993. Characterization of a DMSP-degrading bacterial isolate from the Sargasso Sea. Arch. Microbiol. *160*: 312-318.
26. Levasseur, M., M. D. Keller, E. Bonneau, D. D'Amours and W. K. Bellows. 1994. Oceanographic basis of a DMS-related Atlantic cod (*Gadus morhua*) fishery problem: blackberry feed. Can. J. Fish. Aquat. Sci. *51*: 881-889.
27. Maw, G. A. and V. du Vigneaud. 1948. Dimethyl-β-propiothetin, a new methyl donor. J. Biol. Chem. *174*: 381-382.
28. Nakajima, K. 1993. Dimethylthetin- and betaine-homocysteine methyltransferase activities from livers of fish, chicken, and mammals. Nippon Suisan Gakkaishi *59*: 1389-1393.
29. Paquet, L., B. Rathinasabapathi, H. Saini, L. Zamir, D. A. Gage, Z.-H. Huang and A. D. Hanson. 1994. Accumulation of the compatible solute 3-dimethylsulfoniopropionate in sugarcane and its relatives, but not other gramineous crops. Aust. J. Plant Physiol. *21*: 37-48.
30. Patchett, R. A., A. F. Kelly and R. G. Kroll. 1994. Transport of glycine-betaine by *Listeria monocytogenes*. Arch. Microbiol. *162*: 205-210.
31. Peddie, B. A., M. Lever, C. M. Hayman, K. Randall and S. T. Chambers. 1994. Relationship between osmoprotection and the structure and intracellular accumulation of betaines by *Escherichia coli*. FEMS Micro. Lett. *120*: 125-132.
32. Perroud, B. and D. Le Rudulier. 1985. Glycine betaine transport in *Escherichia coli*: osmotic modulation. J. Bact. *161*: 393-401.
33. Pocard, J.-A., T. Bernard, L. T. Smith and D. Le Rudulier. 1989. Characterization of three choline transport activities in *Rhizobium meliloti*: modulation by choline and osmotic stress. J. Bact. *171*: 531-537.
34. Roulier, M. A., B. Palenik and F. M. M. Morel. 1990. A method for the measurement of choline and hydrogen peroxide in seawater. Mar. Chem. *30*: 409-421.
35. Sherr, E. B. and B. F. Sherr. 1993. Preservation and storage of samples for enumeration of heterotrophic protists. *In* P. F. Kemp, B. F. Sherr, E. B. Sherr and J. J. Cole (ed.), Handbook of Methods in Aquatic Microbial Ecology. Pp. 207-212. Lewis Publishers, Boca Raton.
36. Smith, L. T., J.-A. Pocard, T. Bernard and D. Le Rudulier. 1988. Osmotic control of glycine betaine biosynthesis and degradation in *Rhizobium meliloti*. J. Bact. *170*: 3142-3149.
37. Snipes, W., A. Keith and P. Wanda. 1974. Active transport of choline by a marine pseudomonad. J. Bact. *120*: 197-202.
38. Taylor, B. F. and D. C. Gilchrist. 1991. New routes for aerobic biodegradation of dimethylsulfoniopropionate. Appl. Environ. Microbiol. *57*: 3581-3584.
39. Visscher, P. T., M. R. Diaz and B. F. Taylor. 1993. Enumeration of bacteria which cleave or demethylate dimethylsulfoniopropionate in the Caribbean Sea. Mar. Ecol. Prog. Ser. *89*: 293-296.
40. Visscher, P. T. and B. F. Taylor. 1994. Demethylation of dimethylsulfoniopropionate to 3-mercaptopropionate by an aerobic marine bacterium. Appl. Environ. Microbiol. *60*: 4617-4619.
41. Welsh, D. T. and R. A. Herbert. 1993. Identification of organic solutes accumulated by purple and green sulphur bacteria during osmotic stress using natural abundance ^{13}C nuclear magnetic resonance spectroscopy. FEMS Micro. Ecol. *13*: 145-150.
42. Wolfe, G. V. and R. P. Kiene. 1993. Radioisotope and chemical inhibitor measurements of dimethyl sulfide consumption rates and kinetics in estuarine waters. Mar. Ecol. Prog. Ser. *99*: 261-269.
43. Wolfe, G. V., E. B. Sherr and B. S. Sherr. 1994. Release and consumption of DMSP from *Emiliania huxleyi* during grazing by *Oxyrrhis marina*. Mar. Ecol. Prog. Ser. *111*: 111-119.
44. Zimmermann, R., R. Iturriaga and J. Becker-Birck. 1978. Simulataneous determination of the total number of aquatic bacteria and the number thereof involved in respiration. Appl. Environ. Microbiol. *36*: 925-935.
45. Zweifel, U. L. and C. Hagstrøm. 1995. Total counts of marine bacteria include a large fraction of non-nucleoid-containing bacteria (ghosts). Appl. Environ. Microbiol. *61*: 2180-2185.

N-TERMINAL AMINO ACID SEQUENCES AND COMPARISON OF DMSP LYASES FROM *PSEUDOMONAS DOUDOROFFII* AND *ALCALIGENES* STRAIN M3A

Mark P. de Souza and Duane C. Yoch

Department of Biological Sciences
University of South Carolina,
Columbia, South Carolina 29208

SUMMARY

A comparative *in vitro* study of the DMSP lyases from an estuarine isolate, *Alcaligenes* strain M3A and a marine organism, *Pseudomonas doudoroffii* has been carried out. The enzymes in both organisms were induced to high levels by aeration, but unlike *Alcaligenes*, the *P. doudoroffii* enzyme was easily inactivated *in vivo* by processes not yet understood. The enzymes from both organisms bound to phenyl-Sepharose CL-4B and could then be purified by hydrophobic, anion exchange and gel filtration chromatography. A minor form (isozyme) of DMSP lyase (DL-2) was detected in *Alcaligenes*; its size and subunit composition were similar to the major isoform (DL-1) as was its K_m for DMSP. Polyclonal antibodies raised against the *Alcaligenes* DMSP lyase were equally reactive against both the *Alcaligenes* and *P. doudoroffii* enzymes. Western blots of SDS-polyacrylamide gels together with gel filtration analysis showed that both enzymes were monomers of 48 kDa. The purified DMSP lyases had a similar K_m (1 to 2 mM) and v_{max} (ca. 500 units/mg protein) for DMSP. Methyl-3-mercaptopropionate (MMPA) was inhibitory to DMSP lyase activity *in vivo* and *in vitro* in both organisms. Cyanide and p-chloromercuri-benzoate inhibited *P. doudoroffii* DMSP lyase activity *in vivo* but not *in vitro*; activity in *Alcaligenes* was unaffected. Based on results of kinetic and inhibitor studies, a working model of DMS production, which includes a postulated DMSP-binding protein in *P. doudoroffii*, but not *Alcaligenes*, has been presented. The N-terminal amino acid sequences of the DMSP lyases purified from *Alcaligenes* and *P. doudoroffii* had 75% homology to each other in the first 20 amino acid residues, with 90% homology in the first 10 residues suggesting that the N-terminal region of DMSP lyase in these two marine bacteria is highly conserved. The N-terminal amino acid sequences of these enzymes showed no significant degree of homology with any existing protein in the database.

Biological and Environmental Chemistry of DMSP and Related Sulfonium Compounds
edited by Ronald P. Kiene et al., Plenum Press, New York, 1996

INTRODUCTION

DMSP degradation by microbes in marine environments has been shown to occur via two pathways, one of which involves an initial demethylation to methyl-3-mercaptopropionate (MMPA) (19,20,29), and the other pathway which cleaves DMSP to produce DMS, acrylate and a proton, by a reaction catalyzed by DMSP lyase (15-18,20,21). This report will focus on the enzymology of the DMSP lyase pathway. DMSP degradation to DMS has been reported under both aerobic and anaerobic conditions in salt marsh sediments (17,20,21,29), microbial mats (30), estuarine, oceanic, and coastal waters (15,16,18). A number of pure cultures of bacteria (7,10,15,22,29,33,35) and algae (5,6,11,14,25,26,28) possess DMSP lyase activity which until recently has only been studied *in vivo*. DMSP lyase activity has been studied *in vitro* in the following algae: *Ulva lactuca* (9), *Polysiphonia lanosa* (5,25), and *Gyrodinium cohnii* (13). This enzyme has only recently been purified from a salt marsh bacterial isolate, which was identified as an *Alcaligenes*-like organism (7). The physiology of DMS production by DMSP lyase has recently been compared and contrasted in *Alcaligenes* and a marine species, *P. doudoroffii* (3,22) with regard to their induction characteristics, parameters that affect enzyme turnover, and their *in vivo* kinetic constants (8). This paper compares the *in vitro* biochemical characteristics of DMSP lyase from *Alcaligenes* and *P. doudoroffii*, and is the first to report the N-terminal amino acid sequence of DMSP lyases. These results will provide the foundation for further research in investigating the still unknown aspects of the biochemistry and molecular biology of DMS production from DMSP and *in situ* population sizes of DMSP lyase producers in the marine environment.

MATERIALS AND METHODS

Induction of DMSP Lyase

Alcaligenes was grown in 8-liter batch cultures with acrylate as an inducer and carbon source in the basal salts medium described earlier (7). *P. doudoroffii* was grown in Tryptic Soy Broth (TSB, Difco) in 8-liter or 1-liter cultures and then induced with 1 mM DMSP. Both required high levels of aeration for maximum rates of DMSP lyase induction. After the cells were fully induced, *Alcaligenes* was centrifuged at 8600 x g for 10 min whereas *P. doudoroffii* was harvested by taking the centrifuge up to 11000 x g and then stopping it. This was necessary because *P. doudoroffii* lost a considerable amount of its activity during prolonged centrifugation. DMSP lyase activity, both *in vivo* and *in vitro* was measured by gas chromatographic determination of DMS (7).

Purification and Characterization of DMSP Lyase

The procedure used to purify the *Alcaligenes* DMSP-lyase has been described previously (7) and the *P. doudoroffii* enzyme was purified using a modification of this method. Notable changes included a 40% ammonium sulfate cut instead of a 55% cut and a subsequent gradient of 40 to 0% ammonium sulfate in 50 mM phosphate buffer (pH 8) on a phenyl-Sepharose CL-4B column instead of a 55 to 0% gradient. Due to the still unexplained instability of the *P. doudoroffii* enzyme *in vivo* (8), the amount of electrophoretically pure protein recovered was small and therefore was only used to determine its molecular mass, estimate its kinetic constants (K_m and v_{max}) for DMSP and N-terminal amino acid sequence.

Other experiments were performed using partially purified protein in fractions obtained after gradient elution from the phenyl-Sepharose column.

Determination of the Molecular Mass and Subunit Composition

Pure DMSP lyase from *P. doudoroffii* was chromatographed on a Sephadex G-100S gel filtration column on which the *Alcaligenes* enzyme (molecular mass = 48 kDa, ref. 7) had previously been chromatographed. This procedure allowed us to estimate the molecular weight of the native *P. doudoroffii* enzyme. Subunit composition was determined by sodium dodecyl sulfate polyacrylamide gel electrophoresis (SDS-PAGE) as described earlier for the *Alcaligenes* enzyme (7).

Western Blots

The enzymes were electroblotted from 10% polyacrylamide SDS gels to a polyvinylidine difluoride membrane (PVDF, Bio-Rad) and immuno-stained using polyclonal -*Alcaligenes* DMSP lyase rabbit antiserum (1:1000) and a goat-antirabbit horseradish peroxidase detection system (Bio-Rad) using chloro-naphthol as a chromogenic substrate to identify the bands recognized by the antibody. The techniques of electroblotting and immuno-staining DMSP lyase were essentially those of Sambrook et al. (27).

Determination of the Kinetic Constants of DMSP Lyase for DMSP

The kinetic parameters (K_m and v_{max} for DMSP) of DMSP lyase were determined using Lineweaver-Burke plots and DMSP at concentrations between 20 and 2000 μM. The data reported here are the mean of at least two experiments.

Effect of Inhibitors on *In vitro* DMSP Lyase Activity

Enzyme preparations were incubated with the inhibitors (at 5 mM) shown in Table 2 for 10 min, after which DMSP (2.5 mM) was added and enzyme activity measured.

Effect of pH on *In vitro* DMSP Lyase Activity

The pH optima for the isozymes of DMSP lyase from *Alcaligenes*, and for *P. doudoroffii* were determined for both the purified and partially purified enzyme as described previously (8).

Sequencing of DMSP Lyase

Following electrophoresis and electroblotting onto a PVDF membrane, the bands of *Alcaligenes* and *P. doudoroffii* DMSP lyase were cut from the air-dried membrane and used directly in the autosequencer. The amino-terminal sequences of DMSP lyase from *Alcaligenes* and *P. doudoroffii* were determined using a pulsed liquid protein sequencer (Applied Biosystems model 477A) equipped with an on-line Biosystems model 120A PTH analyzer. Sequences were analyzed using the FASTA program with the GCG software system provided by the manufacturer.

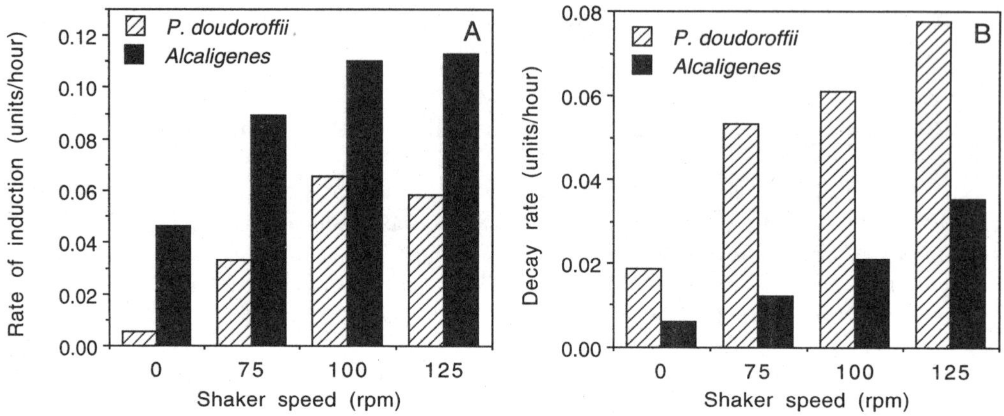

Figure 1. Production and degradation of DMSP lyase in cell cultures of *Alcaligenes* and *P. doudoroffii*. (A) Rate of induction and (B) decay constants of DMSP lyase activity as a function of aeration of the culture. Cultures were induced with 1 mM DMSP and kinetics of DMS production were monitored every 1.5 hours.

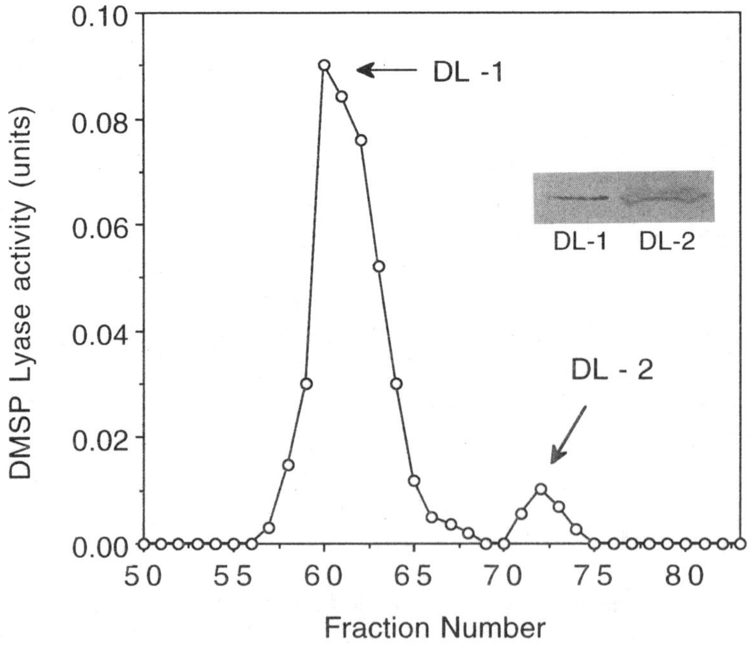

Figure 2. Phenyl-Sepharose elution profile of *Alcaligenes* DMSP lyase activity. A 40% ammonium sulfate supernatant containing the enzyme was applied to the 1 x 15 cm column and eluted with 300 ml of a 40 to 0% ammonium sulfate gradient. The gradient ended at fraction 50 after which 50 mM phosphate buffer (pH 8) was used for the rest of the elution. The volume of each fraction was 4 ml. The inset shows a Western blot of DL-1 and DL-2 probed with the antibody to DL-1.

RESULTS

Induction and Purification of DMSP Lyase

The rate of induction of DMSP lyase in both *Alcaligenes* and *P. doudoroffii* cultures was greatly stimulated by increased culture aeration (Fig. 1A). A shaker speed of 100 rpm saturated the culture with oxygen at approximately 250 μM for *Alcaligenes* in phosphate buffer, and 220 μM for *P. doudoroffii* in sea water. Once induced, the *Alcaligenes* enzyme was quite stable, but cultures of *P. doudoroffii* showed a rapid loss of activity with increased aeration (Fig. 1B). It was therefore difficult to "capture" the *P. doudoroffii* enzyme in a stable, active form *in vitro*. However, when a crude extract with activity was obtained, the enzyme was fairly stable and from those preparations we were able to purify to homogeneity the DMSP lyase from this organism. The enzymes from both organisms were in the soluble fraction and showed similar chromatography characteristics; they bound to phenyl-Sepharose CL-4B when the gel was equilibrated with 40% ammonium sulfate, and were released with buffer containing no ammonium sulfate (Fig. 2 and ref 7). Two forms of the *Alcaligenes*-DMSP lyase with different hydrophobic characteristics eluted from the phenyl-Sepharose column after a gradient of 40 to 0% ammonium sulfate (Fig. 2). The major isozyme that was purified earlier (7) is referred to as DL-1; the minor form of the enzyme is called DL-2. The inset to Fig. 2 shows a Western blot of *Alcaligenes* DL-1 and DL-2, and indicates that they have identical molecular weights. Only one isoform of the enzyme could be detected in *P. doudoroffii* at this time using a similar purification protocol.

Molecular Mass and Subunit Characteristics

The DMSP lyases from *Alcaligenes* and *P. doudoroffii* comigrated on a Sephadex G-100S column (Fig.3) indicating they have similar molecular masses; the mass of the *Alcaligenes* DMSP lyase is known to be 48 kDa (8). With their identical migration on

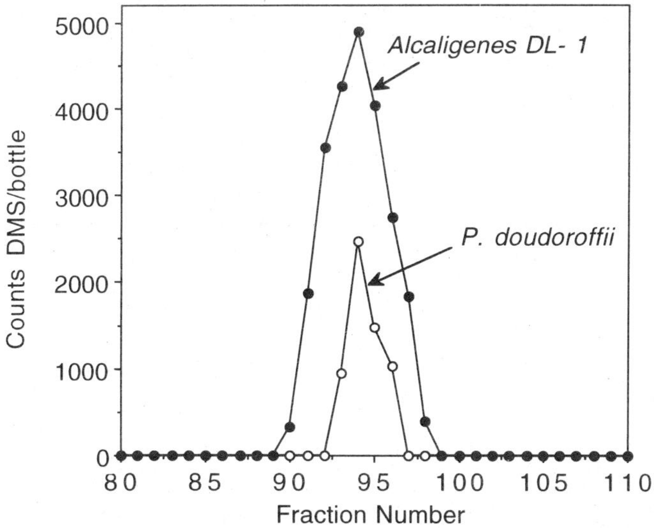

Figure 3. Gel filtration data of *Alcaligenes* M3A and *P. doudoroffii* DMSP lyases on a G100S Sephadex column. The proteins were chromatographed on separate runs. The *Alcaligenes* enzyme was used as the standard since its molecular mass (48 kDa) is known (7).

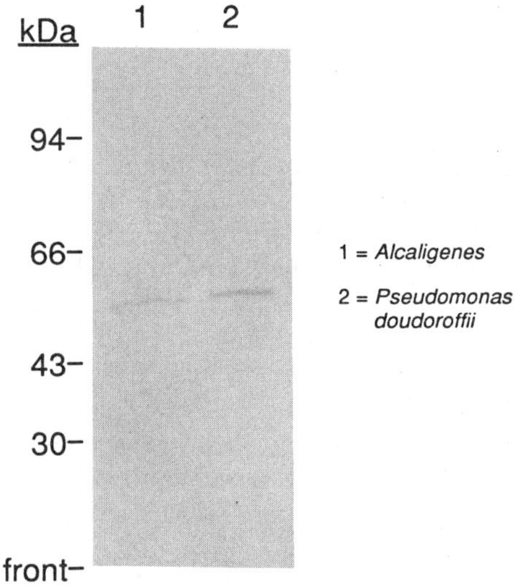

Figure 4. Western blot of *Alcaligenes* and *P. doudoroffii* DMSP lyases. Antibody to the *Alcaligenes* DMSP lyase (DL-1) was used to probe the blot. Molecular mass markers are drawn on the left of the figure.

SDS-PAGE as seen by Western blots (Fig.4), it further indicates that the DMSP lyases from these organisms are both monomeric proteins with similar molecular masses of 48 kDa. These enzymes also have a similar migration pattern on native gels (data not shown). The antibody against the *Alcaligenes* DMSP -lyase responded to the *P. doudoroffii* enzyme at the same dilution at which it reacted with the pure *Alcaligenes* enzyme, indicating the presence of similar antigenic determinants.

Kinetic Characteristics

When the kinetic constants were determined in cell extracts or with the pure enzyme, DMSP lyase from *P. doudoroffii* and *Alcaligenes* had a similar K_m for DMSP that was 1 to 2 mM (Table 1). Note however, that while the *in vitro* K_m values are similar, the *P. doudoroffii* K_m for DMSP *in vivo* is several orders of magnitude lower (< 20μM); the significance of this fact is discussed below. The v_{max} of *Alcaligenes* DMSP lyase was higher than the *P.*

Table 1. Summary table of the kinetic characteristics of DMSP lyase in *Alcaligenes* and *Pseudomonas doudoroffii*

Parameter		*Alcaligenes*	*P. doudoroffii*
K_m (mM)	Cell suspension[1]	1.41	< 20 μM
	Cell extract[2]	0.97	1.40
	Pure enzyme (DL-1)	2.02	1.82
	Isozyme (DL-2)	2.34	NA
v_{max} (units/mg)[3]	Cell suspension[1]	1.09	0.32
	Cell extract[2]	26.75	3.75
	Pure enzyme (DL-1)	408	504

1. The *in vivo* kinetic constants for the *Alcaligenes* enzyme are taken from reference 8.
2. Cell extracts are fractions obtained after sonication, ammonium sulfate precipitation, and hydrophobic chromatography as described previously (7).
3. A unit of DMSP lyase activity is the amount of enzyme that catalyzes the release of 1 μmole of DMS from DMSP in one minute.

Table 2. Effect of inhibitors on DMSP lyase activity in cell-free extracts of *Pseudomonas doudoroffii* and *Alcaligenes* strain M3A

	% Inhibition	
Inhibitor (at 5 mM)	*Alcaligenes*	*P. doudoruffii*
3-mercaptopropionate (MPA)	1	22
Methyl-3-mercaptopropionate (MMPA)	68	74
Methionine	8	22
Glycine betaine	21	18
Propionate	16	14
Homocysteine	15	24
2-mercaptoacetate	31	1

Dimethylsulfoxide, S-methylmethionine, acrylate, acrylamide, methionine sulfoxide, methacrylate, dimethylglycine, dimethylsulfonioacetate (all at 5 mM) showed less than 10% inhibition of DMSP lyase activity. Cyanide, azide, arsenate, and *p*-chloromercuribenzoate (all at 1 mM) did not inhibit DMSP lyase activity *in vitro* in either organism.

doudoroffii enzyme in cell culture and cell extracts but the purified enzymes had similar values of 400-500 units/mg protein (Table 1).

The K_m for DMSP for the minor isozyme (DL-2) of DMSP lyase from *Alcaligenes* was similar to that of its major isozyme, DL-1. The v_{max} could not be compared since DL-2 was not purified to electrophoretic homogeneity.

Optimum pH and Temperature for DMSP Lyase Activity

DMSP lyase activity in both cell cultures and cell extracts of *Alcaligenes* had pH optima at 6.5 and 8.5 (Fig. 5A). The two pH optima are not explained however by the optima of the isozymes, as both the major (DL-1) and minor (DL-2) isoforms of the *Alcaligenes* enzyme had pH optima at 8.5 (Fig. 5B). DMSP lyase activity in *P. doudoroffii* determined *in vitro* had a pH optimum of 8 (data not shown). DMSP lyase activity both *in vivo* and *in vitro* had temperature optima at 37°C (data not shown).

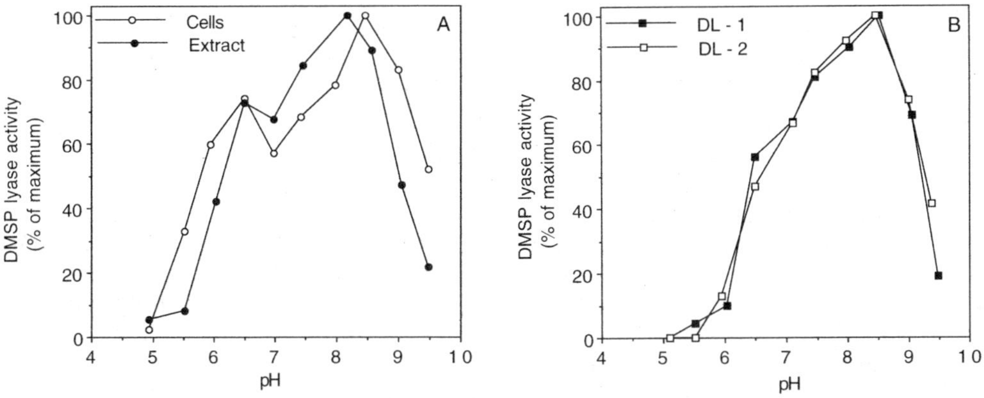

Figure 5. pH optima for activity of *Alcaligenes* M3A DMSP lyases in: (A) cell cultures and crude extracts and (B) pure DMSP lyase (isozyme, DL-1) and the partially purified minor form (isozyme, DL-2).

Effect of Inhibitors

Of the several inhibitors tested, MMPA was the only compound that strongly inhibited DMSP lyase activity *in vitro* both in *P. doudoroffii* and *Alcaligenes* (Table 2).

Effect of Alternative Substrates

None of the following DMSP analogues, tested at 5 mM, functioned as alternative substrates to release DMS or methanethiol during *in vitro* assays of DMSP lyase from either *P. doudoroffii* or *Alcaligenes*: dimethylsulfonioacetate (DMSA), dimethylsulfoxide (DMSO), methionine, methionine sulfoxide, S-methylmethionine, homocysteine, and MMPA.

Amino Acid Sequences

When the N-terminal amino acid sequences of the DMSP lyases from *Alcaligenes* and *P. doudoroffii* were compared to all other proteins in the database (in April 1995) there was no homology greater than 25 to 30% observed, and that was only for 4 or 5 amino acid sequences. There was 75% homology between the *Alcaligenes* and *P. doudoroffii* enzyme over the N-terminal end of these enzymes, while the first 10 amino acid residues were 90% homologous (Fig.6).

DISCUSSION

DMS, the most abundant volatile sulfur compound in seawater (1), is thought to be produced mainly from DMSP by DMSP lyases present in marine environments (See Introduction). Due to an ocean-air flux, DMS is a major source of sulfur to the atmosphere (1), where, following its oxidation, it may even influence climate change (2,24,34). The parameters that influence production, function, and maintenance of DMSP lyases will also affect the production of DMS, and therefore its contribution to the global sulfur cycle. There is a consensus emerging that marine bacteria play an essential, perhaps dominant role in this process.

We have studied the characteristics of DMSP lyases *in vivo* (8) and *in vitro* (7, and this study) from two marine bacteria, *Alcaligenes* strain M3A and *P. doudoroffii*. The *Alcaligenes* enzyme is very stable (Fig. 1) and was easily purified (7), but DMSP lyase activity in *P. doudoroffii* was rapidly lost *in vivo* and therefore the enzyme from this organism could not be purified in large amounts. It is particularly labile when grown in large (1 to 8 liter), aerated, batch cultures. It is possible that DMSP has to be present constantly in the cellular environment in order for the enzyme to be maintained, but this has not yet been tested.

The enzymes isolated from *P. doudoroffii* and *Alcaligenes* are both monomeric, 48 kDa molecules (Figs 3 and 4) which have a high degree of homology in their N-terminal amino acid sequence (Fig. 6) and similar pH and temperature optima. Cell-free preparations of other DMSP lyase-containing bacterial isolates, however, have shown by Western blot analysis a heterogeneity in both molecular mass and the number of proteins that react with the *Alcaligenes* antibody (de Souza and Yoch, unpubl.). The different induction characteristics (Fig. 1) and kinetic parameters (Table 1) in the two organisms, and the presence of an isozyme in *Alcaligenes* (Fig. 2) indicate phenotypic differences in these two organisms that relate to DMS evolving characteristics.

	1	2	3	4	5	6	7	8	9	10	11	12	13	14	15	16	17	18	19	20
Alcaligenes	NH_3-Ala-	Gln-	Phe-	Gln-	His-	Gln-	Asp-	Asp-	Val-	Lys-	Pro-	Ala-	Ala-	Ile-	Ser-	Ala-	Glu-	Glu-	Gly-	Lys-
	|	|	|	|		|	|	|	|	|	|	|		|		|			|	|
Pseudomonas doudoroffii	NH_3-Ala-	Gln-	Phe-	Gln-	Ser-	Gln-	Asp-	Asp-	Val-	Lys-	Pro-	Ala-	Ser-	Ile-	Asp-	Ala-	Trp-	Ser-	Gly-	Lys-

	21	22	23	24	25	26	27	28	29	30	31	32	33	34	35	36	37	38
Alcaligenes	Gly-	Lys-	Leu-	Val-	Asp-	Glu-	Gln-	Phe-	Gln-	Glu-	Ala-	Gln-	Lys-	Asn-	Asn-	Glu-	Ala-	Leu-

75% homology between first 20 residues
90% homology between first 10 residues

Figure 6. Comparison of N-terminal sequences of DMSP lyase from *Alcaligenes* strain M3A and *P. doudorofii*. The residue numbers are depicted above the three letter amino acid symbol.

By developing an oligonucleotide probe (or probes), based on the conserved region of the N-terminal amino acid sequences of DMSP lyase from these and other bacterial and algal isolates, we should be able to clone the gene for DMSP lyase. Once the latter is accomplished, the gene probe can be used to examine the size, location and role of bacterial populations involved in DMS evolution from marine environments.

The two isozymes in *Alcaligenes* had similar pH optima, molecular masses, and K_m for DMSP. The isozyme DL-2 shows < 10% of the activity of DL-1; either DL-2 is produced in much lower amount or its specific activity (which has not yet been determined) is much lower. The molecular basis for the differences in these isozymes is not known except that they elute differently on a hydrophobic chromatography column.

Inhibition by MMPA of DMSP lyase activity from both organisms could have regulatory significance because MMPA is a breakdown product of the demethylation pathway of DMSP degradation (19,20). It is possible that bacteria that demethylate DMSP *in situ*, to produce MMPA, compete for endogenous DMSP by inhibiting DMSP uptake and cleavage in bacteria which contain DMSP lyase.

Cyanide and azide are well known inhibitors of electron transport and PCMB is a cell-impermeant modifier of sulfydryl groups. These compounds (all at 1 mM) strongly inhibited the *P. doudoroffii* DMSP lyase, but not the *Alcaligenes* enzyme, *in vivo* (8). However, these enzyme inhibitors did not affect DMSP lyase activity *in vitro* in either organism suggesting that *P. doudoroffii* probably actively transports DMSP into the cell, and may have a DMSP binding protein on its surface. A DMSP binding protein on the surface of *P. doudoroffii*, is consistent with the much higher apparent affinity (lower K_m) of its *in vivo* DMSP lyase activity for DMSP, compared to that seen in *Alcaligenes* (Table 1). In *Alcaligenes*, however, the DMSP lyase is probably periplasmic and a DMSP-binding protein may therefore not be required. A model which summarizes these results and speculation about a putative binding protein in *P. doudoroffii* is shown in Figure 7. A similar model has been proposed for the transport of DMSP and its subsequent intracellular cleavage in strain LFR (23).

The physiological and biochemical data presented here may be used to make ecological inferences about these two organisms. DMSP lyase from *Alcaligenes* either in its pure form or in cell cultures has similar K_m values for DMSP of 1 to 2 mM (Table 1). One might surmise that *Alcaligenes* strain M3A is found in phytoplankton blooms, *Spartina alterniflora* leaves or roots, or associated with macroalgae like *Ulva lactuca* where high concentrations of DMSP would be expected to be released during their senescence and decay process. *Alcaligenes* can also use acrylate to induce its enzyme and grow on this molecule, unlike *P. doudoroffii* (8). As maximum rates of DMSP lyase induction by acrylate also require

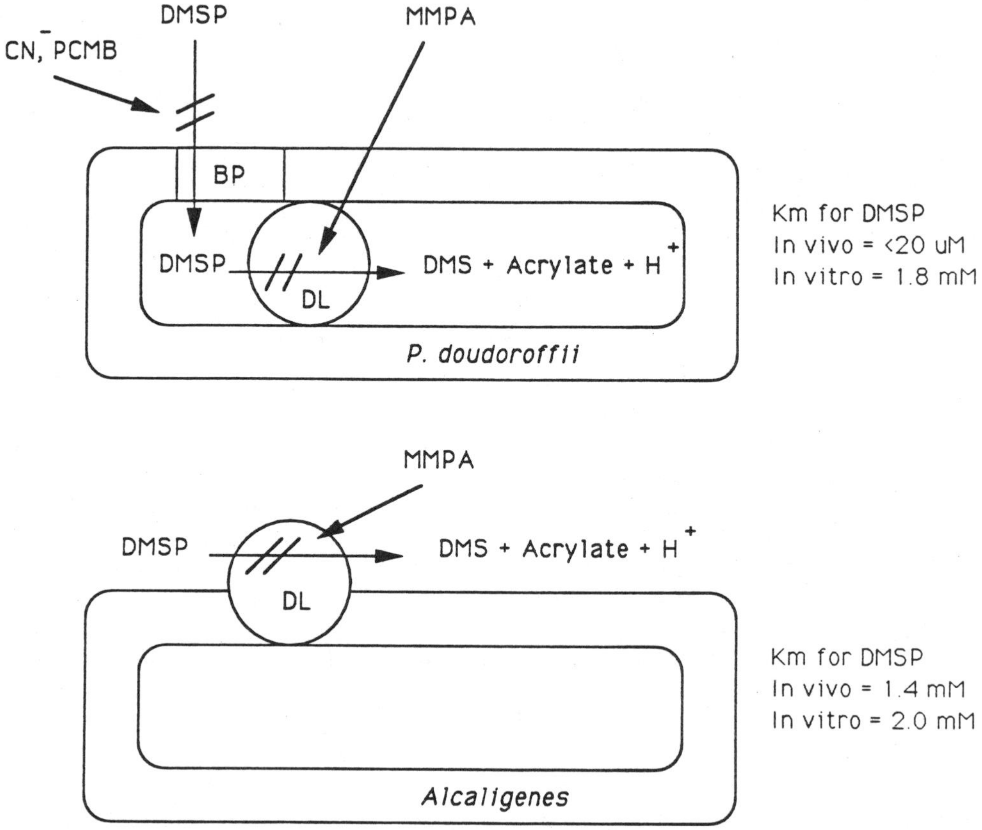

Figure 7. A model summarizing the results of this work and that in ref. 8 with regard to DMS evolution via DMSP lyase in *Alcaligenes* and *P. doudoroffii*. DL = DMSP lyase, BP = putative binding protein, MMPA = methyl-3-mercaptopropionate, PCMB = p-chloromercuribenzoate.

millimolar concentrations (8), one can envision that this molecule might arise from the activity of other DMSP lyase producers whose affinity for DMSP may be much higher than its own. *P. doudoroffii*, strain LFR (23), or phytoplankton with DMSP lyases (28) would be examples of such organisms. *P. doudoroffii* has an upper limit of 20 μM as its *in vivo* K_m for DMSP, probably making it a part of the water column flora. It can presumably metabolize endogenous sea water concentrations of DMSP [2-200 nM (4,12,31,32)] which it may concentrate by the PCMB (and CN^-)-sensitive component on its surface before cleavage by the cytosolic DMSP lyase.

REFERENCES

1. Andreae, M.O. 1985. The emission of sulfur to the remote atmosphere, p. 5-25. *In:* J.N. Galoway, R.J. Charlson, M.O. Andreae, and H. Rodhe (eds.), The biogeochemical cycling of sulfur and nitrogen in the remote atmosphere. D. Reidel Publishing Co., Boston.
2. Bates, T.S., and J.D. Cline. 1985. The role of the ocean in a regional sulfur cycle. J. Geophys. Res. *90*:9168-9172.
3. Baumann, L, P. Baumann, M. Mandel, and R.D. Allen. 1972. Taxonomy of aerobic marine eubacteria. J. Bacteriol. *110*:402-429.

4. Belviso, S., S.K. Kim, B. Krajka, B.C. Nguyen, N. Mihalopoulos, and P. Buat-Ménard. 1990. Production of dimethylsulfoniumpropionate (DMSP) and dimethylsulfide (DMS) by a microbial food web. Limnol. Oceanogr. *35*:1810-1821.
5. Cantoni, G.L., and D.G. Anderson. 1956. Enzymatic cleavage of dimethylpropiothetin by *Polysiphonia lanosa*. J. Biol. Chem. *222*:171-177
6. Challenger, F., and Simpson, M.I. 1948. Studies on biological methylation. Part XII. A precursor of the dimethyl sulfide evolved by *Polysiphonia fastigata*. Dimethyl-2-carboxyethylsulfonium hydroxide and its salts. J. Chem Soc. *3*:1591-1597.
7. de Souza, M.P., and D.C. Yoch. 1995. Purification and characterization of dimethylsulfoniopropionate lyase from an *Alcaligenes*-like dimethyl sulfide-producing marine isolate. Appl. Environ. Microbiol. *61*:21-26.
8. de Souza, M.P., and D.C.Yoch. 1995. Comparative physiology of dimethyl sulfide production by DMSP lyase in two marine strains *Pseudomonas doudoroffii* and *Alcaligenes* M3A. Appl. Environ. Microbiol. *61*: 3986-3911
9. Diaz, M.R., and B.F. Taylor. 1994. Comparison of dimethylsulfoniopropionate (DMSP) lyase activity in a prokaryote and a eukaryote, abstr. N-18, p. 319. *In:* Abstracts of the 94th General Meeting of the American Society for Microbiology 1994. American Society for Microbiology, Washington, D.C.
10. Diaz, M.R., P.T. Visscher, and B.F. Taylor. 1992. Metabolism of dimethylsulfoniopropionate and glycine betaine by a marine bacterium. FEMS Microbiol. Lett. *96*:61-66
11. Dickson, D.M., R.G. WynJones, and J.Davenport. 1980. Steady state osmotic adaptation in *Ulva lactuca*. Planta *150*:158-165.
12. Iverson, R.L., F.L. Nearhoof, and M.O. Andreae. 1989. Production of DMSP and DMS by phytoplankton in estuarine and coastal waters. Limnol. Oceanogr. *34*:53-67.
13. Kadota, H., and Y. Ishida. 1968. Effect of salts on enzymatical production of dimethyl sulfide from *Gyrodinium cohnii*. Bull. Jpn. Soc. Sci. Fish. *34*:512-518.
14. Karsten, U., C. Wiencke, and G.O. Kirst. 1990. The ß-dimethylsulfoniopropionate (DMSP) content of macroalgae from Antarctica and southern Chile. Bot. Mar. *33*:143-146.
15. Kiene, R.P. 1990. DMS production from DMSP in coastal seawater samples and bacterial cultures. Appl. Environ. Microbiol. *56*:3292-3297.
16. Kiene, R.P. 1992. Dynamics of DMS and DMSP in oceanic water samples. Mar. Chem. *37*:29-52.
17. Kiene, R.P. and D.G. Capone. 1988. Microbial transformations of methylated sulfur compounds in anoxic salt sediments. Microb. Ecol. *15*:275-291.
18. Kiene, R.P., and S.K. Service. 1991. Decomposition of dissolved DMSP and DMS in estuarine waters: Dependence on temperature and substrate concentration. Mar. Ecol. Prog. Ser. *76*:1-11.
19. Kiene, R.P., and B.F. Taylor. 1988. Demethylation of dimethylsulfoniopropionate and production of thiols in anoxic marine sediments. Appl. Environ. Microbiol. *54*:2208-2212.
20. Kiene, R.P., and B.F. Taylor. 1989. Metabolism of acrylate and 3-mercaptopropionate, p. 222-230. *In:* E.S. Saltzman and W.J. Cooper (eds.), Biogenic Sulfur in the Environment. Am. Chem. Soc. Symp. Series No. 393. Am. Chem Soc., Washington D.C.
21. Kiene, R.P., and P.T. Visscher. 1987. Production and fate of methylated sulfur compounds from methionine and dimethylsulfoniopropionate in anoxic salt marsh sediments. Appl. Environ. Microbiol. *53*:2426-2434.
22. Ledyard, K.M., E.F. DeLong and J.W.H. Dacey. Characterization of a DMSP-degrading bacterial isolate from the Sargasso Sea. Arch. Microbiol. *160*:312-318.
23. Ledyard, K.M., and J.W.H. Dacey. 1994. Dimethylsulfide production from dimethylsulfoniopropionate by a marine bacterium. Mar. Ecol. Prog. Ser. *110*:95-103.
24. Legrand, M.R., R.J. Delmas, and R.J. Charlson. 1988. Climate forcing implications from Vostok ice-core sulphate data. Nature *334*:418-420.
25. Nishiguchi, M.K. 1994. Identification and detection of dimethylpropiothetin dethiomethylase (4.4.1.3) from macro and micro algae using biomolecular techniques, abstr. 022L-13, p. 98. *In* Abstracts of the ASLO Ocean Science Meeting 1994. Eos, Transactions, American Geophysical Union, Washington, D.C.
26. Reed, R.H. 1983. Measurement and osmotic significance of ß-dimethylsulfoniopropionate in marine macroalgae. Mar. Biol. Lett. *4*:173-181.
27. Sambrook, J., E.F Fritsch, and T. Maniatis. 1989. Molecular cloning: a laboratory manual, 2nd ed. Cold Spring Harbor Laboratory, Cold Spring Harbor, NY.
28. Stefels, J., and W.H.M. van Boekel. 1993. Production of DMS from dissolved DMSP in axenic cultures of the marine phytoplankton species *Phaeocystis* sp. Mar. Ecol. Prog. Ser. *97*:11-18.
29. Taylor, B.F. and D.C. Gilchrist. 1991. New routes for aerobic biodegradation of dimethylsulfoniopropionate. Appl. Environ. Microbiol. *57*:3581-3584

30. Visscher, P.T., and H. van Gemerden. 1991. Production and consumption of dimethylsulfoniopropionate in microbial mats. Appl. Environ. Microbiol. *57*:3237-3242.
31. Wakeham, S.G., and J.W.H. Dacey. 1989. Biogeochemical cycling of dimethyl sulfide in marine environments, p. 152-166. *In:* E.S. Saltzman, and W.J. Cooper (eds.), Biogenic Sulfur in the Environment, Am. Chem. Soc. Symp. Series No. 393. Am. Chem. Soc., Washington D.C.
32. Wakeham, S.G., B.L. Howes, J.W.H. Dacey, R.P. Schwarzenbach and J. Zeyer. 1987. Biogeochemistry of dimethylsulfide in a seasonally stratified coastal salt pond. Geochim. Cosmochim. Acta *51*:1675-1684.
33. Wagner, C., and E.R. Stadtman. 1962. Bacterial fermentation of dimethyl-ß-propiothetin. Arch. Biochem. Biophys. *98*:331-336.
34. Wigley, T.M.L. 1989. Possible climate change due to SO_2 derived cloud condensation nuclei. Nature *339*:365-367
35. Wolfe, G.V., E.B. Sherr, and B.F. Sherr. 1994. Release and consumption of DMSP from *Emiliana huxleyi* during grazing by *Oxyrrhis marina*. Mar. Ecol. Prog. Ser. *111*:111-119.

26

INTRIGUING FUNCTIONALITY OF THE PRODUCTION AND CONVERSION OF DMSP IN *PHAEOCYSTIS* SP

Jacqueline Stefels,[1] Winfried W. C. Gieskes,[1] and Lubbert Dijkhuizen[2]

[1] Department of Marine Biology
[2] Department of Microbiology
University of Groningen
PO Box 14, 9750 AA Haren, The Netherlands

SUMMARY

In many areas of the ocean the distribution and conversion of dimethylsulfoniopropionate (DMSP) may to a high degree be influenced by the activity of the Prymnesiophyte alga *Phaeocystis* sp.: it not only produces DMSP in large amounts, but is also able to convert it enzymatically into dimethylsulfide (DMS) and acrylate. Characteristic properties of DMSP-lyase in *Phaeocystis* sp. indicate that the enzyme is different from lyase enzymes in other organisms studied. During a spring bloom in Dutch coastal waters, DMSP-lyase activity was strongly correlated with *Phaeocystis* sp. cell numbers and potentially capable of producing DMS in excess of abiotic loss factors. In *Phaeocystis* sp. cells, DMSP makes an important contribution to the intracellular osmotic potential, with concentrations of approximately 150 mM. Upon salinity shocks, however, short term regulation of its internal levels was not observed. Although a slow adaptation of the DMSP production in *Phaeocystis* sp. cells may affect intracellular concentrations on a long term, it is concluded that DMSP-lyase is not involved in the short term osmotic adaptation of the cell. DMSP is a structural component of the cell, being produced continuously in the light as well as in the dark. DMSP-lyase activity facilitates the release of DMSP from the cell, with some intriguing beneficial effects.

INTRODUCTION

With respect to the sources and sinks of dimethylsulfoniopropionate (DMSP) in the oceans, the Prymnesiophyte alga *Phaeocystis* sp. is an interesting species for many reasons. This colony-forming microalga is well known for its high intracellular-DMSP concentrations of approximately 150 mM (22, 36). It is one of the few microalgae that benefits strongly from eutrophication (25, 30), resulting in almost uni-algal plankton blooms. Dense spring

Biological and Environmental Chemistry of DMSP and Related Sulfonium Compounds
edited by Ronald P. Kiene et al., Plenum Press, New York, 1996

blooms of *Phaeocystis* sp. have been reported from polar (2, 7, 14, 33, 42), temperate (3, 8, 21) and subtropical areas (1, 13, 17), illustrating its importance in global DMSP production.

Grazing by macro- and micro-zooplankton has been found to be a key process in the release and conversion of DMSP from marine phytoplankton cells other than *Phaeocystis* (6, 26, 43), although the loss of algal DMSP could not always be recovered in the form of dimethylsulfide (DMS). *Phaeocystis* sp. colonies are, in many cases, of poor nutritional value for grazers (18). In culture experiments-even an increase in *Phaeocystis* sp. biomass was observed due to reduced grazing pressure when microzooplankton was offered as an alternative food source to the main grazer (18). Low grazing pressure implies that grazers will not act as a DMSP-sink and that more DMSP is available as substrate for algal and bacterial enzymes.

Algal DMSP that is released in the water due to cell lysis or grazing may be converted into DMS and acrylate by DMSP-lyase. Demethylation of DMSP is another important pathway (24). After cleavage, DMS is to a large extent subject to bacterial consumption (23). This pathway is now thought to be of major importance as a sink for DMS. An intriguing aspect of *Phaeocystis* sp. blooms is that this organism itself is able to convert DMSP extracellularly into DMS and acrylate (36) and that large numbers of bacteria are only found during late stages of these blooms (37, 39, 41). Whether or not the number of DMSP-degrading bacteria follow the same trend as the bloom-associated bacteria in general remains to be investigated.

The relative importance of the different metabolic pathways for DMSP degradation during the course of a bloom is dependent on the development of the bacterio-, zoo- and phytoplankton communities; this will ultimately determine the fraction of the DMSP-sulfur pool which becomes available for exchange with the atmosphere. Mesocosm experiments indicate that only a fraction of the DMSP-sulfur derived from a *Phaeocystis* sp. bloom may reach the atmosphere (29). Community structure variations within *Phaeocystis* sp. blooms exhibit a potential for substantial shifts in DMS emissions to the atmosphere. When bacterial

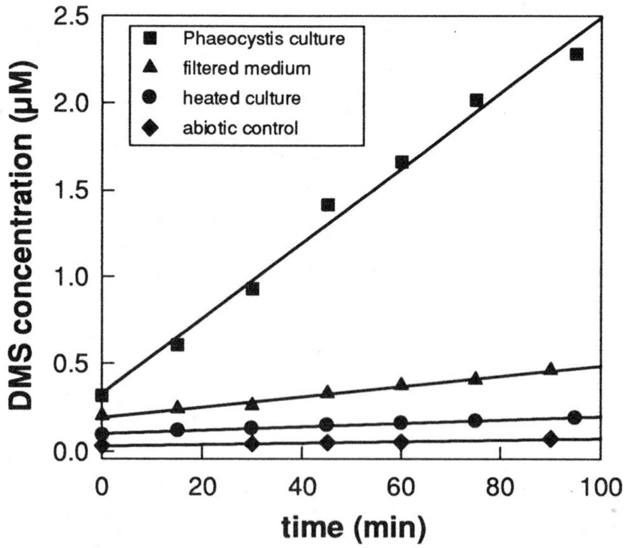

Figure 1. DMS concentration in the medium as a function of time after addition of 10 µM DMSP to a batch culture sample from the end-exponential growth phase (3×10^8 cells l^{-1}), the filtered medium, a heated culture and fresh medium (abiotic control) (36).

activity and grazing pressure are low, *Phaeocystis* sp. blooms may result in a major conversion of DMSP to DMS, as it produces both DMSP-lyase and its substrate DMSP. In order to develop a better insight in the role of *Phaeocystis* sp. in the marine sulfur cycle we present here a review of work done in our laboratory on DMSP-lyase of this organism including some additional new data, and discuss its potential activity in the field and its function in DMSP-metabolism.

DMSP-LYASE IN *PHAEOCYSTIS* SP

The ability of *Phaeocystis* sp. to cleave extracellular, dissolved DMSP into DMS and acrylate was first observed in experiments with axenic cultures of this organism (36). After heating of the cultures for 45 min at 60°C, the cleavage reaction could not be restored, indicating that this conversion was an enzymatic process (Fig. 1), and that DMSP-lyase activity was associated with the cells. In further studies, an enzyme assay was developed to measure DMSP-lyase activity in extracts of natural plankton assemblages. This assay was used during the spring bloom of 1993 off the Dutch coast in order to test the phytoplankton samples for their DMSP-lyase activity in relation to species composition (35). The survey was done at an early stage of the bloom when there was a shift along the coast from a diatom-dominated to a *Phaeocystis*-dominated phytoplankton population. A highly significant correlation was observed between *Phaeocystis* sp. abundance and DMSP-lyase activity, whereas no correlation existed with any of the other species, total diatom numbers, total diatom biomass and total protein content (Fig. 2). Using conservative estimations of the DMS-production by *Phaeocystis* sp., potential DMS-production rates during the spring bloom could be calculated. These estimations were based on the *in vivo* DMS-production rate measured in *Phaeocystis* sp. cultures at the *in situ* sea water temperature. Using a mixed water column depth of 5 m, the calculated DMS-production rate in nearshore waters ranged from 47 to 131 $\mu mol\ m^{-2}\ d^{-1}$ with *Phaeocystis* sp. cell numbers ranging between 8 to 26 × $10^6\ l^{-1}$. Compared with abiotic loss factors for DMS, such as exchange with the atmosphere and photochemical oxidation, calculated for the same area and period, these production rates are 1.5 to 4.5 times higher (35). Bacterial DMS consumption could constitute an additional sink. The strong correlation of the DMSP-lyase activity with *Phaeocystis* sp. abundance suggests that during the early stages of a bloom *Phaeocystis* sp. is a dominant, if not the main, DMS producer.

LOCATION OF DMSP-LYASE IN *PHAEOCYSTIS* SP

A first approach to elucidate the function of DMSP-lyase is to locate the enzyme in relation to the cell. Although unequivocal evidence for the location can only be obtained in labelling studies, a first indication is obtained by comparing the characteristic properties of the enzyme in whole cells and in cell free extracts. An important parameter in this respect is the pH in the reaction mixture. Intracellular pH is expected to be influenced only slightly by external pH when using non-permeable buffers. A comparable reactivity of DMSP-lyase with pH in whole cells and extracts is therefore a strong argument for an extracellular location of the enzyme. Indeed it was found for axenic *Phaeocystis* sp. cultures that whole cells and extracts displayed comparable pH profiles (34). Both preparations also displayed similar activity-salinity profiles. Another strong indication for the extracellular location of the enzyme was obtained from inhibition experiments with whole cells of *Phaeocystis* sp. using *p*-chloromercuribenzenesulfonic acid (*p*CMBS), a nonpermeable thiol-reagent which acts only on the outside of the cell (40). At 2 mM, *p*CMBS caused a 94% inhibition of lyase

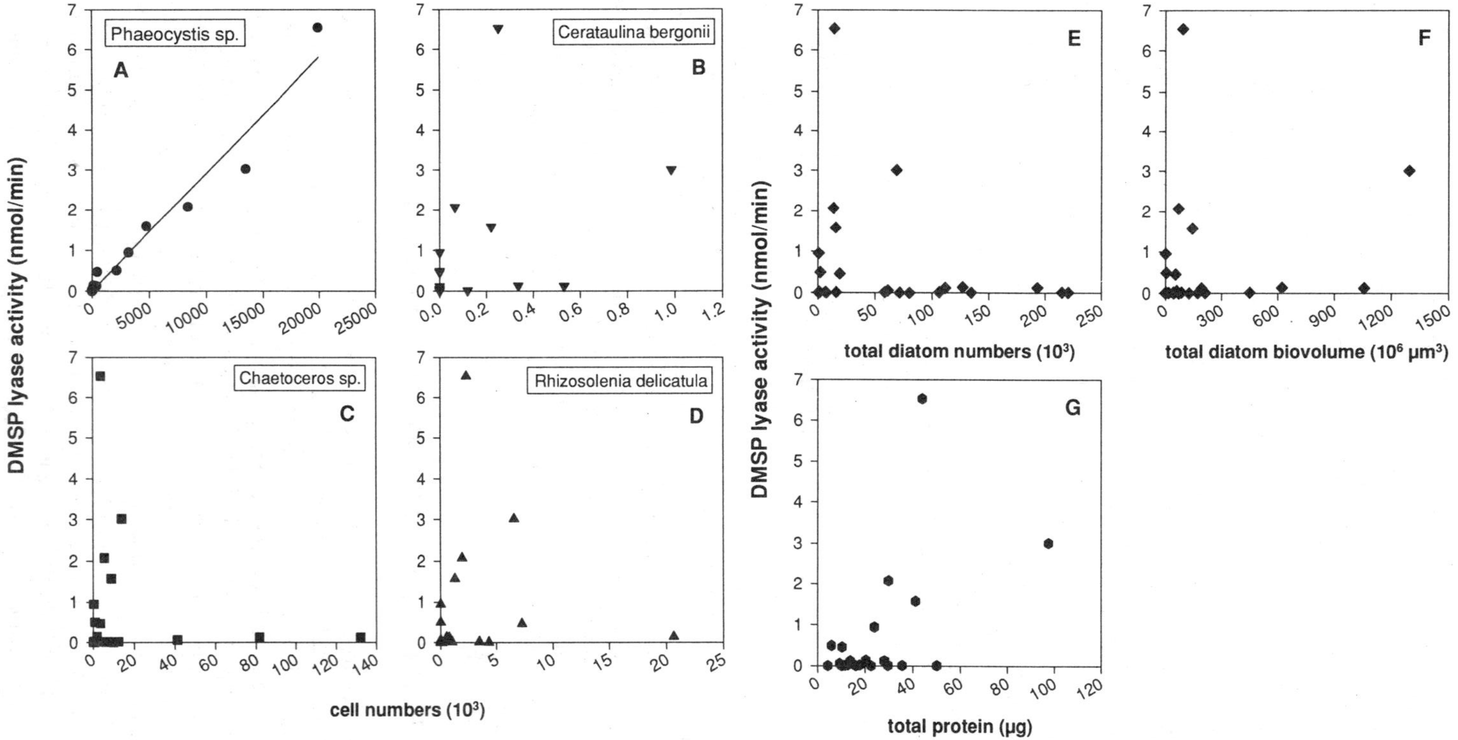

Figure 2. DMSP-lyase activity in 1 ml crude extracs prepared from concentrated particulate material from natural samples versus several parameters measured in these samples: numbers of *Phaeocystis* sp. (A), *Cerataulina bergonii* (B), *Chaetoceros* sp. (C), *Rhizosolenia delicatula* (D), total diatom numbers (E), total diatom biovolume (F) and total protein content (G). A substrate concentrtion of 250 μM was added to the extract for the assay. The equation of the regression line for *Phaeocystis* sp. is $y = 0.2905 \times 10^{-6} x + 0.0105$, $r^2 = 0.9660$, n = 23. All other graphs showed non-significant relationships (35).

Table 1. Comparison of DMSP-lyase properties in several organisms. (–) indicates decrease in specific activity with salinity, (+) indicates increase in specific activity with salinity

Organism	pH optimum	Salinity response	References
Polysiphonia lanosa	5.1		(4)
Enteromorpha clathrata	6.2-6.4	+	Steinke et al. (this issue)
Gyrodinium cohnii	6-6.5	++	(20)
Alcaligenes sp.	8		(9)
Phaeocystis sp.	>10	–	(34)

activity (34). The membrane-bound nature of DMSP-lyase in *Phaeocystis* sp. was confirmed in cell fractionation experiments in which 50 to 80 % of total activity was present in the membrane fraction (34). Association with an intracellular membrane appeared unlikely considering the already high intracellular DMSP concentrations (ca. 150 mM) and the observed instantaneous conversion of extracellular DMSP by whole cells, even in the low micromolar ranges (Fig. 1). For the same reason an internal location of the lyase on the plasmalemma was considered unlikely. Such a location - within the same compartment as its substrate DMSP - would necessitate a strict control of enzyme activity, thereby prohibiting the possibility of converting such low amounts of external DMSP as was observed.

Comparison of some characteristics of DMSP-lyases in various organisms reveals that the DMSP-lyase in *Phaeocystis* sp. displays rather unique properties (Table 1). The inhibition of DMSP-lyase by thiol reagents and its high pH-optimum indicate that cysteine residues are involved in catalysis at the active site (34). This hypothesis is supported when considering the cleavage reaction of DMSP into DMS and acrylate. The reaction will be catalysed when a proton of the second carbon atom of DMSP is bound to a proton acceptor. Under alkaline conditions the deprotonated thiol group of a cysteine residue can act as proton acceptor, thus resulting in an alkaline pH optimum of the cleavage reaction. The alkaline optimum of the extracellular DMSP-lyase in *Phaeocystis* sp. is physiologically relevant in view of the relatively high pH of sea water (pH 8.2) and in bloom situations (pH 9) (19).

FUNCTIONS OF DMSP AND DMSP-LYASE IN *PHAEOCYSTIS* SP

In most research on the physiological function of DMSP much attention has been paid to a possible role in algal osmotic adjustment (10-12, 38; Kirst, this issue). Compatibility of DMSP with algal metabolism was indeed confirmed when studying its effects on enzyme activities in extracts of *Tetraselmis subcordiformis* (16). In this organism, mannitol is the most important organic solute, the concentration of which is readily regulated upon salinity shocks. Mannitol increases in concentration, due to instantaneous production, after a transfer from low to high salinity, and decreases in concentration - probably due to excretion - upon transfer from high to low salinity (10). Other compounds, among others DMSP, displayed a lag-phase of several hours upon salinity upshocks, but were readily released from the cells when they experienced downshocks (10).

In the case of *Phaeocystis* sp. it is obvious that the high intracellular concentration of DMSP contributes to the osmotic potential of the cytoplasm. This does not necessarily imply an active regulation of the cellular DMSP-content upon salinity changes. In general, aside from a possible effect on cell volume which results in a change in concentrations of all solutes, salinity changes may affect intracellular DMSP

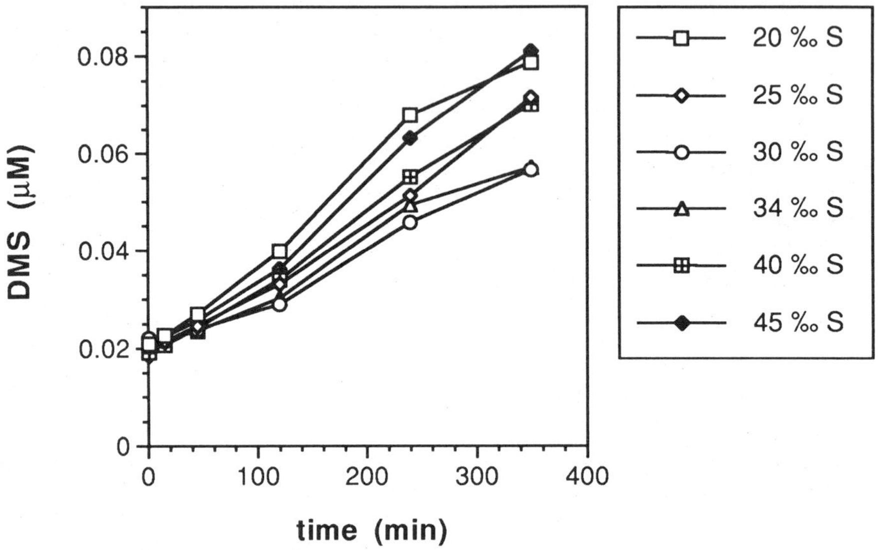

Figure 3. DMS production in salinity shocked axenic *Phaeocystis* sp. cultures over a 6 h period. Results of one experiment are shown; previous experiments gave comparable results. Salinity of the original culture was 34 PSU. At t = 0 h, subsamples were mixed with medium from different salinities to obtain the indicated final values. Data were obtained from headspace analyses of the cultures using a gas chromatograph; cultures were regularly checked for bacterial contamination using Hoechst dye no. 33258 with epifluorescence microscopy.

levels through two different mechanisms. (1) Salinity changes may affect the production of DMSP. Upon a downshock production ceases and with subsequent growth DMSP-concentration decreases by dilution. An upshock may stimulate production. (2) Salinity changes may affect the removal of DMSP through regulated excretion or through a combination of excretion with induced changes in DMSP-lyase activity. Thus, a downshock may stimulate excretion and possibly DMSP-lyase activity. An increased DMSP-lyase activity will facilitate DMSP transport across the membrane by removing DMSP from the outside, thereby preserving a maximal DMSP gradient. An upshock may inhibit excretion and/or DMSP-lyase activity.

To investigate whether *Phaeocystis* sp. displays such mechanisms, shock experiments were performed with axenic *Phaeocystis* sp. cultures in which both the DMSP-lyase activity and the DMS and DMSP-production were measured. A culture grown at 34 ppt salinity (PSU) was subdivided in equal portions and subjected to a salinity range from 20 to 45 PSU final concentration, by dilution with sea water of different salinities. Further extension of the range was not possible, because growth of *Phaeocystis* sp. batch cultures ceased outside this range (Stefels unpubl.). No dissolved DMSP was formed during previous shock experiments within this range, therefore, only DMS production during a 6 h period is shown (Fig. 3). Cultures that had experienced no or only slight salinity changes (the 30 and 34 PSU cultures) displayed the smallest increase in DMS-production, measuring 1.7% of total particulate DMSP after the 6 h period. Cultures subject to large up- or down- shocks (the 20 and 45 PSU cultures) released 2.5% of particulate DMSP as DMS. Apparently, DMS-production had no relation with the direction of change, and may have been caused by the lysis of a small percentage of cells that did not survive either the up- or downshocks. The

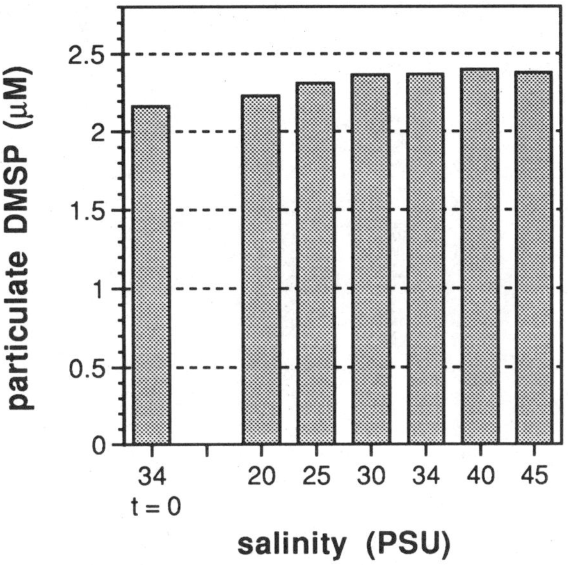

Figure 4. Particulate-DMSP concentrations (in μmol/l culture) in salinity shocked *Phaeocystis* sp. cultures after an incubation period of 6 h. Results of one experiment are shown; previous experiments gave comparable results. The concentration of particulate DMSP in the original culture at t = 0 h is given. Particulate DMSP was measured as DMS after addition of NaOH to the cultures (final concentration: 1M) using headspace analyses after a 12 hour incubation period.

results indicated that upon downshocks intracellular DMSP-concentration was not regulated by active excretion and conversion into DMS.

After the 6 h period, the cells were harvested and analyzed for their DMSP content (Fig. 4). The increase in particulate DMSP-concentration in the culture, when comparing the original culture with the control culture (34 PSU), indicated that growth had occurred during the 6 h period. Only a slight inhibition of DMSP production was observed in the 20 and 25 PSU cultures, but no difference was observed between all other cultures. Also, the effect of salinity changes on DMSP-lyase activity was minimal, resulting in only slight differences between the cultures: a gradual 30% loss of specific activity occurred going from 20 to 45 PSU (34). Although concentrations of intracellular DMSP in the shocked cultures may have differed more distinctly due to changes in cell volume, the results did not indicate active regulation of the DMSP content per cell in short term salinity changes. DMSP content will change in the long term due to gradual shifts in production rates (Stefels unpubl.). The degradation of DMSP through DMSP-lyase, however, can not be regarded as an active mechanism for osmotic regulation.

Carbon:sulfur (C:S) ratios of whole cells may vary considerably over the diel cycle (5). In *Dunaliella tertiolecta*, a non-DMSP producing species, this ratio varied between 300 and 580 on a molar basis. Using this ratio for whole cells of *Phaeocystis* sp. without the contribution of DMSP, the percentage of DMSP-sulfur of total sulfur can be calculated as 71 to 83%. Cysteine and methionine are precursors of DMSP biosynthesis in plants (Hanson and Gage, this volume). The concentrations of these amino acids in the cell are usually strictly regulated (15). Nevertheless DMSP concentrations in *Phaeocystis* sp. are high, comparable to glutathione in higher plants which is proposed to function as a storage and transport form for sulfur (15). A 24 h survey of DMSP concentration and cell numbers in a *Phaeocystis* sp. batch culture revealed a steady increase of the intracellular-DMSP concentration in time, both during the light and the dark period (Fig. 5). Cell division occurred during the dark period. This continuous production of DMSP is comparable to the synthesis of proteins: in general, production of proteins during the light period extends in the dark on the expense of carbohydrates, whereby sulfate can be assimilated from the medium during the dark cycle, if needed (5, 28). This observation is important for the estimation of biological DMS and

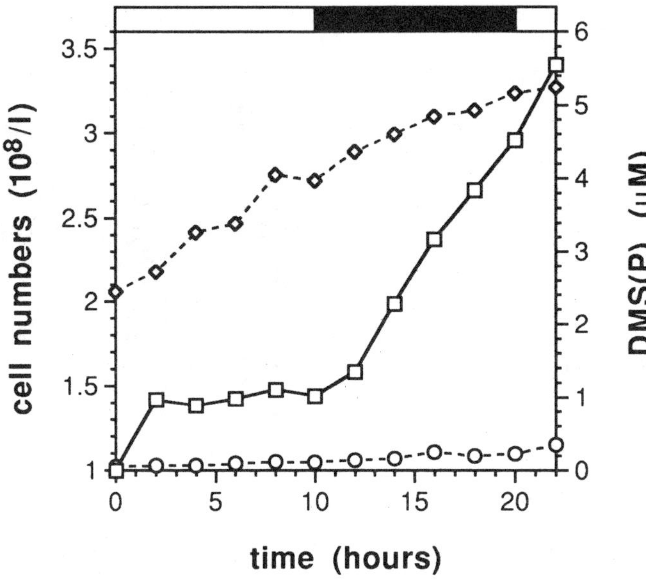

Figure 5. Cell numbers (❑), total DMSP + DMS (◇) and dissolved DMSP + DMS (❍) in a *Phaeocystis* sp. batch culture during a 24 h period. Light and dark periods are indicated.

DMSP conversion rates in natural samples (23, 24). In many incubation experiments no dark-production of DMSP has been reported to occur. Our 24 h experiment clearly demonstrated that this is not valid for habitats were *Phaeocystis* sp. is present.

CONCLUSIONS

Kirst (this issue) showed that DMSP has all the necessary characteristics, like glycine betaine, to act as a compatible solute in cellular osmotic regulation. Its high concentration in *Phaeocystis* sp. cells suggests that it is of vital importance to the cell's integrity. The significant role of sulfate assimilation and incorporation in *Phaeocystis* sp. may have its evolutionary origin in the relatively high concentrations of sulfate in the marine environment. In contrast, nitrogen is frequently limiting growth in these habitats, and the ability to produce organic sulfur containing solutes, which can compensate for nitrogen containing solutes, may be of competitive advantage. Indeed, there appears to be a general trend of DMSP-containing species (like Prymnesiophytes and dinoflagellates) succeeding non-DMSP containing species (like diatoms) during the course of blooms. Thus, algae may benefit by retaining high intracellular DMSP concentrations, even though DMSP is not involved in short-term osmotic adaptations.

Evidence from our research suggests an extracellular location for DMSP-lyase in *Phaeocystis* sp., converting dissolved DMSP into DMS and acrylate. Assuming that the cleavage of DMSP is not a secondary effect of an unknown enzyme, the question remains: what the role is of DMSP-lyase? We hypothesize that intracellular concentrations are regulated by release of DMSP rather than by its production. Removing DMSP extracellularly by cleavage into DMS and acrylate will facilitate this release as DMSP concentration gradients across the membrane are maximized. This is especially relevant for *Phaeocystis* sp., as the mucus layer, surrounding colonial cells, will increase the diffusive boundary layer around these cells. The robustness of this boundary layer may allow the formation of microzones around the cells in which elevated levels of excretion products are observed (27), which, in the case of DMSP, would be unfavourable. In addition to the maintenance of a DMSP gradient, the cell may simultaneously profit from the release of acrylate and protons upon cleavage. The released protons can be used for nutrient uptake. A possible build up of acrylate within the microzone may result in concentrations and conditions with antibacterial activity, as opposed to the more dilute conditions in seawater, were acrylate may serve as a substrate for bacteria (31, 32). Our hypothesis of regulation of intracellular DMSP by means of an extracellular DMSP-lyase is supported by the observation of a daily DMS production measuring a few percent of total particulate DMSP in healthy growing cultures, whereas dissolved DMSP was hardly detected (36; see also Fig. 3). The location of the DMSP-lyase in *Phaeocystis* sp. suggests an association with a transport protein in the plasma membrane. Further research into the physiological function of DMSP-lyase should be undertaken.

ACKNOWLEDGMENTS

The authors thank Marc Staal for his help with the 24 h experiment, and Wim van Boekel and Maria van Leeuwe for critically reading the manuscript. This work was financially supported by the Commission of the European Community, Environment Programme, contract no. EV5V - CT93 - 0326.

REFERENCES

1. Al-Hasan, R. H., A. M. Ali and S. S. Radwan. 1990. Lipids, and their constituent fatty acids, of *Phaeocystis* sp. from the Arabian Gulf. Mar. Biol. *105:*9-14.
2. Barnard, W. R., M. O. Andreae and R. L. Iverson. 1984. Dimethylsulfide and *Phaeocystis poucheti* in the southeastern Bering sea. Cont. Shelf Res. *3:*103-113.
3. Cadée, G. C. and J. Hegeman. 1986. Seasonal and annual variation in *Phaeocystis pouchetii* (Haptophyceae) in the westernmost inlet of the Wadden Sea during the 1973 to 1985 period. Neth. J. Sea Res. *20:*29-36.
4. Cantoni, G. L. and D. G. Anderson. 1956. Enzymatic cleavage of dimethylpropiothetin by *Polysiphonia lanosa*. J. Biol. Chem. *222:*171-177.
5. Cuhel, R. L., P. B. Ortner and D. R. S. Lean. 1984. Night synthesis of protein by algae. Limnol. Oceanogr. *29:*731-744.
6. Dacey, J. W. H. and S. G. Wakeham. 1986. Oceanic dimethylsulfide: production during zooplankton grazing. Science. *233:*1314-1316.
7. Davidson, A. T. and H. J. Marchant. 1992. The biology and ecology of *Phaeocystis* (Prymnesiophyceae), p. 1-45. *In* F. E. Round and D. J. Chapman (ed.), Progress in Phycological Research, 8. Biopress Ltd., Bristol, England.
8. Davies, A. G., I. de Madariaga, B. Bautista, E. Fernández, D. S. Harbour, P. Serret and P. R. G. Tranter. 1992. The ecology of a coastal *Phaeocystis* bloom in the north-western English Channel in 1990. J. Mar. Biol. Ass. U.K. *72:*691-708.
9. de Souza, M. P. and D. C. Yoch. 1995. Purification and characterization of dimethylsulfoniopropionate lyase from an *Alcaligenes*-like dimethyl sulfide-producing marine isolate. Appl. Environ. Microbiol. *61:*21-26.
10. Dickson, D. M. J. and G. O. Kirst. 1986. The role of ß-dimethylsulphoniopropionate, glycine betaine and homarine in the osmoacclimation of *Platymonas subcordiformis*. Planta *167:*536-543.
11. Dickson, D. M. J. and G. O. Kirst. 1987a. Osmotic adjustment in marine eukaryotic algae: the role of inorganic ions, quaternary ammonium, tertiary sulphonium and carbohydrate solutes. I. Diatoms and a Rhodophyte. New Phytol. *106:*645-655.
12. Dickson, D. M. J. and G. O. Kirst. 1987b. Osmotic adjustment in marine eukaryotic algae: the role of inorganic ions, quaternary ammonium, tertiary sulphonium and carbohydrate solutes. II. Prasinophytes and Haptophytes. New Phytol. *106:*657-666.

13. Estep, K. W., P. G. Davis, P. E. Hargraves and J. M. Sieburth. 1984. Chloroplast containing microflagellates in natural populations of north Atlantic nanoplankton, their identification and distribution: including a description of five new species of *Chrysochromulina* (Prymnesiophyceae). Protistologica *20:*613-634.
14. Gibson, J. A. E., R. C. Garrick, H. R. Burton and A. R. McTaggart. 1990. Dimethylsulfide and the alga *Phaeocystis pouchetii* in antarctic coastal waters. Mar. Biol. *104:*339-346.
15. Giovanelli, J. 1990. Regulatory aspects of cysteine and methionine biosynthesis, p. 33-48. *In* H. Rennenberg, C. Brunold, L. J. de KoK and I. Stulen (ed.), Sulfur nutrition and sulfur assimilation in higher plants: fundamental environmental and agricultural aspects, SPB Academic Publishing, The Hague, The Netherlands.
16. Gröne, T. and G. O. Kirst. 1991. Aspects of dimethylsulfoniopropionate effects on enzymes isolated from the marine phytoplankter *Tetraselmis subcordiformis* (Stein). J. Plant Physiol. *138:*85-91.
17. Guillard, R. R. L. and J. A. Hellebust. 1971. Growth and the production of extracellular substances by two strains of *Phaeocystis poucheti*. J. Phycol. *7:*330-338.
18. Hansen, F. C., M. Reckermann, W. C. M. Klein Breteler and R. Riegman. 1993. *Phaeocystis* blooming enhanced by copepod predation on protozoa: evidence from incubation experiments. Mar. Ecol. Prog. Ser. *102:*51-57.
19. Hinga, K. R. 1992. Co-occurrence of dinoflagellate blooms and high pH in marine enclosures. Mar. Ecol. Prog. Ser. *86:*181-187.
20. Ishida, Y. 1968. Physiological studies on the evolution of dimethylsulfide. Mem. Coll. Agric. Kyoto Univ. *94:*47-82.
21. Jones, P. G. W. and S. M. Haq. 1963. The distribution of *Phaeocystis* in the eastern Irish Sea. J. Cons. perm. int. Explor. Mer. *28:*9-20.
22. Keller, M. D., W. K. Bellows and R. R. L. Guillard. 1989. Dimethyl sulfide production in marine phytoplankton, p. 167-182. *In* E. S. Saltzman and W. J. Cooper (ed.), Biogenic sulfur in the environment, Symp. Ser. 393. American Chemical Society, Washington DC.
23. Kiene, R. P. and T. S. Bates. 1990. Biological removal of dimethyl sulfide from sea water. Nature *345:*702-705.
24. Kiene, R. P. and S. K. Service. 1991. Decomposition of dissolved DMSP and DMS in estuarine waters: dependence on temperature and substrate concentration. Mar. Ecol. Prog. Ser. *76:*1-11.
25. Lancelot, C., G. Billen, A. Sournia, T. Weisse, F. Colijn, M. J. W. Veldhuis, A. Davies and P. Wassmann. 1987. *Phaeocystis* blooms and nutrient enrichment in the continental coastal zones of the North Sea. Ambio *16:*38-46.
26. Malin, G., P. S. Liss and S. M. Turner. 1994. Dimethyl sulfide: production and atmospheric consequences, p. 303-320. *In* J. C. Green and B. S. C. Leadbeater (ed.), The Haptophyte Algae, Systematics Association, Special Volume no. 51. Clarendon Press, Oxford.
27. Mitchell, J. G., A. Okubo and J. A. Fuhrman. 1985. Microzones surrounding phytoplankton form the basis for a stratified marine microbial ecosystem. Nature *316:*58-59.
28. Morris, I. 1981. Photosynthesis products, physiological state, and phytoplankton growth, p. 83-102. *In* T. Platt (ed.), Physiological bases of phytoplankton ecology. Can. Bull. Fish. Aquat. Sci. *210.*
29. Quist, P., R. L. J. Kwint, T. A. Hansen, L. Dijkhuizen and K. J. M. Kramer. submitted. Turnover of dimethylsulfoniopropionate and dimethylsulfide in the marine environment - The role of bacteria. Mar. Ecol. Prog. Ser.
30. Riegman, R., A. A. M. Noordeloos and G. C. Cadée. 1992. *Phaeocystis* blooms and eutrophication of the continental zones of the North Sea. Mar. Biol. *112:*479-484.
31. Sieburth, J. M. 1960. Acrylic acid, an "antibiotic" principle in *Phaeocystis* blooms in Antarctic waters. Science *132:*676-677.
32. Sieburth, J. M. 1961. Antibiotic properties of acrylic acid, a factor in the gastrointestinal antibiosis of polar marine animals. J. Bacteriol. *82:*72-79.
33. Smith Jr., W. O., L. A. Codispoti, D. M. Nelson, T. Manley, E. J. Buskey, H. J. Niebauer and G. F. Cota. 1991. Importance of *Phaeocystis* blooms in the high-latitude ocean carbon cycle. Nature *352:*514-516.
34. Stefels, J. and L. Dijkhuizen. 1995. Characteristics of DMSP-lyase in *Phaeocystis* sp. (Prymnesiophyceae). Mar. Ecol. Prog. Ser. in press.
35. Stefels, J., L. Dijkhuizen and W. W. C. Gieskes. 1995. DMSP-lyase activity in a spring phytoplankton bloom off the Dutch coast, related to *Phaeocystis* sp. abundance. Mar. Ecol. Prog. Ser. *123:*235-243.
36. Stefels, J. and W. H. M. van Boekel. 1993. Production of DMS from dissolved DMSP in axenic cultures of the marine phytoplankton species *Phaeocystis* sp. Mar. Ecol. Prog. Ser. *97:*11-18.
37. Thingstad, F. and G. Billen. 1994. Microbial degradation of *Phaeocystis* material in the water column, p. 55-65. *In* C. Lancelot and P. Wassmann (ed.), Ecology of *Phaeocystis*-dominated ecosystems. J. Mar. Syst. *5.*

38. Vairavamurthy, A., M. O. Andreae and R. L. Iverson. 1985. Biosynthesis of dimethyl sulfide and dimethylpropiothetin by *Hymenomonas carterae* in relation to sulfur source and salinity variations. Limnol. Oceanogr. *30:*59-70.
39. van Boekel, W. H. M., F. C. Hansen, R. Riegman and R. P. M. Bak. 1992. Lysis-induced decline of a *Phaeocystis* spring bloom and coupling with the microbial foodweb. Mar. Ecol. Prog. Ser. *81:*269-276.
40. van Iwaarden, P. R., A. J. M. Driessen and W. N. Konings. 1992. What we can learn from the effects of thiol reagents on transport proteins. Biochim. Biophys. Acta. *1113:*161-170.
41. Verity, P. G., T. A. Villareal and T. J. Smayda. 1988. Ecological investigations of blooms of colonial *Phaeocystis pouchetii.* II. The role of life-cycle phenomena in bloom termination. J. Plankton Res. *10:*749-766.
42. Wassmann, P., M. Vernet, B. G. Mitchell and F. Rey. 1990. Mass sedimentation of *Phaeocystis pouchetii* in the Barents Sea. Mar. Ecol. Prog. Ser. *66:*183-195.
43. Wolfe, G. V., E. B. Sherr and B. F. Sherr. 1994. Release and consumption of DMSP from *Emiliania huxleyi* during grazing by *Oxyrrhis marina*. Mar. Ecol. Prog. Ser. *111:*111-119.

27

DMSP LYASE IN MARINE MACRO- AND MICROALGAE

Intraspecific Differences in Cleavage Activity

Michael Steinke, Claudia Daniel, and Gunter O. Kirst

University of Bremen–Marine Botany FB2
P.O. Box 330440, 28334 Bremen, Germany

SUMMARY

The enzymatic cleavage of dimethylsulfoniopropionate (DMSP) to dimethylsulfide (DMS) was investigated in twenty-one strains of marine macro- and microalgae, representing seven algal classes. The enzymes involved in this cleavage are DMSP lyases, producing DMS from DMSP. All algal strains tested were able to synthesize and accumulate various levels of intracellular DMSP but only twelve strains showed DMSP lyase activity. It was possible to identify subgroups of strong and weak DMS producers. The first subgroup included three *Enteromorpha* species (*E. clathrata*, *E. intestinalis*, *E. compressa*) and *Phaeocystis sp.* with specific activities in crude cell extracts ranging from 7 to over 100 nmol DMS min^{-1} (mg cell protein)$^{-1}$. The second subgroup was composed of a sub-antarctic strain of *Acrosiphonia arcta*, *Polysiphonia lanosa*, two strains of *Emiliania huxleyi*, *Acrosiphonia sonderi*, *Ulva lactuca* and *Enteromorpha bulbosa*. In this subgroup activity ranged from 0.01 to 0.2 nmol DMS min^{-1} (mg cell protein)$^{-1}$. No DMSP lyase was detectable in a sub-arctic strain of *Acrosiphonia arcta*, *Acrosiphonia sonderi*, *Monostroma arcticum*, *Prasiola crispa*, *Polysiphonia urceolata*, *Ascoseira mirabilis*, *Laminaria saccharina* and *Tetraselmis subcordiformis*. Non-optimal assay conditions and bacterial contamination may have affected rates in some samples, but the results suggest the widespread presence of DMSP lyase among algal taxa, and also raises the possibility that closely-related species may have quite different lyase activities or function.

INTRODUCTION

The spontaneous breakdown rate of dissolved dimethylsulfoniopropionate (DMSP) to dimethylsulfide (DMS) and acrylic acid is known to be very slow in seawater (3). Since DMS is the dominant sulfur gas found in marine surface waters (1), the understanding of biological processes leading to DMS production is of major importance.

Presently, it is believed that there are two major biological mechanisms whereby DMS is produced in aerobic seawater: microbial (primarily bacterial) activity may cleave dissolved DMSP (free DMSP in the seawater) or particulate DMSP (from decaying algal material and fecal pellets) (4, 16), and enzymes associated with or released from algal cells may cleave the DMSP to DMS (2; 23). The enzymes involved in this process are DMSP lyases, a group of carbon-sulfur lyases, which are classified as dimethylpropiothetin-dethiomethylases. Several reports of bacterial DMSP lyases have examined the production of DMS in seawater and bacterial cultures (6, 15, 19). Recently, a DMSP lyase from a DMS producing marine bacterium was purified and characterized (5).

DMSP is a prominent sulfur compound in various groups of algae (12, 14) but not all taxa produce and accumulate DMSP intracellularly in amounts high enough to be quantified by existing methods. The intracellular concentrations of DMSP vary widely in individual species. Even within one species, the concentrations may be subject to physiological or environmental control (7; 8; 11; see also Keller and Korjeff-Bellows, this volume). Although much is known about the occurrence of DMSP in algae, little is known about the ability of algae to degrade DMSP. The Rhodophytes *Polysiphonia lanosa* (2) and *Polysiphonia paniculata* (21), the Prymnesiophyte *Phaeocystis sp.* (23), and the Chlorophyte *Enteromorpha clathrata* (Steinke and Kirst, in press) are the only currently known algal species to exhibit DMSP lyase activities. Since extracts of *Crypthecodinium* (*Gyrodinium*) *cohnii*, a heterotrophic dinoflagellate, showed DMS production due to lyase activity (10), there is evidence that dinophytes may also be able to cleave DMSP.

In this study we used an enzyme assay developed for DMSP lyase measurements in *Enteromorpha clathrata* (Steinke and Kirst, in press) to screen twenty-one strains of marine macro- and microalgae for the ability to enzymatically cleave DMSP to DMS. The data presented provide a preliminary overview on DMSP lyase distribution in different classes of marine algae.

MATERIALS AND METHODS

Plant Material and Culture Conditions

The algae used in this study originated from different sources and were maintained under conditions listed in Table 1. All cultures were unialgal with the exception of *Scrippsiella sp.* (Dinophyceae). This culture was contaminated with a small (< 10 μm) chlorophyte flagellate.

The macroalgae were kept in 500 or 1000 ml glass beakers, containing 0.2 μm filtered North Sea water. The seawater was aerated, enriched with nutrients (PES enrichment; 22) and adjusted to a salinity of 32 practical salinity units (PSU; equivalent to parts per thousand) with distilled water. This culture medium was changed at least once a month to avoid nutrient limitation. Light was supplied by fluorescent tubes at a light-dark rhythm of 18:6 hours.

Microalgae were cultivated in Erlenmeyer flasks using three types of media (Table 1), adjusted to a salinity of 32 to 34 PSU. *Phaeocystis sp.* cultures were placed on a gyratory shaker at 80 rpm and consisted of non-colonial, single cells when harvested. The culture with *Tetraselmis subcordiformis* was aerated, whereas *Emiliania huxleyi*, *Scrippsiella sp.* and *Calciodinellum operosum* were manually stirred once a day. Light was supplied by fluorescent tubes at a light-dark rhythm of 18:6 hours.

Table 1. Origin and culture conditions of algal strains.

Class	Species	Origin of isolate		Culture conditions		
		Strain #	Location and year of collection	Temp.	Light	Medium [a]
Macroalgae						
Chlorophyceae	*Enteromorpha clathrata*	1086	Disko-Island (Greenland) 1990	0	15	NS (PES)
	Enteromorpha intestinalis	1088	Disko-Island (Greenland) 1990	0	11	NS (PES)
	Enteromorpha bulbosa	1002	King-George-Island (Antarctica) 1986	0	13	NS (PES)
	Enteromorpha bulbosa	1002	King-George-Island (Antarctica) 1986	5	35	NS (PES)
	Enteromorpha compressa	1134	Kongsfjord (Svalbard) 1991	5	25	NS (PES)
	Acrosiphonia arcta	1120	Disko-Island (Greenland) 1990	5	33	NS (PES)
	Acrosiphonia arcta	1071	King-George-Island (Antarctica) 1987	5	26	NS (PES)
	Acrosiphonia sonderi	1130	Kongsfjord (Svalbard) 1991	5	27	NS (PES)
	Acrosiphonia sonderi	1132	Kongsfjord (Svalbard) 1991	5	29	NS (PES)
	Monostroma arcticum	1139	Kongsfjord (Svalbard) 1991	5	27	NS (PES)
	Prasiola crispa	1040	King-George-Island (Antarctica) 1986	5	34	NS (PES)
Ulvophyceae	*Ulva lactuca*	1128	Disko-Island (Greenland) 1990	5	35	NS (PES)
Rhodophyceae	*Polysiphonia urceolata*	2093	Disko-Island (Greenland) 1990	5	21	NS (PES)
	Polysiphonia lanosa	Poly M	Belfast (Irland) 1994	10	6	NS (PES)
Phaeophycae	*Ascoseira mirabilis*	3068	King-George-Island (Antarctica) 1987	0	6	NS (PES)
	Laminaria saccharina	3123 & 3124	Kongsfjord (Svalbard) 1991	5	32	NS (PES)
Microalgae						
Prasinophyceae	*Tetraselmis subcordiformis*	161-1a	New Haven, CT (USA) 1952	25	80	AQUIL (f/2)
Prymnesiophyceae	*Phaeocystis sp.*	Phaeo B	North Sea (Germany)	10	60	AQUIL (K)
	Emiliania huxleyi	Emil D	North Sea (Norway)	15	70	ANT (f/50)
	Emiliania huxleyi	G 1779	(oceanic)	15	70	ANT (f/50)
Dinophyceae	*Scrippsiella sp.*	Scripps		20	70	WIMEX (f/2)
	Calciodinellum operosum	Calcio 74	Naples (Italy) 1995	20	70	WIMEX (f/2)

a) Various media and additional enrichments were used for the cultivation of algae: NS = North Sea water; AQUIL = artificial seawater (prepared after 20); ANT = antarctic water; WIMEX = commercial seasalt mixture; PES = Provasoli enrichment factor (prepared after 22); K = K-enrichment factor (prepared after 13); f/2 and f/50 = dilutions of the f-enrichment factor (prepared after 9).

Preparation of Extracts

Macroalgae: 1 to 2 g fresh weight of algal tissue was briefly rinsed in ice-cold distilled water to remove external salts and carefully blotted dry, then homogenized with a pestle and mortar under liquid nitrogen. All subsequent preparations were carried out at 4°C unless stated otherwise. The homogenate was suspended in 3 ml extraction buffer containing 100 mM 2-[N-morpholino] ethanesulfonic acid (MES), 13 mM calcium chloride dihydrate, 10% (v/v) glycerol and 0.5% (v/v) detergent (Tween 80; polyoxyethylenesorbitan monooleate), adjusted with NaOH to pH 6.2. The extract was incubated on ice for 30 min to dissolve membranes, followed by centrifugation three times for 5 min at 1200 x g. The supernatant was collected and the pellet resuspended in 2 ml extraction buffer after each centrifugation step. The supernatants were pooled and centrifuged 5 min at 4000 x g. The liquid phase of the crude cell-free extract was used in the enzyme assay to test for DMSP lyase activity.

Microalgae: Cultures were harvested during late exponential growth phase (6 to 700 10^6 cells, depending on the strain investigated) by centrifugation at 3500 x g for 10 min at 4°C. The culture with *Scrippsiella sp.* was centrifuged at 700 x g for 5 min at 4°C. Most (80%) of the contaminating flagellates remained in the supernatant and were removed by this procedure. The centrifuged microalgae were resuspended in extraction buffer (described above) and homogenized with a French Press at 138 MPa (equivalent to 20,000 psi). Microscopic observation indicated that >90% of cells were destroyed during this step. The resulting crude extracts, containing cell debris, were incubated on ice for 30 min to allow membranes to dissolve. No attempt was made to further purify the extracts because microalgal biomass was low compared to macroalgal samples.

Macro- and micro-algal crude extracts were purged of gaseous DMS (derived from enzymatic conversion of cellular DMSP) with compressed air for 30 min at 0°C. 1-ml aliquots of the purged crude extracts were stored in Eppendorf micro test tubes at -80°C to be assayed for DMSP lyase activity the day following the extraction. This allowed simultaneous assay of many extracts, following their preparation. Protein concentrations were determined after all enzyme assays were carried out (see below). Another 500 µl subsample was heated for 60 min at 95°C to destroy enzymatic activity and this was used as a control.

Enzyme Assays

Prior to the enzyme assay, the extracts of *Enteromorpha clathrata*, *Enteromorpha intestinalis*, *Enteromorpha compressa* and *Phaeocystis sp.* were diluted by 1:100 with extraction buffer to reduce DMSP lyase activity in the assay. 10-µl extraction buffer containing 30 mM dithiothreitol (DTT) was transferred into a gas tight screw-capped glass vial (volume 1.2 ml) and equipped with a Teflon-coated silicone septum. The potential production of DMS from algal DMSP was assayed after adding 270 µl of the extract to the test buffer after incubating for 10 min at 27°C. Usually, only small amounts of DMS were detectable in the extracts, but the extract of *Scrippsiella sp.* produced large amounts (see results). After this pre-test for DMS, the reaction was started by adding 20 µl extraction buffer containing dissolved DMSP to a final concentration of 2 mM. The DMSP used in all experiments was prepared according to Larher *et al.* (18). 100 to 300 µl of the gas phase (headspace) was sampled over 10 to 60 minutes, and injected with a gas-tight syringe into a Shimadzu 9 A gas chromatograph equipped with a FPD detector. Operating conditions are given in Karsten *et al.* (12). The detection limit was about 8 pmol DMS, or approximately 80 nM DMS in a 100 µl headspace sample. Rates were calculated as the difference in DMS concentrations before DMSP addition, and after incubation with DMSP. Production rates in heated controls, when observed, were subtracted to give enzyme activities. Specific DMSP lyase activities were then calculated based on protein concentration (see following section) in the algal extracts.

Protein Determinations

Protein concentration in the extracts was determined spectrophotometrically at 750 nm using a commercial test assay (Bio-Rad detergent compatible protein assay). Relative absorption was quantified by comparing with a series of protein standards with known concentrations of bovine serum albumin in extraction buffer.

DMSP Determinations

The intracellular DMSP concentrations in the microalgae (*Tetraselmis sp.* clone OPT4, *Phaeocystis sp.* clone 677-3 and *Emiliania huxleyi* clone BT6) were taken from the literature (14) and therefore, do not exactly match strains investigated in this study. The macroalgal DMSP concentrations were determined according to the method described by Karsten *et al.* (12).

RESULTS AND DISCUSSION

All algal species tested were able to synthesize and accumulate DMSP intracellularly. The range of this accumulation varied greatly from 0.01 to 87 mmol (kg fresh weight)$^{-1}$

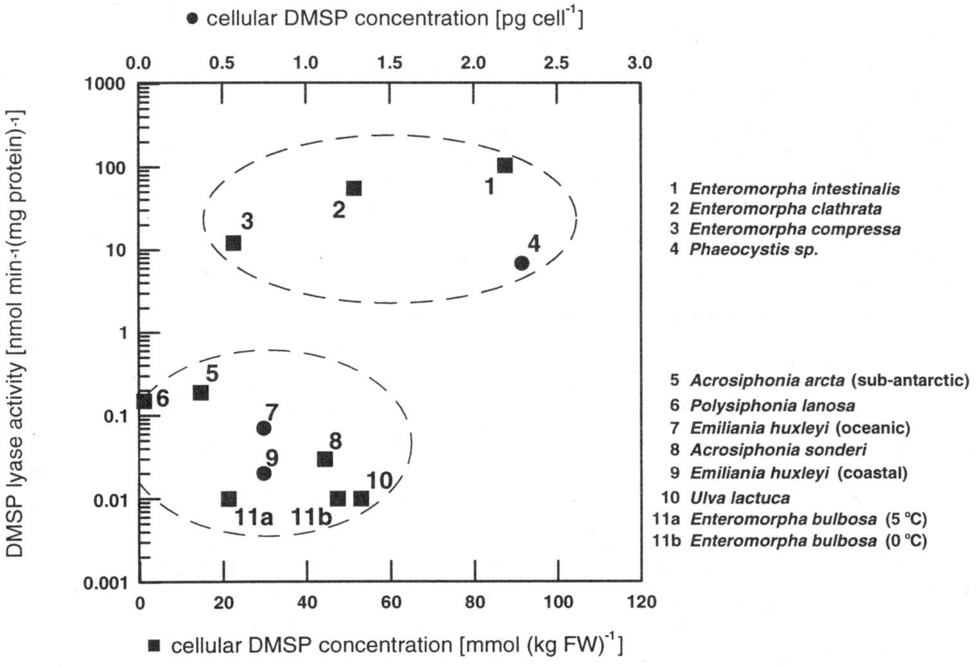

Figure 1. DMSP lyase activities vs. intracellular DMSP concentrations in various algal strains. DMSP concentrations for microalgae (●) are expressed as pg cell^{-1} on the top x-axis (data from 14) and for macroalgae (■) as mmol (kg freshweight)$^{-1}$ on the bottom x-axis. See text for strains without DMSP lyase activity detectable.

(macroalgae) or 0.75 to 2.29 pg cell^{-1} (microalgae). Initially, we choose *Ascoseira mirabilis* and *Laminaria saccharina* (both Phaeophyta) as non-DMSP containing controls but traces of DMSP (DMS released upon NaOH treatment) were detected in both species (0.07 mmol DMSP (kg freshweight)$^{-1}$).

Despite the presence of DMSP in all strains tested, only twelve of twenty-one strains investigated in this study showed measurable DMSP cleavage by means of DMSP lyases. There was no relationship between DMSP lyase activity and intracellular DMSP concentration of the species investigated (Figure 1).

There appeared to be three groups that showed different levels of DMSP lyase activity. In one group no DMSP lyase was detectable (e.g. DMS production in the samples did not exceed that in heated controls). This group included a sub-arctic strain of *Acrosiphonia arcta*, *A. sonderi* (strain 1132), *Monostroma arcticum*, *Prasiola crispa*, *Polysiphonia urceolata*, *Ascoseira mirabilis*, *Laminaria saccharina* and *Tetraselmis subcordiformis*. Another group was composed of weak DMS producers. The species in this group showed activities ranging from 0.01 to 0.2 nmol DMS min^{-1} (mg cell protein)$^{-1}$. A sub-antarctic strain of *Acrosiphonia arcta*, *A. sonderi* (strain 1130), *Polysiphonia lanosa*, both strains of *Emiliania huxleyi*, *Ulva lactuca* and *Enteromorpha bulbosa* belonged into this group. A final group of strong DMS producers showed production rates at least two orders of magnitude higher, from 7 to over 100 nmol DMS min^{-1} (mg cell protein)$^{-1}$. This group was represented by three *Enteromorpha* species (*E. intestinalis*, *E. clathrata* and *E. compressa*) and *Phaeocystis sp*. The strain of *Scrippsiella sp.* appeared to have strong DMSP lyase activity as well. However, we could not quantify production because we were not able to reduce initial DMS concentrations - which arose from the enzymatic conversion of DMSP readily available in the crude extract. This endogenous DMS production made it impossible to quantify any additional DMS production which might have occurred due to the experimental addition of 2 mM DMSP.

It is likely that the distinction between weak and strong producers is to some extent an artefact of our small sample size. We expect that other strains might yield rates in between these two groups, giving a continuum of rates. In fact , *C. operosum* showed a DMSP lyase activity of 2 nmol DMS min^{-1} (mg cell protein)$^{-1}$. It is not included in figure 1 because no estimate of intracellular DMSP concentration was available from the literature.

It is also likely that measured lyase activities were affected by our assay conditions. As already pointed out, we used an enzyme assay which was optimized to test for DMSP lyase in *Enteromorpha clathrata* . Requirements for optimal function of DMSP lyases in other species may be significantly different. For example, different pH optima are reported for several DMSP lyases. The pH optimum for the DMSP lyase from *Polysiphonia lanosa* was 5.1 (2), whereas that from *Crypthecodinium cohnii* was 6 to 6.5 (10), similar to the pH optimum in *Enteromorpha clathrata* (Steinke and Kirst, in press). It is highly likely that our assay conditions were not optimal for some of the weak DMS producers, yielding underestimates of their rates. We also froze algal extracts prior to assay, so that many extracts could be prepared and then tested simultaneously. It is possible that the lyase activities in some species were reduced by cold storage, while others were not. However, our assay conditions were mild, so it is unlikely our underestimates were drastic. We also emphasize the dramatic differences between the strong and weak producers - over 100-fold change in activity - rather than the specific rates. Nonetheless, reaction conditions should be adjusted optimally for each organism tested, and these data must be regarded as preliminary until further tests can be conducted. We note, however, that three species of *Enteromorpha* showed high activities, but *E. bulbosa* was a very weak DMS producer. This suggests, despite the limitations of our method, that closely related species may either have very different concentrations of DMSP lyase, or functionally different enzymes which may require different assay conditions.

While non-optimal assay conditions may have yielded underestimates of the rates in the group of weak lyase activity, bacterial contamination by DMSP-lysing strains could have resulted in overestimates, or false positives. Our algal material was not axenic, but before enzyme extractions, cultures were checked by light microscopy for bacterial contamination. In all algal cultures numbers of bacteria were low. Nonetheless, high bacterial lyase activities could have contributed significantly to DMS production, especially among the weak producers. Using the v_{max} of 1.09 μmol DMS min^{-1} (mg cell protein)$^{-1}$ for the DMSP lyase of strain M3A, an *Alcaligenes*-like bacterium (5), and an approximate protein concentration of 70 μg cell protein per 10^9 cells in M3A (De Souza, pers. comm.) the numbers of bacteria necessary to generate the low activities measured in the group of weak DMS producers may be calculated: 1.5×10^6 to 3×10^6 cells ml^{-1} of strain M3A would be necessary to explain the activities found in this group. We believe such a density would have been detected by microscopy. Furthermore, rinsing the algal tissue in distilled water and blotting it dry probably reduced the number of bacteria attached to the surface of the algal thalli. We therefore believe that bacterial contribution to our results was, in most cases, modest. However, in the strains of *Emiliania huxleyi* which tested positive, concentrating the cultures by centrifugation may have resulted in bacteria numbers high enough to explain the DMSP lyase activity measured.

CONCLUSION

The results of this study, though preliminary, suggest that DMSP lyase is probably widespread among marine algae and that levels of DMSP lyase activity show high intraspecific variability. It is also likely that related strains may show very different DMSP lyase activities, as observed for *Enteromorpha spp*. Whether these differences are due to enzyme concentration, enzyme expression, functional enzyme variation, or environmental cues, we cannot yet say.

Some of the observed variability is undoubtedly due to non-optimal assay conditions, or possible contamination by bacteria. Ideally, all assays should be optimized for each strain, and conducted on axenic material. This is an extreme undertaking and few such studies on comparative enzymology exist even for well-characterized enzymes. We wish to emphasize not the specific rates, but rather the broad pattern of algal DMSP lyase distribution and variability. We believe that algal DMS production should receive more attention because it has been shown that bacterial degradation of extracellular, dissolved DMSP produces relatively little DMS (17) and demethylation of dissolved DMSP may be a more important sink for DMSP than cleavage to DMS (24). Therefore, even slow conversion of the intracellular DMSP pool by algal enzymes might contribute significantly to DMS production.

ACKNOWLEDGMENTS

The authors thank Bärbel Bolt, Monika Kirsch, Christina Langreder, Dr. Christine Maggs, Dr. Linda Medlin, Doris Meyerdierks, Ursula Wellbrock and Dr. Christian Wiencke for the provision of algal cultures. The authors appreciate helpful suggestions on the manuscript by Dr. Ron Kiene and Dr. Gordon Wolfe as well as two anonymous reviewers. Participation of Michael Steinke in this symposium was made possible by funding through the University of Bremen and the University of South Alabama. This study was financially supported through the European Community Project "Role and significance of biological processes in DMS release from ocean to the atmosphere: a close examination of the black box".

REFERENCES

1. Bates, T. S., B. K. Lamb, A. Guenther, J. Dignon and R. E. Stoiber. 1992. Sulfur emissions to the atmosphere from natural sources. J. Atmos. Chem. *14*: 315-337.
2. Cantoni, G. L. and D. G. Anderson. 1956. Enzymatic cleavage of dimethylpropiothetin by *Polysiphonia lanosa*. J. Biol. Chem., *22*: 171-177.
3. Dacey, J. W. H. and N. V. Blough. 1987. Hydroxide decomposition of dimethylsulfoniopropionate to form dimethylsulfide. J. Geophys. Res. Lett. *14*: 1246-1249.
4. Dacey, J. W. H. and S. G. Wakeham. 1986. Oceanic dimethylsulfide: production during zooplankton grazing on phytoplankton. Science *233*: 1314-1316.
5. De Souza, M. P. and D. C. Yoch. 1995. Purification and characterization of dimethylsulfoniopropionate lyase from an *Alcaligenes*-like dimethyl sulfide-producing marine isolate. Appl. Environ. Microbiol. *61*: 21-26.
6. Diaz, M. R., P. T. Visscher and B. F. Taylor. 1992. Metabolism of dimethylsulfoniopropionate and glycine betaine by a marine bacterium. FEMS Microbiol. Lett. *96*: 61-66.
7. Dickson, D. M. J. and G. O. Kirst. 1986. The role of dimethylsulfoniopropionate, glycine betaine and homarine in the osmoacclimation of *Platymonas subcordiformis*. Planta *167*: 536-543.
8. Edwards D. M., R. H. Reed and W. D. P. Stewart. 1988. Osmoacclimation in *Enteromorpha intestinalis*: long-term effects of osmotic stress on organic solute accumulation. Mar. Biol. *98*: 467-476.
9. Guillard, R. R. L. 1975. Culture of phytoplankton for feeding marine invertebrates, p. 29-60. *In* W. L. Smith and M. H. Chanvey (eds.), Culture of marine invertebrate animals. Plenum Press, New York.
10. Kadota, H. and Y. Ishida. 1968. Effect of salts on enzymatical production of dimethyl sulfide from *Gyrodinium cohnii*. Bull. J. Soc. Sci. Fish. *34*: 512-518.
11. Karsten, U. and G. O. Kirst. 1989. Intracellular solutes, photosynthesis and respiration of the green alga *Blidingia minima* in response to salinity stress. Botanica acta *102*: 123-128.
12. Karsten, U., C. Wiencke and G. O. Kirst. 1990. The ß-dimethylsulfoniopropionate content of macroalgae from Antarctica and Southern Chile. Bot. Mar. *33*: 143-146.
13. Keller, M. D., R. C. Selvin, W. Claus and R. R. L. Guillard. 1987. Media for the culture of oceanic ultraphytoplankton. J. Phycol. *23*: 633-638.
14. Keller, M. D., W. K. Bellows and R. R. L. Guillard. 1989. Dimethylsulfide production in marine phytoplankton, p. 167-182. *In* E. S. Saltzman and W. J. Cooper (eds.), Biogenic sulfur in the environment. American Chemical Society Symp. Ser., Washington, DC, Vol. 393.
15. Kiene, R. P. 1990. DMS production from DMSP in coastal seawater samples and bacterial cultures. Appl. Environ. Microbiol. *56*: 3292-3297.
16. Kiene, R. P. 1992. Dynamics of dimethyl sulfide and dimethylsulfoniopropionate in oceanic water samples. Mar. Chem. *37*: 29-52.
17. Kiene, R. P. and S. K. Service. 1991. Decomposition of dissolved DMSP and DMS in estuarine waters: dependence on temperature and substrate concentration. Mar. Ecol. Progr. Ser. 76: 1-11.
18. Larher, F., J. Hamelin and G. R. Stewart. 1977. L'acide Diméthylsulfonium-3 Propanoïque de *Spartina anglica*. Phytochem. *16*: 2019-2020.
19. Ledyard, K. M. and J. W. H. Dacey. 1994. Dimethylsulfide production from dimethylsulfoniopropionate by a marine bacterium. Mar. Ecol. Progr. Ser. *110*: 95-103.
20. Morel, F. M. M., J. G. Rueter, D. M. Anderson and R. R. L. Guillard. 1979. Aquil: a chemically defined phytoplankton culture medium for trace metal studies. J. Phycol. *15*: 135-141.
21. Nishiguchi, M. K. and L. J. Goff. 1995. Isolation, purification, and characterization of DMSP lyase (dimethylpropiothetin dethiomethylase (4.4.1.3)) from the red alga *Polysiphonia paniculata*. J. Phycol. *31*: 567-574.
22. Provasoli, L. 1968. Media and prospects for cultivation of marine algae, p. 47-74. *In* A. Watanabe and A. Hattori (eds.), Cultures and collections of algae. Jap. Soc. Plant Physiol., Tokyo.
23. Stefels, J. and W. H. M. van Boekel. 1993 Production of DMS from dissolved DMSP in axenic cultures of the marine phytoplankton species *Phaeocystis spec.* Mar. Ecol. Progr. Ser. *97*: 11-18.
24. Visscher, P. T., M. R. Diaz and B. F. Taylor. 1992. Enumeration of bacteria which cleave or demethylate dimethylsulfoniopropionate in the Caribbean Sea. Mar. Ecol. Progr. Ser. *89*: 293-296.

KINETICS OF DMSP-LYASE ACTIVITY IN COASTAL SEAWATER

Kathleen M. Ledyard[1] and John W. H. Dacey[2]

[1] Marine Science Institute
University of California
Santa Barbara, California 93106
[2] Biology Department
Woods Hole Oceanographic Institution
Woods Hole, Massachusetts

SUMMARY

The concentration dependence of dissolved-DMSP loss and DMSP-lyase activity was determined at several locations in Monterey Bay over a period of two weeks in spring 1995. Kinetic parameters of DMSP loss and DMS production varied over a wide range and were not obviously correlated with ambient DMSP levels or overall biomass. Apparent half-saturation values of DMSP-lyase activity ranged from 480 nM to 11.3 μM, with corresponding maximal velocities of 5-156 nmol L^{-1} h^{-1}. Kinetic parameters of DMSP loss were also variable, with K_m, v_{max} values of 23 nM, 4.9 nmol L^{-1} h^{-1} and 900 nM, 60 nmol L^{-1} h^{-1} found in two experiments. Based on a comparison of the apparent v_{max} of DMSP loss and DMSP-lyase activity when the kinetics of both were assessed simultaneously, the proportion of DMSP loss resulting in DMS production ranged from low to 100%. These results suggest that DMSP degradation in productive coastal waters can be mediated by extremely different microbial consortia over short temporal and spatial scales.

INTRODUCTION

Production of dimethylsulfide (DMS) from dissolved dimethylsulfoniopropionate (DMSP) is a well-documented process in natural seawater and in pure cultures of marine bacteria and algae (2, 4, 6, 7, 8, 10, 11, 13, 17, 20). DMSP can also be demethylated to form 3-methiolpropionate which can undergo either demethylation to 3-methiolpropionate (MPA) or demethiolation to methanethiol (14, 16, 17, 19). There is evidence as well that DMSP may be accumulated by marine bacteria, apparently without short-term metabolism (22, 24). Knowledge of the extent to which DMSP is converted to DMS versus other products is critical to models of DMS efflux from seawater, but as yet we have a poor understanding of the factors that control this quantity.

The concentration dependence of both DMSP uptake and DMSP-lyase activity has been examined in a variety of marine environments, ranging from estuarine to coastal and oligotrophic seawater (7, 8, 10), as well as in a marine bacterial isolate from the Sargasso Sea (9). DMS production rates measured to date in seawater range up to 24 nmol L^{-1} h^{-1}, and in several cases little or no saturation of activity has been observed at DMSP(d) concentrations far in excess of ambient levels (8, 10). Another striking feature of these measurements is the discrepancy between DMSP loss and DMS production rates at undersaturating concentrations of DMSP. Rates of loss often significantly exceed cleavage rates (7, 8, 10), indicating either a decoupling between the uptake and lyase system (i.e. intracellular location of the lyase) in microbial populations, or high levels of demethylation or storage, or some combination of these. Differentiating between these pathways by measurement of volatile sulfur products is not straightforward in seawater incubations, since neither storage nor successive demethylation of DMSP results in DMS or methanethiol formation. Simultaneous measurement of DMSP loss and DMS production kinetics may provide a way to distinguish between these alternatives (10). In this study we examined the kinetics of DMSP loss and DMSP-lyase activity during the ONR-Marine Boundary Layer (MBL) West Coast Experiment in Monterey Bay (April-May 1995).

MATERIALS AND METHODS

Analytical

DMS analysis has been described elsewhere (9), and was carried out with the following modifications. DMS was preconcentrated at room temperature on approximately 0.5 g gold wool (Degussa; Hanau, Germany) packed in 1/8" stainless steel tubing. Following thermal desorption, the trap was cooled in preparation for the following run by passing compressed air (~10 psi) between the trap and surrounding 1/4" stainless steel sheath for 1 min. Purified air was used as the carrier and sparging gas. Sampling and analysis were fully automated except when incubations involved micromolar-level DMSP additions, in which case sample (0.1-1 mL) was manually introduced into the sparger; the subsequent steps were otherwise automated.

DMS was measured in unfiltered water. Dissolved DMSP (DMSP(d)) was operationally defined as that which passes a Whatman GF/F glass fiber filter (nominal pore size 0.7 μm). For analysis of DMSP(d) in incubation experiments, 4-6 mL sample was filtered under gentle vacuum into 10-mL serum bottles. After being purged of DMS with purified air for 5 min, KOH was added to a final concentration of 1 N. DMSP was analyzed as DMS following alkaline cleavage (21), either by sparging 100-μL aliquots of liquid sample or by on-column injection of 0.25-1 mL headspace gas. Particulate DMSP (DMSP(p)) was not measured in incubation experiments. For analysis of dissolved and particulate DMSP in depth profile samples, 50-100 mL seawater was filtered. GF/F filters were suspended in 25 mL 1 N KOH in 30-mL serum bottles and allowed to react overnight.

Sample Collection

Seawater was collected aboard the Moss Landing ship *Point Sur* in acid-cleaned 10-L Niskin bottles mounted on a Sea-Bird 12-place carousel rosette equipped with a Sea-Bird 9/11+ conductivity-temperature-depth (CTD). Niskin bottles were fitted with silicon O-rings and internal bands.

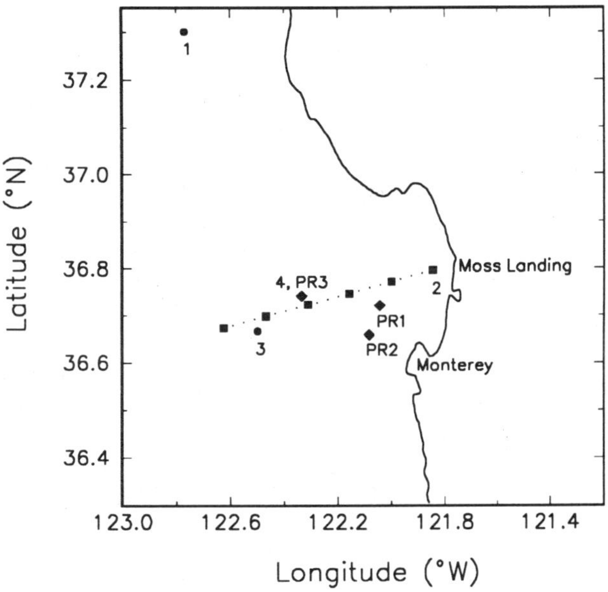

Figure 1. Locations of incubation experiments, profiles and lateral transect stations in Monterey Bay. ●, incubation experiments: 1, 4/19/95; 2, 4/27/95; 3, 4/28/95; 4, 5/1/95. ◆, profiles: PR1, 4/21/95; PR2, 4/22/95; PR3, 5/1/95. ■, transect stations. Experiment 2 was conducted at the easternmost transect station, and experiment 4 was carried out at the same station as profile 3 (PR3).

Incubation Experiments

Over a period of two weeks, four experiments were carried out to assess the kinetics of DMS production (DMSP-lyase activity); locations are given in Fig. 1. In two of these experiments (1 and 4), the kinetics of DMSP(d) loss were also measured. Seawater (1 L) was dispensed through 75-μm Nitex mesh into 1-L foil-wrapped Teflon bottles capped with silicone stoppers. Before incubations, bottles were sequentially rinsed with 10% HCl, nanopure water (Barnstead) and screened seawater. Sample bottles were held at the ambient temperature of surface (3 m) seawater by means of a flowing seawater bath. In experiment 2, incubations were carried out in 125-mL glass serum bottles using 100-mL volumes of seawater. DMSP(d) additions were made as previously described (9); added DMSP concentrations ranged from 5 nM to 10 μM. No attempts were made to selectively inhibit DMS consumption in incubation bottles. Ambient rates of DMS consumption determined in separate experiments using dimethyldisulfide to inhibit consumption of DMS (23) ranged from undetectable to 0.26 nmol L^{-1} h^{-1} (K. Ledyard and J. Dacey, unpubl.) and were assumed to have a negligible effect on DMS production rates measured in response to DMSP additions.

Rates and Kinetic Analysis

Rates were calculated by taking the linear fit of timecourses of DMSP(d) disappearance or DMS production. Rates were based on 4-6 hours' data (with approximately one timepoint acquired per sample per hour). Error bars in figures represent the error of the calculated linear fit. Kinetic parameters were determined by linearizing rate data by means of Eadie-Hofstee (single-reciprocal) plots (5). Following a previous convention (10), the term apparent half-saturation value (K_{mapp}) is used in place of half-saturation constant (K_m),

Table 1. Ambient DMS, DMSP(d) and DMSP(p) for experiments 1-4, and kinetic parameters of DMSP loss and DMSP-lyase activity. Errors represent the calculated error of the linear fit.

Experiment	Date (nM)	DMS (nM)	DMSP loss DMSP(d) (nM)	DMSP(p)	DMSP cleavage K_m (μM)	V_{max} (nmol L^{-1} h^{-1})	K_{mapp} (μM)	V_{max} (nmol L^{-1} h^{-1})
1	19 Apr 95	2.9	11.5	---[1]	0.91 ± 0.09	60 ± 2	---[2]	---[2]
2[3]	27 Apr 95	12.6	10.5	118	---[1]	---[1]	6.3 ± 1.5	156 ± 14
3	28 Apr 95	4.0	6.8	---[1]	---[1]	---[1]	11.3 ± 1.8	63 ± 3
4[4]	1 May 95	4.3	4.2	32.1	0.023 ± 0.002[3]	4.9 ± 0.2[3]	0.48 ± 0.17	4.6 ± 0.7

[1] Not measured.
[2] Kinetic parameters were not determined; rates could not be linearized on a single-reciprocal plot.
[3] DMSP concentration range assayed: 11 nM - 8.4 μM.
[4] DMSP concentration range assayed: 3 nM - 0.8 μM. The DMSP loss rate at ambient DMSP(d) was not included in the linearization.

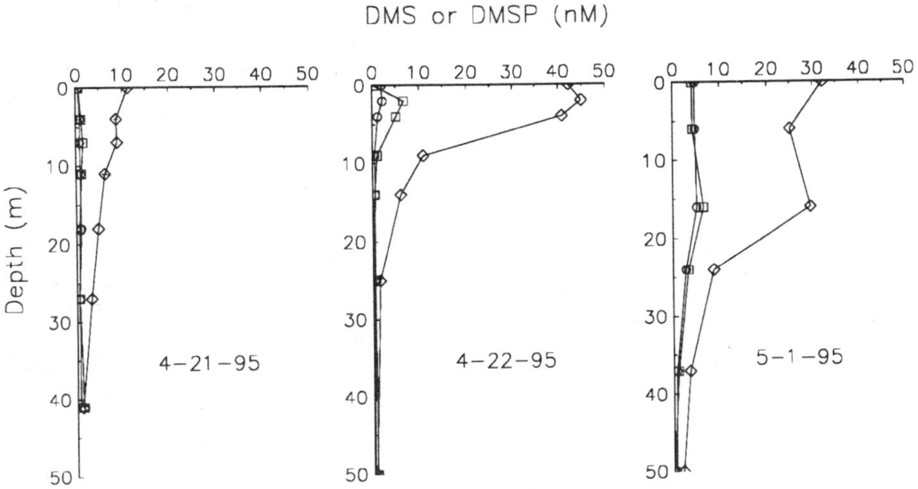

Figure 2. Vertical profiles of DMS (O), DMSP(d) (□) and DMSP(p) (◊) in Monterey Bay.

in acknowledgment of the possibility that DMSP lyase activity may be predominantly intracellular in natural populations, and therefore not directly related to extracellular DMSP(d) concentrations.

Vertical Profiles and Lateral Transect of DMS and DMSP

Water samples were collected as described above and dispensed into 1-L Teflon bottles with no headspace. DMS and DMSP were analyzed as above. Replicate analyses of the same samples generally agreed to better than 10%. Where shown, error bars on profile measurements represent the range of duplicate bottles drawn from the same Niskin. Profile and transect locations are displayed in Fig. 1. Profiles 1 and 2 were collected on April 21 and 22, respectively, while following a drifter deployed near the center of the Bay; profile 3 was taken on May 1 while following a separate drifter.

Reagents and Chemicals

DMS was purchased from Fluka Chemical Corp., Ronkonkoma, NY. DMSP-HBr was synthesized from DMS and bromopropionic acid as described by Challenger and Simpson (1). All other reagents were of analytical grade.

RESULTS

DMS and DMSP Distributions

Vertical profiles taken at locations near those of three of the incubation experiments showed that throughout the period of sampling DMSP(p) was considerably higher than either DMS or DMSP(d), on the order of 10-40 nM (Fig. 2). DMS concentrations were usually comparable to, and occasionally higher than, those of DMSP(d) (see also Table 1). In a lateral transect across the Bay on April 26, DMSP(p) decreased inland from approximately 60 to 20 nM, then increased again with proximity to the coast, with levels of 120 nM recorded at

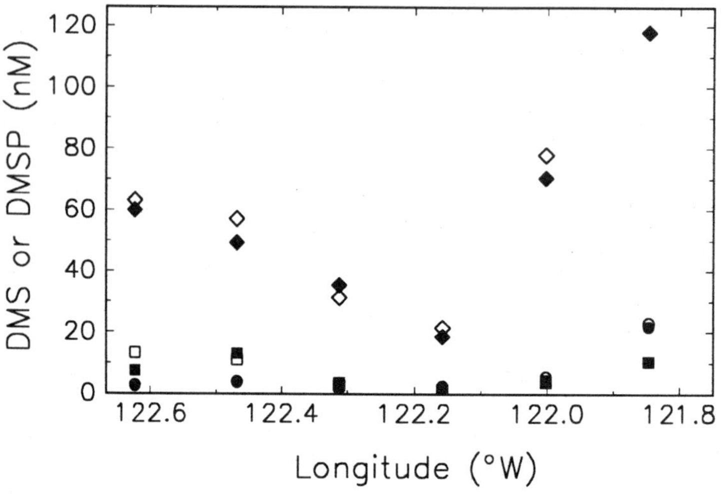

Figure 3. Lateral transect of DMS (●,○), DMSP(d) (■,□) and DMSP(p) (◆,◇) across Monterey Bay on April 26. Closed and open symbols represent measurements at 2 and 5 m water depth, respectively. Samples for particulate and dissolved DMSP at the easternmost station were collected on April 27.

the easternmost station (Fig. 3). This station is directly outside Moss Landing and receives runoff from a slough; however at the time samples were collected the tide was high and slough influence was assumed to be minimal. As in vertical profiles, DMSP(d) and DMS

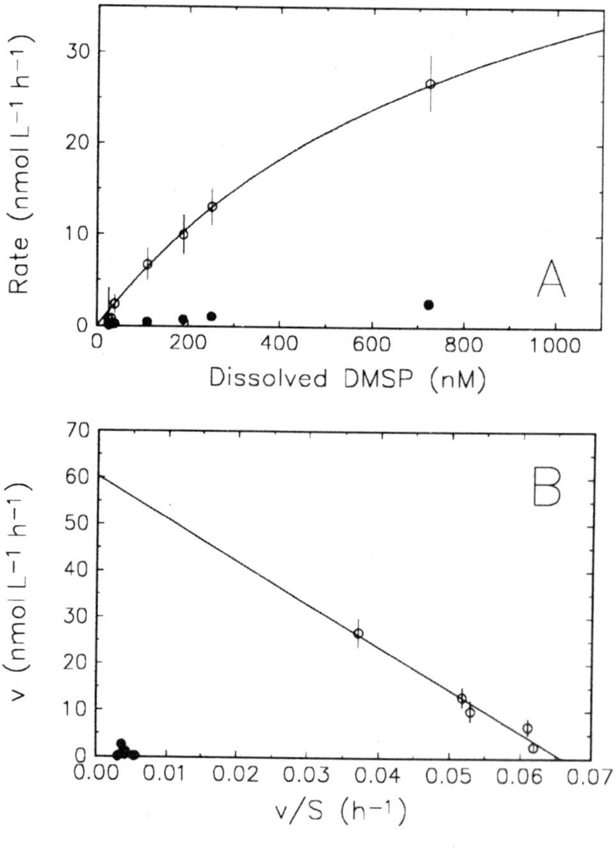

Figure 4. Kinetics of DMSP loss (○) and cleavage (●) on April 19 (experiment 1). (a) Saturation curves. (b) Single reciprocal plots of rate data. For DMSP loss, $K_m = 0.91 \pm 0.09$ μM, $v_{max} = 60 \pm 2$ nmol L^{-1} h^{-1}. Linearization is based on rates for DMSP >40 nM (5 rates). Kinetic parameters could not be determined for DMSP cleavage.

followed the general pattern of DMSP(p) but were present at lower levels. Ambient concentrations of DMS and DMSP in incubation experiments are provided in Table 1.

Kinetic Parameters of DMSP Loss and DMSP-Lyase Activity

On April 19 (experiment 1), at a northern station in the Bay (37°18.03′ N, 122°46.14′ W), a K_m of 0.91 ± 0.09 μM and v_{max} of 60 ± 2 nmol L^{-1} h^{-1} were found for DMSP(d) loss. Corresponding DMS production rates in the concentration range assayed could not be linearized, but were a small fraction (4-21%) of rates of DMSP loss (Fig. 4).

The apparent K_m and v_{max} of DMSP-lyase activity at an inland station just outside Moss Landing (36°47.80′ N, 121°50.72′ W) on April 27 (experiment 2) were 6.3 ± 1.5 μM and 156 ± 14 nmol L^{-1} h^{-1} (Table 1). DMS levels of 20 nM had been measured in surface waters at the same station the previous evening (Fig. 3); at the time of the experiment they had decreased to about half this value.

The following day, April 28 (experiment 3), the kinetics of DMSP cleavage were examined at a western station in the Bay (36°39.99′ N, 122°29.95′ W). The DMSP concentration range assayed (≤10 μM) was similar to that of the previous experiment, and despite the difference in water regime (see below), the apparent K_m and v_{max} of DMSP-lyase activity were comparable as well: 11.3 ± 1.8 μM and 63 ± 3 nmol L^{-1} h^{-1}. These kinetic parameters were calculated on the basis of rates measured in the 2-10 μM DMSP range (4 rates); rates measured at lower concentrations could not be linearized on a single-reciprocal plot (Fig.

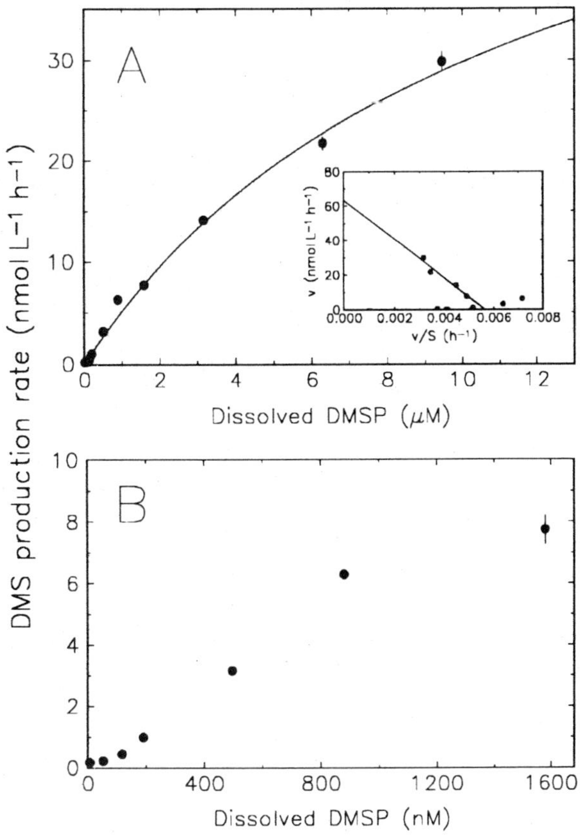

Figure 5. Kinetics of DMSP cleavage on April 28 (experiment 3). (a) Saturation curve showing the full DMSP concentration range, and single reciprocal plot. K_{mapp} = 11.3 ± 1.8 μM, v_{max} = 63 ± 3 nmol L^{-1} h^{-1}. Linearization is based on rates for DMSP concentrations >2 μM (4 rates). (b) Expanded view of DMSP concentration range below 2 μM; a sigmoidal dependence of rate on substrate concentration is evident in this range.

5a). In fact, a sigmoidal dependence of DMS production rate on DMSP(d) was noted at DMSP(d) concentrations <2 μM (Fig. 5b).

On May 1 (experiment 4) the kinetics of both DMSP loss and DMSP-lyase activity were assessed at a location slightly inland from the previous station (36°44.54′ N, 122°20.16′ W). This station and the previous station were assumed to lie within the California current, based on the low salinity of their surface water (32.91 and 32.75‰, respectively, as opposed to values well in excess of 33‰ in upwelled water closer to the coast). The half-saturation value and v_{max} of DMSP(d) loss were strikingly lower than those measured two weeks earlier: 23.1 ± 1.9 nM and 4.9 ± 0.2 nmol L^{-1} h^{-1}. The K_{mapp} of DMSP-lyase activity was also much lower than in the two previous experiments (480 ± 166 nM), and the apparent v_{max} the same within the error of the fit as that of DMSP loss (4.6 ± 0.7 nmol L^{-1} h^{-1}) (Table 1).

DISCUSSION

While some progress has been made in describing the concentration dependence of DMSP(d) removal and cleavage in seawater (7,8,10), our data on this subject are still very limited. Unexpectedly low-affinity kinetics of both removal and cleavage have been noted, with activity failing to saturate over a DMSP concentration range ≤1 μM, in some cases resulting in an inability to estimate kinetic parameters. Ledyard and Dacey (10) have observed that the magnitude of rates of both processes, as well as their apparent half-saturation values, may increase with proximity to summer in coastal and oligotrophic seawater, although these data were not sufficient to establish a seasonal dependence. The results presented here document extreme short-term temporal and spatial variability in the kinetics of DMSP loss and cleavage in Monterey Bay, CA, and indicate that the factors controlling the kinetic parameters of these processes are complex.

The apparent half-saturation values of DMSP cleavage measured in Monterey Bay in this study ranged between 0.5 and 11 μM over a period of two weeks. The difficulties of modeling the kinetics of any process in a natural population aside, the meaning of a half-saturation value for DMSP cleavage is especially problematic. If DMSP cleavage occurs at the cell surface in a given organism, one would expect DMSP loss and DMS production (DMSP cleavage) to be kinetically synonymous; DMSP removal and degradation are the same process. However, rates of DMS production cannot be meaningfully related to environmental levels of dissolved DMSP if the lyase is intracellular, as appears to be the case in the Sargasso Sea isolate strain LFR (9) and possibly the marine bacterium *Pseudomonas doudoroffii* as well (3). In strain LFR the apparent K_m of DMSP cleavage is 640 nM, several hundred nanomolar higher than the K_m for DMSP uptake. On the other hand, DMSP lyase activity appears to be extracellular in a *Phaeocystis* sp., with a K_m of 11.7 μM at 20°C (13, 14), and in an *Alcaligenes*-like salt marsh isolate, with a K_m of 2.02 mM (2, 3).

In light of the uncertainty over localization of DMSP lyase, it is difficult to interpret the high (relative to ambient DMSP(d)) K_{mapp} values for DMSP cleavage found in experiments 2 and 3. They may indicate that this process was mediated by organisms such as *Phaeocystis*. Haptophytes including *Phaeocystis* and motile zooids of *Phaeocystis* made up a small proportion of phototrophs present during experiments 1, 3 and 4 (data were not available for 2), but were more abundant in experiment 3 (9.1% of total autotrophic carbon) than in either 1 or 4 (1.4 and 1.3%, respectively). *Phaeocystis* colonies were tentatively identified in experiments 3 and 4; cells were large (90 pg C $cell^{-1}$) and colonies unusually compact (K. R. Buck, pers. comm.). If organisms with extracellular DMSP-lyase were dominant DMSP cleavers in experiment 3, the kinetics of DMSP loss would be expected to mirror those of DMS production; unfortunately DMSP loss kinetics were not examined on this occasion. In this regard it is interesting that in the few instances in which DMSP loss and cleavage have been monitored simultaneously in natural seawater (7, 8, 10, this work), the kinetics of these processes are never identical.

Maximal velocities of DMS production in Monterey Bay ranged from 4.6 to 156 nmol L^{-1} h^{-1}. In experiment 1, for which kinetic parameters for DMSP cleavage could not be determined (Fig. 4), overall rates were low (<4 nmol L^{-1} h^{-1}). Ledyard and Dacey (9) have previously observed that in a marine bacterium, possessing an intracellular DMSP-lyase, the maximal velocities of uptake and cleavage are identical, although the apparent half-saturation values are strikingly different. Thus in natural waters divergent K_{mapp} values for DMSP removal and cleavage (and, by the same token, non-stoichiometric DMSP removal and DMS production) need not necessarily indicate that DMSP is being degraded via pathways other than cleavage, although there is evidence that this is often the case (18, 19, see also Kiene, this volume). Rather, the relative $v_{\max}$ values of DMSP removal and cleavage may be a more reliable indicator of the comparative importance of demethylation (or accumulation without degradation) and cleavage (10). In the two experiments in which DMSP loss and DMS production were assessed simultaneously, the $v_{\max}$ of DMSP cleavage as a percentage of the $v_{\max}$ of DMSP(d) removal ranged from low (experiment 1, Fig. 4) to virtually 100% (experiment 4, Table 1), suggesting that DMS can shift from being a minor to a major degradation product of DMSP over short time and space scales.

Two estimates of the K_m of DMSP removal in Monterey Bay yielded 0.9 μM and 23 nM. The latter value is on the order of the highest DMSP(d) levels observed in surface water during the cruise (~17 nM). DMSP(d) concentrations as high as 1 μM have been observed in temperate coastal seawater (15), but were not typical of Monterey Bay at the time of this study. While the relation between DMSP cleavage activity and DMSP(d) is potentially complex, one would expect the K_m of DMSP removal (signifying either uptake or extracellular degradation, or both) to reflect ambient DMSP(d) levels. In this regard the high K_m for DMSP removal in experiment 1 is puzzling, though not unprecedented. A K_m for DMSP removal of 157 nM was reported for Vineyard Sound seawater, where ambient DMSP(d) is usually <40 nM (10). The lack of a consistent relation between half-saturation values of DMSP removal and ambient DMSP(d) in this and other work suggests that we may not fully understand the microscale distribution of DMSP(d) in seawater.

Frequently, rates estimated in unamended seawater in these experiments did not fall on a regression line in Eadie-Hofstee plots with rates estimated from additions. This may be indicative of coupling between biological DMS production and consumption at *in situ* substrate levels, resulting in artificially low rates of DMS production. However, this hypothesis does not explain the sigmoidal dependence of rate on DMSP(d) observed in experiment 3 (Fig. 5b). The origin of this concentration dependence, which was not observed in other experiments, is not understood. The effect could be due to mediation of DMSP-lyase activity by predominantly different microorganisms at low and high DMSP levels. However, it may not be necessary to invoke a mixed population to explain this observation. Sigmoidal concentration dependence of DMSP-lyase activity has also been observed in strain LFR (9), where it may reflect the interaction between DMSP uptake, accumulation, and induction of DMSP lyase, or may possibly indicate that the lyase has more than one binding site (5).

No clear dependence of kinetic parameters on *in situ* levels of DMSP was noted in this study (Table 1). During the sampling period strong upwelling prevailed along the coast, and diatom blooms occurred over much of the eastern portion of the Bay. In incubation experiments 1, 3 and 4, total carbon ranged from 76 (experiment 3) to 186 μg C L^{-1} (experiment 4), with centric diatoms dominating in 1 and 4, where rates of DMSP cleavage were lowest. In experiment 3, *Chlorella*-like eukaryotic picoplankton and cryptomonads dominated (63% of autotrophic carbon), and haptophytes (including *Phaeocystis*) comprised almost 10% of autotrophic biomass, instead of on the order of 2% in the other two experiments (K. R. Buck, pers. comm.). Unfortunately, no information is available for experiment 2, and *Phaeocystis*, although a minor component of the algal assemblage, cannot be ruled out as a possible source of the low-affinity DMSP-lyase kinetics observed in experiments 2 and 3. Algal species composition does not explain the low- and high-affinity kinetics of DMSP(d) removal observed, respectively, in experiments 1 and 4, in both of which diatoms comprised ~90% of total carbon. Although the kinetics of DMSP-lyase activity in experiment 3 invite speculation on the role of *Phaeocystis* in turnover of dissolved DMSP in natural waters, *Phaeocystis* cannot be invoked to explain low-affinity kinetics of DMSP-lyase activity in oligotrophic environments such as the Sargasso Sea (10).

The kinetic parameters of DMSP removal and cleavage in Monterey Bay exhibited extreme local variability and did not appear to be a simple function of phytoplankton species composition, overall biomass levels or ambient concentrations of DMS or DMSP. Under bloom conditions in an upwelling area considerable lateral heterogeneity in nutrient status, bacterial and algal assemblages, and other characteristics can be expected, and presumably these were factors in the observed variability of DMSP loss and cleavage kinetics. Elucidating the role of prokaryotes in these processes will be challenging, as we are at this stage unable to resolve specific DMSP-cycling components of the heterotrophic bacterial population without manipulating their growth conditions. However, it is likely that with a larger and higher-resolution data set, correlations between kinetic parameters of DMSP loss and cleavage and possible controlling factors such as DMSP stocks, algal assemblage and temperature will emerge.

ACKNOWLEDGMENTS

All of the algal assemblage and carbon content data in this study were generously provided by K. Buck. We thank R. Dugdale and F. Chavez for the opportunity to participate in this cruise, and the officers and crew of the RV *Point Sur* for their assistance. This work was supported by NSF grants OCE-9411497 and OCE-9102532 and NASA grant NAGW-4291.

REFERENCES

1. Challenger, F. and M. I. Simpson. 1948. Studies on biological methylation. Part XII. A precursor of the dimethyl sulfide evolved by *Polysiphonia fastigiata*. Dimethyl-2-carboxyethylsulphonium hydroxide and its salts. J. Chem. Soc. *1948*: 1591-1597.
2. de Souza, M. P. and D. C. Yoch. 1995. Purification and characterization of dimethylsulfoniopropionate lyase from an *Alcaligenes*-like dimethylsulfide-producing marine isolate. Appl. Environ. Microbiol. *61*: 21-26.
3. de Souza, M. P. and D. C. Yoch. 1996. N-terminal amino acid sequences and comparison of DMSP lyases from *Pseudomonas doudoroffii* and *Alcaligenes* strain M3A. *In:* Environmental biological and chemistry of DMSP and related sulfonium compounds, R. Kiene, P. Visscher, M. Keller, G. Kirst, eds., Plenum Press, New York.
4. Diaz, M. R., P. T. Visscher and B. F. Taylor. 1992. Metabolism of dimethylsulfoniopropionate and glycine betaine by a marine bacterium. FEMS Microbiol. Let. *96*: 61-66.
5. Fersht, A. 1985. Enzyme structure and mechanism. W. H. Freeman and Company, New York.
6. Kiene, R. P. 1990. Dimethyl sulfide production from dimethylsulfoniopropionate in coastal seawater samples and bacterial cultures. Appl. Environ. Microbiol. *56*: 3292-3297.
7. Kiene, R. P. 1992. Dynamics of dimethyl sulfide and dimethylsulfoniopropionate in oceanic water samples. Mar. Chem. *37*: 29-52.
8. Kiene, R. P. and S. K. Service. 1991. Decomposition of dissolved DMSP and DMS in estuarine waters: dependence on temperature and substrate concentration. Mar. Ecol. Prog. Ser. *76*: 1-11.

9. Ledyard, K. M. and J. W. H. Dacey. 1994. Dimethylsulfide production from dimethylsulfoniopropionate by a marine bacterium. Mar. Ecol. Prog. Ser. *110*: 95-103.
10. Ledyard, K. M. and J. W. H. Dacey. In press. Microbial cycling of DMSP and DMS in coastal and oligotrophic seawater. Limnol. Oceanogr.
11. Ledyard, K. M., E. F. DeLong and J. W. H. Dacey. 1993. Characterization of a DMSP-degrading bacterial isolate from the Sargasso Sea. Arch. Microbiol. *160*: 312-318.
12. Stefels, J., W. W. C. Gieskes and L. Dijkhuizen. 1996. Intriguing functionality of the production and conversion of DMSP in *Phaeocystis* sp. *In:* Environmental biological and chemistry of DMSP and related sulfonium compounds, R. Kiene, P. Visscher, M. Keller, G. Kirst, eds., Plenum Press, New York.
13. Stefels, J. and W. H. M. van Boekel. 1993. Production of DMS from dissolved DMSP in axenic cultures of the marine phytoplankton species *Phaeocystis* sp. Mar. Ecol. Prog. Ser. *97*: 11-18.
14. Taylor, B. F. and D. C. Gilchrist. 1991. New routes for aerobic degradation of dimethylsulfoniopropionate. Appl. Environ. Microbiol. *57*: 3581-3584.
15. Turner, S. M., G. Malin and P.S. Liss. 1989. Dimethyl sulfide and (dimethylsulfonio)propionate in European coastal and shelf waters, p. 183-200. *In*: E. S. Saltzman and W. J. Cooper (eds.), Biogenic sulfur in the environment. Am. Chem. Soc. Symp. Ser. 393, Washington, D. C.
16. van der Maarel, M. J. E. C., P. Quist, L. Dijkhuizen and T. A. Hansen. 1993. Anaerobic degradation of dimethylsulfoniopropionate to 3-*S*-methylmercaptopropionate by a marine *Desulfobacterium* strain. Arch Microbiol. *160*: 411-412.
17. Visscher, P. T., M. R. Diaz and B. F. Taylor. 1992. Enumeration of bacteria which cleave or demethylate dimethylsulfoniopropionate in the Caribbean Sea. Mar. Ecol. Prog. Ser. *89*: 293-296.
18. Visscher, P. T., R. P. Kiene and B. F. Taylor. 1994. Demethylation and cleavage of dimethylsulfoniopropionate in marine intertidal sediments. FEMS Microbiol. Ecol. *14*: 179-190.
19. Visscher, P. T. and B. F. Taylor. 1994. Demethylation of dimethylsulfoniopropionate to 3-mercaptopropionate by an aerobic marine bacterium. Appl. Environ. Microbiol. *60*: 4617-4619.
20. Wakeham, S. G., B. L. Howes and J. W. H. Dacey. Biogeochemistry of dimethylsulfide in a seasonally stratified coastal salt pond. Geochim. Cosmochim. Acta *51*: 1675-1684.
21. White, R. H. 1982. Analysis of dimethyl sulfonium compounds in marine algae. J. Mar. Res. *4*: 173-181.
22. Wolfe, G. V. 1996. Accumulation of dissolved DMSP by marine bacteria and its degradation via bacterivory *In:* Environmental biological and chemistry of DMSP and related sulfonium compounds, R. Kiene, P. Visscher, M. Keller, G. Kirst, eds., Plenum Press, New York.
23. Wolfe, G. V. and R. P. Kiene. 1993. Effects of methylated, organic, and inorganic substrates on microbial consumption of dimethyl sulfide in estuarine waters. Appl. Environ. Microbiol. *59*: 2723-2726.
24. Wolfe, G. V., E. B. Sherr and B. F. Sherr. 1994. Release and consumption of DMSP from *Emiliania huxleyi* during grazing by *Oxyrrhis marina*. Mar. Ecol. Prog. Ser. 111: 111-119.

29

TURNOVER OF DISSOLVED DMSP IN ESTUARINE AND SHELF WATERS OF THE NORTHERN GULF OF MEXICO

Ronald P. Kiene

Department of Marine Sciences
University of South Alabama
LSCB 25, Mobile, Alabama 36688
Dauphin Island Sea Lab
Dauphin Island, Alabama 36543

SUMMARY

The turnover rates of dissolved dimethylsulfoniopropionate (DMSP(d)) were determined in a variety of coastal water samples collected from the Northern Gulf of Mexico during a 14 month period spanning 1993 and 1994. A kinetic approach was used in which the rate constant for DMSP(d) consumption was determined from time course data of DMSP(d) concentrations after spike additions of 40 to 50 nM. These rate constants were multiplied by the *in situ* concentration of DMSP(d) to obtain the turnover rate. Rates measured were variable and ranged from 2 to 122 $nM{\cdot}d^{-1}$ during the study period. No seasonal pattern was observed nor was there any significant relationship between turnover rates (or rate constants) with temperature, salinity, or concentrations of dimethylsulfide (DMS), particulate DMSP (DMSP(p)) or DMSP(d). Based on differential filtration experiments, the turnover of DMSP(d) was mainly attributed to submicron-sized organisms, probably bacteria. DMSP was degraded to both dimethylsulfide (DMS) and methanethiol (CH_3SH) and the turnover of DMSP(d) is likely to be a major source of these compounds in Gulf of Mexico waters.

INTRODUCTION

Dimethylsulfoniopropionate (DMSP; $[(CH_3)_2S^+CH_2CH_2COO^-]$) is an organic sulfur compound which is distributed throughout the euphotic zone of the oceans (7, 15, 36). Marine phytoplankton appear to be the main source of DMSP in the marine water column but only certain species produce large amounts of this osmotic solute (16, 17, 39). The distribution of DMSP, is therefore, heterogeneous and, in most cases, not strictly tied to algal biomass (31).

The marine biogeochemistry of DMSP is of interest, in part, because this compound is the main precursor of dimethylsulfide (DMS), the major volatile sulfur compound found in the surface ocean (1, 3, 15, 21, 30, 36). DMS emissions from the ocean contribute nearly half of the global biogenic sulfur emission to the atmosphere (1) and these emissions may also play a role in modulating global climate through formation of sulfur-containing aerosols (2, 4, 8, 32).

The release and degradation of DMSP, as well as the production of DMS are closely linked with food web activities (6, 9, 14, 21, 41). Bacterial degradation of the dissolved DMSP pool (DMSP(d)) (operationally defined as that which passes a GF/F or equivalent glass fiber filter) is thought to be a major pathway leading to DMS formation (19, 25, 36). While degradation of DMSP appears to be the main source of DMS in sea water, not all DMSP is degraded to DMS (6, 25). A significant fraction of the DMSP(d) degraded in marine surface waters appears to be converted to highly reactive methanethiol (CH_3SH) (18). In addition to its important role as a precursor of DMS and CH_3SH, DMSP may also represent an important source of carbon for marine bacterioplankton (21, 37). For these reasons, it is important to know how fast the different pools of DMSP (dissolved and particulate) turn over. To date, the only study in which determined the rates of DMSP(d) turnover under natural conditions in the marine water column was presented by Ledyard and Dacey (28).

In the present study, the rates of DMSP(d) turnover in Gulf of Mexico waters were estimated with a kinetic approach. The results indicate a large potential for DMSP(d) consumption and relatively rapid, but variable, rates of turnover in the waters sampled. The potential problems with the kinetic approach and its potential utility are discussed.

MATERIALS AND METHODS

Sample Collection and Processing

Most water samples used during this study were collected over a 14 month period from a pier on the east end of Dauphin Island, AL. This site is located in the Northern Gulf of Mexico near the mouth of Mobile Bay (30° 20′ N, 88° 10′ W). One set of samples was collected by boat approximately 10 km south of the mouth of Mobile Bay at a site termed the Sea Buoy and another was collected 50 km south of Dauphin Island in the open Gulf of Mexico. In still other cases, water was collected from shore sites located near Fort Walton Beach, FL and in Santa Rosa Sound, FL. Water from these location was generally higher in salinity and lower in suspended solids than the Mobile Bay water. Unless otherwise indicated, samples were collected from the surface by bucket or carboy and dispensed immediately into 1 liter or 250 ml Teflon bottles. Water samples were stored in the dark and returned to the laboratory within 1 h where they were used immediately for experimental incubations. During the incubations, the bottles were maintained within 1 degree of *in situ* temperature and kept in the dark, except during sub-sampling (< 2 min duration) when they were exposed to room light. The water samples were not shaken during the incubations, but were gently inverted several times before subsamples were removed for sulfur compound analysis.

Experimental Design

In order to determine the kinetics of DMSP(d) degradation, spike additions of DMSP (40-50 nM final concentrations) were made to freshly collected water samples. Controls which received no additions were run in parallel. The concentrations of DMSP(d), particulate DMSP (DMSP(p)) and DMS were monitored in water samples over time courses which

DMSP (DMSP(p)) and DMS were monitored in water samples over time courses which lasted from 5 to 30 h depending on the experiment. The first sample for DMSP(d) was taken within 2 min of the addition and this time point was designated as time zero. Changes in sulfur pools in unamended controls were minor compared to those observed in DMSP-treated samples. In two experiments, different amounts of DMSP(d) were added to a series of bottles from the same batch of seawater to give a range of initial DMSP concentrations. The highest concentration used was 230 nM. In these cases, the results are presented graphically as the ratio of the concentration (C) measured at each time point divided by the initial concentration (C_o), which normalizes each treatment to a common initial value. Because the consumption kinetics for DMSP(d) were fast (see results), and a rapid sampling schedule was necessary, single bottles were often run for each treatment. Duplicate treatments were occasionally run and replication for DMS and DMSP(d) measurements was usually better than 10%.

Calculation of Kinetic Rate Constants and DMSP Turnover Rates

Apparent first order rate constants (k) for the loss of DMSP(d) were estimated by plotting the natural log of the DMSP(d) concentration vs time and taking the slope of this line as k. The loss of DMSP from the dissolved pool represented degradation since particulate pools were constant or declined slightly during the dark incubations. All rate constants reported here are from experiments in which the DMSP(d) loss data obeyed apparent first order kinetics and in which the added DMSP(d) was equal to or less than 50 nM. In 3 out of the 20 experiments conducted during this investigation it was evident that DMSP(d) decreased linearly (zero order kinetics) rather than with an exponential pattern, perhaps suggesting that DMSP consumption was saturated in these 3 cases. The DMSP(d) turnover rates for these experiments are not included here, so that only the data for experiments showing apparent first order consumption are compared. Endogenous DMSP(d) was less than 11 nM in all cases; often less than 6 nM. The turnover rate of DMSP(d) in $nM{\cdot}d^{-1}$ was estimated by multiplying the endogenous DMSP(d) concentration determined at the start of the incubations by the loss rate constant according to:

$$\frac{d\,[DMSP\,(d)]}{dt} = -k\,[DMSP\,(d)]$$

where k is the first order loss rate constant (d^{-1}) and [DMSP(d)] is the concentration of DMSP(d) in nM.

Differential Filtration

In order to determine the size class of organisms responsible for consuming DMSP(d) several experiments were carried out in which subsamples of the initial batch of seawater were filtered through either 0.2 or 1.0 µm polycarbonate membrane filters (47 mm, Nucle-pore) before the addition of DMSP(d). A glass filtration tower was used and the vacuum was kept below 10 mm Hg. Despite the relatively low vacuum, the filtration of whole seawater though these membrane filters resulted in considerably higher DMSP(d) concentrations in the filtrates as compared to unfiltered water samples. This resulted most likely from lysis of DMSP-containing algal cells during the filtrations.

DMSP and DMS Time Series

The concentrations of DMSP(p), DMSP(d) and DMS were determined (see below for methods) in water samples collected from the Dauphin Island East End Pier over the 14 month period considered in this study. Some of the samples in this time series represent the initial values obtained in water samples used for incubations. Temperature and salinity data were also collected for these water samples. Salinity was determined by refractometer.

Analytical Methods

DMS and DMSP(p) were measured by purge and trap gas chromatography as described previously (25). The procedure for DMSP(d) analysis was modified somewhat from that used previously. A 20 ml subsample was removed from the incubation bottle and allowed to drip through a 47 mm Gelman AE (1.0 μm nominal pore size) filter held in a glass filter tower. The filter was used for DMSP(p) determinations while the filtrate was used for DMSP(d). After all of the sample had passed the filter, approximately 5 ml of the filtrate was placed in a small open sparge tube and bubbled with He (100 ml·min^{-1} for 2 min) to remove DMS. After the He flow was turned off, 1 ml of the sample was removed by pipette and placed in a 14 ml serum vial. One ml of 5 N NaOH was added to this vial and it was sealed quickly with a Teflon-faced butyl rubber septum. DMSP in the water sample was decomposed quantitatively to DMS (and acrylic acid) in the NaOH. After 30 min, the reaction was complete and the sample could be analyzed, though they were often analyzed the next day (< 24 h). The DMS in the vials was measured by sweeping the headspaces of the serum vials into a cryotrap and subsequently into a gas chromatograph as described in Kiene and Gerard (23). Standards were prepared using the same liquid volumes as the samples. This approach yielded excellent precision (typically better than 5%) and low detection limits (0.3-0.5 nM) for 1 ml samples.

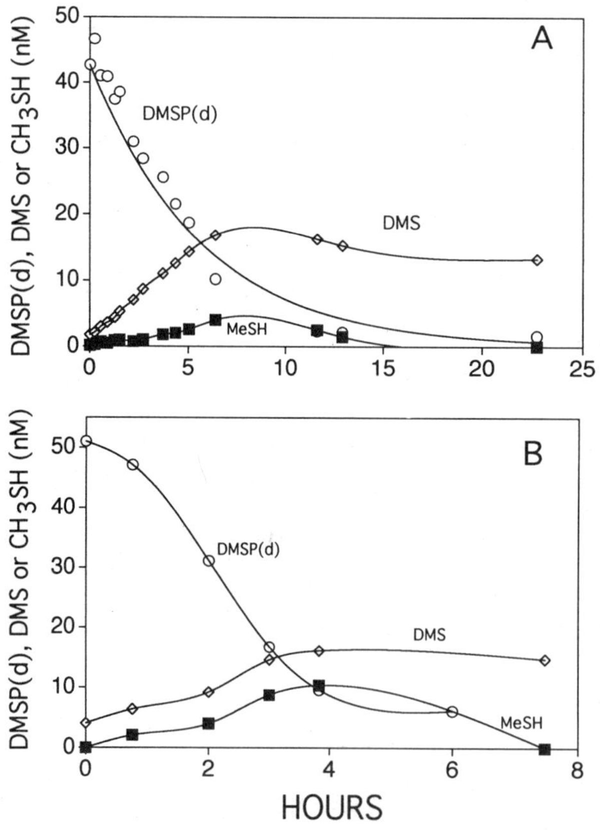

Figure 1. Time courses of DMSP(d), DMS and CH_3SH in water samples collected from the Dauphin Island Pier after the addition of ~40 nM DMSP(d). Panel A: Sample collected on October 26, 1994. The incubation temperature was 21°C and the salinity was 17. The DMSP(d) concentration in parallel unspiked controls ranged between 1.5 and 3.3 nM throughout the incubation. Panel B: Sample collected on September 8, 1994. Incubation temperature was 26°C and the salinity was 19. The DMSP(d) concentration in parallel unspiked controls ranged between 4.5 and 3.8 nM throughout the incubation. Data are from single bottles and all curves are interpolated fits except for DMSP(d) in panel A which is an exponential fit.

Methanethiol (CH_3SH) was identified by cochromatography with authentic CH_3SH and this compound was quantified by using standard curves for DMS. Previous work has shown that the standard curves for CH_3SH and DMS differ by less than 5% (22). DMSP·HCl was obtained from Research Plus Inc. and was used from concentrated stocks which were kept frozen.

RESULTS

After the addition of exogenous DMSP(d) (~50 nM) to freshly collected water samples, the concentrations of DMSP(d) typically declined rapidly and within 4-12 h approached those in unamended samples (Fig. 1). In most cases the net consumption of the added DMSP(d) began immediately and could be modeled with a first order (exponential) decay function (Fig. 1A). Occasionally, a short (< 1 h) lag in the net DMSP(d) consumption rate was observed as illustrated by the data in Figure 1B, but it is not always clear if this represented a true lag or simply variability in the measurements. The consumption of DMSP(d) was accompanied by the production of both DMS and CH_3SH (Fig. 1A & B). The net yields of DMS and CH_3SH varied in different experiments and ranged from 7-64% and 12-65% for CH_3SH and DMS respectively. The results on relative yields of sulfur gases from DMSP degradation are presented in greater detail in (18). For DMSP(d) amendments below ~100 nM, plots of C/C_o vs time showed nearly identical time courses (Fig. 2 A & B), suggesting that the rate of DMSP(d) consumption was not saturated at these concentrations.

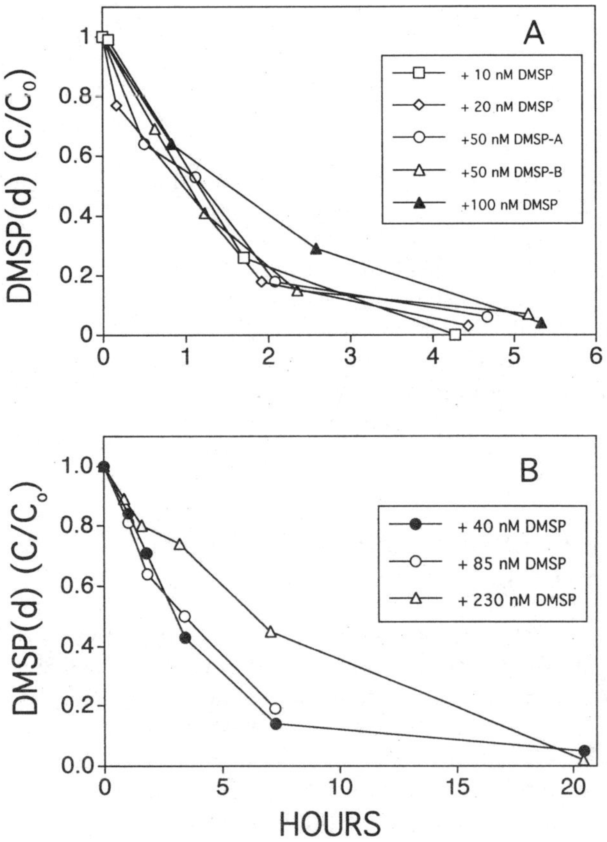

Figure 2. DMSP(d) consumption curves for two experiments (panels A & B) in which DMSP(d) was added at several different concentrations. Data are presented as C/C_0 to illustrate the relative rates of consumption at the different levels of DMSP. Data in Panel A are from samples collected from Santa Rosa Sound (May 11, 1994; 25°C, salinity = 26) and samples in Panel B are from the Dauphin Island Pier (December 7, 1994; 17.5°C, salinty = 11).

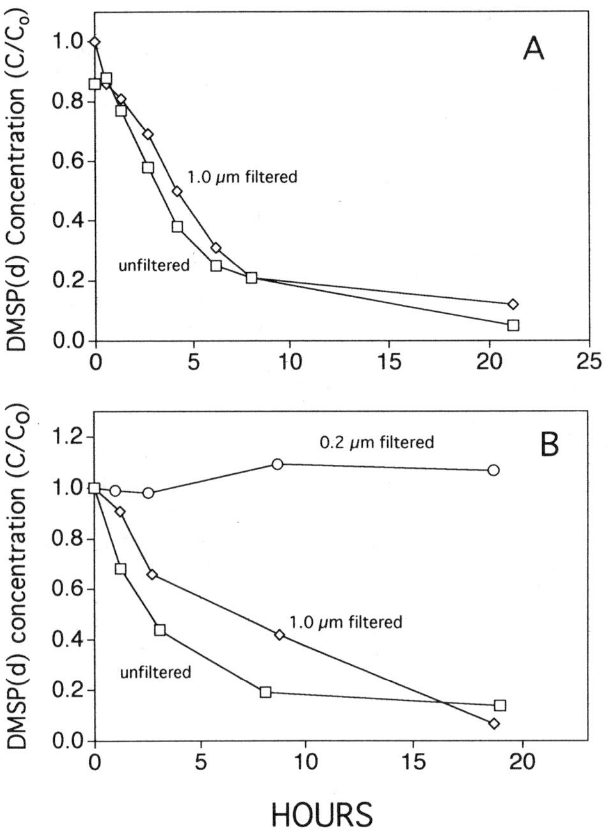

Figure 3. The effect of filtration on consumption of ~40 nM added DMSP(d) in two separate experiments. The indicated samples were filtered through polycarbonate membrane filters of the pore sizes shown. Both experiments used water from the Dauphin Island Pier.

At or above 100 nM, the relative rates of DMSP(d) consumption were often slower (Fig. 2 A & B), suggesting that these substrate concentrations may begin to saturate the DMSP(d) consumption activity.

The removal of DMSP(d) from seawater was primarily due to submicron-sized organisms, as 1.0 μm filtered samples showed only slightly slower consumption rates than unfiltered controls (Fig. 3 A & B). Similar minimal effects have been obtained with Gelman AE glass fiber filtered sea water (nominal retention 1.0 μm) (data not shown). On the other hand, 0.2 μm filtration essentially stopped DMSP consumption (Fig 3B). Both 1.0 μm and AE filter treatments removed > 95% of the particulate DMSP in the samples but allowed passage of > 80% of the bacteria as determined by acridine orange direct counts (data not shown).

Temporal Variability of Dimethyl Sulfur Compounds and DMSP Turnover Rates

The concentrations of DMS, DMSP(p) and DMSP(d) measured in water samples collected over a year long period from the East End Pier on Dauphin Island are presented in Fig. 4. The corresponding temperature and salinity data are shown in Fig. 5. Each pool of sulfur compound displayed considerable temporal variability, but no distinct seasonal pattern was evident in this data set. Correlation analysis did not reveal any significant relationships among the sulfur compounds or between these and salinity or temperature. However, the highest concentrations of DMSP(d) and DMS were observed during a bloom of DMSP-pro-

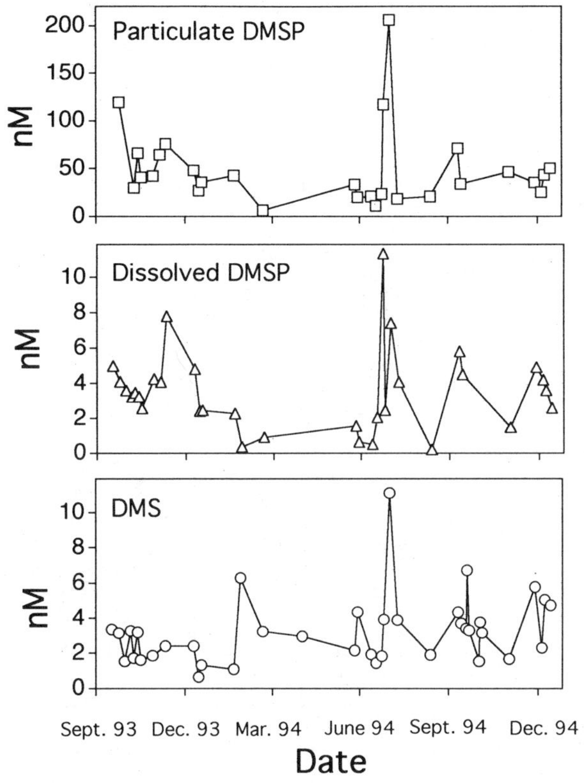

Figure 4. Time series of organic sulfur pools in surface waters collected from the Dauphin Island Pier.

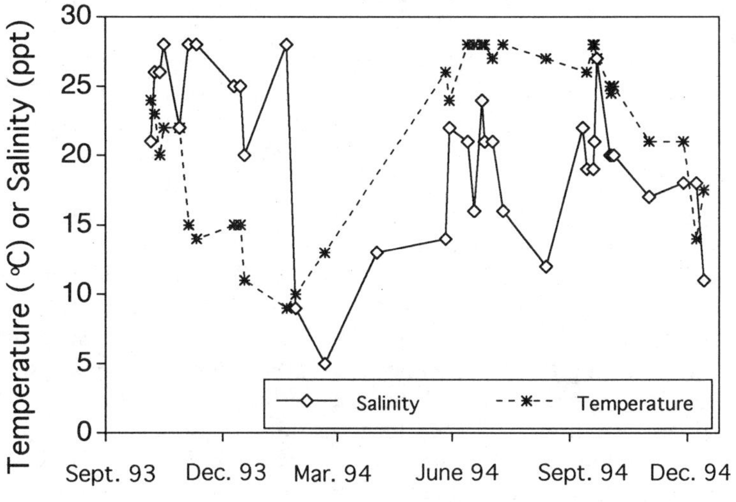

Figure 5. Time series of salinity and temperature in water samples collected from the Dauphin Island East End Pier.

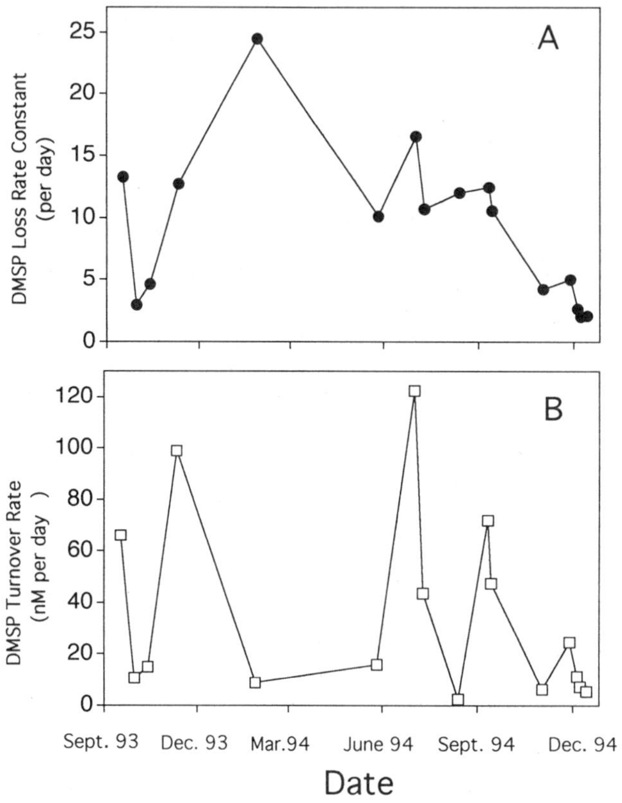

Figure 6. Panel A. DMSP loss rate constants determined in water samples collected from the Dauphin Island Pier over a 14 month period. Each data point represents a separate determination of the rate constant on water collected on that date. Panel B. DMSP turnover rates on different dates during the study period. Turnover rates were obtained by multiplying the rate constants (panel A) by the endogenous DMSP(d) concentrations (Figure 4) on that date.

ducing phytoplankton which occurred in late June, 1994. During this period, DMSP(p) reached 206 nM while DMSP(d) and DMS reached lower maximal concentrations of 11.4 and 11.1 nM respectively. For the entire data set, the average (± std dev.) concentration of DMS was 3.2 ± 2.0 nM (n= 38), while DMSP(d) averaged 3.5 ± 2.4 nM (n=32). The mean DMSP(p) concentration (49.0 ± 41.2 nM, n= 28) was much higher than for either of the other dimethyl sulfur pools.

The first order rate constants for DMSP(d) consumption (calculated from loss curves such as those in Fig.1) in water samples from the Dauphin Island Pier varied from 2.0 to 24.5 $\cdot d^{-1}$ over the study period (Fig. 6A). The turnover rates for DMSP(d) calculated from these rate constants and the *in situ* DMSP(d) concentration are presented in Fig. 6B and ranged from 2.4 to 122 $nM \cdot d^{-1}$. Each of these parameters exhibited considerable temporal variability, although the rate constant showed a narrower range than did the turnover rate. Some of the variability may have been due to variations in temperature, salinity or pools of DMSP(p) (see Fig. 6). However, for the entire data set, no significant correlations were observed when the rate constants or the turnover rates were regressed against DMSP(p), DMS, DMSP(d), temperature or salinity (data not shown). During the late summer and Fall of 1994 the rate constants appeared to decline into the winter period and this was reflected in the turnover rates as well. On several occasions, low DMSP(d) turnover rates resulted from very low *in situ* concentrations of DMSP(d). The highest rate constant (24.5 $\cdot d^{-1}$) was obtained in January, 1994 when the DMSP(d) concentration was very low (0.36 nM). The resulting turnover rate (8.8 $nM \cdot d^{-1}$) was among the lowest observed. The highest DMSP(d) turnover rate obtained (122 $nM \cdot d^{-1}$) was during the bloom in DMSP(p) on June 28,

1994 when the DMSP(d) concentration was 7.4 nM. The highest DMS concentration of the year (11.1nM) was observed on this date as well.

Data on DMSP and DMS concentrations as well as DMSP turnover rates for Gulf of Mexico sites other than the Dauphin Island Pier are presented in Table 1. These data were collected from offshore locations and from shore sites located east of Mobile Bay where salinities are generally higher. Concentrations of DMSP(p), DMSP(d) and DMS fell in the same range as those observed in Mobile Bay, though the average concentrations were somewhat lower than for Mobile Bay. The limited number of kinetically determined turnover rate estimates also showed somewhat lower rates than observed at the Dauphin Island Pier site. For the Santa Rosa Sound samples, there was considerable potential for DMSP(d) turnover as evidenced by a relatively high rate constant of $20 \cdot d^{-1}$, however DMSP(d) was not detectable (< 0.3 nM) and the overall turnover rate was therefore very low (< $6\ nM \cdot d^{-1}$).

DISCUSSION

The turnover of DMSP is of interest because this compound is one of the most abundant reduced sulfur compounds in the surface ocean and it is degraded to important products such as DMS and CH_3SH (Fig. 1). Presently, there is limited information available on the turnover rates of DMSP(d) under *in situ* conditions in seawater (28). The results obtained here suggest that DMSP(d) turnover rates are both rapid and highly variable in Gulf of Mexico Coastal waters.

There is considerable potential for turnover of DMSP(d) in subtropical coastal waters as evidenced by the fact that spike additions of DMSP(d) in the range of 40-50 nM were consumed to near ambient levels within 3 to 10 h in all of the water samples collected during this 14 month study. In 17 out of 20 experiments the consumption of added DMSP(d) appeared to follow first order kinetics (e.g. Fig. 1A), that is DMSP(d) declined exponentially with time. DMSP(d) consumption rates were rarely saturated below substrate concentrations of 100 nM (Fig. 2) despite the fact that *in situ* concentrations were almost always at least 10-fold lower than this (Fig. 4). These findings are consistent with other studies of DMSP(d) consumption in marine coastal waters (25,28) and suggest that DMSP turnover is limited primarily by the availability of DMSP(d), which is maintained at low concentrations throughout the year (Fig. 4). The relatively large loss rate constants ($2\text{-}24.5 \cdot d^{-1}$) determined for DMSP(d) suggest that the *in situ* dissolved DMSP pool turns over rapidly with turnover times of 1-12 h. Such fast turnover is similar to what has been observed for other labile, low molecular weight compounds such as amino acids (12,13).

The turnover rates obtained from the product of kinetically-determined rate constants and the *in situ* concentration of DMSP(d) were quite variable during the study period and ranged from 2 to 122 $nM \cdot d^{-1}$ (mean 35.6 $nM \cdot d^{-1}$, n = 17; Fig. 6B) on different dates. These rates represented a turnover of between 11 to 240% (mean 70%) of the measured particulate DMSP pool per day, assuming the latter was in steady state. No distinct seasonal pattern was evident in the turnover rates nor in the DMSP and DMS pools measured (Fig. 4). The fact that the coastal waters sampled during this study varied considerably with respect to salinity, temperature and other factors made it highly unlikely that simple correlations between turnover rates and other parameters would be significant. The samples from sites less influenced by Mobile Bay (Table 1) had somewhat lower average concentrations of sulfur compounds as well as DMSP(d) turnover rates, but the data set is too small to draw firm conclusions about site differences. The DMSP(d) turnover rate estimates from the present study (2 to 122 $nM \cdot d^{-1}$) are in reasonable agreement with the net rates of DMSP(d) loss (undetectable to 38.4 $nM \cdot d^{-1}$) reported by Ledyard and Dacey for coastal waters of Vineyard

Table 1. Concentrations of DMSP and DMS as well as DMSP turnover rates for sites in the Northern Gulf of Mexico which are not strongly influenced by Mobile Bay.

Location	Date	Salinity	Temperature (oC)	DMS (nM)	DMSP(d) (nM)	DMSP(p) (nM)	DMSP Turnover	
							Rate Constant (d-1)	Turnover rate (nM.d-1)
Santa Rosa Sound	5/11/94	26	25	3.8	<0.3	5.1	20.0	< 6.0
Fort Morgan Beach	7/19/94	28	28	2.5	2.7	18.2	6.1	16.4
Sea Bouy	11/2/94	32	22	1.4	2.6	45.0	2.4	6.3
Gulf of Mexico 50 km south of Dauphin Island (20 m depth)	12/8/94	37	18	3.6	2.1	9.6	1.1	2.3

Sound (28). These authors also have reported considerable temporal and spatial variability in DMSP turnover and in the kinetics of DMSP uptake and lyase degradation (27,28), again consistent with the present study.

At the present time it is not known if the kinetic approach to determining DMSP turnover rates gives valid results. During the same period considered in the present study Kiene and Gerard (in press) used an alternative approach which was based on the inhibition of DMSP(d) uptake by 50 μM glycine betaine (GBT; GBT is a potent short term inhibitor of DMSP degradation, and its addition to seawater causes endogenous DMSP(d) to accumulate (24)). The net rates of GBT-induced DMSP(d) accumulation in Gulf of Mexico waters during 1993-1994 ranged from 4.1 to 27.3 $nM \cdot d^{-1}$ (mean of 12.9 $nM \cdot d^{-1}$). These rates were somewhat lower and less variable than the kinetically determined rates and they were also strongly correlated with incubation temperature (24). Unfortunately direct comparisons of the two approaches were not done during this period. The fact that the two different approaches yielded values which are of the same order of magnitude suggests that the true turnover rates may have been in the range of those determined.

Evaluating the approaches used for DMSP(d) turnover rate determinations is difficult since there are no accepted reference techniques. For the kinetic approach, errors in determination of the *in situ* concentration of DMSP(d) or the rate constant for its removal would lead to errors in the turnover rate. The most likely error in determination of DMSP(d) would be an over estimation of the true dissolved concentration due either to filtration-induced release from the particulate pool, or possible inclusion of DMSP-containing bacterial cells (see Wolfe (40)) in the glass fiber filtrates (29). Either of these situations would result in over-estimation of the turnover rate. Likewise, if the rate constant is over-estimated, due perhaps to induction of DMSP lyase activity (10), this would also cause an over-estimation of the turnover rate. Induction of activity is quite possible, since DMSP was added to concentrations significantly above (about 10- fold) ambient levels, and incubations of several hours were required to determine uptake constants. In some experiments, a lag of 1 h or less in the consumption of DMSP(d) was observed (Fig. 1B) and such a lag could have been missed in some experiments due to the sampling schedule. If this were a common, but often missed feature, the rate constants (and therefore turnover rates) reported here may be too high. Furthermore, it may not be possible to apply the kinetic approach with samples in which the *in situ* DMSP(d) concentration is not in approximate steady state. In most cases, though, the steady state approximation holds reasonably well (R. Kiene, unpublished data). The GBT inhibition method appears to give somewhat lower rates than the kinetic approach, (24) but this too has its limitations. Further studies will be necessary to evaluate these techniques and to determine DMSP(d) turnover rates under a variety of conditions.

DMSP(d) turnover in Gulf of Mexico waters was mainly due to submicron sized organisms, probably bacteria (Fig. 3). Marine bacteria which are capable of metabolizing DMSP have been isolated in a number of studies (10,11,26,38) and DMSP-utilizing bacteria are relatively abundant in the field (37). Other organisms such as the alga *Phaeocystis* sp. may also be capable of degrading DMSP(d) (34), but their importance in Gulf of Mexico waters appears to have been small. DMSP may represent a potentially important carbon substrate for bacterial populations in the marine environment since DMSP carbon may comprise as much as 1-10% of the carbon in living phytoplankton (5,21,33). It is therefore not surprising that the ability to degrade DMSP is widespread among marine aerobic bacteria (26,37). At least 2 functional groups of bacteria are responsible for degrading DMSP in sea water: those which cleave DMSP into DMS and acrylic acid and those which demethylate it to 3-methiolpropionate (11,26,35,37,38). Evidence for the operation of the demethylation pathway in seawater has been circumstantial (6,20,25), but recent findings that CH_3SH is a major product of DMSP (see Fig. 1) confirm the importance of the demethylation pathway (18).

The relatively rapid turnover rates for DMSP(d) reported here are significant because they imply a substantial production of reactive gases such as DMS and CH_3SH (see Fig. 1 and also (18)). DMSP(d) is likely to be a major source of DMS in the water column, but because DMSP conversion efficiency to DMS can be highly variable in different water masses (18,28), it is difficult to predict, from DMSP turnover alone, what that contribution might be. Further studies are needed to investigate the factors which control DMSP turnover and the relative amounts of products produced. The kinetic approach used here, if validated, should be useful in these investigations.

ACKNOWLEDGMENTS

I thank Ghislain Gerard for excellent technical assistance during this study. I also thank Pieter Visscher and especially Kathleen Ledyard for providing critical comments on the manuscript. Funding for this work was provided by NSF grant OCE9203728. This is contribution # 280 of the Dauphin Island Sea Lab.

REFERENCES

1. Andreae, M. O. 1990. Ocean-atmosphere interactions in the global biogeochemical sulfur cycle. Mar. Chem. *30*: 1-29.
2. Andreae, M. O., W. Elbert and S. J. deMora. 1995. Biogenic sulfur emissions and aerosols over the tropical South Atlantic. 3. Atmospheric dimethylsulfide, aerosols and cloud condensation nuclei. J. Geophys. Res. *100*: 11335-11356.
3. Andreae, M. O. and H. Raemdonck. 1983. Dimethyl sulfide in the surface ocean and the marine atmosphere: a global view. Science. *221*: 744-747.
4. Bates, T. S., R. J. Charlson and R. H. Gammon. 1987. Evidence for the climatic role of marine biogenic sulphur. Nature. *329*: 319-321.
5. Bates, T. S. et al. 1994. The cycling of sulfur in surface sea water of the Northeast Pacific. JGR-Oceans. *99*: 7835-7843.
6. Belviso, S. et al. 1990. Production of dimethylsulfonium propionate (DMSP) and dimethylsulfide (DMS) by a microbial food web. Limnol. Oceanogr. *35*: 1810-1821.
7. Burgermeister, S. et al. 1990. On the biogenic origin of dimethylsulfide: relation between chlorophyll, ATP, organismic DMSP, phytoplankton species, and DMS distribution in Atlantic surface water and atmosphere. J. Geophys. Res. *95*: 20607-20615.
8. Charlson, R. J., J. E. Lovelock, M. O. Andreae and S. G. Warren. 1987. Oceanic phytoplankton, atmospheric sulfur, cloud albedo and climate. Nature. *326*: 655-661.
9. Dacey, J. W. H. and S. G. Wakeham. 1986. Oceanic dimethylsulfide: production during zooplankton grazing on phytoplankton. Science. *233*: 1314-1316.
10. de Souza, M. P. and D. C. Yoch. 1995. Purification and characterization of DMSP lyase from an *Alcaligenes*-like dimethylsulfide-producing marine isolate. Appl. Environ. Microbiol. *61*: 21-26.
11. Diaz, M. R., P. T. Visscher and B. F. Taylor. 1992. Metabolism of dimethylsulfoniopropionate and glycine betaine by a marine bacterium. FEMS Microbiol. Lett. *96*: 61-66.
12. Ferguson, R. L. and W. G. Sunda. 1984. Utilization of amino acids by planktonic marine bacteria: importance of clean technique and low substrate additions. Limnol. Oceanogr. *29*: 258-274.
13. Fuhrman, J. 1987. Close coupling between release and uptake of dissolved free amino acids in seawater studied by an isotope dilution approach. Mar. Ecol. Prog. Ser. *37*: 45-52.
14. Gabric, A., N. Murray, L. Stone and M. Kohl. 1993. Modelling the production of dimethylsulfide during a phytoplankton bloom. J. Geophys. Res. *98*: 22805-22816.
15. Iverson, R. L., F. L. Nearhoof and M. O. Andreae. 1989. Production of dimethylsulfonium propionate and dimethylsulfide by phytoplankton in estuarine and coastal waters. Limnol. Oceanogr. *34*: 53-67.
16. Keller, M. D. 1991. Dimethyl sulfide production and marine phytoplankton: The importance of species composition and cell size. Biol. Oceanogr. *6*: 375-382.

17. Keller, M. D., W. K. Bellows and R. R. L. Guillard. 1989. Dimethyl sulfide production in marine phytoplankton, p. 167-182. *In* E. Saltzman andW. J. Cooper (Ed.), Biogenic sulfur in the environment. Am. Chemical Soc., New York.
18. Kiene, R. P. Production of methane thiol from dimethylsulfoniopropionate in sea water in marine surface waters. Submitted to Marine Chemistry.
19. Kiene, R. P. 1990. Dimethyl sulfide production from dimethylsulfoniopropionate in coastal seawater samples and bacterial cultures. Appl. Environ. Microbiol. *56*: 3292-3297.
20. Kiene, R. P. 1992. Dynamics of dimethyl sulfide and dimethylsulfoniopropionate in oceanic seawater samples. Mar. Chem. *37*: 29-52.
21. Kiene, R. P. 1993. Microbial sources and sinks for methylated sulfur compounds in the marine environment, p. 15-33. *In* D. P. Kelly andJ. C. Murrell (Ed.), Microbial growth on C_1 compounds. Intercept Ltd., London.
22. Kiene, R. P. 1996. Microbial cycling of organosulfur gases in marine and freshwater environments, p. *In press*. *In* D. Adams, S. Seitzinger andP. Crill (Ed.), Cycling of reduced gases in the hydrosphere. E. Schweitzerbart'sche Verlagsbuchhandlung (Naglele u. Obermiller), Stuttgart.
23. Kiene, R. P. and G. Gerard. 1994. Determination of trace levels of dimethylsulfoxide (DMSO) in sea water and rainwater. Mar. Chem. *47*: 1-12.
24. Kiene, R. P. and G. Gerard. 1995. Evaluation of glycine betaine as an inhibitor of dissolved dimethylsulphoniopropionate degradation in marine waters. Mar. Ecol. Prog. Ser. *128*: 121-131.
25. Kiene, R. P. and S. K. Service. 1991. Decomposition of dissolved DMSP and DMS in estuarine waters: dependence on temperature and substrate concentration. Mar. Ecol. Prog. Ser. *76*: 1-11.
26. Ledyard, K. M. and J. W. H. Dacey. 1994. Dimethylsulfide production from dimethylsulfoniopropionate by a marine bacterium. Mar. Ecol. Prog. Ser. *110*: 95-103.
27. Ledyard, K. M. and J. W. H. Dacey. 1996. Kinetics of DMSP-lyase activity in coastal seawater, *In:* Biological and environmental chemistry of DMSP and related sulfonium compounds. R. P. Kiene, P. T. Visscher, M. D. Keller and G. O. Kirst (Ed.), Plenum Press, New York. p. 325-336.
28. Ledyard, K. M. and J. W. H. Dacey. *In Press*. Microbial cycling of DMSP and DMS in coastal and oligotrophic seawater. Limnol. Oceanogr.
29. Lee, S., Y. Kang and J. Fuhrman. 1995. Imperfect retention of natural bacterioplankton cells by glass fiber filters. Mar. ecol. Prog. Ser. *119*: 285-290.
30. Lovelock, J. E., R. J. Maggs and R. A. Rasmussen. 1972. Atmospheric dimethyl sulfide and the natural sulfur cycle. Nature. *237*: 452-453.
31. Malin, G., S. Turner, P. S. Liss, P. Holligan and D. Harbour. 1993. Dimethyl sulfide and dimethylsulphoniopropionate in the Northeast Atlantic during the summer coccolithophore bloom. Deep-Sea Res. *40*: 1487-1508.
32. Malin, G., S. M. Turner and P. S. Liss. 1992. Sulfur: The plankton/climate connection. J. Phycol. *28*: 590-597.
33. Matrai, P. A. and M. D. Keller. 1994. Total organic sulfur and dimethylsulfoniopropionate (DMSP) in marine phytoplankton: Intracellular variations. Mar. Biol. *119*: 61-68.
34. Stefels, J., L. Dijkhuizen and W. Gieskes. 1995. DMSP-lyase activity in a spring phytoplankton bloom off the Dutch coast, related to *Phaeocystis* sp. abundance. Mar. Ecol. Prog. Ser. *123*: 235-243.
35. Taylor, B. F. and D. C. Gilchrist. 1991. New routes for aerobic biodegradation of dimethylsulfoniopropionate. Appl. Environ. Microbiol. *57*: 3581-3584.
36. Turner, S. M., G. Malin and P. S. Liss. 1988. The seasonal variation of dimethyl sulfide and dimethylsulfoniopropionate concentrations in nearshore waters. Limnol. Oceanogr. *33*: 364-375.
37. Visscher, P. T., M. R. Diaz and B. F. Taylor. 1992. Enumeration of bacteria which cleave or demethylate dimethylsulfoniopropionate in the Caribbean Sea. Mar. Ecol. Prog. Ser. *89*: 293-296.
38. Visscher, P. T. and B. F. Taylor. 1994. Demethylation of dimethylsulfoniopropionate to 3-mercaptopropionate by an aerobic marine bacterium. Appl. Environ. Microbiol. *60*: 4617-4619.
39. White, R. H. 1982. Analysis of dimethyl sulfonium compounds in marine algae. J. Mar. Res. *40*: 529-535.
40. Wolfe, G. V. 1996. Accumulation dissolved DMSP by marine bacteria and its degradation via bactivory, *In* Biological and environmental chemistry of DMSP and related sulfonium compounds. R. P. Kiene, P. T. Visscher, M. D. Keller and G. O. Kirst (Eds.), Plenum, New York.
41. Wolfe, G. V., E. B. Sherr and B. F. Sherr. 1994. Release and consumption of DMSP from *Emiliania huxleyi* during grazing by *Oxyrrhis marina*. Mar. Ecol. Prog. Ser. *111*: 111-119.

30

ANAEROBIC MICROORGANISMS INVOLVED IN THE DEGRADATION OF DMS(P)

Marc J. E. C. van der Maarel and Theo A. Hansen

Department of Microbiology
Groningen Biomolecular Sciences
Biotechnology Institute
University of Groningen,
NL-9751 NN Haren, The Netherlands

SUMMARY

In anoxic intertidal sediments DMSP can be degraded via an initial cleavage to dimethyl sulfide (DMS) and acrylate or via demethylation. 3-Methiolpropionate is a possible intermediate of the demethylation pathway; it can also be demethiolated to methanethiol (MSH) and acrylate. In freshwater sediments DMS and MSH are mainly formed by the transfer of a methyl group from methoxylated aromatic compounds to sulfide. Both DMS and MSH can further be converted to methane, carbon dioxide and sulfide. In recent years a variety of pure cultures of anaerobic microorganisms that carry out one of the above mentioned conversions have been described in the literature.

INTRODUCTION

Most of the DMSP from marine algae is likely to be degraded in the oxic water phase of the seas. A wide variety of aerobic DMSP and DMS oxidizing bacteria have been isolated in recent years (see Taylor and Visscher, this volume). Less is known about the microorganisms involved in DMS(P) conversion in anoxic environments such as sediments and the gut system of marine animals. Benthic algae and the settling of pelagic algae explain the presence of DMSP in anoxic intertidal marine sediments. Concentrations of DMSP ranging from 1 up to 110 μmol/l sediment have been measured in the surface layer of various intertidal sediments (18, 42, 49). In a microbial mat, a concentration of up to 200 μmol/l sediment has been reported (48). For many years the only data on the anaerobic degradation of DMSP came from work of Wagner and Stadtman (51) on a freshwater isolate. The pioneering work of Kiene and colleagues on the conversion of DMSP (19, 22, 23) and DMS (24, 36) in anoxic marine sediments showed that anaerobically, DMSP can be converted by two different

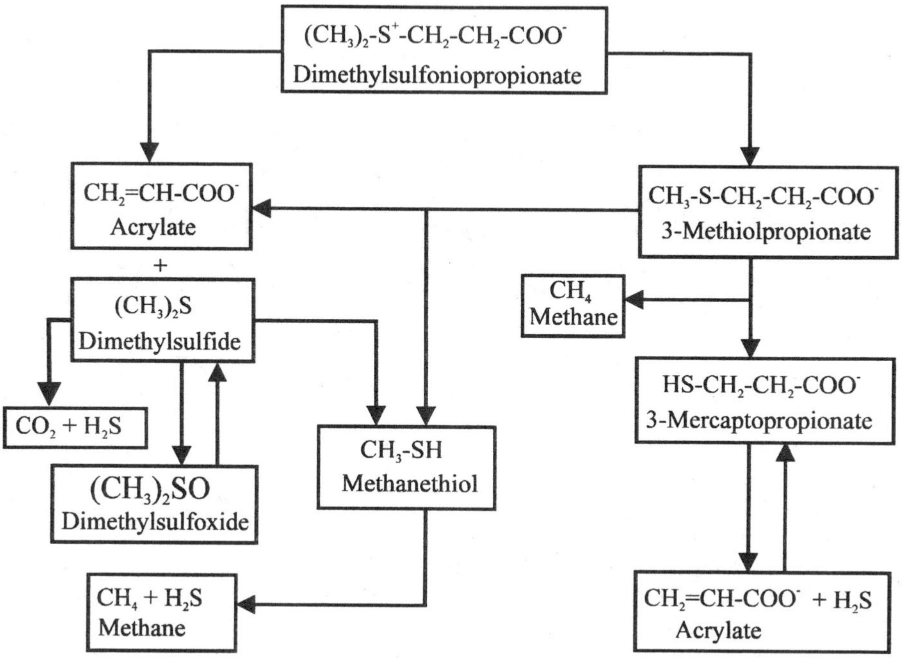

Figure 1. Pathways involved in the conversion of DMSP in anoxic intertidal sediments.

pathways: a cleavage to DMS and acrylate or a demethylation to 3-mercaptopropionate (MPA) with 3-*S*-methyl- mercaptopropionate (MMPA) as a possible intermediate. They also succeeded in the isolation of the first DMS-metabolizing methanogen (24). In this paper an overview will be presented on the anaerobic microorganisms, from marine as well as non-marine origin, that have been shown to convert DMSP, MMPA, MPA, or DMS in pure culture studies. The various processes involved in the anaerobic degradation of DMS(P) which will be discussed below in detail are presented in Fig. 1.

CLEAVAGE OF DMSP

Until now cleavage of DMSP to DMS and acrylate has only been described for one pure culture of an anaerobic bacterium (51). This organism, a strain of *Clostridium propionicum*, was isolated from river mud and ferments acrylate to acetate and propionate (in a 1:2 ratio). It also ferments pyruvate, lactate, and alanine. Lactate is converted via the acrylate pathway in which acrylylCoA is an intermediate (1, 39). An acetate to propionate ratio of 1:2 was also found when acrylate was added to suspensions of anoxic coastal sediments (20). This ratio indicated that fermentative organisms, such as *C. propionicum* , most likely are involved in the fermentation of acrylate and in the cleavage of DMSP to DMS and acrylate.

Recently we isolated a DMSP-cleaving bacterium (strain W218) from anoxic intertidal Wadden Sea sediment; the isolate was identified as a new species of *Desulfovibrio* (van der Maarel et al., submitted), the best described genus of sulfate-reducing bacteria. After cleavage of DMSP strain W218 does not ferment acrylate but it uses acrylate as an alternative electron acceptor which is reduced to propionate. Reduction of acrylate occurs simultaneously with the reduction of sulfate and might give the organism a competitive advantage under conditions of low electron donor concentrations. The amount of energy available per mol of electron donor with acrylate ($E^{0'}$= -217 mV, calculated from (41)) as an electron acceptor is higher than with sulfate ($E^{0'}$=-25 mV, calculated from (41)). Strain W218 has a very active DMSP lyase which has been purified. The v_{max} for DMSP at 30°C was 12.0 mmol.min^{-1}.mg^{-1} protein and the K_m for DMSP 0.9 mM. The enzyme is not active towards *S*-methylmethionine nor MMPA.

DEMETHYLATION OF DMSP

A structural analog of DMSP is betaine (*N*, *N*, *N*, trimethylglycine) that similarly functions as an osmolyte of many organisms (54). The anaerobic conversion of betaine involves two different pathways of which one is analogous to the demethylation pathway of DMSP. This pathway involves a demethylation of betaine to dimethylglycine and was found in the acetogen *Eubacterium limosum* (31) and the sulfate reducer *Desulfobacterium* sp. strain PM4 (15). It was therefore tempting to speculate that the demethylation of DMSP in anoxic sediments, which was observed by Kiene and Taylor (19), might be carried out by such organisms. The demethylation of DMSP to MMPA is indeed catalyted by *Desulfobacterium* strain PM4 (43). Two other *Desulfobacterium* strains, *D. niacini* DSM 2059 and *D. vacuolatum* DSM 3385, and a novel type of CO dehydrogenase-containing sulfate reducer were found to demethylate DMSP(van der Maarel and Jansen, unpublished results).

Desulfobacterium species have the acetylCoA/CO dehydrogenase pathway in which an acetylCoA is cleaved into a hydropteroyltetraglutamate-bound methyl group and a bound carbon monoxide; both C_1-groups are subsequently oxidized to CO_2 (53; Fig 2). Our present working hypothesis concerning the conversion of DMSP by *Desulfobacterium* species involves a methyl group transfer from DMSP to a tetrahydropterin acceptor followed by an oxidation to CO_2 (Fig 2). A methyltransferase system has been described for the transfer of a methyl group by *Holophaga foetida* (26). The acetylCoA/CO dehydrogenase pathway also exists in homoacetogenic bacteria (7, 10), such as *Acetobacterium* sp. and *Eubacterium limosum,* where this pathway is used in the direction of acetylCoA synthesis. DMSP-demethylating acetogens have not been described but it is possible that they exist as suggested by Kiene and Taylor (19).

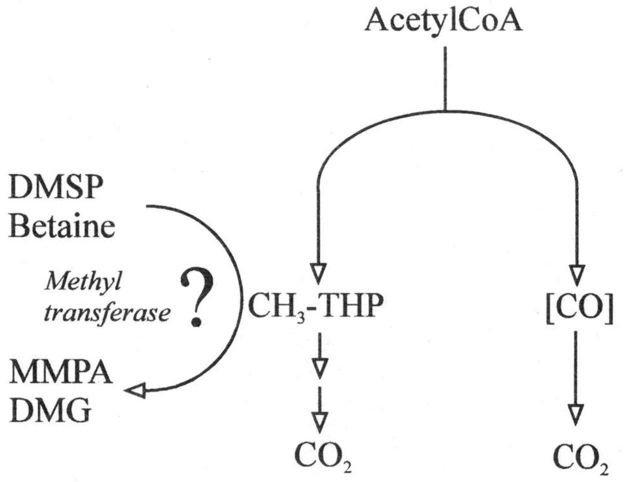

Figure 2. AcetylCoA/CO dehydrogenase pathway as found in *Desulfobacterium* spp. and the possible involvement of a methyltransefrase system in the conversion of DMSP or betaine. Abbreviations: THP, tetrahydropteroyltetraglutamate; DMG, dimethylglycine. Adapted from Widdel and Hansen (53).

CONVERSION OF MMPA

The second step in the demethylation pathway of DMSP is the conversion of MMPA which was found in the marine methanogens *Methanosarcina* sp. strain MTP4, *M. acetivorans*, and *M. siciliae* (42). All of these three strains also utilize methanol, trimethylamine, and DMS and were isolated from marine sediments, an environment where methylated compounds such as TMA are thought to be the predominant precursors for methanogenesis (35). Methanogenic conversion of MMPA was shown to occur in sediment slurries when antibiotics that specifically inhibit Bacteria but not Archaea were added (42). Under non-inhibited conditions MMPA (500μM) was readily converted to methanethiol (MSH) and subsequently methane, indicating that the organisms involved in the demethiolation of MMPA belong to the domain Bacteria.

The biochemistry of MMPA demethylation by methanogens has not been studied but has some interesting aspects. MMPA is structurally very similar to methyl-S-coenzyme M (methyl-S-ethanesulfonic acid) and can serve, albeit poorly, as a substrate for methyl-S-coenzyme M reductase (50), the enzyme that catalyses the last step of methanogenesis (4; Fig 3). It is possible that after uptake MMPA is directly converted by the methyl-S-coenzyme M reductase to methane and the heterodisulfide of MPA and *N*-7-mercaptohexanoylthreoninephosphate (HS-HTP). This heterodisulfide of MPA and HS-HTP would subsequently have to be converted by the enzyme heterodisulfide-reductase to HS-HTP and MPA. It is questionable whether the heterodisulfide reductase is able to reduce the MPA-HTP heterodisulfide since this enzyme seems to be highly specific (12, 13). Of the eleven different methanogens that were tested for growth on MMPA, only three strains were found positive (42). This suggests that the metabolism of MMPA is not due to a lack of specificity of the methyl-S-coenzyme M reductase. Most probably a MMPA:coenzyme M methyltransferase system is involved in the conversion of MMPA. Specific methyltransferases for methanol and TMA have been found to exist in methanogens (4; Fig. 3).

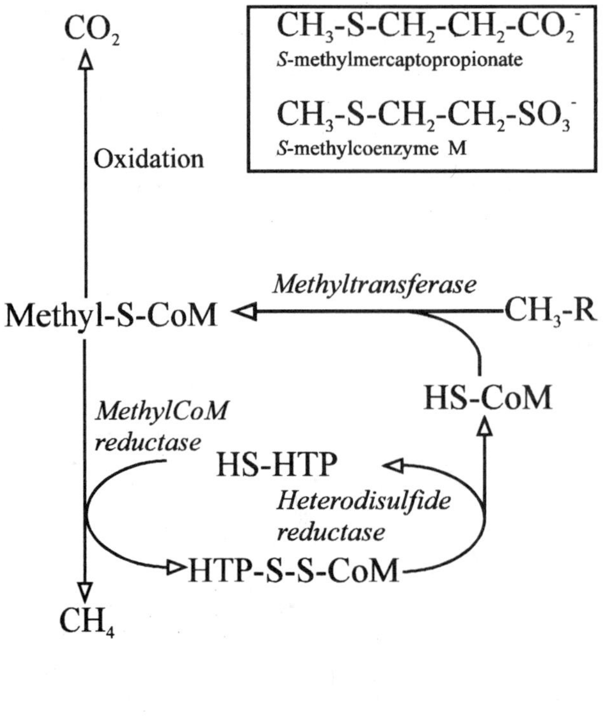

Figure 3. Structural similarity between methyl-S-coenzyme M (methyl-S-CoM) and MMPA and the mechanism of the final step in the formation of methane from methylated C_1-compounds such as methanol or methylamines. Abbreviations: CoM, coenzyme M; HS-HTP, N-7-mercaptohexanoylthreoninephosphte; CH_3-R, methyl containing C_1 compound (R represents the remainder of the C_1 compound). Adapted from Blaut (4).

CONVERSION OF MPA

In addition to microbial formation, MPA can also be formed by the chemical addition of sulfide to acrylic acid (44). Little is known about the microbial degradation of MPA . Kiene and Taylor (19) found that MPA accumulated and was subsequently converted after the addition of DMSP to sediment suspensions. Acetate accumulated transiently when millimolar concentrations of MPA were added to anoxic sediment suspensions (20). They suggested that MPA was degraded biologically but no MPA utilizing isolates have been described.

The phototrophic bacteria *Rhodopseudomonas* sp. strain BB1 and *Thiocapsa roseopersicina* can grow on MPA. It is cleaved to acrylate and sulfide, which was used as an electron donor by both organisms (45). Acrylate was used as a carbon source by the *Rhodopseudomonas* isolate but not by *Thiocapsa roseopersicina*. Both strains also cleaved mercaptomalate to sulfide and fumarate.

PRODUCTION OF DMS AND MSH

In anoxic marine sediments, DMS is readily produced from DMSP (19, 23). Fermentative bacteria such as *Clostridium propionicum* (51) or sulfate reducers such as strain W218 (van der Maarel et al., submitted) can be responsible for this cleavage. MSH is produced in marine sediments from sulfur containing compounds such as L-homocysteine, 2-keto-4-methiolbutyrate, and DL-methionine (22, 23); the bacteria that produce MSH from these compounds have not been isolated. A wide variety of aerobic bacteria produce MSH from methionine or S-methylmethionine; some anaerobes such as certain *Clostridium* species were found to produce MSH from methionine, S-methylcysteine or thioglycolate (17). MSH is also formed transiently during the metabolism of DMS by methanogenic archaea (9). As discussed above, in anoxic marine sediments the demethiolation of MMPA can also result in the formation of MSH (19); the only characteristic that is known about the MMPA-cleaving organisms is that they belong to the domain of the Bacteria (42). Cleavage of DMSP and demethiolation of MMPA both involve the breakage of a C-S bond and result in the formation of a volatile sulfur compound plus acrylate. A major difference between DMSP and MMPA is the positively charged S-atom. It can be speculated that MMPA-demethiolating bacteria utilize the acrylate formed in a similar way as the DMSP-cleaving sulfate reducer strain W218. This strain was not able to grow on MMPA (van der Maarel et al, submitted).

An interesting mechanism that results in the formation of MSH and DMS is the O-demethylation of methoxylated aromatic compounds, such as syringate (3,4,5-trimethoxybenzoate), followed by the methylation of sulfide (8). Methoxylated aromatic compounds are major constituents of plants, mainly as lignin precursors (5). The O-demethylation mechanism has been found to exist in freshwater as well as marine sediments (8). A homoacetogenic bacterium, strain TMBS4, that forms gallate from syringate using the O-demethylation mechanism and subsequently metabolizes gallate to acetate and CO_2, was isolated from anaerobic freshwater sediment (3, 25). Unfortunately, an isolate from marine sediment was lost. Strain TMBS4 has been identified as the new species *Holophaga foetida* (27). The methylation of sulfide using a methyl group of methoxylated aromatic compounds may be a major pathway for the formation of MSH and DMS in freshwater environments, since DMSP is probably present at low concentrations. It cannot be excluded that the DMS-producing O-demethylation mechanism also plays an important role in marine environments, especially in periods when the DMSP concentration is low due to a low algal biomass in open waters.

Besides DMSP and methoxylated aromatic compounds, DMSO can also be a precursor of DMS. DMSO reductase activity has been found in a wide variety of (facultative) anaerobic bacteria from different habitats (11, 29, 52, 56, 57). The DMSO reductase enzyme of *Escherichia coli* shows a broad substrate specificity and is for example able to reduce trimethylamine *N*-oxide, sodium chlorate, and nicotinic acid *N*-oxide (52). No anaerobic marine DMSO reducing bacteria have been described in the literature. Kiene and Capone (21) found that in anoxic marine sediment suspensions DMSO was rapidly reduced to DMS and concluded from inhibitor experiments with molybdate that sulfate-reducing bacteria were not involved in the reduction of DMSO. However, very recently it was demonstrated that several pure cultures of marine sulfate-reducing bacteria can grow by using DMSO as an alternative electron acceptor (Jonkers et al., in preparation). Whether such sulfate reducers are responsible for the observed conversion of DMSO in marine sediment remains to be established.

CONSUMPTION OF DMS

Anaerobic oxidation of DMS to DMSO can be carried out by the several anoxygenic phototrophs (11, 47, 55). In these organisms DMS is used as a photosynthetic electron donor. The anaerobic oxidation of DMS to DMSO does not lead to a substantial removal of DMS from the environment because the reverse process, the reduction of DMSO to DMS as is described in the previous section, also takes place. DMS can also be oxidized by the facultative anaerobe *Thiobacillus* sp. strain ASN-1 that uses nitrate as an electron acceptor; besides DMS it can use a variety of alkyl sulfides including dimethyldisulfide and diethylsulfide (46).

Zinder and Brock (58) already showed that sediments from the freshwater Lake Mendota were able to convert DMS and MSH to methane. However, the methanogen that was responsible for this conversion could not be isolated. They also tested a few pure cultures of methanogens for their ability to utilize DMS without positive results. Later, the methanogenic conversion of DMS was described by Kiene and coworkers (24) for anoxic estuarine sediment and they were the first to isolate a methanogen that can grow on DMS (24). In addition to DMS, this organism also utilizes MSH, methylamines and methanol (36) and it was identified as *Methanolobus taylorii* (34). In recent years a variety of DMS- and MSH utilizing methanogens have been described originating from different habitats (Table 1).

Kiene and coworkers (24) found indications from sediment suspension experiments with low concentrations of radio labelled DMS and specific inhibitors that sulfate-reducing bacteria are directly involved in the oxidation of DMS. Until now, no pure cultures of marine DMS-oxidizing sulfate reducers have been isolated, but DMS oxidation was reported for

Table 1. Methanogens that are able to utilize DMS or MSH and their habitat

Organism	Habitat	Reference
Methanohalophilus oregonense	Alkaline saline lake	28
Methanohalophilus zhilinae	Alkaline saline lake	30
Methanolobus bombayensis	Marine sediment	16
Methanolobus taylorii	Estuarine sediment	34
Methanosarcina acetivorans	Marine mud	32
Methanosarcina siciliae	Oil well	32, 33
Methanosarcina sp. Strain MTP4	Marine mud	9

three thermophilic strains belonging to the genus *Desulfotomaculum*; these strains were isolated from an anaerobic digestor (40). No DMS-oxidizing methanogens could be isolated from the reactor and the authors concluded that sulfate-reducers are responsible for the degradation of DMS and MSH in thermophilic freshwater environments. Thus far our attempts to isolate mesophilic marine DMS-oxidizing sulfate reducers have not been successful (unpublished results), but this aspect deserves further study. Alternative explanations for the results of Kiene and coworkers (24) about the involvement of sulfate reducing bacteria in the oxidation of DMS should not be excluded such as the oxidation of DMS to CO_2 and H_2 (or formate) by methanogens coupled to the transfer of H_2 to sulfate-reducing bacteria. In order to reduce a methyl group from a substrate such as methanol or DMS, methanogens have to oxidize part of the methyl group to CO_2 (4). Provided that the DMS-oxidizing methanogen has hydrogenase activity, part of the reducing equivalent pool will be channelled to hydrogen. Many sulfate reducers are very good hydrogen scavengers (38) and are thought to be the predominant hydrogen consumers in marine sediments (35). The process of interspecies H_2 transfer forces the methanogen to oxidize more of the methyl group to CO_2 and results in a lower CH_4/CO_2 ratio then would be expected on the basis of exclusive methanogenesis from DMS. A similar phenomenon was demonstrated for a coculture of *Methanosarcina barkeri* and *Desulfovibrio vulgaris* growing on methanol or acetate (37) and for cocultures of acetogenic methanol-utilizing bacteria and a *Desulfovibrio* strain (14).

FUTURE PERSPECTIVES

Studies on pure cultures and sediment suspensions have shown that a variety of different microorganisms are involved in the anaerobic conversion of DMS(P). Pure cultures have been described for most of the conversions pathways presented in Fig. 1. Some of these organisms show interesting strategies such as the reduction of acrylate after cleavage of DMSP by the sulfate reducer strain W218 (van der Maarel et al., submitted). Another recent finding was the metabolism of MMPA by methanogens (42). It can be expected that the strains that we have in pure culture at the moment represent only a minority of all the organisms involved in the conversion of DMS, DMSP, or their intermediate degradation products. Microbial ecologists have to cope with the vicious circle that enumerations, enrichments, and isolation of microorganisms are very much dependent on culture conditions and composition of media used and that for the proper knowledge of these conditions a study of pure cultures is often required. The next step will be the elucidation of the enzymes that are involved in the conversions. A start has been made by the purification and characterization of the DMSP lyase from strain W218 which will allow a comparison with the enzyme from the aerobic *Alcaligenes*-like isolate of deSouza and Yoch (6). Another important and challenging aspect will be to determine whether the pure cultures now available can be considered as major players in the conversion of DMS(P) or degradation intermediates. Novel molecular techniques may be useful to address this question such as the use of labelled antibodies and 16S rRNA probes (see 2).

ACKNOWLEDGMENTS

Marc J.E.C. van der Maarel and Theo A. Hansen were supported by The Netherlands Ministry of Housing, Physical Planning, and the Environment (National Research Project NOLK 026/90) and the European Union (contract EV5V-CT93-0326). The authors thank

Lubbert Dijkhuizen, Walter Aukema, Henk Jonkers and Michael Jansen for valuable discussions and permission to use unpublished data.

REFERENCES

1. Akedo, M., C.L. Cooney and A.J. Sinskey. 1983. Direct demonstration of lactate-acrylate interconversion in *Clostridium propionicum*. Biotechnology *1:* 791-794.
2. Amman, R.I., W. Ludwig and K.H. Schleifer. 1995. Phylogentic identification and in situ detection of individual microbial cells without cultivation. Microb. Rev. *59:*143-169.
3. Bak, F., Finster, K. and F. Rothfuβ. 1992. Formation of dimethylsulfide and methanethiol from methoxylated aromatic compounds and inorganic sulfide by newly isolated anaerobic bacteria. Arch. Microbiol. *157:* 529-534.
4. Blaut, M. 1994. Metabolism of methanogens. Ant. Leeuwenhoek *66*: 187-208.
5. Colberg, P.J. 1988. Anaerobic microbial degradation of cellulose, lignin, oligolignols, and monoaromatic lignin derivatives, p. 333-372. *In* A.J.B. Zehnder (ed.), Biology of anaerobic microorganisms. John Wiley & Sons, New York.
6. deSouza, M.P. and D.C. Yoch. 1995. Purification and characterization of dimethylsulfoniopropionate lyase from an *Alcaligenes*-like dimethyl sulfide producing marine isolate. Appl. Environ. Microbiol. *61:* 21-26.
7. Diekert, G. and G. Wohlfarth. 1994. Metabolism of homoacetogens. Ant. Leeuwenhoek *66:* 209-221.
8. Finster, K., G.M. King and F. Bak. 1990. Formation of methylmercaptan and dimethylsulfide from methoxylated aromatic compounds in anoxic marine and fresh water sediments. FEMS Microbiol. Ecol. *74:* 295-302.
9. Finster, K., Y. Tanimoto and F. Bak. 1992. Fermentation of methanethiol and dimethylsulfide by a newly isolated methanogenic bacterium. Arch. Microbiol. *157:* 425-430.
10. Fuchs, G. 1986. CO_2 fixation in acetogenic bacteria: variations on a theme. FEMS Microbiol. Rev. *39:*181-213.
11. Hanlon, S.P., R.A. Holt, G.R. Moore and A.G. McEwan. 1994. Isolation and characterization of a strain of *Rhodobacter sulfidophilus*: a bacterium which grows autotrophically with dimethylsulphide as electron donor. Microbiol. *140:* 1953-1958.
12. Hedderich, R., A. Berkessel and R.K.Thauer. 1989. Catalytic properties of the heterodisulfide reductase involved in the final step of methanogenesis. FEBS Lett. *1:* 67-71.
13. Hedderich, R., A. Berkessel and R.K.Thauer. 1990. Purification and properties of the heterodisulfide reductase from *Methanobacterium thermoautotrophicum*. Eur. J. Biochem. *193:* 255-261.
14. Heijthuijsen, J.H.F.G. and T.A. Hansen. 1986. Interspecies hydrogen transfer in co-cultures of methanol-utilizing acidogens and sulfate-reducing or methanogenic bacteria. FEMS Microbiol. Ecol. *38:* 57-64.
15. Heijthuijsen, J.H.F.G. and T.A. Hansen. 1989b. Anaerobic degradation of betaine by marine *Desulfobacterium* strains. Arch. Microbiol. *152:* 393-396.
16. Kadam, P.C., D.R. Ranade, L. Mandelco and D.R. Boone. 1994. Isolation and characterization of *Methanolobus bombayensis* sp. nov., a methylotrophic methanogen that requires high concentrations of divalent cations. Int. J. Syst. Bacteriol. *44:* 603-607.
17. Kadota, H. and Y. Ishida. 1972. Production of volatile sulfur compounds by microorganisms. Annu. Rev. Microbiol. *53*: 127-143.
18. Kiene, R.P. 1988. Dimethyl sulfide metabolism in anoxic salt marsh sediments. FEMS Microbiol. Ecol. *53:* 71-78.
19. Kiene, R.P. and B.F. Taylor. 1988. Demethylation of dimethylsulfoniopropionate and production of thiols in anoxic marine sediments. Appl. Environ. Microbiol. *54:* 2208-2212.
20. Kiene, R.P. and B.F. Taylor. 1989. Metabolism of acrylate and 3-mercaptopropionate, p. 222-230. *In* E.S. Saltzman and W.J. Cooper (eds.), Biogenic sulfur in the environment. American Chemical Society, Washington DC.
21. Kiene, R.P. and D.G. Capone. 1988. Microbial transformation of methylated sulfur compounds in anoxic salt marsh sediments. Microb. Ecol. *15:* 275-291.
22. Kiene, R.P., K.D. Malloy and B.F. Taylor. 1990. Sulfur-containing amino acids as precursors of thiols in anoxic coastal sediments. Appl. Environ. Microbiol. *56:* 156-161.
23. Kiene, R.P. and P.T. Visscher. 1987. Production and fate of methylated sulfur compounds from methionine and dimethylsulfoniopropionate in anoxic salt marsh sediments. Appl. Environ. Microbiol. *53:* 2426-2434.

24. Kiene, R.P., R.S. Oremland, A. Catena, L.G. Miller and D.G. Capone. 1986. Metabolism of reduced methylated sulfur compounds in anaerobic sediments and by a pure culture of an estuarine methanogen. Appl. Environ. Microbiol. *52:* 1037-1045.
25. Kreft, J.-U. and B. Schink. 1993. Demethylation and degradation of phenylmethylesters by the sulfide-methylating homoacetogenic bacterium TMBS 4. Arch. Microbiol. *159:* 308-315.
26. Kreft, J.-U. and B. Schink. 1994. *O*-demethylation by the homoacetogenic anaerobe *Holophaga foetida* studied by a new photometric methylation assay using electrochemically produced cobalamin. Eur. J. Biochem. *226:* 945-951.
27. Liesack, W., F. Bak, J.-U. Kreft and E. Stackebrandt. 1994. *Holophaga foetida* gen. nov., sp. nov., a new, homoacetogenic bacterium degrading methoxylated aromatic compounds. Arch. Microbiol. *162:* 85-90.
28. Liu, Y., D.R. Boone and C. Choy. 1990. *Methanohalophilus oregonense* sp. nov., a methylotrophic methanogen from an alkaline, saline aquifer. Int. J. Syst. Bacteriol. *40:* 111-116.
29. Lorenzen, J., S. Steinwachs and G. Unden. 1994. DMSO respiration by the anaerobic rumen bacterium *Wolinella succinogenes*. Arch. Microbiol. *162:* 277-281.
30. Mathrani, I.M., D.R. Boone, R.A. Mah, G.E. Fox and P.P. Lau. 1988. *Methanohalophilus zhilinae* sp. nov., an alkalophilic, halophilic, methylotrophic methanogen. Int. J. Syst. Bateriol. *38:* 139-142.
31. Müller, E., K. Fahlbusch, R. Walther and G. Gottschalk. 1981. Formation of *N,N*,-dimethylglycine, acetic acid, and butyric acid from betaine by *Eubacterium limosum*. Appl. Environ. Microbiol. *42:* 439-445.
32. Ni, S., C.R. Woese, H.C. Aldrich and D.R. Boone. 1994. Transfer of *Methanolobus siciliae* to the genus *Methanosarcina*, naming it *Methanosarcina siciliae*, and emendation of the genus *Methanosarcina*. Int. J. Syst. Bacteriol. *44:* 357-359.
33. Ni, S. and D.R. Boone. 1991. Isolation and characterization of a dimethyl sulfide-degrading methanogen, *Methanolobus siciliae* HI 350, from an oil well, characterization of *M. siciliae* T4/M^T, and emendation of *M. siciliae*. Int. J. Syst. Bacteriol. *41:* 410-416.
34. Oremland, R.S. and D.R. Boone. 1994. *Methanolobus taylorii* sp. nov., a new methylotrophic estuarine methanogen. Int. J. Syst. Bacteriol. *44:* 573-575.
35. Oremland, R.S. and S. Polcin. 1982. Methanogenesis and sulfate reduction: competitive and non-competitive substrates in estuarine sediments. Appl. Environ. Microbiol. *44:* 1270-1276.
36. Oremland, R.S., R.P. Kiene, I. Mathrani, Whiticar, M.J. and D.R. Boone. 1989. Description of an estuarine methylotrophic methanogen which grows on dimethyl sulfide. Appl. Environ. Microbiol. *55:* 994-1002.
37. Phelps, T.J., R. Conrad and J.G. Zeikus. 1985. Sulfate-dependent interspecies H_2 transfer between *Methanosarcina barkeri* and *Desulfovibrio vulgaris* during coculture metabolism of acetate or methanol. Appl. Environ. Microbiol. *50: 589-594.*
38. Robinson, J.A. and J.M. Tiedje. 1984. Competition between sulfate-reducing and methanogenic bacteria for H_2 under resting and growing conditions. Arch. Microbiol. *137:* 26-32.
39. Schweiger, G. and W. Buckel. 1985. Identification of acrylate, the product of dehydration of (R)-lactate catalysed by cell-free extracts from *Clostridium propionicum*. FEBS Lett. *185:* 253-256.
40. Tanimoto, Y. and F. Bak. 1994. Anaerobic degradation of methylmercaptan and dimethyl sulfide by newly isolated thermophilic sulfate-reducing bacteria. Appl. Environ. Microbiol. *60:* 2450-2455.
41. Thauer, R.K., K. Jungermann and K. Decker. 1977. Energy conservation in chemotrophic bacteria. Bact. Rev. *41*: 100-180.
42. van der Maarel, M.J.E.C., M. Jansen and T.A. Hansen. 1995. Methanogenic conversion of 3-*S*-methyl-mercaptopropionate to 3-mercaptopropionate. Appl. Environ. Microbiol. *61*: 48-51.
43. van der Maarel, M.J.E.C., P. Quist, L. Dijkhuizen and T.A. Hansen. 1993. Anaerobic degradation of dimethylsulfoniopropionate to 3-*S*-methylmercaptopropionate by a marine *Desulfobacterium* strain. Arch. Microbiol. *160*: 411-412.
44. Variavamurthy, A. and K. Mopper. 1987. Geochemical formation of organosulphur compounds (thiols) by addition of H_2S to sedimentary organic compounds. Nature *329:* 623-625.
45. Visscher, P.T. and B.F. Taylor. 1993a. Organic thiols as organolithotrophic substrates for growth of phototrophic bacteria. Appl. Environ. Microbiol. *59:* 93-96.
46. Visscher, P.T. and B.F. Taylor. 1993b. Aerobic and anaerobic degradation of a range of alkyl sulfides by a dentrifying marine bacterium. Appl. Environ. Microbiol. *59:* 4083-4089.
47. Visscher, P.T. and H. van Gemerden. 1991. Photoautotrophic growth of *Thiocapsa roseopersicina* on dimethyl sulfide. FEMS Microbiol. Lett. *81:* 247-250.
48. Visscher, P.T., P. Quist and H. van Gemerden. 1990. Methylated sulfur compounds in microbial mats: *in situ* concentrations and metabolism by a colorless sulfur bacterium. Appl. Environ. Microbiol. *57:* 1758-1763.
49. Visscher, P.T., R.P. Kiene and B.F. Taylor. 1994. Demethylation and consumption of dimethylsulfonio-propionate in marine intertidal sediments. FEMS Microbiol. Ecol. *14:* 179-190.

50. Wackett, L.P., J.F.Honek, T.P. Begley, V. Wallace, W.H. Orme-Johnson and C.T. Walsh. 1987. Substrate analogues as mechanistic probes of methyl-S-coenzyme M reductase. Biochemistry. *26:* 6012-6018.
51. Wagner, C. and E.R. Stadtman. 1962. Bacterial fermentation of dimethyl-β-propiothetin. Arch. Biochem. Biophys. *98:* 331-336.
52. Weiner, J.H., R.A. Rothery, D. Sambasivaroa and C.A. Trieber. 1992. Molecular analysis of dimethylsulfoxide reductase: a complex iron-sulfur molybdoenzyme of *Escherichia coli*. Bioch. Bioph. Acta *1102:* 1-18.
53. Widdel, F. and T.A. Hansen. 1991. The dissimilatory sulfate- and sulfur-reducing bacteria, p. 583-624. *In* A. Balows, H.G. Trüper, M. Dworkin, W. Harder and K.-H. Schleifer (ed.), The prokaryotes, 2nd edition, Springer Verlag, New York.
54. Yancey, P.H., M.E. Clark, C. Hand, R.D. Bowlus and G.N. Somero. 1982. Living with water stress: evolution of osmolyte systems. Science *217:* 1214-1222.
55. Zeyer, J., P. Eicher, S.G. Wahekam and R.P. Schwarzenbach. 1987. Oxidation of dimethyl sulfide to dimethyl sulfoxide by phototrophic purple bacteria. Appl. Environ. Microbiol. *53:* 2026-2032.
56. Zinder, S.H. and T.D. Brock. 1978. Dimethyl sulfoxide as an electron acceptor for anaerobic growth. Arch Microbiol. *116:* 35-40.
57. Zinder, S.H. and T.D. Brock. 1978. Dimethyl sulfoxide reduction by micro-organisms. J. Gen. Microbiol. *105:* 305-313.
58. Zinder, S.H. and T.D. Brock. 1978. Production of methane and carbon dioxide from methane thiol and dimethyl sulfide by anaerobic lake sediments. Nature *273:* 226-228

31

DIMETHYLSULFONIOPROPIONATE AS A POTENTIAL METHANOGENIC SUBSTRATE IN MONO LAKE SEDIMENTS

Pieter T. Visscher,[1] Janet R. Guidetti,[2] Charles W. Culbertson,[2] and Ronald S. Oremland[2]

[1] Department of Marine Sciences
University of Connecticut
Groton, Connecticut 06340
[2] Water Resources Division
U.S. Geological Survey
Menlo Park, California 94025

SUMMARY

A high concentration of dimethylsulfoniopropionate (DMSP) was found in the water column (0.1-1.8 μM particulate plus dissolved) of Mono Lake, CA, an alkaline, hypersaline waterbody. The dense *Artemia monica* population contained high levels of DMSP (1.7-2.5 $mmol.g^{-1}$ wet weight), presumably as an osmolyte. Death of these brine shrimp caused accumulation of DMSP along the shoreline of the lake, where concentrations peaked at 7-13 $\mu mol.cm^{-3}$ sediment. DMSP was also associated with the phototrophic microbial population in microbial mats close to the shoreline. Chemical hydrolysis of DMSP caused by the high pH value of the water (9.7-10.0) competed with biological consumption. Flux chamber experiments suggested that part of the dimethylsulfide (DMS) generated by hydrolysis escaped to the atmosphere. Vertical profiles of DMSP and DMS in the sediment correlated well. Methane and DMS also had similar distributions. Additional inhibitor studies showed that a major biological sink for DMS(P) is methanogenesis, although monooxygenase-containing bacteria also contributed to its consumption.

INTRODUCTION

Dimethylsulfoniopropionate (DMSP) is found in marine sediments, often originating from phototrophic organisms (17, 25). Salt marsh sediments, microbial mats, diatom mats and carbonate sediments may contain DMSP (8, 27), the concentration of which frequently displays a correlation with the organic content or chlorophyll a (Chl*a*) concentration (25, 27). Amended DMSP is rapidly consumed by the microbial population in water samples and

Biological and Environmental Chemistry of DMSP and Related Sulfonium Compounds
edited by Ronald P. Kiene et al., Plenum Press, New York, 1996

sediment slurries via demethylation or cleavage (9, 12, 25, 27). Degradation results in production of 3-methiolpropionate (MMPA), 3-mercaptopropionate (MPA), methanethiol (CH_3SH) or dimethylsulfide (DMS).

A variety of bacterial isolates, obtained from marine environments, have been found to contain DMSP lyase activity (2, 3, 4, 9, 13). In addition, pure culture studies revealed demethylation and demethiolation of DMSP (21, 22, 23, 28) under oxic and anoxic conditions, resulting in CH_3SH and MPA production. Metabolic pathways of and organisms involved in DMS consumption are better documented than those of DMSP, and lead ultimately to CO_2, CH_4, or dimethyl sulfoxide (20).

DMSP degradation pathways have been established for coastal marine sediments and microbial mats. Hypersaline environments, such as Mono Lake, are expected to contain high concentrations of osmolytes, such as K^+, glycine betaine, sucrose, DMSP etc. (6). In this study, we found the presence of DMSP in the water column and near-shore sediments of Mono Lake and determined some of its sources and fates in benthic communities on the Northeastern shore.

MATERIALS AND METHODS

Mono Lake is a hypersaline (9.1%), meromictic, alkaline (pH 9.7-10.0) lake in California, USA. Sulfide levels in the deep water column can reach 15 to 40 mM, but surface waters are usually well oxygenated. Primary productivity, sustained by green algae, diatoms and cyanobacteria, averages at 100 mg $C.m^{-2}.h^{-1}$. *Artemia monica* is the dominant zooplankton species (1, 6).

Sediment samples were taken in July and December 1994 near the Northeastern shoreline. Depth profiles of oxygen and sulfide were measured with microelectrodes (24), dissolved CH_4 and DMS were measured with a gas diffusion probe as described by Rothfuss et al. (18). Chl*a* was determined spectrophotometrically after methanol extraction (19). Water and plankton samples were collected at the center of the lake with Niskin bottles and a 60 μm-mesh Nitex plankton net, respectively. DMSP was measured in these samples and in depth profiles in sediment cores and porewaters. All assays were done in triplicate.

Slurry experiments were performed in 57-ml (anoxic) and 160-ml (oxic conditions) serum bottles closed with black butyl rubber stoppers. Sediment cores were kept in the dark at 4°C after they were collected. Slurries were prepared by mixing 1 part of sediment with 2 parts filter-sterilized (0.2 μM) Mono Lake water (for details, see (27)). The headspace of anoxic incubations were flushed with O_2-free N_2 for 10 minutes prior to additions. The following inhibitors were employed: $CHCl_3$ (0.5 mM), methyl butyl ether (0.5 mM), dimethyl ether (DME; 5% vol/vol), sodium molybdate (10 mM), bromoethanesulfonate (BES; 5 mM), CH_3F (1% vol/vol), and glutaraldehyde (0.5% vol/vol). Slurry experiments were either carried out with three replicates or repeated twice to ensure replication.

Samples for water column DMSP(P) were collected by gently filtering a known volume through a Whatman GF/F glass fiber filter. The filter was placed in a small serum vial which was sealed with a teflon-faced septum. DMSP was measured as DMS by headspace analysis after alkaline hydrolysis (25, 26). DMSP in organisms and in sediments was determined similarly, by placing a known weight of material in a sealed serum vial and treating this with NaOH. Sediment DMS and CH_4 samples obtained with the gas diffusion probe were collected in stoppered tubes which contained a saturated salt solution. Gas analysis was carried out on a gas chromatograph (Hewlett Packard 5730A, HNU Systems 301 or Shimadzu GC14A) with flame ionization or photo-ionization detection (26).

RESULTS

The concentration of total DMSP in the water column of Mono Lake ranged from 0.1 to 1.8 µM, with higher values close to the water surface, where *Artemia* sp. densities were the highest. The DMSP content in *Artemia*, collected during a plankton tow, ranged from 1.7 to 2.5 mmol.g^{-1} dry weight. Large amounts of these brine shrimp accumulate at the Northeastern shoreline, which is also reflected in a DMSP(P) concentration of up to 0.9 mmol.g^{-1} sediment. Porewater profiles of DMSP (7 to 13 µM) and DMS (150 to 600 µM) peaked at 5 and 10 mm depth, respectively (Fig. 1). Elevated DMS concentrations typically coincided with elevated methane concentrations. Near the shoreline, cyanobacteria colonized the sediment (sub)surface, but the oxygen concentration profile did not show a pronounced maximum. At the sediment surface, the cyanobacterial population was actively grazed upon by brine flies (*Ephydra hians*), resulting in a brown appearance. Approximately 10 m away from the shoreline, a green microbial mat developed, with a Chl*a* content of 80 to 230 µg.cm^{-3}. An oxygen concentration of approximately 350 µM showed that this population was actively photosynthesizing (Fig. 1). Both mats had a sulfide maximum of 450 to 650 µM at approximately 30 mm depth. Sulfate reduction rates peaked at 1 and 1.5 µmol.cm^{-3}.d^{-1} in the brown and green mats, respectively (L.G. Miller, unpublished results).

Preliminary flux chamber studies indicated that dimethyl ether (5%) and CH_3F (1%) resulted in accumulation of DMS in the gas phase above the sediment, while methylfluoride addition had no effect on the methane concentration in the chambers. Sediment temperatures

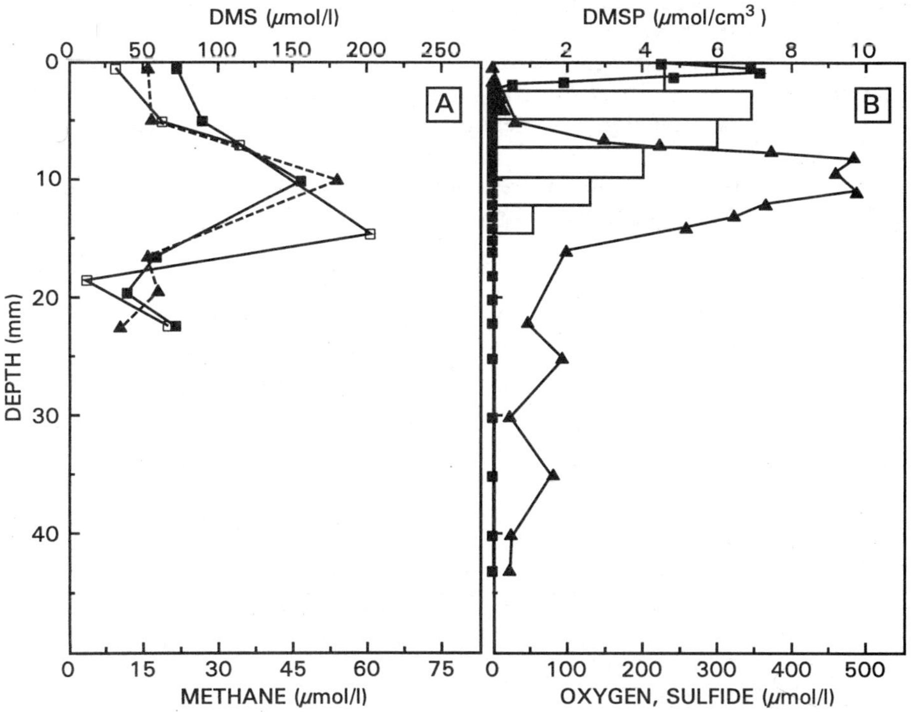

Figure 1. Depth profile of (A) DMS (solid lines) and CH_4 (dashed line) and (B) O_2 (squares), sulfide (triangles) and DMSP (bars) (right panel). Profiles were measured in July 1994 in the green mat (see text).

Table 1. DMSP turnover in slurries prepared from Mono Lake sediments, collected in July 1994. Slurries (20 ml) were incubated for 12 hours at room temperature in the dark. Glut = glutaraldehyde, MBE = methyl butyl ether. Mean of three replicate bottles and standard deviation (in parenthesis) are given

Treatment	DMSP added (μmol.bottle^{-1})	DMSP consumed (μmol.bottle^{-1})	DMS produced (μmol.bottle^{-1})	CH_4 produced (μmol.bottle^{-1})
$-O_2$	0	0 (0)	0 (0)	3.5 (0.4)
$-O_2$	5	4.2 (0.2)	0.4 (0.2)	12.2 (1.1)
$-O_2$ + BES	5	2.2 (0.5)	1.7 (0.4)	0.4 (0.5)
$-O_2$ + MoO_4^{2-}	5	3.8 (0.1)	0.5 (0.1)	10.7 (3.6)
$-O_2$ +Glut	5	0.9 (0.3)	1.1 (0.2)	0 (0)
$+O_2$	5	5.0 (0.1)	0.2 (0.4)	0 (0)
$+O_2$ +MBE	5	4.7 (0.4)	3.1 (0.4)	0 (0)
$+O_2$ + $CHCl_3$	5	4.8 (0.2)	0.2 (0.2)	0 (0)
$+O_2$ + CH_3F	5	3.8 (0.2)	2.3 (0.7)	0 (0)
$+O_2$ + Glut	5	1.1 (0.2)	0.8 (0.1)	0 (0)

in July reach values > 30°C at 5 mm depth during the afternoon, which is close to the boiling point of DMS.

Slurry experiments with mat samples collected in July 1995 showed rapid consumption of amended DMSP under oxic and anoxic conditions. Anoxic DMSP metabolism was inhibited by BES and, to a lesser extent, MoO_4^{2-} addition, whereas abiotic DMSP consumption was the slowest (Table 1). In the slurry which received DMSP only, CH_4 production was almost 4 times higher than in the control. Under oxic conditions, CH_3F addition inhibited biological DMSP consumption, and the killed control displayed the lowest consumption rate.

When DMS was added to anoxic slurries, BES and MoO_4^{2-} both inhibited its consumption (Table 2). Aerobic DMS consumption was inhibited by methyl butyl ether, but not by $CHCl_3$. This indicated involvement of organisms containing a monooxygenase and not a methyltransferase (26).

Similar slurry experiments with sediments collected in December 1994 display a less clear picture (Fig. 2). Oxygen and sulfide depth profiles showed peaks, suggesting that

Table 2. DMS turnover in slurries prepared from Mono Lake sediments, collected in July 1994*

Treatmen	DMS consumed (μmol.bottle^{-1})	CH4 produced (μmol.bottle^{-1})
$-O_2$	2.3 (0.2)	3.8 (0.3)
$-O_2$ + BES	0.2 (0.1)	0 (0)
$-O_2$ + MoO_4^{2-}	0.9 (0.4)	1.2 (0.6)
$+O_2$	1.7 (0.4)	0 (0)
$+O_2$ + $CHCl_3$	2.1 (0.1)	0 (0)
$+O_2$ + MBE	0.2 (0)	0 (0)

*Slurries (20 ml) received 5 μmol DMS per bottle and were incubated for 8 hours at room temperature in the dark. MBE = methyl butyl ether. Killed slurries did not consume DMS; untreated anoxic slurries produced 0.1 μmol CH_4.bottle^{-1}. Means of three replicate and standard deviation (in parenthesis) are given.

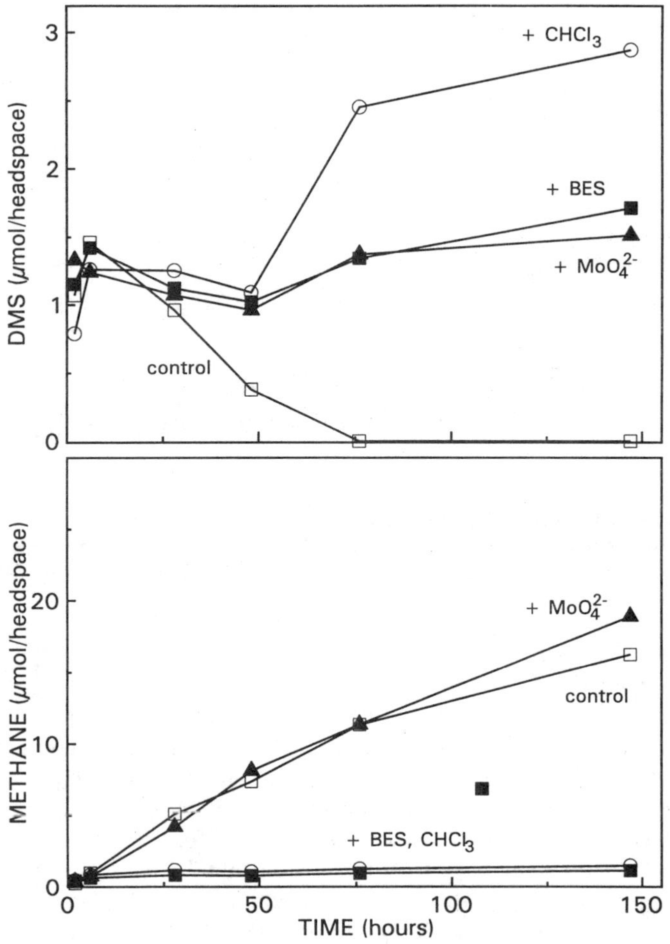

Figure 2. Time course of DMS consumption (upper panel) and CH_4 production (lower panel) in anoxic sediment slurries from Mono Lake (sediment samples obtained in December 1994). At t=0, DMS was added to the bottles (approximately 5 $\mu mol.bottle^{-1}$).

microbial actively was still present, despite low sediment temperatures (0 to 14°C during the course of a day). In $CHCl_3$ treatments, the DMS concentration doubled during the course of the experiment (150 h). Addition of MoO_4^{2-} and BES both had the same effect, resulting in an unchanged DMS concentration. CH_4 formation was observed in the control but not in the MoO_4^{2-} treatment. Similarly, DMSP addition stimulated CH_4 formation in the control and $MoO_4^{=}$ treatment, but not when BES was added (Fig. 3). Transient accumulation of DMS was observed in all treatments.

DISCUSSION

The high concentration associated with the *Artemia* population in Mono Lake, suggests that this organism may use DMSP as an osmolyte. A lower salinity of the surface water (67-72 ppt) in the summer of 1995 coincided with lower specific DMSP content in the brine shrimp (Visscher et al., unpublished results). *Nitszschia* spp, which may be present in high density in this lake (6), or any of the other phytoplankton species, may be responsible for producing DMSP (7). Prevailing winds from the Southwest cause accumulation of *Artemia* on the Northeastern shoreline. The ambient pH of 9.7 to 10.0 leads to slow chemical

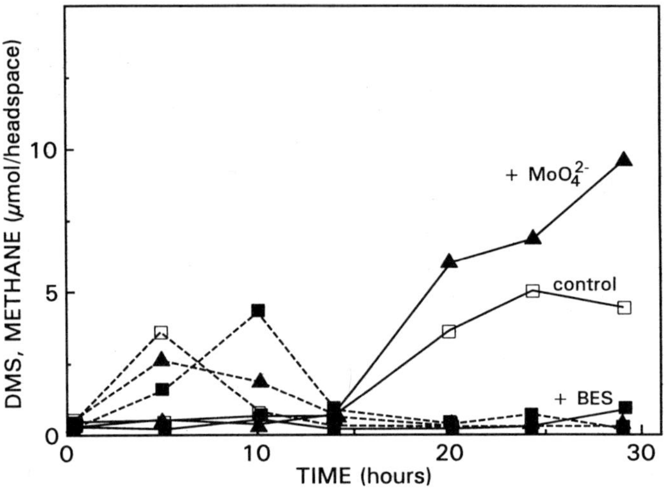

Figure 3. Production of DMS and CH_4 from amended DMSP (7 μmol.bottle^{-1}) in anoxic sediment slurries from Mono Lake (December 1994). Symbols: open squares represent the controls, triangles indicate MoO_4^{2-} additions, solid squares indicate BES additions. Dashed lines indicate DMS, solid lines indicate CH_4 concentrations.

hydrolysis of DMSP to yield DMS (25). However, slurry experiments did suggest that live sediments have a higher rate of DMSP degradation than killed sediments, indicating a role of the microbial population. The similarity between CH_4 and DMS profiles and the stimulation of CH_4 production upon addition of DMS(P) suggests that methanogenesis is one of the sinks of DMS(P). Although other sinks for DMS(P) exist, experiments in the summer of 1995 employing ^{14}C-DMS indicated that $^{14}CH_4$ was a major product (Visscher et al., unpublished results).

Methane and CO_2 were produced from CH_3SH and DMS in lake sediments (29). Addition of methylated sulfur compounds (DMS, CH_3SH, dimethyl disulfide) to a variety of aquatic habitats, including Mono Lake sediments, stimulated methane production (10), which lead to the hypothesis that DMS is a non-competitive substrate for methanogens at high concentrations. Methane production also was stimulated by the addition of DMSP in salt marsh sediments (11) and marine microbial mats (25). Oremland et al. (16) reported high rates of methane formation and oxidation, as well as a flux of CH_4 from the water column and sediments around Mono Lake.

Methanogenic cultures that utilize DMS have been isolated from an estuary (10, 15), a salt marsh (5) and the marine sediments (14). Alternatively, MMPA may be used by marine methanogens (23). MMPA, which is the first product of DMSP demethylation, was not quantified in this study. Finally, the product of MMPA demethiolation, CH_3SH, may sustain CH_4 formation (5, 15).

Slurry and flux chamber experiments suggest that in addition to methanogens, sulfate reducers and monooxygenase-containing bacteria play a role in DMSP turnover as well. Presently, only one pure culture of a marine sulfate reducer that can grow on DMSP has been described (22). Similarly, methanotrophic isolates, which may be involved in DMS(P) oxidation based on results with CH_3F additions, have not been found in hypersaline environments. Further studies are underway to determine the role of the different functional groups in DMS(P) degradation. Interestingly, due to the physical and chemical characteristics of Mono Lake, solely chemical decomposition of DMSP to DMS and a rapid volatilization due to high sediment temperatures during the summer can cause a flux of DMS to the atmosphere.

ACKNOWLEDGMENTS

We thank Ralf Conrad and colleagues for providing the gas diffusion probe, Barrie Taylor for the gift of DMSP and Ron Kiene for lending his Keithley picoammeter. The technical assistance of Larry Miller, Tracey Connell, Bush Boy Dowdle, Caleb Sasson and Russel Hudson, and Thelma and Louise is much appreciated. Larry Miller is also thanked for sharing unpublished data on sulfate reduction rates.

REFERENCES

1. Connell, T.L. 1993. The chemical evolution of shallow groundwater along the Notheastern shoreline of Mono Lake, California. Msc thesis, U.C. Santa Cruz, 197 pp.
2. de Souza, M.P., and D.C, Yoch. 1995. Purification and characterization of dimethylsulfoniopropionate lyase from an *Alcaligenes*-like dimethyl sulfide-producing marine isolate. Appl. Environ. Microbiol. 61:21-26.
3. Dacey, J.W.H., and N.V. Blough. 1987. Hydroxide decomposition of DMSP to form DMS. Geophys. Res. Lett. 14:1246-1249.
4. Diaz, M.R., P.T. Visscher, and B.F. Taylor. 1992. Metabolism of dimethylsulfoniopropionate and glycine betaine by a marine bacterium. FEMS Microbiol. Lett. 96:61-66.
5. Finster, K., Y. Tanimoto, and F. Bak. 1992. Fermentation of methanethiol and dimethylsulfide by a newly isolated methanogenic bacterium. Arch. Microbiol. 157:425-430.
6. Javor, B. 1989. Hypersaline environments: Microbiology and biogeochemistry. Springer Verlag, Heidelberg, Germany, 328 pp.
7. Keller, M.D, W.K. Bellows, and R.R.L. Guillard. 1989. Dimethyl sulfide production in marine phytoplankton. In: Biogenic sulfur in the environment. E.S. Saltzman and W.J. Cooper (eds), Am. Chem. Soc., Washington DC, pp167-182.
8. Kiene, R.P. 1988. Dimethylsulfide metabolism in salt marsh sediments. FEMS Microbiol. Ecol. 53:71-78.
9. Kiene, R.P. 1990. DMS production from DMSP in coastal seawater samples and bacterial cultures. Appl. Environ. Microbiol. 56:3292-3297.
10. Kiene, R.P., R.S. Oremland, A. Catena, L.G. Miller, and D.G. Capone. 1986. Metabolism of reduced sulfur compounds by anaerobic sediment and a pure culture of an estuarine methanogen. Appl. Environ. Microbiol. 52:1037-1045.
11. Kiene, R.P., and P.T. Visscher. 1987. Production and the fate of methylated sulfur compounds from methionine and dimethylsulfoniopropionate in anoxic salt marsh sediments. Appl. Environ. Microbiol. 53:2426-2434
12. Kiene, R.P, and B.F. Taylor. 1988. Demethylation of dimethylsulfoniopropionate and production of thiols in anoxic marine sediments. Appl. Environ. Microbiol. 54:2208-2212.
13. Ledyard, KM., E.F. DeLong, and J.H.W. Dacey. 1993. Characterization of a DMSP-degrading bacterial isolate from the Sargasso Sea. Arch. Microbiol. 160:312-318.
14. Ni, S., and D.R. Boone. 1991. Isolation and characterization of a dimethyl sulfide-degrading methanogen, *Methanolobus siliciae* HI350, from an oil well, characterization of *M. siliciae* T4/M^T, and emendation of *M. siliciae*. Int. J. Syst. Bacteriol. 41:410-416.
15. Oremland, R.S., R.P. Kiene, I. Mathrani, M.J. Whiticar, and D.R. Boone. 1989. Description of an estuarine methylotrophic methanogen which grows on dimethyl sulfide. Appl. Environ. Microbiol. 55:994-1002.
16. Oremland, R.S., L.G. Miller, C.W. Culbertson, S.W. Robinson, R.L. Smith, D.R. Lovley, M.J. Whiticar, G.M. King, R.P. Kiene, N. Iversen and M. Sargent. 1993. Aspects of the biogeochemistry of methane in Mono Lake and the Mono Basin of California. In: Biogeochemistry of global change: radiatively active trace gases. R.S. Oremland (ed), Chapman & Hall, New York NY, pp 704-741
17. Pakulski, J.D., and R.P. Kiene. 1992. Foliar release of dimethylsulfoniopropionate from *Spartina alterniflora*. Mar Ecol. Prog. Ser. 81:277-287.

18. Rotfuss, F., P. Frenzel and R. Conrad. 1994. Gas diffusion probe for measurement of CH4 gradients. In: Microbial mats: structure, development and environmental significance. L.J. Stal and P. Caumette (eds). Springer Verlag, Heidelberg, Germany, pp 167-172.
19. Stal, L.J., H. van Gemerden, and W.E. Krumbein. 1984. Simultaneous assay of chlorophyll and bacterio-chlorophyll in natural microbial communities. J. Microb. Meth. 2:295-306.
20. Taylor, B.F. 1993. Bacterial transformations of organic sulfur compounds in marine environments. In: Biogeochemistry of global change: radiatively active trace gases. R.S. Oremland (ed), Chapman & Hall, New York NY, pp 745-781
21. Taylor, B.F. and D.C. Gilchrist. 1991. New routes for the aerobic biodegradation of dimethylsulfonio-propionate. Appl. Environ. Microbiol. 57:3581-3584.
22. Van der Maarel, M.J.E.C., P. Quist, L. Dijkhuizen and T.A. Hansen. 1993. Anaerobic degradation of dimethylsulfoniopropionate to 3-S-methylmercaptopropionate by a marine *Desulfobacterium* strain. Arch. Microbiol. 160:411-412.
23. Van der Maarel, M.J.E.C., M. Jansen, and T.A. Hansen. 1995. Methanogenic conversion of 3-S-methyl-mercaptopropionate to 3-mercaptopropionate. Appl. Environ. Microbiol. 61:48-51.
24. Visscher, P.T., J. Beukema, and H. van Gemerden. 1991. In situ characterization of sediments: measurements of oxygen and sulfide profiles with a novel combined needle electrode. Limnol. Oceanogr. 36:1476-1480.
25. Visscher, P.T. and H. van Gemerden. 1991. Production and consumption of dimethylsulfoniopropionate in marine microbial mats. Appl. Environ. Microbiol. 57:3237-3242.
26. Visscher, P.T., and B.F. Taylor. 1993. A new mechanism for the aerobic catabolism of dimethyl sulfide. Appl. Environ. Microbiol. 59:3784-3789.
27. Visscher, P.T., R.P. Kiene and B.F. Taylor. 1994. Demethylation and cleavage of dimethylsulfoniopropionate in marine intertidal sediments. FEMS Microbiol. Ecol. 14:1879-190.
28. Visscher, P.T., and B.F. Taylor. 1994. Demethylation of dimethylsulfoniopropionate to 3-mercapto-propionate by an aerobic marine bacterium. Appl. Environ. Microbiol. 60:4617-4619.
29. Zinder, S.H.,, and T.D. Brock. 1978. Production of methane and carbon dioxide from methane thiol and dimethyl sulfide by anaerobic lake sediments. Nature 273:226-228.

32

THE ROLE OF OXYGENIC PHOTOTROPHIC MICROORGANISMS IN PRODUCTION AND CONVERSION OF DIMETHYLSULFONIOPROPIONIATE AND DIMETHYLSULFIDE IN MICROBIAL MATS

S. A. van Bergeijk and L. J. Stal

Laboratory for Microbiology
University of Amsterdam, Nieuwe Achtergracht 127
NL-1018 WS Amsterdam, The Netherlands

SUMMARY

The dimethylsulfoniopropionate (DMSP) content of several strains of benthic marine cyanobacteria and diatoms was determined. We were unable to detect this compound in any of the cyanobacterial strains even though some of these had been isolated from cyanobacteria-dominated (sub)tidal sediments in which we had measured considerable amounts of DMSP. The diatom *Cylindrotheca closterium* contained an average concentration of about 4 mmoles DMSP (g Chl*a*)$^{-1}$ and a strain of *Navicula* sp. contained approximately 30 µmoles DMSP (g Chl*a*)$^{-1}$. DMSP production by diatoms seems to be highly species-specific but it provides a potential source for the DMSP encountered in the sediment. The role of cyanobacteria in the transformation of DMSP and DMS was limited. The strains that were tested were not able to oxidize DMS during anoxygenic photosynthesis. Cyanobacteria are probably not able to cleave DMSP enzymatically to DMS and acrylate, however the rise in pH they cause as a result of the photosynthetic CO_2 fixation may lead to the enhanced chemical hydrolysis of DMSP. A strain of the cyanobacterium *Phormidium* sp. reduced DMSO to DMS during fermentation under anoxic dark conditions. This is another potential source of DMS in coastal marine sediments.

INTRODUCTION

Dimethyl sulfide (DMS) has been identified as an important component of the global sulfur cycle and it is assumed that this compound is a major factor in the transport of sulfur from the oceans to the continent (13). It has been estimated that ± 50% of the total flux of

Biological and Environmental Chemistry of DMSP and Related Sulfonium Compounds
edited by Ronald P. Kiene et al., Plenum Press, New York, 1996

sulfur to the atmosphere consists of DMS (10), and because DMS is also assumed to play an important role in global climate regulation, it is not surprising that much research concentrates on processes that are involved in the turnover of this compound. About 95% of the DMS emitted into the atmosphere originates from the marine environment (10). The main source of atmospheric DMS is the oceans. Most of the DMS originates from the degradation of dimethylsulfoniopropionate (DMSP), which is produced by a variety of marine phototrophic organisms (9,33). DMSP can be enzymatically cleaved to acrylate and DMS. This reaction is carried out either by bacteria (2) after release of DMSP in the environment or by the DMSP-producing organisms themselves (7,23). DMS that is produced in the water column is either further metabolized or it escapes into the atmosphere. In order to be able to predict fluxes of DMS, all sources and sinks must be known and quantified. However, there are still many uncertainties in this area.

Although there is no doubt that the oceans are the major source of DMS, concentrations of this compound and its precursor and degradation products are extremely low. These low concentrations hamper studies of processes involved in the production and conversion of DMS. Marine intertidal sediments are often characterized by the occurrence of dense communities of phototrophic microorganisms, notably cyanobacteria and diatoms. Such communities are also known as microbial mats. Compared to the oceans the total surface of these coastal sediments is small, but, nevertheless, they produce considerable amounts of volatile sulfur compounds. On the basis of the few data that are available, it was calculated that up to 10% of the sulfur emissions can be attributed to coastal marine sediments (24). In marine microbial mats sulfur cycling is important with respect to energy flow and interactions among organisms (27). Microbial mats represent dynamic, small-scale ecosystems with a high biomass and biological activity, in which the environmental impact on the flux of DMS can be studied and modeled.

The discovery of the presence of considerable amounts of DMSP in microbial mats as well as in the mat-forming cyanobacterium *Microcoleus chthonoplastes* (28,31) motivated us to study the contribution of the oxygenic phototrophic organisms (cyanobacteria and diatoms) to the production and consumption of DMSP and DMS in coastal marine sediments.

MATERIALS AND METHODS

Field Samples

Sediment samples were taken with stainless steel cores (17-mm inner diameter); in August 1994 from a microbial mat in a subtidal pool in the Étang du Prévost, a lagoon at the Mediterranean coast of France, and in February 1995 from a biofilm of diatoms at an intertidal sand flat (the Cocksdorp) on the island of Texel in The Netherlands. The cores were stored in the dark on ice and taken to the laboratory, where they were analyzed within 24 hours. For the determination of vertical profiles of DMS(P) and pigments, the sediment cores were cut into 2 or 5-mm slices. For the determination of pigments, slices were frozen, freeze-dried and stored until analysis. For the determination of DMS(P), the sediment slices were transferred to 10-ml glass Chrompack vials. ASN III-medium (18) with 2.5% formaldehyde was added to a final volume of 3 ml and the vials were sealed with teflon coated rubber septa and aluminum crimp caps. These teflon-coated septa were used in all experiments, except for the DMSO-reduction experiments. For these experiments butyl rubber stoppers were used which proved to be more efficient in maintaining anoxic conditions. Total hydrolyzable DMS was measured in the sediments from the island of Texel, whereas DMS and DMSP were determined in the microbial mat samples from Étang du Prévost. DMS and DMSP were measured as described below.

Cultures and Experiments

Cyanobacteria. Isolates of several strains of mat-forming cyanobacteria (Table 1) were provided by Dr. M. Villbrandt (Laboratory for Microbiology, University of Amsterdam). *Microcoleus chthonoplastes* strain 11 was isolated from a marine microbial mat on the island of Mellum (Germany) by Stal and Krumbein (21) and was obtained in pure culture by Visscher and Van Gemerden (28). The non-heterocystous strains of *M. chthonoplastes* 11, *Phormidium* sp., and *Oscillatoria* sp. were grown in ASN III medium. The heterocystous strains of *Anabaena* sp. and *Nostoc* sp. were cultured in a 1:1 mixture of nitrate-free ASN III and BG 11 medium (17), which had a salinity of approximately 17 ppt. Cultures were grown in Erlenmeyer flasks in an orbital shaking incubator, at an irradiance of ± 35 μmol m^{-2} s^{-1} (white light) and ± 20°C.

DMSP content. For the determination of the DMSP content 10 to 100 ml of culture was harvested at the end of the exponential growth phase and was transferred into 30-ml vials (final volume 10 ml). Chl *a* was measured in 1 ml subsamples. The sample was centrifuged for 10 minutes at 4000 rpm, the supernatant was removed and replaced by 1 ml of pure methanol and Chl *a* was extracted overnight in the dark at ± 4°C.

DMS-oxidation experiment. DMS oxidation was tested in 5-ml cultures of *M. chthonoplastes* 11, *Phormidium* sp., *Anabaena* sp. and *Nostoc* sp. which were inoculated in 20 ml of fresh medium in 30-ml glass bottles. Subsequently, 1 mM of DMS and 5 μM of 3-(3,4-dichlorophenyl)1,1-dimethylurea (DCMU) -from an ethanolic stock solution- were added. DCMU inhibits oxygenic photosynthesis. The vials were incubated on a shaking incubator at ± 10 μmol m^{-2} s^{-1} and ± 20 °C. DMS was determined gas chromatographically by headspace analysis. Chl*a* was determined as described above.

DMSO-reduction experiment. To test whether DMSO was reduced, 25-ml samples of a stationary culture of *Phormidium* sp. were transferred to 60-ml bottles. The bottles were wrapped with aluminium foil and were purged with Argon for 10 minutes and then 1 mM of DMSO was added. DMS and Chl*a* were determined as described above. For the determination of DMSO and storage polyglucose (total glucose) samples of 1 ml were taken with a sterile plastic syringe and were stored at -20°C until analysis.

Table 1. DMSP content of benthic marine cyanobacteria and diatoms

Cyanobacteria	Isolation site	DMSP content (μmol DMSP (g Chl*a*)$^{-1}$)
M. chthonoplastes 11	Mellum (Ger.)	ND
Phormidium sp.	Prévost (Fr.)	ND
Oscillatoria sp.	Prévost (Fr.)	ND
Anabaena sp.	Arcachon (Fr.)	ND
Nostoc sp.	Texel (Nl.)	ND
Diatoms		
C. closteriu	Eems-Dollard (Nl.)	± 153
Navicula sp. 1295	Texel (Nl.)	± 3

ND = not detected. For the detection limit, see Materials and Methods section.

DMSP-cleavage experiment. To test whether DMS was produced from DMSP, 5-ml samples of cultures of *M. chthonoplastes* 11, *Phormidium* sp. and *Nostoc* sp. were inoculated in 20 ml of fresh medium in 60-ml bottles. Subsequently, ± 0.5 mM of DMSP was added and the vials were incubated on a shaking incubator at an irradiance of ± 10 µmol m^{-2} s^{-1} and ± 20 °C. DMS and Chl*a* were determined as described above. For the determination of DMSP 1 ml-samples were taken and stored at -20 °C. until analysis.

Diatoms. Diatoms were cultured in "f" medium (5). The strains that were used, were *Cylindrotheca closterium*, provided by N. Staats (Laboratory for Microbiology, University of Amsterdam) and *Navicula* sp., isolated from a diatom biofilm on intertidal sediments of the island of Texel. Cultures were grown in Erlenmeyer flasks, which were bubbled with air in order to avoid CO_2 limitation. The diatoms were grown on a layer of pure sea sand (Merck) on the bottom of the Erlenmeyer flasks. The cultures were incubated at an irradiance of ± 40 µmol m^{-2} s^{-1} and ± 20 °C.

DMSP Content. The DMSP content of diatoms was determined as described for cyanobacteria.

Effect of salinity. To test the effect of the salt concentration on the DMSP content, the salinity of the medium was adjusted by adding different amounts of NaCl. The salinities applied were 38, 56, 74 and 94 ppt. Growth of the cultures was followed as the optical density (750 nm) using a spectrophotometer (Novaspec II, Pharmacia). At the end of the exponential growth phase the cells were harvested and Chl*a* and DMSP were determined.

Determination of DMS and DMSP, Pigments and Total Glucose. DMS was determined by headspace analysis: 250 µl of headspace was taken with a gas-tight Hamilton syringe and was injected in a gas chromatograph (Chrompack,CP 9000), equipped with a wide bore column (Poraplot U; 0.53 mm, 25 m, Chrompack, The Netherlands) and a flame ionization detector. The temperatures of the detector, injector and oven were 200, 175 and 150°C, respectively. The flow of air was 250 ml min^{-1}, the flow of H_2 was 25 ml min^{-1} and the flow of the carrier gas, N_2, was ± 5.5 ml min^{-1}. The retention time of DMS was ± 4.8 minutes. DMSP and DMSO were measured indirectly as DMS after alkaline hydrolysis and reduction, respectively. Samples were transferred to glass Chrompack vials, which were closed. DMS was measured in the headspace. For the determination of DMSP, 10 M NaOH was added to the vials through the septa to a final concentration of at least 2 M. The samples were kept overnight in the dark at ± 4°C and DMS was measured again. The detection limit for DMS, using 30-ml vials with a volume of 14 ml of culture and NaOH, was 0.1 µM. For the determination of DMSO a saturated $SnCl_2$ solution (29) was added to the vials. They were kept for at least 90 minutes at 55 °C and DMS was measured again. The Chl*a* concentration in the culture samples was determined by measuring the extinction of the methanol extracts at 665 nm using a spectrophotometer (Novaspec II, Pharmacia). The extinction coefficient used was 74.5 ml mg^{-1} cm^{-1} (15). For the determination of pigments, the sediment samples were extracted with 90% acetone and the extracts were analyzed by HPLC (ISCO, model 2350, V^4 absorbance detector), as described by Mantoura and Llewellyn (14). Polyglucose storage products were measured by using Boehringers GOD reagent (32), with absorbance being determined at 420 nm.

Chemicals. All chemicals that were used were purchased from E. Merck AG, Darmstadt, BRD and Sigma Chemical Co., St. Louis, Mo., USA, except for DMSP, which was produced and kindly provided by the Department of Microbiology, University of Groningen, The Netherlands.

RESULTS

In the diatom film of Texel, the DMSP concentration (total hydrolyzable DMS) was highest in the upper 2 mm of the sediment (Fig. 1A). In the microbial mats of Étang du Prévost, DMS and DMSP concentrations were also highest in the upper 2 mm of the sediment and amounted to 7 and 140 nmoles cm^{-3} sediment, respectively. The vertical distribution of DMSP corresponded well with profiles of Chl*a* (Fig. 1B). This suggested that DMSP production is associated with the layer of oxygenic phototrophs. None of the cyanobacteria tested contained detectable amounts of DMSP (Table 1). We tested several strains, isolated from microbial mats of the islands of Texel (The Netherlands) and Mellum (Germany) and from the coastal lagoons of Arcachon and Étang du Prévost (France). At these sites (except Mellum for which no data are available), high amounts of DMSP were found. In contrast, two species of diatoms that were tested contained considerable amounts of DMSP. When normalized to Chl*a*, *Cylindrotheca closterium* contained almost 2 orders of magnitude more

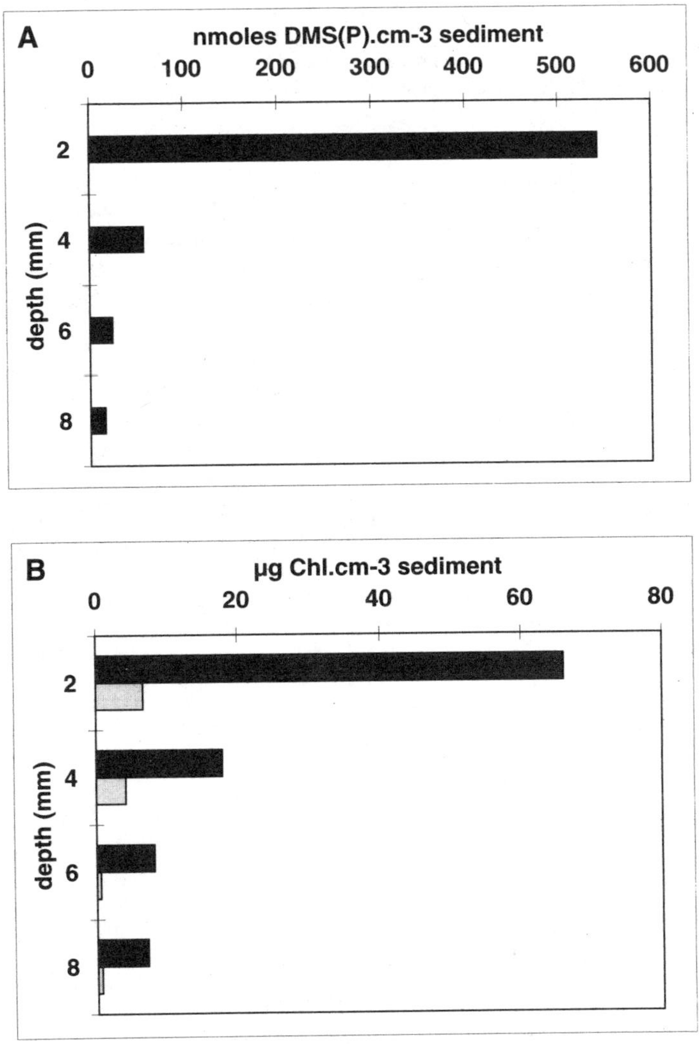

Figure 1. Depth profiles of total hydrolyzable DMS (A) and chlorophyll (B) content of the intertidal sediment (diatom film) of Texel (The Netherlands). Dark grey bars = Chl*a*; light grey bars = Chl*c*.

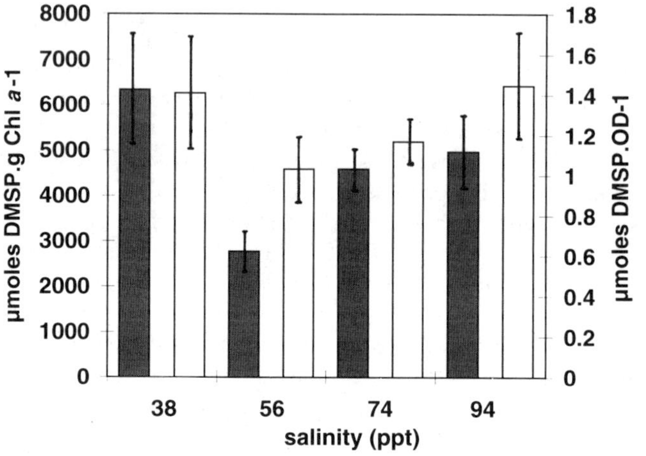

Figure 2. DMSP content of *C. closterium* grown at different salinities. Bars represent means of triplicate measurements of one batch culture (with standard deviation). Dark grey bars = DMSP (g Chl*a*)$^{-1}$; light grey bars = DMSP OD^{-1}.

DMSP than *Navicula* sp. In *C. closterium* no correlation of DMSP content and the salinity of the growth medium was found (Fig. 2). When normalized to Chl*a*, DMSP content was lowest at a salinity of 56 ppt. However, the ratio Chl*a* and biomass (optical density) varied with salinity, and no significant differences were found in DMSP content when expressed on the basis of biomass.

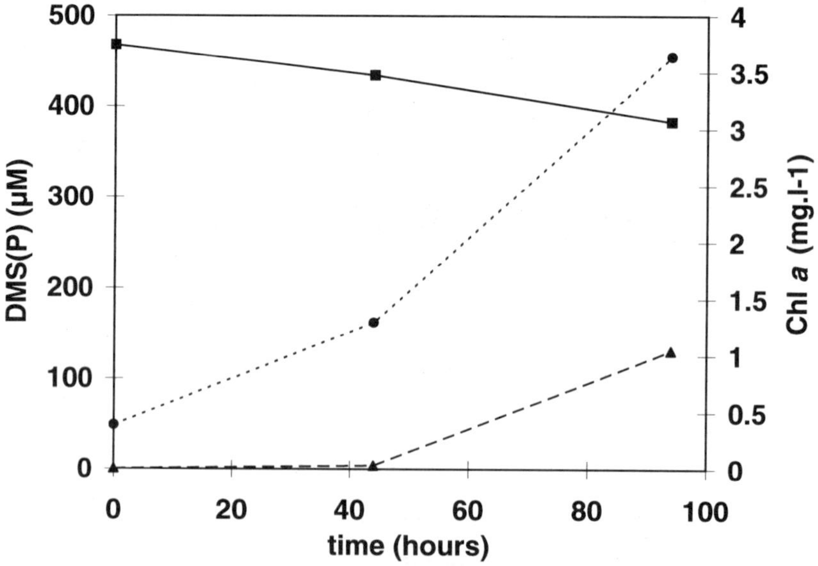

Figure 3. Cleavage of DMSP in a batch culture of *M. chthonoplastes* 11. ± 0.5 mM of DMSP was added to 25 ml of culture in closed 60-ml bottles. Symbols represent means of duplicate batch cultures. Initial pH was 7.7; final pH was 10.1. Symbols: —■— = DMSP; —▲— = DMS; ··●·· = Chl*a*.

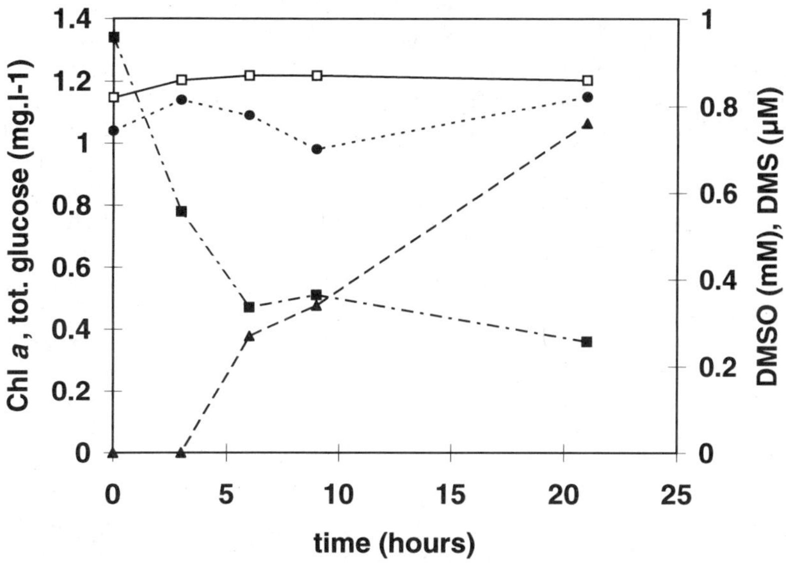

Figure 4. DMSO-reduction by *Phormidium* sp. ± 1 mM of DMSO was added to 25 ml of culture in closed 60-ml bottles, which were incubated under dark anoxic conditions. Symbols represent means of duplicate batch cultures. Symbols: —□— = DMSO; —▲— = DMS; ··●··= Chla; --■-- = total glucose.

M. chthonoplastes 11 produced DMS from DMSP (Fig. 3). The increase of Chl*a* with time showed that this culture was growing. As a result, the pH increased to > 10 due to CO_2 fixation. While 140 µM of DMS were formed in the experiment with *M. chthonoplastes* only 3.5 µM were produced in the sterile control, in which the pH was 7.7. Also, in slow growing cultures of other species, in which the pH did not increase much, only small amounts of DMS were detected (12 µM for *Nostoc* sp. with a final pH of the medium of 7.9).

None of the strains tested oxidized DMS. During the time course of the experiment (4 days) the amount of DMS and of Chl*a* remained constant, which indicated that no growth occurred and that DMS was not used as an electron donor for anoxygenic photosynthesis.

In Fig. 4 it is shown that DMS is produced from the reduction of DMSO at the expense of storage glucose under anoxic conditions in the dark. After 21 h of incubation 1 mg l^{-1} of storage polyglucose was degraded while 0.7 µM of DMS was produced. The concentration of Chl*a* remained constant during the course of the experiment. In abiotic controls (ASN III and DMSO), no DMS was detected.

DISCUSSION

Although DMSP has been reported to occur in cyanobacteria (28,33) we were unable to detect it in any of the strains we investigated. Surprisingly, we were also unable to detect DMSP in *Microcoleus chthonoplastes* strain 11. Visscher and Van Gemerden (28) reported previously the occurrence of DMSP in this organism. The detection limit of the method we used was 5 times lower than the one used by Visscher and Van Gemerden (28). With a protein content of the cultures of 100 mg l^{-1}, amounts as low as 1 µmol of DMSP (g protein)$^{-1}$ could have been measured. By concentrating the cultures 10 times, even 100 nmol of DMSP (g protein)$^{-1}$ would have been detected, which is 2 orders of magnitude lower than the 37.3 µmol (g protein)$^{-1}$ found for *M. chthonoplastes* strain 11.

C. Vogt (pers. comm.) has tested several strains of cyanobacteria and has found concentrations of DMSP that are in the same order of magnitude as found by Visscher and Van Gemerden (28). At present we can not offer a satisfactory explanation for this discrepancy. We would like to emphasize that the method used for the determination of DMSP is indirect and the gas chromatographic measurement of DMS using a flame ionisation detector is also less specific because it is sensitive to the methyl group rather than to sulfur. The column we used has a better separation than packed columns, which allowed us to distinguish DMS from other compounds which are sometimes produced during the alkaline hydrolysis of DMSP. Furthermore, we are confident about our method since we were able to detect DMSP in 2 species of diatoms. The need for the application of a specific method for the detection of DMSP is obvious and such experiments are in progress. Of course, it can not be excluded that the cyanobacteria we have tested produce higher amounts of DMSP under natural conditions than under laboratory conditions, or that they even stop to produce DMSP after several transfers into fresh medium, but we feel that this is unlikely.

The DMSP that was detected in the sediments could originate from benthic diatoms (9). The two strains we tested contained DMSP. Diatoms are often present in cyanobacterial mats in vast numbers, and were also observed in samples from the subtidal pool of Étang du Prévost. The large amount of DMSP in the diatom film on the island of Texel suggests that diatoms were responsible for its production. However, the DMSP content of *Navicula* sp. 1295 does not account for the amount measured in the field. In the upper layer of the sediment 400 nmol DMSP and 70 μg Chl*a* cm^{-3} were present. Seventy μg Chl*a* in *Navicula* sp. 1295 corresponds with 2.5 nmol DMSP, which is about 2 orders of magnitude too low to explain the sediment content. There are several possible explanations for the difference that we observed. Theoretically, the method used to determine the DMSP content in the sediment may have overestimated it, since the total hydrolyzable DMS concentration was measured. Another interesting possibility is the occurrence of vast amounts of DMSP accumulated extracellularly. Preliminary data do not support this. In intertidal areas deposition of DMSP-containing macroalgae such as *Ulva* sp. and *Enteromorpha* sp., may be an additional source of DMSP. However, for this possibility no data are available. Moreover, macroalgae were not observed in the samples from the sediments containing the diatom biofilm on the island of Texel. It is also possible that other diatom species are present in the sediment that produce larger amounts of DMSP. For instance, comparing the DMSP content of the biofilm with that produced by *C. closterium* (Table 1), it is obvious that this species could be responsible for it. However, this species has not been observed in the diatom biofilm studied here, and information about the DMSP content of the sediment where this species was isolated is currently lacking.

The experiment shown in Fig. 2 reveals even an average DMSP content in *C. closterium* of 5 mmol (g Chl*a*)$^{-1}$ which is about 3 times the amount we measured originally (Table 1). It is possible that different light conditions affected the DMSP content of this diatom. Karsten et al.(8) demonstrated that the DMSP content of macroalgae increased with increasing daylength and light intensity.

It has been conceived that DMSP may serve as an osmoregulator in organisms that contain it. We investigated this possibility in the diatom *C. closterium* and conclude that DMSP does not play a role as osmoregulator. Growth at increased salinities did not result in significant higher DMSP content but a higher ratio of Chl*a* : biomass was found. This was probably due to an increased chlorophyll requirement to cover a higher energy demand to cope with osmotic stress.

DMSP present in the sediment can be degraded biologically or chemically. It can either be demethylated (12,26,30), a process which does not result in the formation of DMS or it can be cleaved to acrylate and DMS (28,30). DMS production is often associated with the presence of cyanobacteria (3,4). Our results suggest that cyanobacteria are indirectly responsible for the cleavage of DMSP (Fig. 3). The formation of DMS in the cultures was most likely the result of the chemical degradation of extracellular DMSP and not of the enzymatic cleavage by a DMSP-lyase. There are no reports of the presence of this enzyme in cyanobacteria and there was a strong correlation between the amount of DMS formed and the pH of the culture. In the light, cyanobacteria and other photoautotrophic organisms such as diatoms will cause a rise of pH in the sediment which will accelerate the chemical cleavage of DMSP resulting in the formation of DMS (28).

DMS is degraded by a variety of aerobic and anaerobic bacteria (11,29,31). The colorless sulfur bacterium strain T5 (31) oxidized DMS to CO_2 and SO_4^{2-} and the purple sulfur bacterium *Thiocapsa roseopersicina* used DMS as electron donor in anoxygenic photosynthesis, producing dimethylsulfoxide (DMSO) from it. Many cyanobacteria are capable of sulfide dependent anoxygenic photosynthesis (1) but none of the strains we tested were capable of oxidizing DMS. Oxidation of DMS results in formation of either DMSO or CO_2 and H_2S.

Many organisms, such as for instance colorless sulfur bacteria (10) and purple non-sulfur bacteria (17), can use DMSO as an alternative electron acceptor for anaerobic respiration. Our results demonstrate that *Phormidium* sp. is also capable of reducing DMSO. Cyanobacteria are metabolically versatile organisms (19). Many species have now been shown to switch to fermentation under anoxic dark conditions. A variety of fermentation pathways have been observed in cyanobacteria (6, 19), and some species are capable of reducing elemental sulfur. It seems unlikely that this organism used DMSO as a terminal electron acceptor in respiratory electron transport. Oxidation of 1 mg l^{-1} storage polyglucose to CO_2 using DMSO as electron acceptor would yield 144 μM of DMS, however, only 0.7 μM was found. It is therefore more likely that DMSO served as an electron sink during fermentation. This is also the case in cyanobacteria such as *M. chthonoplastes* strain 11 and *Oscillatoria limosa* that use elemental sulfur as electron sink, resulting in the formation of sulfide (19,20). The use of an electron sink increases the ATP yield during fermentation (19).

Summarizing, it can be concluded that benthic cyanobacteria probably play only a minor role in the production and degradation of DMSP and DMS in intertidal marine sediments. The strains tested do not produce DMSP in significant amounts under laboratory conditions, but may indirectly cause the chemical decomposition of DMSP by increasing the pH as a result of photosynthetic activity. Cyanobacteria also did not oxidize DMS, but were capable of producing it as the result of the reduction of DMSO.

ACKNOWLEDGMENTS

This research was subsidised in part by the European Community ENVIRONMENT programme, contract N° EV5V-CT92-0080 and by the Netherlands Organization of Scientific Research (NWO) and the National Research Programme (NOP).

REFERENCES

1. Cohen, Y., B.B. Jørgensen, N.P. Revsbech and R. Poplawski. 1986. Adaptation to hydrogen sulfide of oxygenic and anoxygenic photosynthesis among cyanobacteria. Appl. Environ. Microbiol. *51*: 398-407.

2. De Souza, M.P. and D.C. Yoch. 1995. Purification and characterization of dimethylsulfoniopropionate lyase from an *Alcaligenes*-like dimethyl sulfide-producing marine isolate. Appl. Environ. Microbiol. *61*: 21-26.
3. Granroth, B. and T. Hattula. 1976. Formation of dimethyl sulfide by brackish water algae and its possible implication for the flavor of baltic herring. Finn. Chem. Lett.: 148-150.
4. Gries, C., T.H. Nash III and J. Kesselmeier. 1994. Exchange of reduced sulfur gases between lichens and the atmospere. Biogeochem. *26*: 25-39.
5. Guillard, R.R.L. and J.H. Ryther. 1962. Studies on marine planktonic diatoms. I. *Cyclotella nana* Hustedt and *Detonula confervacea* (Cleve) Gran. Can. J. Microbiol. *8*: 229-239.
6. Heyer, H., L.J. Stal and W.E. Krumbein. 1989. Simultaneous heterolactic and acetate fermentation in the marine cyanobacterium *Oscillatoria limosa* incubated anaerobically in the dark. Arch. Microbiol. *151*: 558-564
7. Kadota, H. and Y. Ishida. 1972. Production of volatile sulfur compounds by microorganisms. Ann. Rev. Microbiol. *26*: 127-138.
8. Karsten, U., C. Wiencke and G.O. Kirst. 1990. The effect of light intensity and daylength on the ß-dimethylsulphoniopropionate (DMSP) content of marine green macroalgae from Antarctica. Pl. Cell Env. *13*: 989-993.
9. Keller, M.D., W.K. Bellows and R.R.L. Guillard. 1989. Dimethylsulfide production in marine phytoplankton, p. 167-200. *In* E.S. Saltzman and W.J. Cooper (eds.), Biogenic Sulfur in the Environment, American Chemical Society, Washington DC.
10. Kelly, D.P. and N.A. Smith. 1990. Organic sulfur compounds in the enironment: Biogeochemistry, microbiology, and ecological aspects. Adv. Microbial Ecol. *11*: 345-385.
11. Kiene, R.P. 1988. Dimethyl sulfide metabolism in salt marsh sediments. FEMS Microbiol. Ecol. *53*: 71-78.
12. Kiene, R.P. and B.F. Taylor. 1988. Demethylation of dimethylsulfoniopropionate and production of thiols in anoxic marine sediments. Appl. Environ. Microbiol. *54*: 2208-2212.
13. Lovelock, J.E., R.J. Maggs and R.A. Rasmussen. 1972. Atmospheric dimethylsulfide and the natural sulfur cycle. Nature *237*: 452-453.
14. Mantoura, R.F.C. and C.A. Llewellyn. 1983. The rapid determination of algal chlorophyll and carotenoid pigments and their breakdown products in natural waters by reverse-phase high-performance liquid chromatography. Anal. Chim. Acta *151*: 297-314.
15. McKinney, G. 1941. Absorption of light by chlorophyll solutions. J. Biol. Chem. *140*: 315-322
16. Moezelaar, R. 1995. Fermentation in the cyanobacteria *Microcystis aeruginosa* and *Microcoleus chthonoplastes*. PhD thesis, University of Amsterdam.
17. Richardson, D.J., G.F. King, D.J. Kelly, A.G. McEwan, S.J. Ferguson and J.B. Jackson. 1988. The role of auxiliary oxidants in maintaining redox balance during growth of *Rhodobacter capsulatus* on propionate or butyrate. Arch. Microbiol. *150*: 131-137.
18. Rippka, R., J. Deruelles, J.B. Waterbury, M. Herdman and R.Y. Stanier. 1979. Generic assignment, strain histories and properties of pure cultures of cyanobacteria. J. Gen. Microbiol. *111*: 1-61.
19. Stal, L.J. 1991. The metabolic versatility of the mat-building cyanobacteria *Microcoleus chthonoplastes* and *Oscillatoria limosa* and its ecological significance. Alg. Stud. *64*: 453-467.
20. Stal, L.J. 1991. The sulfur metabolism of mat-building cyanobacteria in anoxic marine sediments. Kieler Meeresforsch., Sonderh. *8*: 152-157.
21. Stal, L.J. and W.E. Krumbein. 1985. Isolation and characterization of cyanobacteria from a marine microbial mat. Bot. Marina *28*: 351-365.
22. Stal, L.J. and W.E. Krumbein. 1986. Metabolism of cyanobacteria in anaerobic marine sediments. Deuxième Colloque International de Bactériologie Marine. Actes de Colloques *3*: 301-309. Gerbam, Ifremer, Brest, France.
23. Stefels, J. and W.H.M. van Boekel. 1993. Production of DMS from dissolved DMSP in axenic cultures of the marine phytoplankton species *Phaeocystis* sp.. Mar. Ecol. Prog. Ser. *97*: 11-18.
24. Steudler, P.A. and B.J. Peterson. 1984. Contribution of gaseous sulfur from salt marshes to the global sulfur cycle. Nature *311*: 455-457.
25. Vairavamurthy, A, M.O. Andreae and R.L. Iverson. 1985. Biosynthesis of dimethylsulfide and dimethylpropiothetin by *Hymenomonas carterae* in relation to sulfur source and salinity variations. Limnol. Oceanogr. *30*: 59-70.
26. Van der Maarel, M.J.E.C., P. Quist, L. Dijkhuizen and T.A. Hansen. 1993. Anaerobic degradation of dimethylsulfoniopropionate to 3-*S*-methylmercaptopropionate by a marine *Desulfobacterium* strain. Arch. Microbiol. *160*: 411-412.
27. Van Gemerden, H. 1993. Microbial mats: a joint venture. Mar. Geology *113*: 3-25.

28. Visscher, P.T. and H. van Gemerden. 1991. Production and consumption of dimethylsulfoniopropionate in marine microbial mats. Appl. Environ. Microbiol. *57*: 3237-3242.
29. Visscher, P.T. and H. van Gemerden. 1991. Photo-autotrophic growth of *Thiocapsa roseopersicina* on dimethyl sulfide. FEMS Microbiol. Lett. *81*: 247-250.
30. Visscher, P.T., R.P. Kiene and B.F. Taylor. 1994. Demethylation and cleavage of dimethylsulfoniopropionate in marine intertidal sediments. FEMS Microbiol. Ecol. *14*: 179-190.
31. Visscher, P.T., P. Quist and H. van Gemerden. 1991. Methylated sulfur compounds in microbial mats: in situ concentrations and metabolism by a colorless sulfur bacterium. Appl. Environ. Microbiol. *57*: 1758-1763.
32. Werner, W., H.-G. Rey and H. Wielinger. 1970. Über die Eigenschaften eines neuen Chromogens für die Blutzuckerbestimmung nach der GOD/POD-methode. Z. Analyt. Chem. *252*: 224-228.
33. White, R.H. 1982. Analysis of dimethyl sulfonium compounds in marine algae. J. Marine Res. *40*: 529-536.

33

PRODUCTION OF DIMETHYLSULFIDE AFTER DEPOSITION OF INCREASING AMOUNTS OF *EMILIANIA HUXLEYI* ONTO SEDIMENTS IN MARINE MICROCOSMS

Ronald Osinga, Johanna J. Minnaard, Wilma E. Lewis, and Fleur C. van Duyl

Netherlands Institute for Sea Research, P.O. Box 59
1790 AB Den Burg (Texel), The Netherlands

SUMMARY

Rapid release of dimethylsulfoniopropionate (DMSP) and subsequent production of dimethylsulfide (DMS) may occur after deposition of DMSP-containing algae onto sediments, potentially causing a temporal accumulation of DMS. This relation between sedimentation and DMS formation was studied by supplying increasing amounts of the marine microalga *Emiliania huxleyi* to anoxic marine sediment microcosms, resulting in initial DMSP concentrations of 950 nM, 1600 nM and 5300 nM. In all experiments, rapid formation of DMS was observed, the highest concentrations were reached after 2 to 5 days. The DMS concentrations remained high for more than 5 days, suggesting a slow response of anaerobic DMS consuming bacteria. In a control experiment, in which the algae were kept in suspension, the release of DMS was an order of magnitude lower, and more gradual. It was therefore concluded that sedimentation of DMSP containing algae to anoxic sediments can lead to emission of DMS to the water column. The ratio between DMS produced and DMSP added was highest at the highest algal density. This may indicate an increasing importance of the cleavage of DMSP under increasing substrate concentrations. Hence, benthic DMS formation after sedimentation of algae is most likely to occur in eutrophic, coastal areas, where large amounts of algae are deposited onto very reduced sediments.

INTRODUCTION

The organosulfur compound dimethylsulfoniopropionate (DMSP) has gained increasing scientific attention, since it is the most important biogenic precursor for the proposed "anti-greenhouse gas" dimethylsulfide (DMS) (5, 23). Microbial conversion of

Biological and Environmental Chemistry of DMSP and Related Sulfonium Compounds
edited by Ronald P. Kiene et al., Plenum Press, New York, 1996

DMSP to DMS and microbial degradation of DMS are key processes in the regulation of the flux of DMS to the atmosphere (10, 12).

Different physiological types of bacteria are involved in the microbial degradation of DMSP and DMS (9). These processes may not always be closely coupled: periods of rapid DMS formation from DMSP may temporarily lead to high DMS concentrations in seawater and elevated fluxes of DMS into the atmosphere, as was found by Kwint & Kramer (15) in the Dutch Coastal Zone.

Very high DMS concentrations of more than 300 nM were observed during pelagic mesocosm experiments, in which a bloom of the DMSP producing Prymnesiophyte *Phaeocystis sp.* was simulated (19, 20). These high concentrations persisted only for a few days. Osinga et al. (19) suggested the following mechanism to explain this phenomena: due to sedimentation and a rapid breakdown of the deposited algae on the seafloor, intracellular DMSP will be released. Subsequently, rapid bacterial turnover of the released DMSP will cause high levels of DMS. However, *Phaeocystis sp.* possesses a very active DMSP-lyase system itself (21), and the activity of this enzyme could as well have been responsible for the rapid formation of DMS in these mesocosm-experiments.

In this study, the formation of DMS after deposition of another Prymnesiophyte, *Emiliania huxleyi*, was measured in gas-tight marine sediment microcosms. *E. huxleyi* is also an important producer of DMSP (8, 17), but in contrast to *Phaeocystis sp.*, it shows little if any DMSP-lyase activity (22).

Kiene & Service (12), who found a close coupling between the microbial degradation of DMSP and DMS, hypothesized that DMS will accumulate when the concentration of dissolved DMSP reaches a certain threshold. An increase of the amount of DMSP above this threshold concentration may thus further enhance the accumulation of DMS. This was studied by monitoring the concentration of DMS after deposition of increasing amounts of cultured *E. huxleyi* onto the microcosm surface.

Both DMSP and DMS can be degraded aerobically as well as anaerobically (23). In sediments with a high organic matter input, anaerobic bacterial processes will predominate over aerobic processes (1). Large amounts of algae were used in this study, simulating a high organic matter input. Hence, we focused on the fate of DMSP under anaerobic conditions.

MATERIAL AND METHODS

The microcosms used for the experiments were perspex cores (94 mm diameter) that could be closed with removable perspex lids containing an o-ring. These lids had a conical shape, and two openings on top that could be closed with screw-caps containing a teflon-coated septum. The microcosms consisted of a sediment phase of at least 10 cm depth, and a water phase of 10.5-14.5 cm (0.75-1.0 l). For the sediment phase, sandy sediment with a low organic matter content was used, which had been collected in the southern North Sea. This sediment had been preserved in the lab for three years at 12°C in darkness, and was therefore not likely to contain significant amounts of organosulfur compounds. A magnetic stirrer was mounted inside the microcosm, approximately 5 cm above the sediment surface.

To study the effects of deposition of algae, the microcosms were supplied with monospecific, non-axenic cultures of *Emiliania huxleyi*. The algae were cultured on a nutrient-rich F2 medium (6), to which 150 μM ammonium was added. The cultures were grown in Erlenmeyer flasks at a temperature of 15°C, under a light regime of 14 hours light and 10 hours darkness. Cell densities in the cultures were followed by microscopic counts. Shortly before the addition to the microcosms, the cultures were gently flushed with nitrogen gas for 15 minutes to remove oxygen. Immediately thereafter, a sample was taken from the cultures for determination of the total DMSP content (particulate and dissolved).

In three separate experiments, cultures with cell densities of 1.5 x 10^5 cells.ml^{-1}, 4.0 x 10^5 cells.ml^{-1} and 2.0 x 10^6 cells.ml $^{-1}$ were supplied to the microcosms. In order to check whether the microcosms remained anoxic, the oxygen concentration in the microcosm with 4.0 x 10^5 cells.ml^{-1} at the end of the incubation was measured using the standard Winkler method.

The microcosms were not stirred. Since the average sinking velocity of *E. huxleyi* cells is approximately 1 m per day (P. van der Wal, personal communication), most of the algae will therefore have been deposited onto the sediment surface a few hours after the start of the incubation. However, to make sure that the observed effects are indeed due to deposition, a control experiment was performed with an addition of 4.0 x 10^5 cells.ml^{-1}, which was continuously stirred in order to slow down the deposition rate.

The additions of 1.5 x 10^5, 4.0 x 10^5 and 2.0 x 10^6 cells/ml resulted in initial DMSP concentrations in the water phase of respectively 950 nM, 1600 nM and 5300 nM. A microcosm supplied with autoclaved F2 medium only was used as a starved control.

Samples for DMS measurements were taken from the microcosms with a 10 ml injection syringe, using the openings in the lids. The sample size was 10 ml. Just before sampling, the microcosms were gently stirred for half a minute to ensure a homogeneous distribution of DMS through the water column. Visually, no resuspension of algal material was observed due to this stirring. Resuspension was therefore assumed to be low. Immediately after sampling, 10 ml of 0.2 µm filtered aged seawater was supplied to the microcosm. This was done to prevent the formation of a gas phase in the microcosms. In the experiments with 1.5 x 10^5 and 4.0 x 10^5 cells.ml^{-1}, samples for DMS analysis were taken until 10 days after the start of the incubation, while the experiment with 2.0 x 10^6 cells/ml continued for 27 days. The DMS concentration in the starved control was monitored for five subsequent days.

In order to compare a semi-natural sediment with an artificial bottom, an additional experiment was performed in which a silicon rubber stopper was used as artificial bottom. To this microcosm, 2.0 x 10^6 cells/ml were added (5300 nM DMSP), and the DMS concentration was monitored for 27 days.

DMS is a very volatile compound. The loss of DMS due to adsorption and/or diffusion processes was estimated from a control microcosm with distilled water, to which 170 nM DMS (Aldrich Chemicals) was added. A microcosm with a silicon stopper instead of sediment was tested in this way as well. The DMS concentration in these microcosms was measured on three subsequent days.

DMS samples were preconcentrated following the procedures described by Kwint & Kramer (14). Pre-concentrated samples were analyzed for organic sulfur compounds on a Varian 3500 gas chromatograph, according to Lindqvist (16). DMSP in the algal cultures was measured as DMS after incubation of the samples for 24 hours at room temperature with 1 N NaOH, for chemical conversion (in an equimolar ratio) of DMSP to DMS (2). For calculating the DMSP concentration, the DMS concentration measured in an untreated sample was subtracted from the value obtained in the NaOH treated sample.

RESULTS

In the control experiments with distilled water and 170 nM DMS, a slow decrease in the concentration of DMS was observed during the first two days. Thereafter, the DMS concentration did not decrease further (Fig. 1). The daily loss of DMS was in general less than 10%, both in the system with sediment and in the system with the silicon stopper.

The results from the experiments with increasing *E. huxleyi* densities are presented in Fig. 2, showing the concentration of DMS in the water column during a period of 10 days. In all cases, addition of the algae resulted in a short period of rapid DMS formation after the first day, followed by a period of persisting high levels of DMS, during which the concentration did not further increase. The concentration of DMS in the experiments with 4.0 x 10^5 cells.ml^{-1} and 2.0 x 10^6 cells.ml^{-1} started to decrease after day 5.

The DMS concentration in the starved control (Fig 2A) remained at levels around or below the detection limit. In the stirred control experiment with 4.0 x 10^5 cells.ml^{-1} (Fig. 3), the concentration of DMS also remained low, and did not show the sharp increase between day 1 and day 3 as was observed in the experiment that had not been stirred.

Addition of increasing amounts of *E. huxleyi* cells resulted in increasing DMS concentrations in the water column (Figs. 2B, 2C and 2D). Moreover, with increasing algal densities, the DMS concentration that was reached after 3 to 5 days showed a relative increase to the amount of DMSP that had been added: when 1.5 x 10^5 cells.ml^{-1} were added, only 5% of the initial DMSP had been recovered as DMS, while 100% had been recovered in the experiment with 2.0 x 10^6 cells.ml^{-1} (Table 1). No oxygen could be detected in the water column of the microcosm with 4.0 x 10^5 cells.ml^{-1} at the end of the incubation. This indicates that the sediment microcosms remained anoxic throughout the whole experimental period.

Fig. 4 shows the results of a sediment microcosm compared to a microcosm with a silicon stopper. Remarkable differences were found between the two systems: the sediment microcosm exhibited a maximal DMS concentration that was almost one order of magnitude higher than the maximal concentration in the microcosm with the silicon stopper. In the latter system, the DMS concentration decreased rapidly after day 5, and dropped to values around zero after day 15. In contrast, DMS in the sediment microcosm did not show an obvious decline until day 17, and elevated concentrations were still found at the end of the incubation period.

In most of the DMS samples that were analysed, some dimethyldisulfide (DMDS) was detected as well. In the experiment comparing the sediment microcosm and the silicon stopper microcosm, obvious differences in the concentration of DMDS were observed. Therefore, the concentration of this compound is also plotted in Fig. 4.

In contrast to DMS, the concentrations of DMDS were highest in the silicon stopper microcosm, where DMDS and DMS were found in nearly equal concentrations. The highest DMDS concentrations were found after day 5, when DMS had already started to decrease. In the sediment microcosm, the maximal DMDS concentration was 6-fold lower than in the silicon stopper microcosm, and was equal to only 1 or 2% of the DMS concentration. However, the values were still more than an order of magnitude higher than the concentrations measured in the starved control (data not shown).

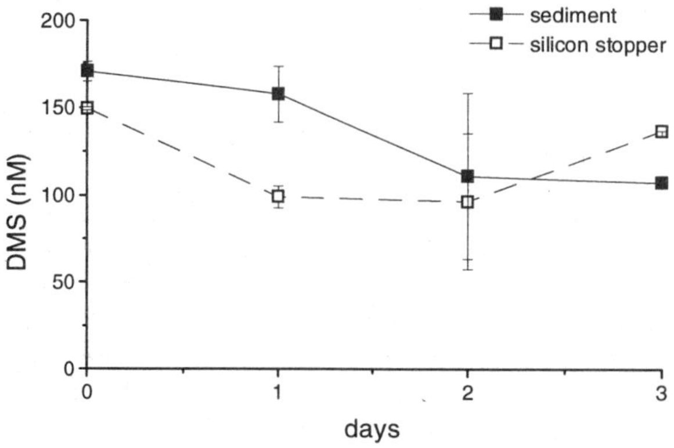

Figure 1. Loss of DMS due to adsorption and diffusion from a microcosm with sediment and from a microcosm with a silicon stopper. Error bars indicate the range between duplicate samples.

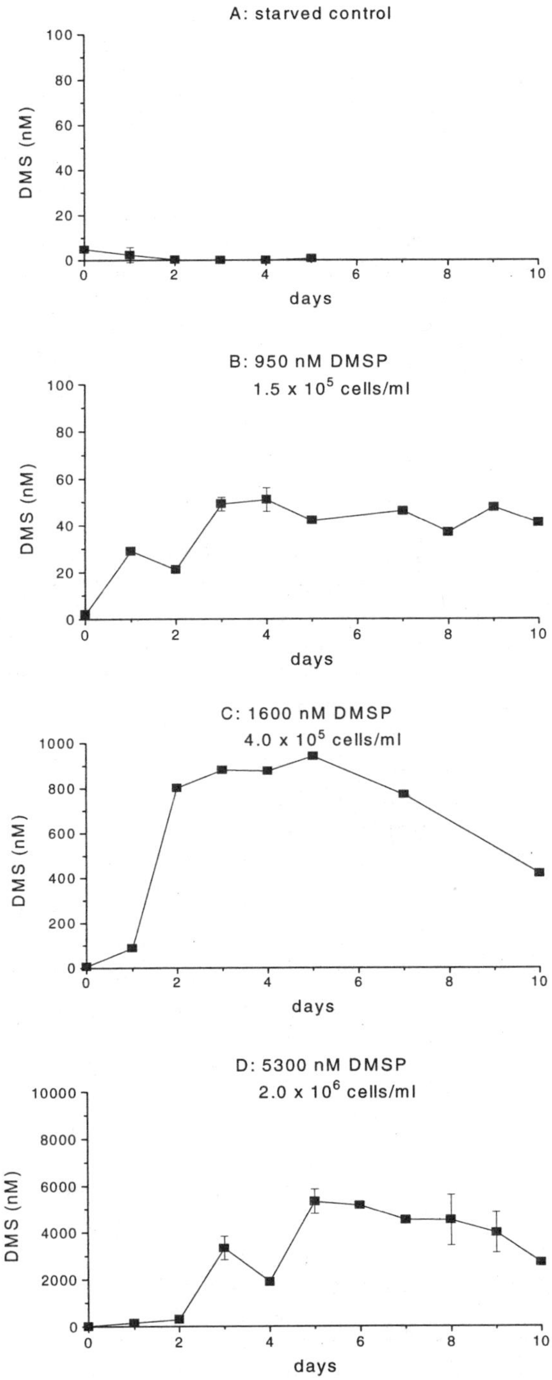

Figure 2. DMS concentrations in the water column of microcosms supplied with different densities of *Emiliania huxleyi*. A. Starved control. B. 1.5 x 10^5 cells.ml^{-1} (950nM DMSP). C. 4.0 x 10^5 cells.ml^{-1} (1600 nM DMSP). D. 2.0 x 10^6 cells.ml^{-1} (5300 nM DMSP). Error bars indicate the range between duplicate samples. Note the differences in scale on the Y-axis in figures 2A/2B (up to 100 nM), 2C (up to 1000 nM) and 2D (up to 10000 nM).

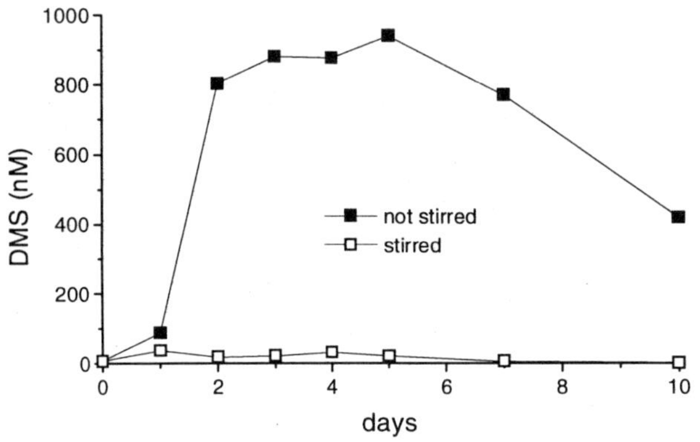

Figure 3. DMS concentrations in the water column of a continuously stirred microcosm and a microcosm that was not stirred. 4.0 x 10^5 cells.ml^{-1} (1600 nM DMSP) were added to these microcosms.

DISCUSSION

The control experiments suggest that abiotic loss of DMS mainly occurs during the first two days after DMS formation. The abiotic loss during the first two days may be due to adsorption to the microcosm wall: DMS will be adsorbed until the wall is saturated. The stabilization of the DMS concentration in the control experiments after two days, and the long period with relatively stable DMS concentrations in the sediment microcosm with 2.0 x 10^6 cells.ml^{-1} indicate that losses due to diffusion out of the system are low. Nevertheless, the total abiotic loss after long incubation periods may be considerable.

In addition to the abiotic losses, the DMS concentration will also slowly decrease due to the daily dilution of the microcosm water with filtered seawater after sampling. This decrease will be around one percent per day.

Our results showed that under anoxic conditions, sedimentation events of DMSP-containing algae will lead to DMS emissions from the sediment surface into the water column. DMS formation in the continuously stirred microcosm was low, indicating that the observed effects in the non-stirred microcosms are indeed due to processes occurring on the sediment surface. The fact that the DMS formation in the microcosms showed a lag-phase like pattern suggests the involvement of bacteria. It has been shown that benthic heterotrophic bacteria can respond very rapidly to deposition of algae (4,7). This gives further support for the ideas of Osinga et al.(19) about the mechanism that leads to accumulation of DMS after deposition of DMSP containing algae: a strong release of DMSP from the rapidly decaying algae, followed by a rapid conversion of the dissolved DMSP to DMS.

The persisting high levels of DMS in the sediment microcosms after the initial increase suggest a low microbial consumption rate of DMS in these microcosms. There are

Table 1. Comparison of the amount of DMSP added and the amount of DMS accumulated after 5 days in the experiments with different densities of *Emiliania huxleyi*

Cell density (cells.ml^{-1})	1.5 x 10^5	4.0 x 10^5	2.0 x 10^6
DMSP at day 0 (nM)	950	1600	5300
DMS at day 5 (nM)	51	940	5325
DMS/DMSP	0.05	0.59	1.00

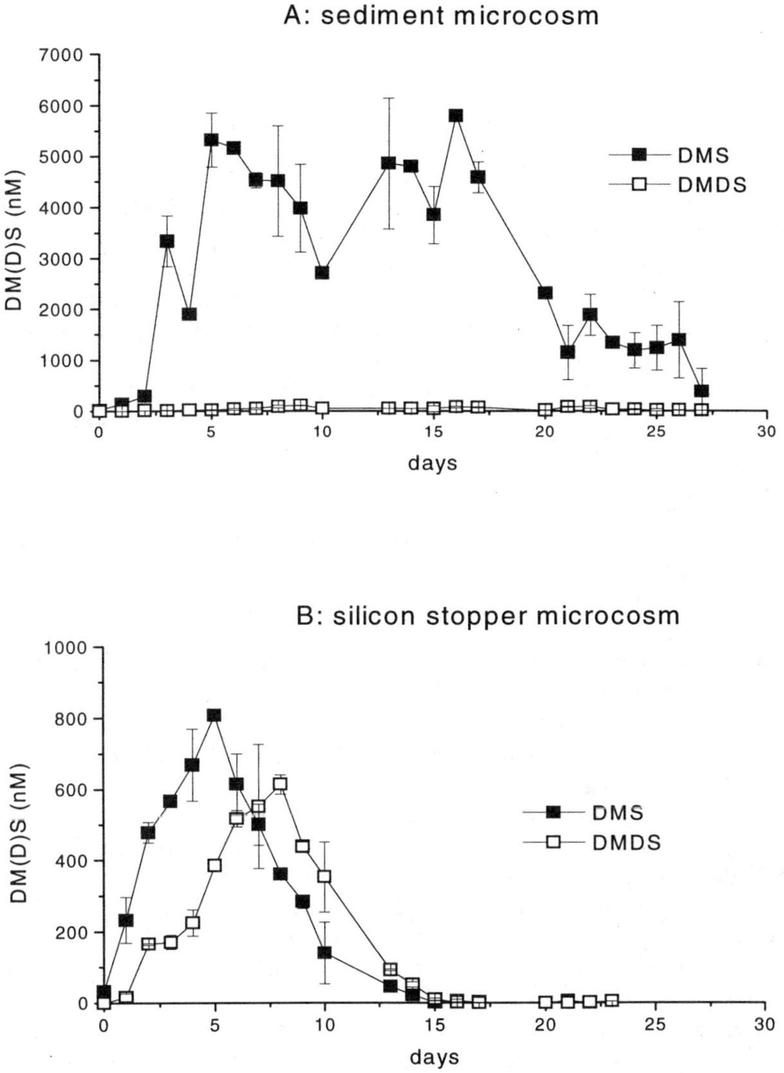

Figure 4. Concentrations of DMS and DMDS in microcosms with sediment (4A) and with a silicon stopper (4B). Error bars indicate the range between duplicate samples. 2.0 x 10^6 cells.ml^{-1} (5300 nM DMSP) were added to these microcosms.

two possible explanations for the discrepancy between this slow DMS consumption and the rapid DMS production.

The sediment in the microcosms had been starved for three years. It is therefore not likely that this sediment contained a large, metabolically active population of anaerobic DMSP and DMS degrading bacteria. However, a high production of DMS from DMSP does not necessarily require a large bacterial population. It has been reported that some anaerobic DMSP consumers have very active DMSP-lyases (3,25). A small bacterial population may thus still be able to convert a lot of DMSP to DMS. The subsequent anaerobic breakdown of DMS apparently requires the growth of a larger bacterial population than present in the microcosms.

It may as well be possible that many active DMSP and DMS consumers have been introduced in the microcosms with the added cultures of *E. huxleyi*. Since these are presumably aerobic bacteria, DMS degradation by these bacteria will be blocked by the anoxic conditions, but the bacterial DMSP-lyases may still be active. Hence, DMS will still be produced rapidly, but will not be immediately degraded in the anoxic microcosms: time would be required for a population of anaerobic DMS consumers to grow.

Particularly at the lowest DMSP concentration, a substantial part of the DMSP was not immediately cleaved to DMS and acrylate. The unrecovered DMSP may either not have been degraded (perhaps due to adsorption on sediment particles) or may have been degraded via another pathway. Demethylation to 3-methiolpropionate and 3-mercaptopropionate is the only other pathway that has been described for the biological degradation of DMSP. This process occurs both under oxic conditions (24) and under anoxic conditions (13). Kiene and Taylor (13) suggested that this demethylation pathway is quantitatively more important at lower DMSP concentrations, while cleavage becomes the predominant pathway at higher DMSP concentrations. This is in agreement with observations by Visscher et al.(26), who found high numbers of cleaving bacteria in an environment with high a DMSP concentration, and high numbers of demethylators in environments with a low DMSP concentration. Recently, De Souza and Yoch (3) determined high K_m values for bacterial DMSP-lyase activity in various environments, ranging from 30 μM in anoxic sediment slurries to 2 mM in aerobic seawater. This shows that cleaving bacteria are physiologically adapted to high substrate concentrations.

An increasing dominance of the cleavage pathway with respect to the demethylation pathway at increasing substrate concentrations could explain the observed differences in the microcosms with increasing amounts of *E. huxleyi*. Future microcosm research should therefore also focus on demethylation products in order to estimate the importance of this pathway under different conditions.

The differences between the microcosm with the silicon stopper and the sediment microcosm may suggest at first sight that the presence of sediment has an important effect on DMSP metabolism. However, the differences may as well have been caused by effects of the silicon stopper. Silicon rubber has a high permeability for DMS (11). DMS may be absorbed by, or diffuse through the stopper, but the control experiment with DMS in distilled water did not show a strong effect. The rapid removal of DMS in the silicon stopper microcosm must therefore be attributed primarily to biological degradation.

The increased formation of DMDS in the microcosm with the silicon stopper points to another possible effect of the silicon stopper: oxygen may have diffused through the stopper into the microcosm. In the marine environment, DMDS is mainly produced under oxic conditions by the oxidation of methanethiol (MSH), which is an intermediate in the degradation of DMS (23). The fact that most DMDS was formed when DMS had already started to decrease suggests that DMDS was indeed formed from degradation products of DMS. Interestingly, Wolfe & Kiene (27) recently reported on the inhibiting properties of DMDS on DMS consumption. These authors found that 100 nM DMDS was already sufficient to block the DMS consumption by more than 90 %. The DMDS concentration in the silicon stopper microcosm increased to 600 nM, but DMS consumption seemed not to be inhibited. In contrast to the study of Wolfe and Kiene (27), who used DMS concentrations less than 10 nM, the DMS concentration in our experiment exceeded the concentration of DMDS. This suggests that perhaps DMDS inhibits DMS consumption because it works as a competitive substrate.

A control experiment revealed that oxygen indeed penetrated easily through the silicon stopper: a microcosm filled with anoxic distilled water was incubated in darkness at 12°C. Winkler measurements showed that the water in the microcosm was saturated with oxygen already after 24 hours. The presence of oxygen may very well explain the difference in DMS formation between the sediment microcosm and the silicon stopper microcosm. Under these more oxic circumstances, the DMS-consuming bacteria present in the *E. huxleyi* culture will continue their metabolic activity. This makes the period of unbalance between DMS production and DMS consumption that occurs after the deposition of the algae shorter and less intensive when compared to the anoxic-sediment microcosm. The higher accumulation of DMS in the sediment microcosms is thus an effect of the anoxic conditions in the overlying water rather than an effect of processes occurring in the sediment. Nedwell et al. (18) showed that high concentrations of DMSP and DMS in North Sea sediments did not lead to large emissions of DMS into the water column. Hence, high DMS concentrations in the water column are more likely to be caused by processes occurring on the sediment surface than by processes in the sediment.

In summary, our experiments showed that deposition of DMSP containing algae can be followed by a temporal accumulation of DMS, which confirms the ideas of Osinga et al. (19). Accumulation of DMS is especially likely to occur in anoxic environments that are suddenly exposed to a high input of DMSP. The accumulation is due to a low initial potential for degradation of DMS, and to the importance of the cleavage pathway at high DMSP concentrations. Situations like this will mainly take place in eutrophic coastal areas. Although these areas are of little significance to the global sulfur cycle, the phenomena occurring in these areas may affect the local climate: periods of DMS accumulation may lead to enhanced DMS emission to the atmosphere. This may locally stimulate cloud formation and it may lead to an increased amount of acid precipitation.

ACKNOWLEDGMENTS

We thank Rik Kwint for technical assistance with the DMS analyses. Anna Noordeloos kindly provided cultures of *Emiliania huxleyi*. This study was funded by the European Community, project CT-93-0326.

REFERENCES

1. Canfield, D.E., B.B. Jørgensen, H. Fossing, R. Glud, J. Gundersen, N.B. Ramsing, B. Thamdrup, J.W. Hansen, L.P. Nielsen and P.O.J. Hall. 1993. Pathways of organic carbon oxidation in three continental margin sediments. Mar. Geol. *113*: 27-40.
2. Dacey, J.W.H. and N.V. Blough. 1987. Hydroxide decomposition of dimethylsulfoniopropionate to form dimethylsulfide. Geophys. Res. Lett. *14*: 1246-1249.
3. De Souza, M.P. and D.C. Yoch. 1995. Purification and characterization of dimethylsulfoniopropionate lyase from an *Alcaligenes*-like dimethyl sulfide-producing marine isolate. Appl. Environ. Microbiol. *61*: 21-26.
4. Graf, G., R. Schultz, R. Peinert and L.-A. Meyer-Reil. 1983. Benthic response to sedimentation events during autumn to spring at a shallow-water station in the Western Kiel Bight. I. Analysis of processes on a community level. Mar. Biol. *77*: 235-246.
5. Gröne, T. 1995. Biological production and consumption of dimethylsulfide (DMS) and dimethylsulfoniopropionate (DMSP) in the marine epipelagic zone: a review. J. Mar. Res. *6*: 191-209.
6. Guillard, R.R.L. 1975. Culture of phytoplankton for feeding marine invertebrates. In: W.L. Smith, M.H. Chanley (Eds.), Culture of marine invertebrate animals. Plenum Press, New York, pp. 26-60.
7. Hansen, L.S. and T.H. Blackburn. 1992. Effects of algal bloom deposition on sediment respiration and fluxes. Mar. Biol. *112*: 147-152.
8. Keller, M.D., W.K. Bellows and R.R.L. Guillard. 1989. Dimethylsulfide production in marine phytoplankton. In: E.S. Saltzmann, W.J. Cooper (Eds.), Biogenic sulfur in the environment. ACS Symp. Ser. 393, Washington DC, pp. 167-182.
9. Kiene, R.P. 1992. Dynamics of dimethylsulfide and dimethylsulfoniopropionate in oceanic seawater samples. Mar. Chem. *37*: 29-52.

10. Kiene, R.P. and T.S. Bates. 1990. Biological removal of dimethylsulfide from sea water. Nature *345*: 702-705.
11. Kiene, R.P. and D.C. Capone. 1988. Microbial transformations of methylated sulfur compounds in anoxic salt marsh sediments. FEMS Microb. Ecol. *15*: 275-291.
12. Kiene, R.P. and S.K. Service. 1991. Decomposition of dissolved DMSP and DMS in estuarine waters: dependence on temperature and substrate concentration. Mar. Ecol. Prog. Ser. *76*: 1-11.
13. Kiene, R.P. and B.F. Taylor. 1988. Demethylation of dimethylsulfoniopropionate and production of thiols in anoxic marine sediments. Appl. Environ. Microbiol. *54*: 2208-2212.
14. Kwint, R.L.J. and K.J.M. Kramer. 1995. DMS production by plankton communities. Mar. Ecol. Prog. Ser. *121*: 227-237.
15. Kwint, R.L.J. and K.J.M. Kramer. 1995. The annual cycle of the production and fate of DMS(P) in a marine coastal system. Submitted to Mar. Ecol. Prog. Ser.
16. Lindqvist, F. 1989. Sulfur-specific detection in air by photoionization in a multiple detector gas chromatographic system. J. High Res. Chrom. *12*: 628-631.
17. Matrai, P.A. and M.D. Keller. 1993. Dimethylsulfide in a large scale coccolithophore bloom in the Gulf of Maine. Cont. Shelf Res. *13*: 831-843.
18. Nedwell, D.B., M.T. Shabbeer and R.M. Harrison. 1994. Dimethyl sulfide in North Sea waters and sediments. Estuar. Coast. Shelf Sci. *39*: 209-217.
19. Osinga, R., R.L.J. Kwint, W.E. Lewis, G.W. Kraay, J.D. Lont, H.J. Lindeboom and F.C. van Duyl. 1995. Production and fate of dimethylsulfide and dimethylsulfoniopropionate in pelagic mesocosms: the role of sedimentation. Mar. Ecol. Prog. Ser. In Press.
20. Quist, P., R.L.J. Kwint, T.A. Hansen, L. Dijkhuizen and K.J.M. Kramer. 1995. Turnover of dimethylsulfoniopropionate and dimethylsulfide in the marine environment - a mesocosm experiment. Submitted to Mar. Ecol. Prog. Ser.
21. Stefels, J. and W.H.M. van Boekel. 1993. Production of DMS from dissolved DMSP in axenic cultures of the marine phytoplankton species *Phaeocystis sp*. Mar. Ecol. Prog. Ser. *97*: 11-18.
22. Steinke, M., C. Daniel and G.O. Kirst. 1996. DMSP lyase in marine macro- and microalgae: interspecific differences in cleavage activity. This volume.
23. Taylor, B.F. 1993. Bacterial transformations of organic sulfur compounds in marine environments. In: R.S. Oremland (Ed.). Biogeochemistry of global change. (Radiatively active trace gases). Chapman & Hall, New York, pp. 745-781.
24. Taylor, B.F. and D.C. Gilchrist. 1991. New routes for aerobic biodegradation of dimethylsulfoniopropionate. Appl. Environ. Microbiol. *57*: 3581-3584.
25. Van der Maarel, M.J.E.C. and T.A. Hansen. 1996. Anaerobic microorganisms involved in the degradation of DMS(P). This volume.
26. Visscher, P.T., R.P. Kiene and B.F. Taylor. 1994. Demethylation and cleavage of dimethylsulfoniopropionate in marine intertidal sediments. FEMS Microb. Ecol. *14*: 179-190.
27. Wolfe, G.V. and R.P. Kiene. 1993. Effects of methylated, organic, and inorganic substrates on microbial consumption of dimethyl sulfide in estuarine waters. Appl. Environ. Microbiol. *59*: 2723-2726.

34

DMSP, DMS AND DMSO CONCENTRATIONS AND TEMPORAL TRENDS IN MARINE SURFACE WATERS AT LEIGH, NEW ZEALAND

P. A. Lee[1] and S. J. de Mora[2]

[1] Chemistry Department
University of Auckland
Private Bag 92019
Princes Street, Auckland
New Zealand
[2] Département d'océanographie
Université du Québec à Rimouski
300 allée des Ursulines
Rimouski, Québec
Canada G5L 3A1

SUMMARY

Marine waters at the Leigh Marine Laboratory (36° 17′ S, 174° 48′ E), situated on the north-east coast of the North Island of New Zealand, were investigated on four occasions during 1993. Surface sampling was conducted at 1 or 2 h intervals for 24 or 48 h. Dissolved DMSO was determined in all cases. During the later two field trips, dissolved DMS, total DMSP (DMSP(t)) and chlorophyll *a* concentrations were also measured. DMS and DMSP(t) exhibited no apparent diurnal trends with respect to concentration. The DMS content was positively correlated to that of DMSP(t), but with differing slopes (3 in late August and 1 in mid November). The results for DMSO are tantalising rather than conclusive. Concentrations show a marked seasonal change, with lowest values observed during winter. The DMSO content also exhibited obvious diurnal variations, with a maximum usually in the afternoon and minimum values just before sunrise. Night-time loss processes for DMSO have no profound influence on the ambient DMS levels. Positive correlations are apparent for DMSO with respect to DMS and with respect to sunlight intensity. Given that DMSO concentrations generally exceed those of DMS by an order of magnitude, the photo-oxidation of DMS seems unlikely to be the sole source of DMSO. The direct biosynthesis of DMSO may account in part for its abundance in the marine environment.

Biological and Environmental Chemistry of DMSP and Related Sulfonium Compounds
edited by Ronald P. Kiene et al., Plenum Press, New York, 1996

INTRODUCTION

Since the first field measurements of dimethylsulfide (DMS) in sea water were made (28), the ubiquity of DMS has been established in many aqueous marine environments (3, 13, 26, 30, 42). Such investigations were also stimulated by the hypothesis that DMS was involved in moderating global warming (8). DMS is the dominant volatile sulfur compound emitted from the ocean (8) and may represent up to 90% of the sea to air sulfur flux (29). Concentrations of DMS in sea water have been found to range from 0.03 to 193 nmol L^{-1} with an approximate average of 2.9 nmol L^{-1}. The highest concentrations are observed in the most productive waters. DMS exhibits strong temporal and spatial trends (40), with summer values exceeding winter concentrations by up to a factor of 70. Some variation with depth is observed, although not necessarily consistently (2, 9). However, DMS is found to be concentrated in the top 200 m of the ocean and is associated with biological activity in the euphotic zone (10).

The major precursor of DMS in the oceans is dimethylsulfoniopropionate (DMSP). Widely occurring in the marine environment, DMSP is thought to act as a cryoprotectant in polar waters and as an osmoregulator in saline environments (25, 41). Phytoplanktonic DMSP is released to the ocean via two pathways, both producing DMS. Firstly, cell

Table 1. Concentration of DMSO in various environmental media

System	Study		Concentration (nmol L^{-1})
Seawater	Andreae (1980) (1)		< 0.9 - 532
	Bates *et al.* (1994) (5)		< 1.2
	de Mora *et al.* (submitted) (14)		2.9 - 123
	Gibson *et al.* (1990) (16)		< 2.5 - 24.3
	Hatton *et al.* (1994) (18)		3 - 15
	Kiene & Gerard (1994) (23)		1.2 - 25.7
	Lee & de Mora (this study)		< 6.2 - 124
	Ridgeway *et al.* (1992) (33)		< 5 - 24
	Wakeham *et al.* (1987) (42)		up to 21
Rainwater	Andreae (1980) (1)		< 2.5 - 20.3
	Harvey & Lang (1986) (17)		< 0.3 - 20.3
	Ridgeway *et al.* (1992) (33)		
		Coastal Samples	19.7 - 29.3
		Inland Samples	5 - 10
Freshwater	Andreae (1980) (1)		
		River Samples	< 2.5 - 210
		Snow Samples	< 2.5
	de Mora *et al.* (submitted) (14)		
		Snow Samples	1.4 - 2.5
		Melt Water Ponds	0.06 - 185
	Richards *et al.* (1994) (32)		
		Lakes/Wetland Ponds	0 - 180
Aerosols	Harvey & Lang (1986) (17)		69 - 125 pmol m^{-3}
	Watts *et al.* (1987) (44)		1- 62 pmol m^{-3}
	Watts *et al.* (1990) (43)		3 - 12 pmol m^{-3}

senescence and lysis releases DMSP into the ocean where it can then undergo enzymatic cleavage (40). Secondly, zooplankton grazing on phytoplankton can ingest organisms containing intracellular DMSP which is then digested and excreted as DMS (12). Dissolved oceanic DMSP can greatly exceed that of DMS.

Previously little consideration has been given to the role of dimethylsulfoxide (DMSO) in the biogeochemical cycling of sulfur and its interaction with DMS in the aqueous marine environment. Very few measurements of DMSO levels have been made due to the lack of an analytical procedure which was sufficiently sensitive and selective (18, 23). The first measurements of DMSO showed that it occurred throughout the euphotic zone in sea water and in photosynthetically active freshwater systems (1). Since then, observations have shown that DMSO is as ubiquitous as DMS, and present in higher concentrations (Table 1). Laboratory studies have shown that DMS can be oxidised in sea water (6). The losses of DMS and O_2 from a UV-irradiated solution were consistent with DMSO being the oxidation product and it was estimated that DMS photo-oxidation was on the same scale as atmospheric exchange. The rate of bacterial consumption of DMS is on average 10 times faster than the transfer of DMS to the atmosphere, but exhibits a large variation, being in the range 3- 430 (24).

The microbial processes linking DMS and DMSO are not well understood. A photosynthetic purple sulfur bacterium, tentatively identified as *Thiocystis*, has been reported to anaerobically oxidize DMS to DMSO (48) and luminescent marine bacteria may also produce the same result (36). A wide variety of micro-organisms, including prokaryotes and eukaryotes, aerobes and anaerobes, can reduce DMSO to DMS. Under anaerobic conditions, bacteria and yeasts have been observed to reduce DMSO to DMS (49) and in some organisms, DMSO can support anaerobic growth. Aerobic growth occurred only in the case of *Hyphomicrobium* EG and *Hyphomicrobium* S. Purple nonsulfur bacteria are capable of aerobic respiration in the absence of light. This led to the suggestion that any DMSO reaching active microbial communities would be reduced to DMS thus potentially influencing the atmospheric sulfur cycle (49). Any mechanism involving anaerobic bacteria is unlikely to be significant in well oxygenated oceanic surface waters. Luminescent bacteria are, however, very abundant and may be influential.

Using a direct gas chromatographic procedure (15), DMSO concentrations in marine surface waters near Leigh, New Zealand, were examined together with the precursors DMS and DMSP. This study was conducted to test the modified DMSO technique and to allow assessment of the importance of DMSO in the aqueous marine biogeochemical cycling of DMS.

METHODS

Sample Collection

Marine waters at Leigh (36° 17′ S, 174° 48′ E), situated on the north-east coast of the North Island of New Zealand, were investigated on four occasions during 1993. Between 2 and 5 L samples were collected from a rocky point jutting into a channel at a depth of approximately 20 cm. The sample bottles (LDPE Nalgene) were filled slowly to capacity to minimize bubbling effects and to eliminate the headspace. The channel has a water depth of 5 m. The bottom was sandy with rocky outcrops and some macroalgae. The adjacent shore is a relatively narrow (10 m) sandy beach, also with rocky outcrops, and a cliff at the head of the beach.

Chlorophyll *a* and Climate Data

Once the samples were returned to the laboratory, an aliquot (125 mL) was used for immediate DMS and DMSP analyses which were completed within 2 h of sample collection. The remaining sample was then filtered through a 0.45 µm HA Millipore filter, with samples taken in November being filtered through Whatman GF/F filters. The filter papers were stored at 0°C in the dark for subsequent determination of chlorophyll *a* (Chl*a*;35). With the exception of the samples from November, the filter papers were extracted using 90% acetone and the absorbance of the resulting extracts was measured using a spectrophotometer. The samples taken during November were extracted using 95% ethanol with the samples being boiled in 95% ethanol for five minutes and then left to stand for 24 hours at ambient temperatures. The absorbances of the extract solutions were measured at 665nm and 750nm (19). A further 125 mL aliquot of the filtered sea water was used for DMSO analysis, which was also completed within 2 h of sample collection. Sunlight intensity measurements (Kipp and Zonen Pyronometer) and wind velocities were obtained from the Leigh Marine Laboratory climate station.

DMSO

Previously published techniques (15, 31) were modified only slightly for the determination of DMSO dissolved in sea water. The analysis was conducted using a Perkin-Elmer PE8500 gas chromatograph equipped with a flame photometric detector (GC-FPD). Other than filtration, no sample pretreatment was required and no reagents were added. Aliquots of 5 µL of filtered seawater, with three to six replicate injections per sample, were injected directly into the GC. The analytical column comprised a 1.4 m x 2.5 mm teflon tube packed with 15% Free Fatty Acid Phase (FFAP) on 40-60 mesh Chromosorb T. A teflon guard column (125 mm x 2.5 mm) containing identical packing material was installed prior to the analytical column to prevent salts depositing in the analytical column. However, at an injection port temperature of 210°C, all salts were deposited in the glass injector liner and none were found in the guard column. The injector liner was cleaned with Milli-Q (18 MΩ purity) water after every 30 injections. The detector block temperature was set at 210°C. Both DMSO and $DMSO_2$ could be resolved isothermally at 150°C with retention times of approximately 1.8 and 6 minutes, respectively. It should be pointed out that volatile sulfur gases (DMS, CS_2, *etc.*) do not interfere with the analysis because the chromatographic phase is very polar and so they pass through the system essentially at the speed of the carrier gas. The detection limit for DMSO was 1 pg or equivalent to 6 nmol L^{-1} in a sea water sample. Because of the non-linear response of the FPD, introducing a continuous flow of a sulfur gas into the H_2 flame invokes a dramatic improvement in sensitivity (20). Sensitivity was markedly enhanced by using SF_6 in the flame, giving an absolute detection limit of 10 fg or equivalent to 0.06 nmol L^{-1} DMSO. The detection limit for $DMSO_2$ was 20 nmol L^{-1} under normal flame conditions and 0.2 nmol L^{-1} under high sensitivity (SF_6 bleed) flame conditions. The standard deviation for both analyses was 10% at a concentration of 20 nmol L^{-1} (no SF_6) or 0.25 nmol L^{-1} (using SF_6). Milli-Q water was used to provide a blank and the DMSO content in this blank was consistently below detectable levels.

DMS and DMSP(t)

A purge and trap technique was employed for the analysis of DMS (13, 46). A 50 mL aliquot of the unfiltered seawater sample was purged with oxygen-free dry nitrogen (80 mL min^{-1}) for 20 minutes. Trapping took place in a tube containing Molecular Sieve 5A. A tube containing K_2CO_3 was placed in front of the sample tube to prevent water from interfering in the analysis. The sample tube was held at ambient temperature whilst trapping

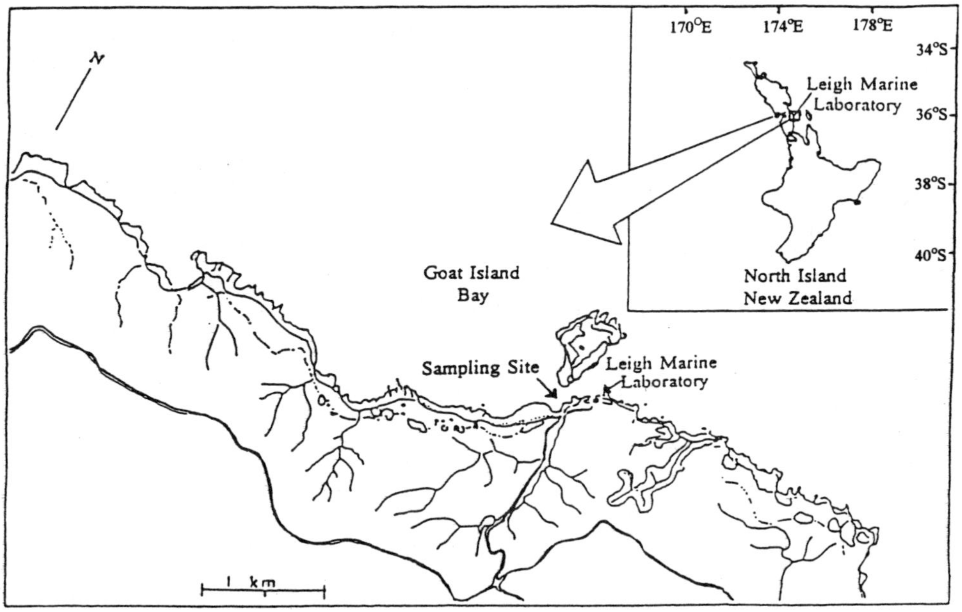

Figure 1. Sampling location at the Leigh Marine Laboratory, New Zealand (36° 17′ S, 174° 48′ E).

took place. Thereafter, the sampling tube was connected to the gas sampling loop of the GC. The tube itself was placed in a block heater and instantly heated to 320°C. After two minutes, the sample valve was opened thereby allowing passage of the volatilized sample onto the chromatographic column. The column used for the volatile sulfur gases was a 2 m x 2.5 mm teflon column packed with Carbopack B/ 1.5% XE60/ 1% H_3PO_4 (Supelco Inc.). Temperature programming (40°C - 140°C) was used to achieve baseline separation of DMS and CS_2. Quantification of the reduced sulfur gases was carried out using permeation tubes (34). The detection limit for DMS was 5 pg or equivalent to 0.8 pmol L^{-1} DMS in solution for a 100 ml water sample.

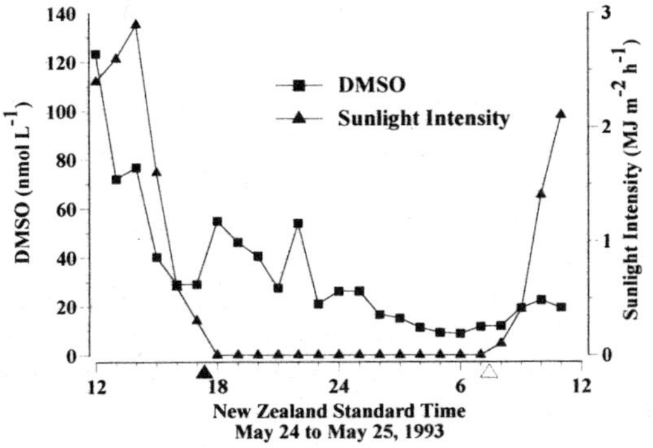

Figure 2. DMSO concentration and sunlight intensity during May 24-25, 1993. Open and closed triangles on this and subsequent graphs indicates sunrise and sunset, respectively.

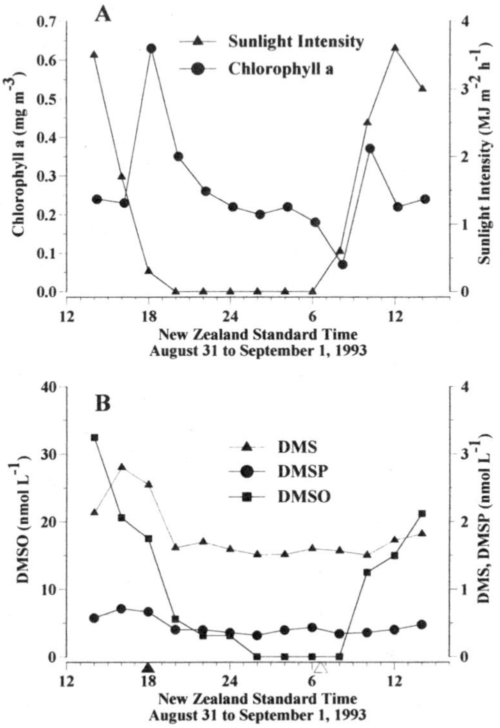

Figure 3. Time series for August 31 - September 1, 1993, showing (a) chlorophyll-*a* and sunlight intensity and, (b) concentrations of DMSO, DMS, and DMSP(t).

DMSP determinations were conducted following DMS stripping (45). After the addition of 3 mL of 2 mol L^{-1} NaOH to hydrolyse the DMSP to DMS, the sample was then reanalysed for DMS as outlined previously. To prevent hydrolysis by residual base, the bubbling chamber was rinsed with 1 mol L^{-1} HCl and then Milli-Q water after each DMSP analysis. The detection limit for DMSP was 0.08 nmol L^{-1} as DMS equivalents

RESULTS

The sampling programme was carried out near the Leigh Marine Laboratory (36° 17′ S, 174° 48′ E, Figure 1), with samples being taken from Goat Island Bay adjacent to the Marine Laboratory. Sunlight intensity data was available from the Leigh Laboratory climate station for all four occasions (Figures 2, 3a, 4a). In all cases, maximum sunlight intensity occurred at approximately 1 pm (New Zealand Standard Time). Chl *a* determinations were only carried out in August (Figure 3a) and November (Figure 4a). No data were collected for planktonic speciation. On both occasions the chlorophyll-*a* measurements show no clear temporal trends and the distribution observed may be due to phytoplankton moving vertically in the water column or water-mass movement resulting in new plankton communities arriving at the sampling site.

DMSO levels in marine surface waters were determined in May, July, August and November 1993. Concentrations of DMSO present in May (Figure 2) showed that the highest values were observed during the early afternoon, shortly after maximum sunlight intensity, and subsequently declined until just prior to the following sunrise. During the second day, such high

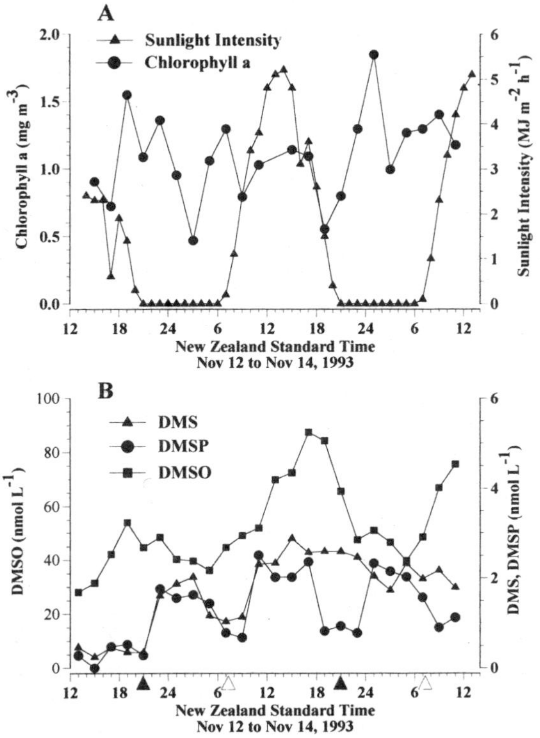

Figure 4. Time series for November 12 - 14, 1993, showing (a) chl *a* and sunlight intensity and, (b) concentrations of DMSO, DMS, and DMSP(t).

DMSO concentrations were not attained. Data from July (*not shown here*) were very erratic. DMSO concentrations were generally very low and no trend could be discerned. For both August (Figure 3b) and November (Figure 4b) a strong diurnal trend was evident. As with the May data, daily DMSO maxima were observed shortly after maximum sunlight intensity and concentrations fell until just before sunrise. However, on these occasions the increase in DMSO concentrations after sunrise was very much larger than for May. Since sampling for November was conducted for 48 hours, the diurnal variations were readily apparent.

Measurements for DMSP(t) and DMS were only undertaken during the sampling sessions in August (Figure 3b) and November (Figure 4b). In both cases there was no apparent temporal trend. For August, maximum concentrations for both compounds were observed during the day and in general, DMS and DMSPT(t) levels exhibited similar tendencies. The situation for November was less clear. Although maximum concentrations were observed during daylight hours, peaks were observed during the first night of sampling. During the first 24 h period, DMS and DMSP(t) concentrations covaried. However, the trends diverged during the second 24 h sampling session

DISCUSSION

DMSP(t) concentrations during August 1993 varied between 0.3 nmol L^{-1} and 0.7 nmol L^{-1} (average 0.45 nmol L^{-1}), and for November 1993 ranged from not detected to 2.6

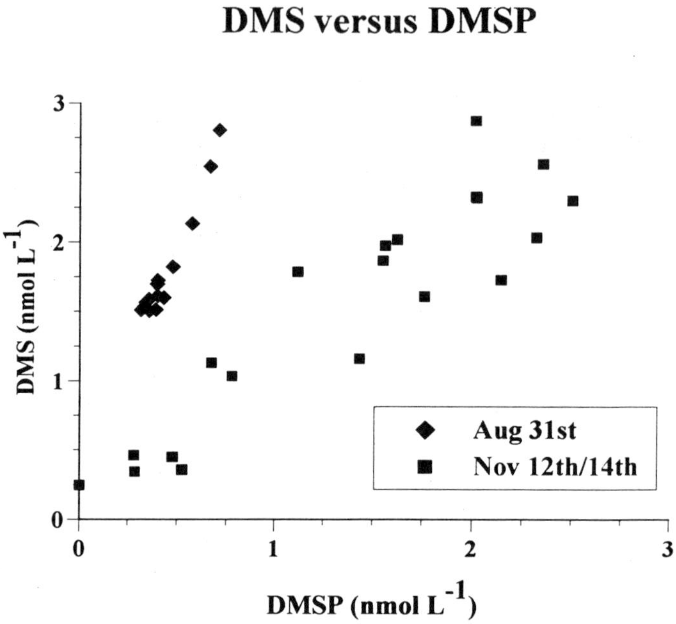

Figure 5. DMS as a function of DMSP(t) for the August and November data sets.

nmol L^{-1} (average 1.29 nmol L^{-1}). Based on the non-parametric test of the Spearman's rank correlation coefficient, DMSP(t) concentrations were not correlated with those of chl *a*. This has been observed previously and relates to zooplankton grazing (12) and planktonic speciation (27). In Goat Island Bay during winter and early spring, diatoms are known to be the predominant planktonic species whereas during late summer and early autumn, dinoflagellates are abundant (37, 38). The coccolithophore, *Emiliania huxleyi*, has also been observed to be dominant during spring-time (37). The DMSP results obtained in this study are consistent with this pattern of planktonic speciation. Diatoms, occurring in winter, are considered to be amongst the poorest producers of DMSP (22), hence low levels of DMSP are observed. Conversely, coccolithophores are better producers of DMSP (22) and this is reflected in the higher levels of DMSP measured in the bay during spring. In particular, *Emiliania huxleyi*, has been shown to be capable of producing significant quantities of DMSP (21) and DMS (27).

Average DMS levels for August were 1.82 nmol L^{-1} and for November were 1.65 nmol L^{-1} with ranges of 1.50 nmol L^{-1} to 2.80 nmol L^{-1}, and 0.25 nmol L^{-1} to 2.87 nmol L^{-1} respectively. DMS has been reported as having a night-time maximum in both surface waters (11, 39) and the atmosphere (7). In our study, DMS was not correlated to sunlight intensity (based on the non-parametric test of the Spearman's rank correlation coefficient), suggesting there was no photochemically mediated DMSP cleavage. Similarly, the lack of a strong negative correlation between DMS and sunlight intensity indicates that DMS photo-oxidation does not control its concentration. DMSP and DMS were found to be significantly correlated in August (Figure 5, Spearman's Rank Correlation Coefficient, P=0.05, ρ=0.909, n=13) and November (Figure 5, ρ=0.869, n=19). The greater slope observed for August as compared to November (Figure 5) may be due to a lower rate of DMS photo-oxidation as a result of lower sunlight intensities and sunlight hours. Another contributing factor may be less efficient bacterial activity (assimilation or oxidation of DMS) during the cooler month. This would allow a more rapid build up of DMS levels at lower DMSP concentrations.

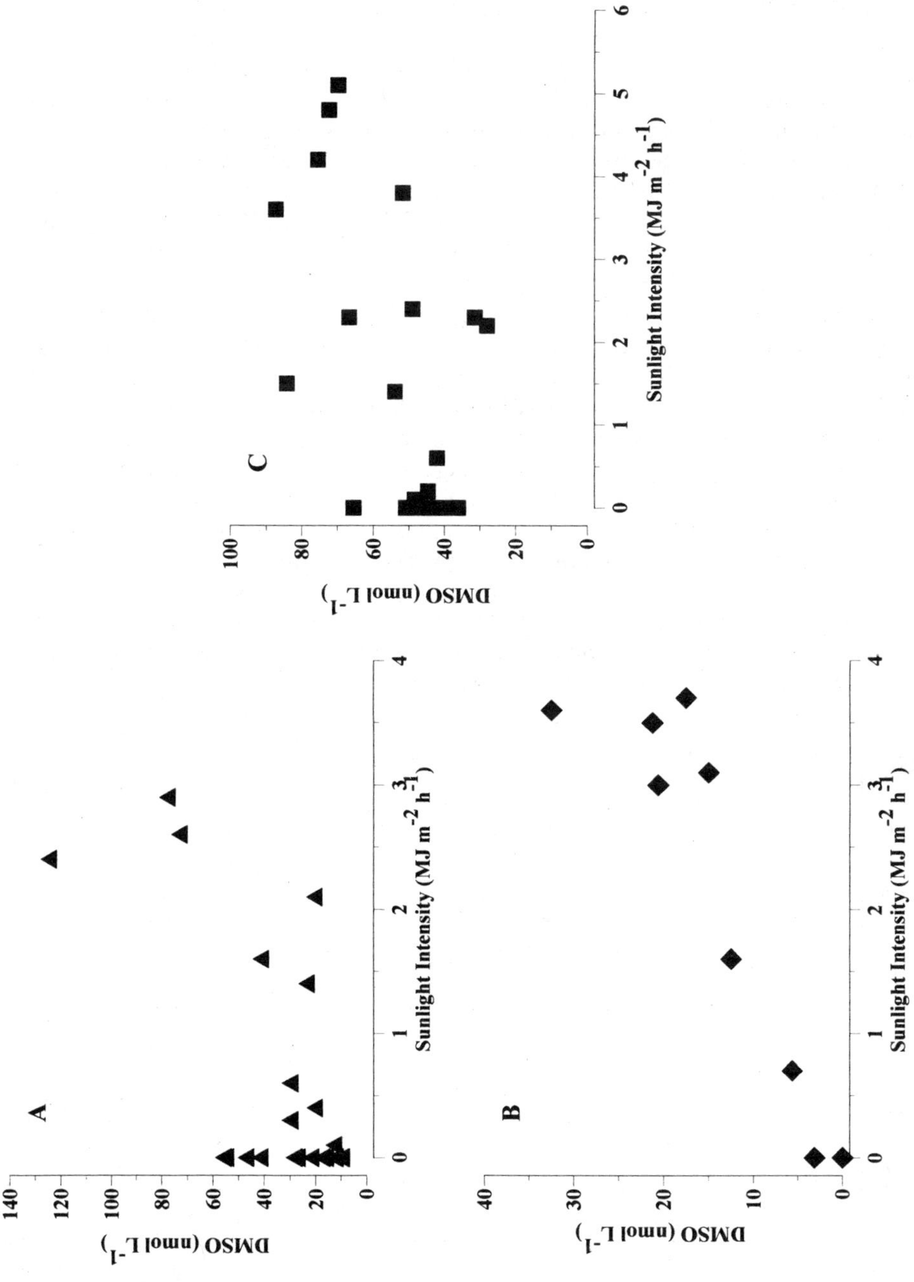

Figure 6. DMSO as a function of sunlight intensity for (a) May, (b) August and (c) November.

Surface seawater samples were found to contain concentrations of DMSO between 6 and 124 nmol L^{-1}. Average monthly values of DMSO found were 33.8 nmol L^{-1} (May), 4.0 nmol L^{-1} (July), 10.1 nmol L^{-1} (August) and 54.2 nmol L^{-1} (November). A comparison of DMSO data can be seen in Table 1, these results representing a notable proportion of the results published for DMSO concentrations in aqueous environments. The levels of DMSO observed during this study were somewhat higher and have a larger range than values previously found, with the exception of the seminal study by Andreae (1), in oceanic environments. As noted previously (18, 23), this original study may have over-estimated DMSO due to the lack of specificity in the reduction of DMSO to DMS. However, comparably high levels of DMSO have been found in hypersaline lakes (14, 32). DMSO concentrations reported here were often found to exceed DMS levels by one to two orders of magnitude (Figures 3b and 4b). Except during the night when DMSO levels fell below the detection limit, it was the dominant sulfur compound of those under consideration. Although the slopes differ, DMSO was significantly correlated with DMS during both August (Spearman's Rank Correlation Coefficient, P=0.05, ρ=0.762, n=13) and November (Spearman's Rank Correlation Coefficient, P=0.05, ρ=0.603, n=24). Data for August show a more rapid increase in DMSO concentrations with DMS compared to November, which may be due to the more efficient degradation of DMSO during November. However, overall DMSO levels were higher during November when a better DMS-producing plankton species was present (coccolithophores *c.f.* diatoms) and increased sunlight was available for both DMS photo-oxidation and enhanced photosynthetic activity.

The most striking feature of the data presented is the strong diurnal trend observed for DMSO concentrations in May (Figure 2), in August (Figure 3b) and November (Figure 4b). No such trend was observed during July when winter concentrations were generally low and very close to the detection limit of the GC-FPD system being used at the time. No study has previously considered the likelihood of a diurnal variation. Although the slopes differ, a significant correlation between sunlight intensity and DMSO occurs for three of the four

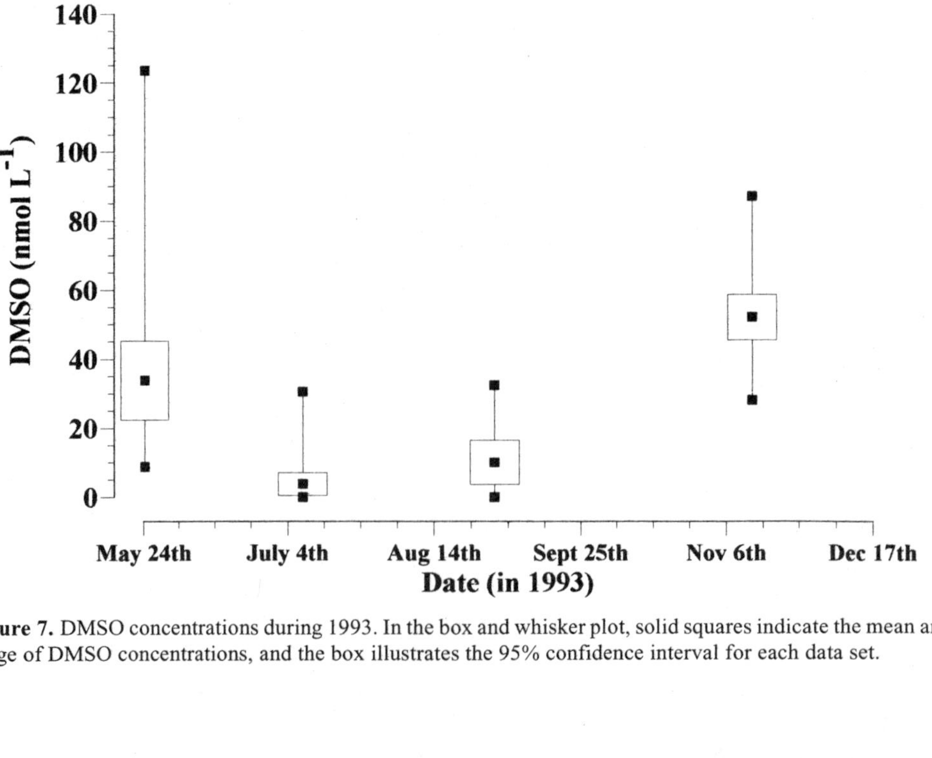

Figure 7. DMSO concentrations during 1993. In the box and whisker plot, solid squares indicate the mean and range of DMSO concentrations, and the box illustrates the 95% confidence interval for each data set.

Table 2. Estimated rates of DMSO production and removal

Date (1993)	DMSO Production Rate	DMSO Destruction Rate
May 24	2.5 nmol L^{-1} h^{-1}	19 nmol L^{-1} h^{-1}
Aug 31	4.5 nmol L^{-1} h^{-1}	3.6 nmol L^{-1} h^{-1}
Nov 12	4.0 nmol L^{-1} h^{-1}	1.2 nmol L-1 h^{-1}
Nov 13	5.7 nmol L^{-1} h^{-1}	5.1 nmol L^{-1} h^{-1}
Nov 14	6.0 nmol L^{-1} h^{-1}	

dates in which sampling was undertaken, May (Figure 6a, Spearman's Rank Correlation Coefficient, P=0.05, ρ=0.423, n=24), August (Figure 6b, Spearman's Rank Correlation Coefficient, P=0.05, ρ=0.918, n=13), November (Figure 6c, Spearman's Rank Correlation Coefficient, P=0.05, ρ=0.531, n=24). There are two known processes which could contribute to the daytime production of DMSO, namely photosensitized oxidation (6) and oxidation by luminescent bacteria (36) of DMS. These aspects are further considered below. The rapid loss of DMSO at night is consistent with the notion that micro-organisms may be using DMSO as an energy source or as a terminal electron acceptor in a dark aerobic growth mode (47, 49).

The DMSO concentrations exhibited a marked seasonal variability (Figure 7). By comparing the 95% confidence intervals for the various data sets, the DMSO levels found in May were significantly different from the concentrations observed in July/August, and significantly different from the values obtained in November. For July and August there was no significant difference between the two sets of DMSO concentrations, although August did exhibit a higher average. Two factors may account for the seasonal dependence. Firstly, during winter, photosynthetic activity is minimal, thereby limiting the production of DMS. Secondly, because both sunlight intensity and sunlight hours are lower during winter, less photochemical oxidation of DMS to DMSO can occur. As apparent for DMSP(t) and Chl *a*, no correlation (based on Spearman's rank correlation coefficients) was evident between DMSO and Chl *a*. This would be expected if DMSO originates with DMSP released from plankton. The two step pathway to DMSO involves DMS, which can evade to the atmosphere or be utilised by bacteria.

Rates of production and loss of DMSO can be estimated from the slopes of DMSO as a function of time (Figures 2,3 & 4). As shown in Table 2, DMSO was produced during the day with rates varying in the range of 2.5 to 5.7 nmol L^{-1} h^{-1}. The highest rates occurred during the spring. The production of DMSO showed a strong dependence upon irradiance. DMSO can be produced via the photosensitized oxidation of DMS (6). However, the reaction rate is insufficient to account for all DMSO production observed here. DMS oxidation by luminescent bacteria represents a second possible source during the day (36), but the relatively low dissolved DMS concentrations would limit the resulting flux. Indeed, it is intriguing that DMS levels, about 1-2 orders of magnitude less concentrated than those of DMSO were not dramatically influenced. As noted above, DMS concentrations were not correlated with the sunlight intensity. This leads to the speculation that there is an additional mechanism, not dependent upon environmental DMS levels. It has long been known that DMSO concentrations naturally present in foodstuffs and beverages can be relatively high (4, 15, 31). The implication here is that there is a biosynthetic pathway for DMSO, which, for organisms in the marine environment, is apparently linked to sunlight intensity, suggesting that photosynthetic processes play a role.

Finally, this study also showed that there was a marked loss of DMSO during the night (Table 2). Loss rates, 1-20 nmol L^{-1} h^{-1}, were more variable than production rates. Again there was no concurrent rise in DMS concentrations, suggesting that DMSO removal resulted from microbial assimilation. DMSO is known to be a viable substrate for some bacteria (49). Presumably such microbial assimilation would also occur during the day, but this study provides no data to support such a claim.

CONCLUSIONS

DMSO, but not $DMSP_T$ and DMS exhibited a strong diurnal variation in marine surface waters. DMSO was found to be significantly positively correlated to DMS and sunlight intensity. Additionally, DMSO varied seasonally, with lowest values present in winter. The dominance of DMSO was evident as concentrations were typically between 1 and 2 orders of magnitude greater than those of DMSP(t) or DMS. The rapid formation and loss processes observed for DMSO appeared to have little impact on DMS content. DMSO apparently is produced during the day. Photo-oxidation of DMS may be a contributing factor, but cannot account for the apparent production of DMSO. DMS oxidation by luminescent bacteria may also play a role. We speculate here that DMSO may be directly synthesised by photosynthetically active organisms. Equivalent rates of removal are observed during the dark, ascribed here to bacterial assimilation as DMS concentrations were not seen to increase sufficiently to account for the DMSO loss.

ACKNOWLEDGMENTS

We thank Mark Tucker and Miriam McArthur of the Chemistry Department, University of Auckland for their assistance in the collection of samples. We appreciate the considerable logistical support provided by Don and Eileen Tucker throughout the study. Thanks are also due to the staff at the Marine Laboratory at Leigh, particularly Brady Doak (skipper of the R.V. Proteus) and Jo Evans.

REFERENCES

1. Andreae, M. O. 1980. Dimethylsulfoxide in Marine and Freshwaters. Limnol. Oceanogr. *25:* 1054-1063.
2. Andreae, M. O. and W. R. Barnard. 1984. The Marine Chemistry of Dimethylsulfide. Mar. Chem. *14*: 267-279.
3. Andreae, M. O., W. R. Barnard and J. M. Ammons. 1983. The Biological Production of Dimethylsulfide in the Ocean and its Role in the Global Atmospheric Sulfur Budget. p. 167-177. *In* Hallberg (ed.), Environmental Biogeochemistry. Ecol. Bull., Stockholm.
4. Bamforth, C. W. and B. J. Anness. 1981. The role of dimethyl sulphoxide reductase in the formation of dimethyl sulphide during fermentations. J. Inst. Brew. *87*: 30-34.
5. Bates, T. S., R. P. Kiene, G. V. Wolfe, P. A. Matrai, F. P. Chavez, K. R. Buck, B. W. Blomquist and R. L. Cuhel. 1994. The cycling of sulfur in surface seawater of the northeast Pacific. J. Geophys. Res. *C99*: 7835-7843.
6. Brimblecombe, P. and D. Shooter. 1986. Photooxidation of dimethylsulfide in aqueous solution. Mar. Chem. *19*: 343-353.
7. Bürgermeister, S., R. L. Zimmermann, H.W. Georgii, H. G. Bingemer, G. O. Kirst, M. Janssen and W. Ernst. 1990. On the Biogenic Origin of Dimethylsulfide: Relation Between Chlorophyll, ATP, Organismic DMSP, Phytophlankton Species, and DMS Distrubution in Atlantic Surface Water and Atmosphere. J. Geophys. Res. *95*: 20,607-20,615.
8. Charlson, R. J., J. Lovelock, M. O. Andreae and S. Warren. 1987. Oceanic phytoplankton, atmospheric sulfur, cloud albedo and climate. Nature *326*: 655.
9. Cline, J. D. and T. S. Bates. 1983. Dimethyl Sulfide in the Equatorial Pacific Ocean: A Natural Source of Sulfur to the Atmosphere. J. Geophys. Res. *10*: 949-952.

10. Cooper, W. J. and P. A. Matrai. 1989. Distribution of Dimethyl Sulfide in the Oceans: A Review. p. 141-151. *In* Saltzman and Cooper (ed.), Biogenic Sulfur in the Environment. American Chemical Society, Washington, D. C.
11. Crocker, K. M., M. E. Ondrusek, R. L. Petty and R. C. Smith. in press. Dimethysulfide, algal pigments and light in an Antarctic *Phaeocystis* sp. bloom. Mar. Biol.
12. Dacey, J. W. H. and S. G. Wakeham. 1986. Oceanic Dimethylsulfide. Production During Zooplankton Grazing on Phytoplankton. Science *233*: 1314-1316.
13. de Mora, S. J., A. Grout and D. Shooter. 1990. The Analysis of Reduced Sulfur Gases in Air and Melt Waters by Gas Chromatography at Bratina Island, 78S. N.Z. Antarctic Rec. *10*: 12-21.
14. de Mora, S. J., P. Lee, A. Grout, C. Schall and K. Heumann. 1996. Aspects of the biogeochemistry of sulfur in glacial melt water ponds on the McMurdo Ice Shelf, Antarctica. Antarctic Sci. 8: 15-22.
15. de Mora, S. J., P. Lee, D. Shooter and R. Eschenbruch. 1993. The analysis and importance of dimethylsulfoxide in wine. Am. J. Enol. Vitic. *44*: 327-332.
16. Gibson, J. A. E., R. C. Garrick, H. R. Burton and A. R. McTaggart. 1990. Dimethylsulfide and the Alga *Phaeocystis Pouchetii* in Antarctic Coastal Waters. Mar. Biol. *104*: 339-346.
17. Harvey, G. R. and R. F. Lang. 1986. Dimethylsulfoxide and Dimethylsulfone in the Marine Atmosphere. Geophys. Res. Letts. *13*: 49-51.
18. Hatton, A. D., G. Malin, A. G. McEwan and P. S. Liss. 1994. Determination of dimethyl sulfoxide in aqueous solution by an enzymelinked method. Anal. Chem. *66*: 4093-4096.
19. HowardWilliams, C., R. Pridmore, M. T. Downes and W. F. Vincent. 1989. Microbial biomass, photosynthesis and chlorophyll *a* related pigments in the ponds of the McMurdo Ice Shelf, Antarctica. Antarctic Sci. *1*: 125-131.
20. Farwell, S. O. and C.J. Barinaga. 1986. Sulfur-selective detection with the FPD: Current enigmas, practical usage, and future directions. J. Chromatogr. Sci. *24*: 483-494.
21. Keller, M. D. 1989. Dimethyl Sulfide production and marine phytoplankton: The importance of species composition and cell size. Biol Oceanogr *6*: 375-382.
22. Keller, M. D., W. K. Bellows and R. R. L. Guillard. 1989. Dimethyl Sulphide Production in Marine Phytoplankton. p. 167-182. *In* Saltzman and Cooper (ed.), Biogenic Sulphur in the Environment, ACS. Symp. Ser. 393. Am. Chem. Soc., Washington, D. C.
23. Kiene, R. and G. Gerard. 1994. Determination of trace levels of dimethylsulfoxide (DMSO) in seawater and rainwater. Mar. Chem. *47*. 1-12.
24. Kiene, R. P. and T. S. Bates. 1990. Biological Removal of Dimethyl Sulphide from Sea Water. Nature *345*: 702-704.
25. Kirst, G. O., C. Thiel, J. Nothnagel, M. Wanzek and R. Ulmke. 1991. Dimethylsulponiopropionate (DMSP) in icealgae and its possible biological role. Mar. Chem. *35*: 381-388.
26. Kirst, G. O., M. Wanzek, R. Haase, S. Rapsomanikis, S. J. de Mora, G. Schebeske and M. O. Andreae. 1993. Ecophysiology of Ice Algae (Antarctica): Dimethylsulfoniopropionate Content and Release of Dimethylsulfide During the Ice Melt. p. 23-36. *In* Restelli and Angeletti (ed.), Dimethylsulfide: Oceans, Atmosphere and Climate. Kluwer Academic Publishers, Dordrecht.
27. Liss, P., G. Malin and S. Turner. 1993. Production of DMS by marine phytoplankton. p. 1-14. *In* Restelli and Angeletti (ed.), Dimethylsulfide: Oceans, Atmosphere and Climate. Kluwer Academic Publishers, Dordrecht.
28. Lovelock, J. E., R. J. Maggs and R. A. Rasmussen. 1972. Atmospheric Dimethyl Sulphide and the Natural Sulphur Cycle. Nature *237*: 452-453.
29. Malin, G., S. M. Turner and P. S. Liss. 1992. Sulfur: The plankton/climate connection. J. Phycol. *28*: 590-597.
30. Nguyen, B. C., B. Bonsang and A. Gaudry. 1983. The Role of the Ocean in the Global Atmospheric Sulfur Cycle. J. Geophys. Res. *88*: 10903-10914.
31. Pearson, T. W., H. J. Dawson and H. B. Lackey. 1981. Natural occurring levels of dimethyl sulphoxide in selected fruits, vegetables, grains, and beverages. J. Agric. Food Chem. *29*: 1089-1091.
32. Richards, S. R., J. W. N. Rudd and C. A. Kelly. 1994. Organic volatile sulfur in lakes ranging in sulfate and dissolved salt concentration over five orders of magnitude. Limnol Oceangr *39*: 562-572.
33. Ridgeway, R. G., D. C. Thornton and A. R. Bandy. 1992. Determination of Trace Aqueous Dimethylsulfoxide Concentrations by Isotope Dilution GC/MS: Application to Rain and Sea Water. J. Atmos. Chem. *1992*: 53-60.
34. Shooter, D., S. J. de Mora, A. Grout, D. J. Wylie and H. ZhiYun. 1992. The Chromatographic Analysis of Reduced Sulfur Gases in Antarctic Waters following the Preconcentration onto Tenax. Inter. J. Environ. Anal. Chem. *47*: 239-249.

35. Strickland, J. D. H. and T. R. Parsons. 1972. A Practical Handbook of Seawater Analysis. Fisheries Research Board of Canada, Ottawa.
36. Taylor, B. F. and R. P. Kiene. 1989. Microbial Metabolism of Dimethyl Sulfide. p.201-221. *In* Saltzman and Cooper (ed.), Biogenic Sulfur in the Environment. American Chemical Society, Washington D. C..
37. Taylor, F. J. 1981. Phytoplankton and Nutrients in Goat Island Bay, New Zealand. Int. Revue ges. Hydrobiol. *66*: 377-406.
38. Taylor, N. J. and F. J. Taylor. 1985. Hydrology and Changes in the Nutrients and Phytoplankton Levels in Goat Island Bay, Northern New Zealand. Int. Revue ges. Hydrobiol. *70*: 173-186.
39. Turner, S. M., G. Malin and P. S. Liss. 1989.p. 183-200 Dimethyl Sulfide and (Dimethylsulfonio)propionate in European Coastal and Shelf Waters. American Chemical Society, Washington D. C.
40. Turner, S. M., G. Malin, P. S. Liss, D. S. Harbour and P. M. Holligan. 1988. The Seasonal Variation of Dimethyl Sulfide and Dimethylsulfoniopropionate Concentrations in Nearshore Waters. Limnol. Oceanogr. *33*: 364-375.
41. Vairavamurthy, A., M. O. Andreae and R. L. Iverson. 1985. Biosynthesis of Dimethylsulfide and Dimethylpropiothetin by Hymenomonas Carterae in Relation to Sulfur Source and Salinity Variations. Limnol. Oceanogr. *30*: 59-70.
42. Wakeham, S. G., B. L. Howes, J. W. H. Dacey, R. P. Schwarzenbach and J. Zeyer. 1987. Biogeochemistry of dimethylsulfide in a seasonally stratified pond. Geochimica Cosmochimica Acta *51*: 1675-1684.
43. Watts, S. F., P. Brimblecombe and A. J. Watson. 1990. Methanesulphonic Acid, Dimethyl Sulphoxide and Dimethyl Sulphone in Aerosols. Atmos. Environ. *24A*: 353-359.
44. Watts, S. F., A. Watson and P. Brimblecombe. 1987. Measurements of the Aerosol Concentrations of Methanesulphonic Acid, Dimethyl Sulphoxide and Dimethyl Sulphone in the Marine Atmosphere of the British Isles. Atmos. Environ. *21*: 2667-2672.
45. White, R. H. 1982. Analysis of Dimethylsulfonium Compounds in Marine Algae. Journal of Marine Research *40*: 529-536.
46. Wylie, D. J. 1993. Atmospheric Chemistry of Biogenic Sulfur in the Southern Hemisphere. PhD. Thesis. Auckland.
47. Yen, H. C. and B. Marrs. 1977. Growth of *Rhodopseudomonas capsulata* under anaerobic dark conditions with dimethyl sulfoxide. Arch. Biochem. Biophys. *181*: 411-418.
48. Zeyer, J., P. Eicher, S. G. Wakeham and R. P. Schwarzenbach. 1987. Oxidation of dimethyl sulfide to dimethyl sulfoxide by phototrophic purple bacteria. Appl. Environ. Microbiol. *53*: 2026-2032.
49. Zinder, S. H. and T. D. Brock. 1978. Dimethyl Sulphoxide reduction by microorganisms. J. Gen. Microbiol. *105*: 335-342.

35

DMSO

A Significant Compound in the Biogeochemical Cycle of DMS

A. D. Hatton, G. Malin, S. M. Turner, and P. S. Liss

School of Environmental Sciences
University of East Anglia
Norwich, NR4 7TJ, United Kingdom

SUMMARY

Dimethylsulfoxide (DMSO) in seawater is thought to arise from photochemical and bacterial oxidation of dimethylsulfide (DMS) and is likely to be a key compound in the marine biogeochemical cycle of sulfur. DMSO was measured using a new highly specific method (12). This involves removal of DMS from samples by purging; DMSO is then reduced to DMS using the enzyme DMSO reductase purified from the photosynthetic bacterium *Rhodobacter capsulatus*. This technique has been used to measure DMSO concentrations during three field campaigns: in coastal and shelf waters of the North Sea during a cold and cloudy April, the equatorial Pacific inside and outside the nutrient rich plume in the vicinity of the Galapagos Islands, and the Arabian Sea along a transect from upwelled eutrophic to oligotrophic waters. Results for DMSO concentrations have been compared with those for DMS and dimethylsulfoniopropionate (DMSP), in surface waters and throughout the water column. These data are used to establish the significance of DMSO in marine waters.

INTRODUCTION

DMSO occurs naturally in marine and freshwater environments (1, 8, 19), in rainwater and in the atmosphere (5, 11). Early work on DMSO suggested that it was produced by phytoplankton and was present at concentrations far in excess of DMS (1, 2). However, this work was carried out before much was known about DMSP production and the method used caused degradation of DMSP to DMS. Current literature suggests that DMSO in seawater arises from photochemical and bacterial oxidation processes and is a key compound in the marine biogeochemical cycle of DMS.

Until recently these oxidation processes have received little attention, even though this may represent an important sink for DMS (21). Brimblecombe & Shooter (6) showed that aqueous solutions of DMS were susceptible to photooxidation by visible light in the

Biological and Environmental Chemistry of DMSP and Related Sulfonium Compounds
edited by Ronald P. Kiene et al., Plenum Press, New York, 1996

presence of photosensitizers. They estimated the global quantity of DMS photooxidized would amount to about 6 Tg(S) a^{-1}. This is significant when compared with the global sea-air flux of DMS of 17 Tg(S) a^{-1} estimated by Bates & Gammon (4). The photochemical oxidation of DMS to DMSO also occurs in the marine atmosphere (3, 5, 14). Due to its low volatility and hygroscopic nature, DMSO would be scavenged from the air by rain (15) and returned to the oceans in precipitation.

In addition to photochemical processes, the formation of DMSO from DMS may be enzymically catalyzed (21). It has been demonstrated that some phototrophic bacteria are capable of oxidizing DMS to DMSO, however, most of the organisms studied to date are obligate anaerobes (10, 23, 25). Zeyer *et al.* (25) showed that enrichment cultures of phototrophic purple bacteria, including a pure strain of a marine *Thiocystis* species, rapidly oxidized up to 10 mM DMS to DMSO. Hansen *et al.* (10) isolated seven, pure bacterial cultures that could utilize DMS as an electron donor for CO_2 fixation, with DMSO as the only product. Recently Hanlon *et al.* (9) isolated a strain of *Rhodobacter sulfidophilus* which grew autotrophically with DMS serving as an electron donor in photosynthesis and respiration, but not as a carbon source. They identified a periplasmic DMS acceptor oxidoreductase, which was distinct from the DMSO reductase found in this bacterium.

Until recently, the few available measurements of DMSO in seawater were thought to be unreliable due to analytical difficulties (12), and hence its role in the marine sulphur cycle has yet to be defined . Various methods have been reported for DMSO analysis in aqueous samples. Direct measurement of DMSO by gas chromatography is generally insufficiently sensitive (but see Lee and deMora, this volume) for the nanomolar concentration range found in environmental samples (7, 17, 18, 24). Chemical reduction of DMSO to DMS with subsequent analysis of the DMS has greater sensitivity, however these methods require the prior removal of DMSP from the sample (8, 13). Recently we have developed a sensitive and highly specific technique for the measurement of DMSO in aqueous solution (12). In this paper we discuss some of the results obtained during three field campaigns to shelf, coastal and open ocean waters using the new technique.

MATERIALS AND METHODS

Measurements of DMSO, DMS, DMSP(p) and DMSP(d)

The analytical system consisted of a Varian 3300 gas chromatograph fitted with a flame photometric detector and a Chromosil 330 column. The GC was operated using a temperature program (40-85°C), and the DMS retention time on this column was 3.4 min. The detector output was monitored on a Hewlett–Packard 3390A reporting integrator. The method for the hydrolysis of DMSP(p) and DMSP(d) to DMS and the system for trapping DMS was identical to that described by Turner *et al.*, (22).

For DMSO analysis the system was also similar to that described by Turner *et al.* (22) with the exception that the purge tube was illuminated using three 60 W daylight bulbs, which initiated the reduction of DMSO. Reduction of DMSO to DMS was achieved using a reducing solution containing 25 μg cm^{-3} DMSO reductase, 30 mM ethylenediaminetetraacetic acid (EDTA), and 540 μM flavin mononucleotide (FMN). Under illumination EDTA forms radicals which reduce FMN to $FMNH_2$, and this acts as an electron donor to DMSO reductase, catalyzing the reduction of DMSO to DMS. DMSO reductase was purified from the cells of *R. capsulatus* strain H123 using the method outlined by McEwan *et al.* (16). This method is described in more detail in Hatton *et al.* (12).

Field Measurements. The DMSO assay was tested under field conditions during three shipboard campaigns: A survey of the region surrounding the Galapagos Islands was conducted onboard the research vessel *Columbus Iselin* (5-24 November 1993). A series of stations were occupied both upstream (east) and down stream (west) of the Islands. Depth profiles were taken to 1500 m. A survey of the southern North Sea was conducted onboard the *Cirolana* (5-10 April 1994). Seawater samples were taken routinely from a depth of approximately 3 m using the non-contaminated ship's continuous pump supply and analyzed immediately. A survey was conducted in the Arabian Sea along a transect from upwelled eutrophic, to oligotrophic waters onboard the *Discovery* (20 August-6 October 1994). Samples for depth profiles were collected from the CTD rosette and underway samples using the non-contaminated ship's continuous pump supply and analyzed immediately.

All water samples collected from the CTD rosette were stored in 500 ml ground glass stoppered bottles with minimal headspace for a maximum of 8 hr in the dark at ambient seawater temperature to minimize possible storage artifacts.

RESULTS AND DISCUSSION

DMSO in Surface Seawater

Fig. 1 shows the concentrations of DMS and DMSO found in near surface waters in the North Sea and the Arabian Sea. From these results we have concluded that the distribution of DMSO in surface seawater closely follows that of DMS. The correlation between DMS and DMSO for the whole data set was $r = 0.77$, showing that a strong relationship exists between the two compounds. DMS was below the detection limit for a few of the North Sea analyses, but, with the exception of one sample, DMSO was readily detected in all of these (Fig. 2). We feel that this correlation helps to support the theory that photooxidation of DMS in surface waters may be an important source of DMSO. Shooter and Brimblecombe (20) suggested that a number of oxidation pathways exist in seawater, but that near the sea surface, photochemical processes probably play the more important role in DMS oxidation. The loss of DMS via photooxidation to DMSO is thought to occur in seawater at rates which are of the same magnitude as the emissions of DMS to the atmosphere (6). Brimblecombe and Shooter (6) also showed that in coastal seawater photooxidation of DMS did not differ considerably between sunshine or overcast conditions. This would help explain the correlation even though the data sets were taken from both the North Sea and the Arabian Sea under sunny and overcast conditions. In the past, the mechanisms for DMS oxidation have received little attention, even though this may represent an important sink for DMS (21). Hydrogen peroxide is thought to be a potential chemical oxidation route for DMS in seawater (26). In addition, it has been shown that the formation of DMSO from DMS may be enzymatically catalyzed (21). A number of works have demonstrated that some phototrophic bacteria are capable of oxidizing DMS to DMSO, however, most of the organisms studied to date were obligate anaerobes (9, 10, 23, 25). DMSP(p) appeared to follow the same general trend as DMSO (Fig. 1), but did not give a significant correlation ($r = 0.34$).

Vertical distribution. A series of depth profiles of DMS, DMSO, DMSP(p) and DMSP(d) concentrations were determined for seawater collected during the equatorial Pacific and the Arabian Sea campaigns. Fig. 3 shows two example depth profiles collected inside and outside the nutrient rich plume in the vicinity of the Galapagos Islands. The DMSP(p) concentrations were consistently higher than DMS, DMSO and DMSP(d) concentrations in the upper water column. The DMSO concentrations were also consistently higher than DMS and DMSP(d) concentrations and were higher than DMSP(p) concentrations at

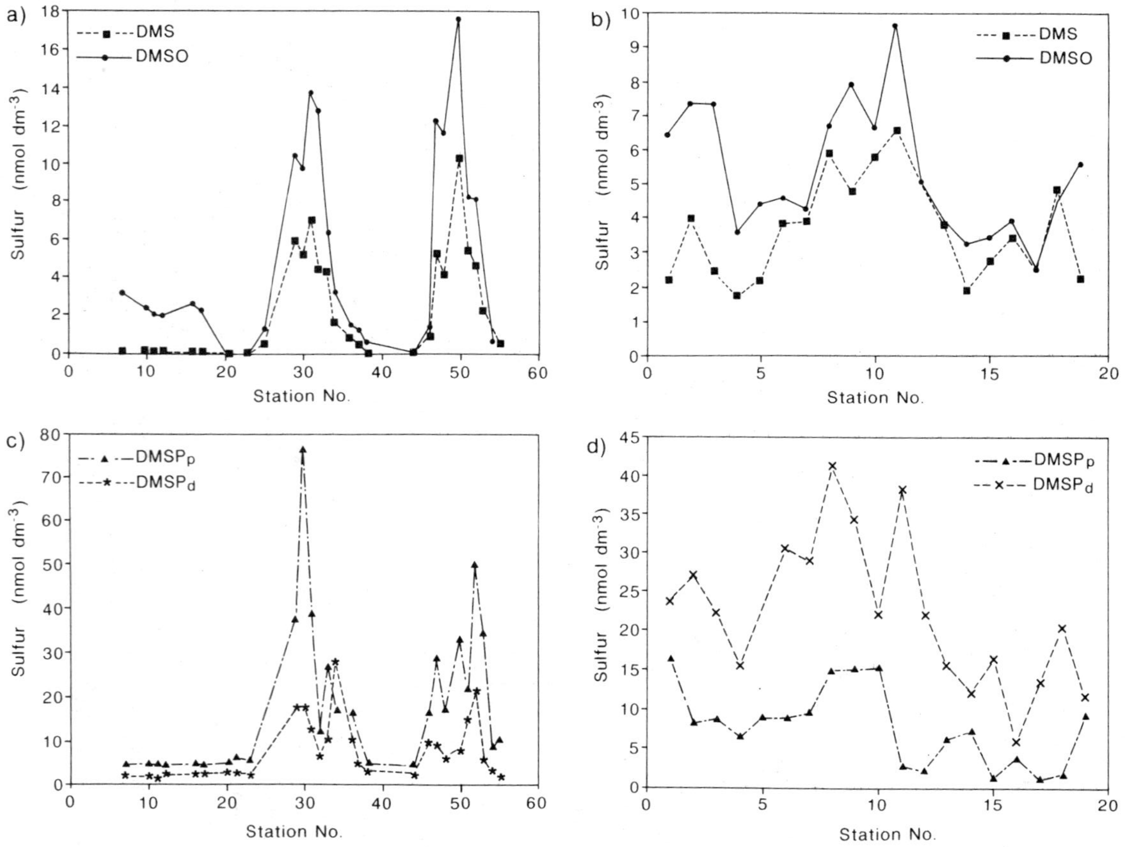

Figure 1. Near-surface concentrations for DMS, DMSO, DMSP(p) and DMSP(d) collected in the North Sea from 52°46N, 01°50E to 54°03N, 02°10E to 52°51N, 03°07E (a &b & c), 5th - 10th April 1994 and Arabian Sea from 19°30N, 58°09E to 16°02N, 62°00E (b & d), 20th Aug - 6th Oct 1994.

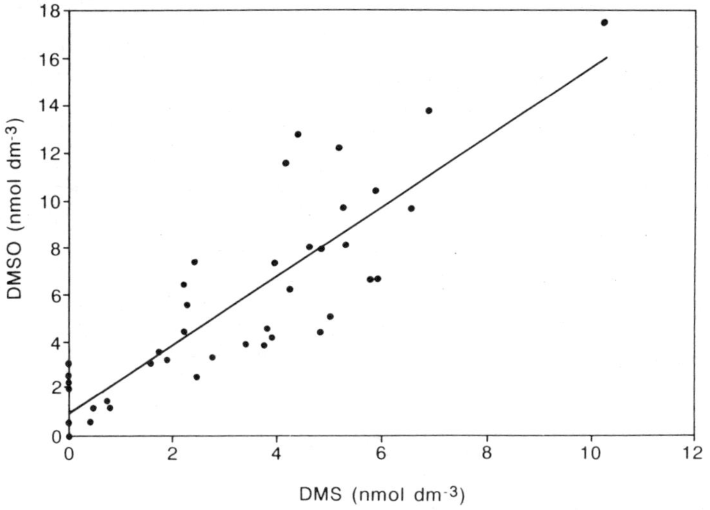

Figure 2. Correlation between DMS and DMSO for data set collected in the North Sea, 5th - 10th April 1994 and Arabian Sea, 20th Aug -6th Oct 1994. Concentrations show a correlation of r = 0.77.

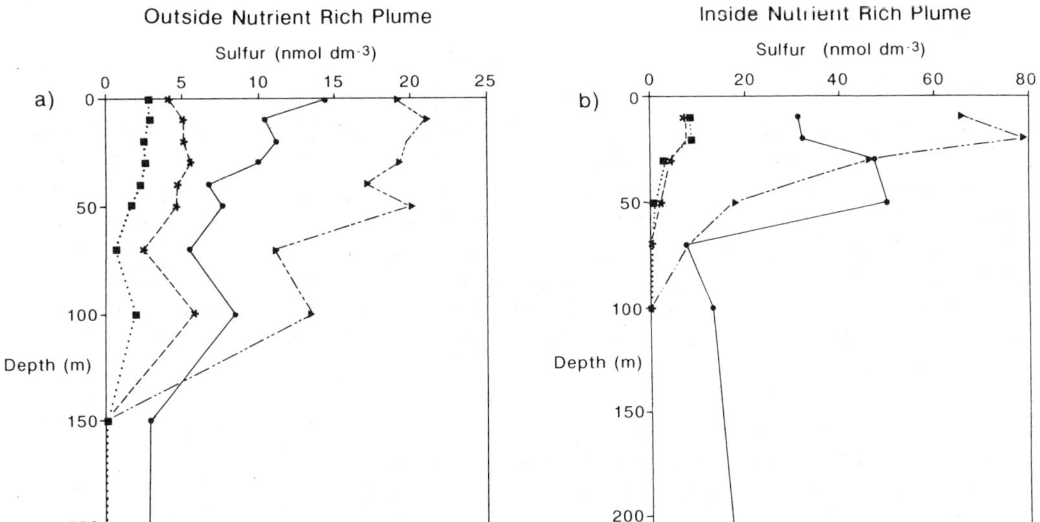

Figure 3. Concentration profiles for DMS, DMSO, DMSP(p) and DMSP(d) for contrasting stations outside (a, 0°30N, 89°0W) and inside (b, 0°23S, 92°60W) the high productivity plume to the west of the Galapagos Islands. Samples were collected during the PlumEx study onboard RV *Columbus Iselin*, equatorial Pacific 5th - 24th Nov 1993. Note the difference in X axis scales.

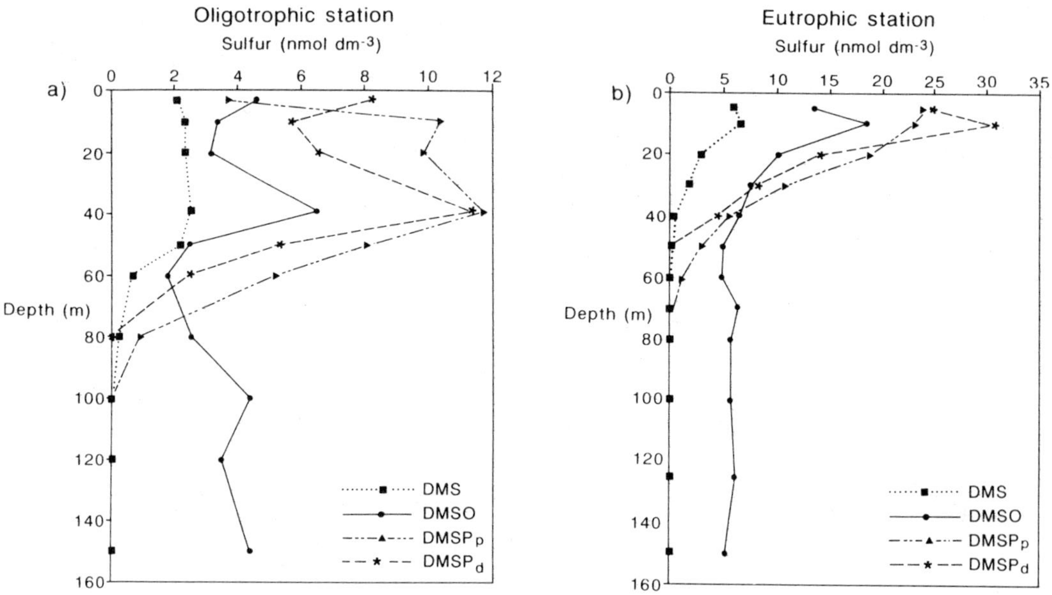

Figure 4. Concentration profiles for DMS, DMSO, DMSP(p) and DMSP(d) for contrasting stations in an oligotrophic (a, 10°14N, 65°33E) and a eutrophic (b, 17°30N, 60°30E) area. Samples were collected during Arabesque 1 onboard RRS *Discovery*, Arabian Sea 20 Aug - 6th Oct 1994. Note the difference in X axis scales.

greater depths. The concentrations of all four compounds were higher within the plume than outside. Fig. 4 shows two example depth profiles collected in upwelled eutrophic to open ocean oligotrophic waters of the Arabian Sea. The DMSP(p) and DMSP(d) concentrations were consistently higher than DMS and DMSO down through the top 50 m of the profile. The DMSO concentrations were consistently higher than DMS and were higher than DMSP(p) and DMSP(d) at greater depths. Concentrations of all four compounds were higher in the eutrophic waters than the oligotrophic regions. Results for biological parameters are now being processed and these will be published with the full biogenic sulphur data set in the future.

DMSO in the Deep Water Column. In both the Pacific and the Arabian Sea, DMSO was found to be ubiquitous throughout the water column. DMSO concentrations never dropped below 1.5 nmol S dm^{-3} at greater depth and were higher throughout the water column in highly productive regions. Fig. 5 shows values for DMSO and DMSP from a series of deep profiles inside and outside the high productivity plume to the west of the Galapagos Islands. Results were integrated to give the average concentration in nmol S dm^{-3} over a 1500 m water column. This shows that when the whole water column is taken into account DMSO levels are significantly higher than those of DMSP(p). Similar results were also observed for deep profiles taken in the Arabian Sea. Recently, Ridgeway *et al.* (19) observed that DMSO was well mixed throughout the water column, and suggested that DMSO had a longer mean lifetime than DMSP or DMS. Our results appear to confirm this observation and therefore indicate that DMSO may act as a substantial sink for DMS in deep waters.

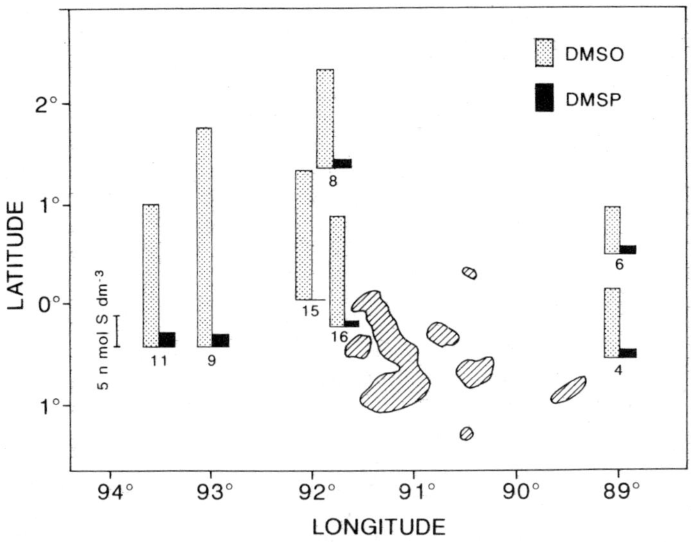

Figure 5. PlumEx, RV *Columbus Iselin*, equatorial Pacific 5th - 24th Nov 1993. Biogenic sulfur compounds were analyzed in a series of deep profiles inside and outside the high productivity plume to the west of the Galapagos Islands. DMSP(p) concentrations reached their minimum detection levels by 200 meters, whereas DMSO was detected in all samples down to 1500 m. The bars represent the average DMSO and DMSP(p) concentrations (nmol S dm^{-3}) over the 1500 m water column.

CONCLUSIONS

Dimethylsulfoxide is present in a wide variety of marine environments, with the highest concentrations in upwelled and highly productive regions. DMSO is significantly correlated with DMS and closely parallels DMSP concentrations in near-surface water. Additionally DMSO appears to be ubiquitous throughout the water column being easily detectable in deep waters.

ACKNOWLEDGMENTS

This work was supported by a NERC/CASE studentship (to ADH) and grants from the NERC, the EEC, and the UEA Research Promotion Fund . We would like to thank Al McEwan, Neil Benson and Wendy Broadgate for their contributions to this project. We also thank the officers, crew, and scientists onboard the research vessels *Columbus Iselin*, *Cirolana* and *Discovery*.

REFERENCES

1. Andreae, M. O. 1980a. Determination of trace quantities of DMSO in aqueous solution. Anal. Chem. *52*: 150-153.
2. Andreae, M. O. 1980b. Dimethylsulphoxide in marine and freshwaters. Limnol. Oceanogr. *25*: 1054-1063.
3. Barnes, I., K. H. Becker, P. Carlier and G. Mouvier. 1987. FTIR study of the DMS/NO2/I2/N2 photolysis system: The reaction of IO radicals with DMS. Int. J. Chem. Kinet. *19*: 489-501.

4. Bates, T. S. and R. H. Gammon. 1985. Oceanic dimethylsulphide and the global atmosphere sulphur cycle. Trans. Am. Geophys. Union. *66*: 1309.
5. Berresheim, H., D. J. Tanner and F. L. Eisele. 1993. Real time measurement of DMSO in ambient air. Anal. Chem. *65*: 84-86.
6. Brimblecombe, P. and D. Shooter. 1986. Photo-oxidation of DMS in aqueous solution. Mar. Chem. *19*: 343-353.
7. deMora, S. J. P. Lee, D. Shooter and R. Eschenbruch. 1993. The analysis and importance of DMSO in wine. Am. J. Enol. Vitic. *44*: 327-332.
8. Gibson, J. A. E., R. C. Garrick, H. R. Burton and A. R. McTaggart. 1990. DMS in the alga *Phaeocystis pouchetii* in antarctic coastal waters. Mar. Biol. *104*: 339-346.
9. Hanlon, S. P., R. A. Holt, G. R. Moore and A. G. McEwan. 1994. Isolation and characterisation of a strain of *Rhodobacter sulphidophilus*: a bacterium which grows autotrophically with DMS as an electron donor. Microbiology *140*: 1953-1958.
10. Hansen, T. A., P. Quist, M. J. E. C. Van der Maarel and L. Dijkhuizen. 1993. Isolation of marine DMS-oxidizing bacteria, p. 37-41. *In* Restelli, G. and Angeletti, G. (eds.), DMS: oceans, atmosphere and climate. Kluwer Academic Publishers, Dordrecht.
11. Harvey, G. R. and R. F. Lang. 1986. Dimethylsulphoxide and dimethylsulphone in the marine atmosphere. Geophys. Res. Lett. *13*: 49-51.
12. Hatton, A. D., G. Malin, A. G. McEwan and P. S. Liss. 1994. Determination of DMSO in aqueous solution by an enzyme-linked method. Anal. Chem. *66*: 4093-4096.
13. Kiene, R. P. and G. Gerard. 1994. Determination of trace levels of dimethylsulfoxide (DMSO) in seawater and rainwater. Mar. Chem. *47*: 1-12.
14. Koga, S. and H. Tanaka. 1993. Numerical study of the oxidation process of dimethylsulfide in the marine atmosphere. J. Atmos. Chem. 17: 201-228.
15. Lovelock, J. E., R. J. Maggs and R. A. Rasmussen. 1972. Atmospheric dimethylsulphide and the natural sulphur cycle. Nature. *237*: 452-453.
16. McEwan, A. G., H. G. Wetzstein, S. J. Ferguson and J. B. Jackson. 1985. Periplasmic location of the terminal reductase in TMAO and DMSO respiration in the photosynthetic bacterium *Rhodopseudomonas capsulata*. Biochim. Biophys. Acta. *806*: 410-417.
17. Ogatoa, M., T. Fujii and Y. Yoshida. 1979. Quantitative determination of urinary DMSO and $DMSO_2$ by GC equipped with FPD. Ind. Health. *17*: 73-78.
18. Paulin, H.J., J. B. Murphy and R. E. Larson. 1966. Determination of DMSO in plasma and cerebrospinal fluid by GLC. Anal. Chem. *38*: 651-652.
19. Ridgeway, R. G., D. C. Thornton and A. R. Bandy. 1992. Determination of trace aqueous DMSO concentrations by isotope dilution GC/MS: application to rain and seawater. J. Atmos. Chem. *14*: 53-60.
20. Shooter, D. and Brimblecombe, P. (1989). Dimethylsulphoxide oxidation in the ocean. Deep-Sea Res. *36*: 577-585.
21. Taylor, B. F. and R. P. Kiene. 1989. Microbial metabolism of dimethylsulfide, p. 202-221. *In* Saltzman, E. S. and Cooper, W. J. (eds.) Biogenic sulfur in the environment. American Chemical Society Symposium Series 393, Washington.
22. Turner, S. M., G. Malin, L. E. Bagander and C. Leck. 1990. Interlaboratory calibration and sample analysis of dimethylsulphide in water. Mar. Chem. *29*: 47-62.
23. Visscher P.T. and H. Van Gemerden. 1991. Photo-autotrophic growth of *Thiocapsa roseopersicina* on dimethylsulfide. FEMS Microbiology Letters. *81*: 247-250.
24. Wong, K. K., G. Mei Wang, J. Dreyfuss and E. C. Schreiber. 1971. Absorption, excretion and biotransformation of DMSO in man and miniature pigs after topical application as an 80 % gel. J. Invest. Dermatol. *56*: 44-48.
25. Zeyer, J., P. Eicher, S. G. Wakeham and R. P. Schwarzenbach. 1987. Oxidation of DMS to DMSO by phototrophic purple bacteria. Appl. Environ. Microbiol. *53*: 2026-2032.

36

AEROBIC DMS DEGRADATION IN MICROBIAL MATS

The Use of a Newly Isolated Species *Methylophaga sulfidovorans* in the Mathematical Description of Sulfur Fluxes in Mats

Jolyn M. M. de Zwart and J. Gijs Kuenen

Kluyverlaboratory for Biotechnology
Delft University of Technology,
Julianalaan 67, 2628 BC Delft, The Netherlands

SUMMARY

A new dimethyl sulfide (DMS)-degrading bacterium, *Methylophaga sulfidovorans,* was isolated from a Dutch intertidal mud flat. This bacterium was obligately methylotrophic, and able to use sulfide as an additional electron donor. Experiments with *M. sulfidovorans* were carried out in continuous and batch cultures in order to measure DMS oxidation under the types of conditions in a microbial mat. As *M. sulfidovorans* appears to be representative of the aerobic DMS oxidizing community in such a mat, its (kinetic) characteristics were used for the mathematical modelling of sulfur fluxes in a microbial mat. The predicted fluxes from a microbial mat sediment to the atmosphere in a 12 hour light period were within the range observed by direct measurements (derived from literature) in the field.

INTRODUCTION

Dimethyl sulfide (DMS) metabolism has been described for a small number of aerobic and anaerobic bacteria from freshwater and marine systems. Interest in freshwater DMS oxidizing bacteria originates primarily from their potential use in biological waste gas treatment systems for gases that contain (organic) sulfur compounds (9, 18, 20, 29). DMS in waste gases arises either as a product of the chemical lignin degradation in the paper industry, or from food manufacturing where DMS is mainly formed from the degradation of sulfur containing amino acids (11). The main precursor of DMS in marine environments is dimethylsulfoniopropionate (DMSP), an osmolyte present in many marine phototrophic organisms (10). Production of DMSP can be estimated at 4.10^{15} g S (DMSP) y^{-1} (15). Bacterial degradation of oceanic DMS is estimated to be about 90% of the total DMS

Biological and Environmental Chemistry of DMSP and Related Sulfonium Compounds
edited by Ronald P. Kiene et al., Plenum Press, New York, 1996

produced (15). Marine DMS oxidizing bacteria therefore play a significant role in the regulation of fluxes from the oceans to the atmosphere, and thus in climate regulation processes based on DMS fluxes (1).

DMS degradation under anoxic conditions has been studied with several methanogenic bacteria from marine systems (12, 16, 17). It has been found that sulfate-reducing bacteria play a role in DMS oxidation in marine sediments (13), but so far only a freshwater-sulfate reducing bacteria able to use DMS as a substrate has been isolated (22).

Aerobic DMS degradation has only been described for a small number of marine bacteria (24, 27). A large part of the research on DMS oxidation in marine systems has concentrated on DMS transformations in salt marsh sediments and microbial mats (13, 14, 25, 30). The physiological characteristics of aerobic DMS-oxidizing bacteria must therefore be examined in greater detail, in order to provide understanding of the role of these bacteria in the oxidation of DMS. In the current study, microbial mats are used as model systems to describe aerobic DMS oxidation.

Microbial Communities in a Microbial Mat

A microbial mat is a laminated microbial ecosystem a few millimetres thick that develops on solid surfaces (23). The driving force behind this ecosystem is photosynthesis in the top layer. Photosynthesis is, in most cases, the only source of organic carbon in mats (2). A microbial mat can, therefore, be regarded as a semi-closed system, with only input of light energy, carbon dioxide and sulfate. Only a few groups of microorganisms with different metabolic activities are therefore theoretically necessary to explain the interactions and activities within such a mat. In our model, five functional groups of microorganisms are considered; oxygenic phototrophs, sulfide-oxidizing bacteria, aerobic heterotrophs, aerobic DMS oxidizing bacteria and sulfate-reducing bacteria. These groups form a commensal relationship in substrate production and consumption. Carbon dioxide is photosynthetically fixed. The organic carbon is then used by the heterotrophic community in the aerobic zone of the mat, and by the sulfate-reducing bacteria in the anaerobic zone of the mat. The sulfide produced by the sulfate-reducers is oxidized with oxygen (produced by the oxygenic phototrophs) by colorless sulfur bacteria in the aerobic zone of the mat. The top layer of the mat contains phototrophs that may contain DMSP (25) from which DMS is produced. A population of DMS-oxidizing bacteria will, therefore, probably be present in the top layer of the mat where most of the DMS is produced. As our interest lies in the role of these organisms in the control of DMS fluxes, the DMS-metabolizing bacteria represent a separate functional group of organisms in our model.

The metabolic processes in a microbial mat are strongly dependent on the day/night cycle, during which pH values, oxygen and sulfide concentrations can reach extreme values. The pH in the top of the microbial mat can rise to above 10 due to the depletion of bicarbonate, which is the natural buffer in marine systems (24). During the night, the pH falls to 7 or 8 because of carbon dioxide production from the dissimilatory processes of the microbial community, and diffusion of carbon dioxide and bicarbonate throughout the mat. The oxygen produced during photosynthesis during the day can result in supersaturation (2 to 3 times the air saturation value) of oxygen (500 $\mu M\ O_2$) at the *in situ* temperature (4). Concentrations of 500 $\mu M\ H_2S$ can be found in microbial mats (26). During the dark period the oxygen may only diffuse slowly from the atmosphere into the mat, and not all of the sulfide will be oxidized. Sulfide can then diffuse upwards and may reach the surface resulting in hydrogen sulfide emissions from the mat.

DMS concentration profiles have been determined in microbial mat sediments, ranging from 250 $\mu mol\ l^{-1}$ at the top 5 mm to almost 0 at 25 mm depth in a specific sediment

(26). Furthermore, measurements have shown a DMS-uptake capacity in sediments, indicating the presence of a microbial DMS oxidizing population (13).

This paper describes the isolation and characterisation of an aerobic, DMS-oxidizing bacterium from a marine microbial mat. The physiological characteristics of the isolate are used in a mathematical model describing sulfur fluxes in a microbial mat. In this general model, both biological activity and transport processes are considered in order to predict DMS release into the atmosphere.

MATERIALS AND METHODS

Bacterial Culture and Natural Samples

Microbial mat sediment samples were obtained from an estuarine intertidal region at the south-west coast of the Netherlands (51° 26.3′ N, 4° 7.5′ E). Samples were obtained in 1993 and 1994. *Methylophaga sulfidovorans* was isolated from these mat samples (7).

Culture Media

Medium contained per liter: 15 or 25 g NaCl, 0.5 g $(NH_4)_2SO_4$, 0.33 g $CaCl_2.6H_2O$, 0.2 g KCl, 1 g $MgSO_4.7H_2O$, 0.02 g KH_2PO_4, 2 g Na_2CO_3, 1 mg $FeSO_4.7H_2O$, 1 ml trace solution, 1 ml vitamin solution. pH was set at 7.5 (± 0.3) with 1 N HCl. The trace element solution was as described by Widdel and Pfennig (28), except for the addition of 0.6 μM $Na_2SeO_4.5H_2O$ (final concentration). The vitamin solution contained per liter: biotin (20 mg), nicotinic acid (200 mg), thiamine (100 mg), *p*-aminobenzoic acid (100 mg), pantothenate (50 mg), pyridoxine.HCl (500 mg), riboflavine (10 mg) and vitamin B_{12} (10 mg). Medium used for purification of strains was supplemented with HEPES buffer (1mM).

Artificial sea water, used for suspension of sediment samples, contained 22.5 g l^{-1} NaCl, 24.6 g l^{-1} $MgSO_4.7H_2O$, 1.5 g l^{-1} KCl and 2.2 g l^{-1} $CaCl_2$. pH was set at 7.5.

Liquid medium containing DMS was kept in glass bottles, which were sealed with butyl rubber or teflon stoppers to avoid DMS loss.

The purity of the cultures was checked by streaking on Brain Heart Infusion plates supplemented with 1.5% NaCl.

Analytical Techniques

DMS from the headspace of the cultures was measured with a gas chromatograph with a sulfur-specific flame photometric detector (20), equipped with a Hayesep R column. Measurements of DMS were accurate (less than 5% variance) down to 0.5 μM in the gas phase. The equilibrium constant for DMS at room temperature was estimated at 13.5 (19).

Oxygen consumption rates of bacterial suspensions were measured with a polarographic Clark-type electrode in a temperature-controlled Biological Oxygen Monitor. Oxygen concentrations in the gas phase were determined by gas chromatography, using a molecular sieve packed column and hard wire detector.

Culture density was determined using a Total Organic Carbon analyzer (Tocamaster 815B). The biomass obtained from continuous cultures was analyzed for its carbon, hydrogen and nitrogen content with a Carlo Erba CHNS-O elemental analyzer. The formula for biomass composition was found to be $CH_{1.75}O_{0.46}N_{0.19}$ for *M. sulfidovorans* which shows that about 50% of the biomass is carbon. Biomass was also measured by dry weight determination.

DMS Consumption in Microbial Mat Sediment Samples

DMS oxidation rates in natural samples were determined in 100 ml serum bottles filled with 20 ml sediment (1 part) and artificial seawater (4 parts), leaving the larger part as headspace. Aerobic and anaerobic headspaces were created by flushing with air or N_2, respectively. The bottles were sealed with butyl rubber septa and teflon stoppers, in order to prevent inward oxygen diffusion and outward DMS diffusion. DMS was added to the samples (< 1 mM DMS) with a syringe from a stock solution (27 mM DMS). As the bottles were not shaken during incubation, it must be assumed that even with an air-flushed headspace, part of the slurry in the flask rapidly became anoxic. The total DMS removal was therefore the sum of the aerobic and anaerobic breakdown of DMS. When the headspace was flushed with nitrogen, only anaerobic DMS degradation could occur.

Most Probable Number (MPN) Counts of Aerobic DMS-Oxidizing Bacteria in Sediment Samples and Isolation Procedure

Dilution series were made from natural samples using bicarbonate buffered mineral medium, and incubated at 25°C. The tubes containing the highest positive dilution were used for enrichment and strain purification. Enrichments were carried out in the same liquid medium. The flasks were inoculated in a desiccator with DMS in the headspace at a concentration that would not exceed 1 mM DMS in the liquid phase. One enrichment culture was selected for further study on the basis of its DMS-oxidizing capacity. As this culture did not grow on agar plates with DMS in the headspace, the aerobic DMS-oxidizing bacterium was isolated using several dilutions to extinction. Culture purity was checked with BHI agar plates and microscopy, and confirmed when the DNA of the culture gave only one visible 16 S rDNA band after PCR amplification.

Physiology and Taxonomy of the Isolate

The substrate range of the isolate was tested in batch cultures at 25°C and 100 rpm. Aerobic growth was tested for the following substrates: DMS, methane, methanol, formate, thiosulfate, hydrogen sulfide, methylamine, dimethylsulfoxide, acrylate, acetate, ethanol, propanol, lactate, galactose, glucose and fructose. Furthermore, denitrification capacity with methanol as electrondonor and nitrate as electronacceptor was tested in a nitrogen flushed chemostat at 25°C.

To determine the kinetics of DMS utilization by the isolate, it was grown in a methanol-limited continuous culture under standard conditions: temperature = 27°C, pH = 7.6, oxygen tension = 50% air saturation and dilution rate = 0.05 h^{-1}. After a steady state on methanol had been reached, the medium supply to the continuous culture was stopped, and the aeration of the chemostat was changed to headspace recirculation, in order to prevent loss of DMS through the outgoing airflow. A pulse of DMS was added to the chemostat culture, resulting in a DMS concentration of 0.8 mM, and the rate of DMS consumption was followed by gas chromatography. The Michaelis–Menten parameters, v_{max} and k_s, were then determined in a Biological Oxygen Monitor and directly estimated from the rate of DMS disappearance in the chemostat.

Biomass yields on DMS were determined in shaking batch cultures, incubated at 25°C at 100 rpm. The flasks were sealed with butyl rubber stoppers to prevent DMS loss. DMS was added with a syringe and did not exceed 0.5 mM in order to prevent DMS toxicity and to prevent oxygen exhaustion. Biomass formation was measured by total organic carbon analysis.

For further identification of the isolate the full (1500 base pairs) 16 S rDNA sequence was determined and compared with known eubacterial sequences at the Deutsche Sammlung für Microorganismen.

Cultivation of *M. sulfidovorans* at Various pH Values, H_2S and O_2 Concentrations

The influence of pH and oxygen tension on the Michaelis–Menten parameters (the maximum oxidation rate v_{max} and the half saturation constant k_s) for DMS-oxidation by *M. sulfidovorans* was determined with chemostat grown cells. After a steady state under standard conditions on methanol had been reached a pulse of about 0.8 mM DMS was added to the chemostat culture in order to adapt the culture to DMS as a substrate. After the DMS had dissappeared from the chemostat (partly through the outgoing airflow and partly because of consumption), a particular variable (pH or oxygen tension) was changed. One hour later, DMS was added to the chemostat (now operated as a batch fermenter with a recirculating headspace) and its disappearance from the culture was followed in time.

To test the effect of hydrogen sulfide on the kinetics of DMS oxidation, *M. sulfidovorans* was cultivated in batch cultures in serum bottles sealed with butyl rubber or teflon septa. The serum bottles (75 ml) were filled with 20 ml mineral medium with $Na_2S.9H_2O$ to final concentrations ranging from 0 to 6 mM. The bottles were inoculated with about 4 mg l^{-1} biomass from a chemostat culture grown under standard conditions. DMS solution was added to give a concentration of 0.5 mM DMS, in order to avoid toxic and inhibitory effects and maintain an aerobic headspace during the experiment. The bottles were incubated at 25 °C. DMS removal rates were determined with gas chromatography. The rate of DMS removal was low, due to the small inoculum, to ensure that the medium remained oxic. The v_{max} was calculated from the initial DMS oxidation. The affinity constant was calculated from the slower DMS oxidation rate obtained with lower DMS concentrations.

During these experiments purity was carefully checked with BHI agar plates and microscopy.

Mathematical Modelling of Fluxes in a Mat

To investigate the interrelationships between microorganisms in a microbial mat a mathematical model has been developed. This model is used to design relevant experiments with *M. sulfidovorans* and to predict the fluxes of DMS to the atmosphere. The model is based on simple Michaelis Menten kinetics and diffusion processes through sediment.

In the model, five functional groups of microorganisms are considered (23). These groups of organisms are listed in table 1 in order of (vertical) appearance in the mat, with the most significant biological reactions carried out by each group (figure 1). It should be

Table 1. List of the functional groups of microorganisms with their metabolic reactions, used in the mathematical model

Functional group	Metabolic reaction*
Oxygenic phototrophs	Light + $H_2O + CO_2 \rightarrow CH_2O + O_2 + (CH_3)_2S$
Chemolithoautotrophs	$H_2S + 2O_2 \rightarrow H_2SO_4$
Methylotrophs	$(CH_3)_2S + 5O_2 \rightarrow 2CO_2 + H_2SO_4 + 2H_2O$
Heterotrophs	$CH_2O + O_2 \rightarrow CO_2 + H_2O$
Sulfate reducers	$H_2SO_4 + 2CH_2O \rightarrow H_2S + 2CO_2 + 2H_2O$

(*)The stoichiometric relations of the reactions are presented, except for the DMS production by the phototrophic bacteria. The DMS formation rate is assumed to be coupled to the average DMSP content of photosynthetic bacteria and the photosynthetic rate, as explained in detail in the text. The functional groups are listed in order of vertical appearance in the model (see also figure 1).

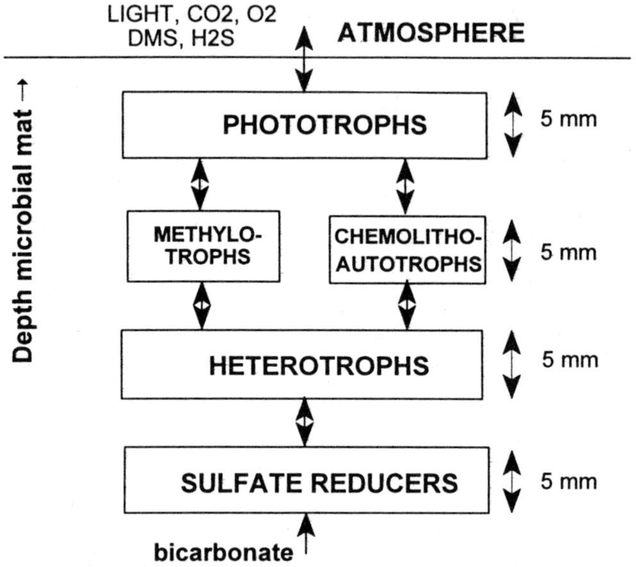

Figure 1. Schematic, simplified representation of a microbial mat. In the model, 5 groups of organisms are assumed to be in separate compartments, situated in different layers of the mat.

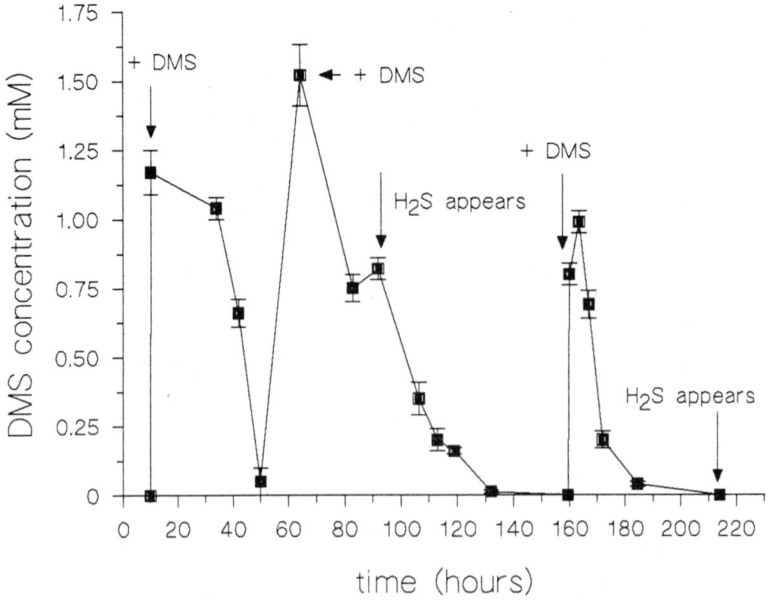

Figure 2. Aerobic and anaerobic DMS consumption in microbial mat sediment. The "aerobic" experiment had a gas phase containing air, but parts of the sediment in the bottle were anaerobic, due to low oxygen diffusion from the headspace to the sediment. This experiment therefore represents a sum of aerobic and anaerobic DMS conversion. In the anoxic experiment, DMS was added to the sediment after the headspace was flushed with N_2.

noted that the anoxygenic phototrophs, for simplicity reasons, have not been included in this model. Such a group could be accommodated in an additional compartment below the compartments of the colorless sulfur bacteria and the DMS-oxidizing bacteria, where the oxygen tension will be low. It can be seen from the five reactions shown in table 1, that all products can be consumed. The reactions take place in separate compartments (figure 1), in different layers, of the mat. These five metabolic reactions are the basic, and, when considering control of DMS production and metabolism, quantitatively the most important biological reactions. Metabolic activity is described with Michaelis–Menten kinetics (v_{max} and k_s). Transport to and from the compartments was described with simple transport equations based on diffusion. A general mass balance was set up in the model for each component in each compartment, and the changes of different components with time were described. Details of this model have been published elsewhere (6).

Calculations over 12-Hour Periods

The model was used to describe changes that take place in a microbial mat during a 12-hour light period. The driving force in the microbial mat is then photosynthesis, during which the pH in the mat will rise because of the depletion of the bicarbonate that buffers the mat. The effect of increasing pH on the Michaelis–Menten parameters for DMS oxidation was determined for *Methylophaga sulfidovorans,* and incorporated in the model. The influence of pH on DMS fluxes from the mat to the atmosphere was simulated using the model.

RESULTS

DMS Consumption in Microbial Mat Sediments. MPN Enumerations with Sediment Samples

Different samples from an intertidal coastal area were tested for their DMS- degrading capabilities under aerobic and anaerobic conditions. Most of the fresh samples (which had been stored in the dark) produced DMS. After 1 to 3 days, the production of DMS ceased, and DMS uptake rates under aerobic and anaerobic conditions could be determined, as shown for one sample in figure 2. As long as oxygen could be detected in the gas phase, the experiment was labelled "aerobic". When the oxygen was depleted (at the end of the aerobic experiment, or during the anaerobic experiments), hydrogen sulfide was detected. Most

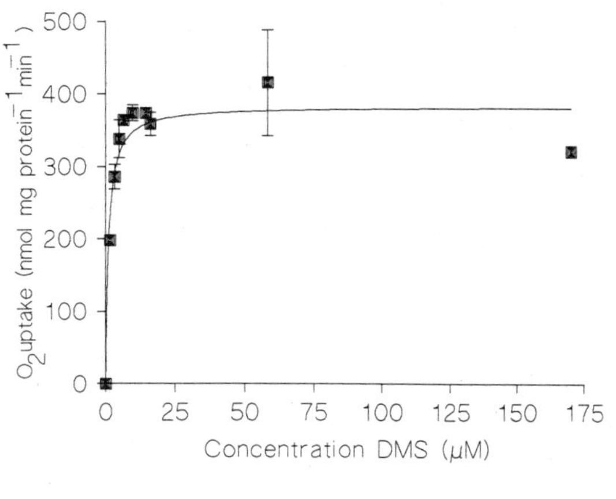

Figure 3. Michaelis Menten-type saturation curve for *Methylophaga sulfidovorans,* of the specific oxygen consumption rate versus the DMS concentration. From this curve the kinetic parameters v_{max} and k_s were estimated with a Line Weaver–Burk Plot: $v_{max} = 383 \pm 16$ nmol O_2 mg^{-1} biomass min^{-1} and $K_s = 1.0 \pm 0.32$ μmol l^{-1} DMS.

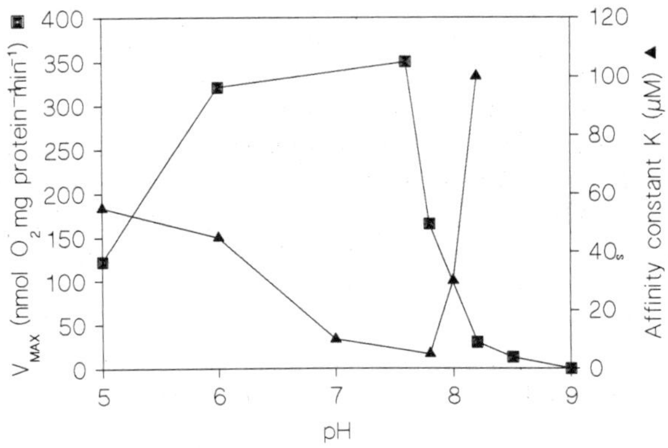

Figure 4. The effect of pH on the kinetic parameters (v_{max} and k_s) for DMS oxidation of *M. sulfidovorans.*

probably the hydrogen sulfide produced in the anoxic part of the sediment had been directly oxidized with oxygen by chemolitho (auto)trophic bacteria present in these sediments. The aerobic DMS-oxidation rate was estimated at 1 (± 0.5) μmol ml^{-1} d^{-1} for the different samples, by subtracting the DMS oxidation rate found under *anoxic* conditions from the rate found under "aerobic" conditions.

The same samples (to which no DMS was added) were used for MPN enumerations using a bicarbonate-buffered mineral medium with DMS in the headspace. The counts obtained ranged from 10^4 to 10^5 cells ml^{-1} sediment.

Physiology and Taxonomy of Isolate

As evident from its limited substrate range, the isolate appeared to be an obligately methylotrophic organism, capable of growth on DMS, methanol, methylamine and dimethylamine. Hydrogen sulfide could be used as additional energy source, but autotrophic growth on H_2S was not possible.

From experiments with cells grown in a methanol-limited chemostat to which DMS was added, the maximum oxidation rate of DMS (v_{max}) was found to be 400 (± 40) nmol O_2 min^{-1} mg^{-1} biomass, and the affinity constant (k_s) was 1.5 (± 0.5) μM DMS (figure 3).

The yield of the isolate was found to be 10.36 (± 1.44) g biomass mol^{-1} DMS (n =15). Standard tests for identification (e.g. API 20 and 20 NE) are partly based on substrate range. All of these tests were negative for the isolate, confirming its limited range of metabolic possibilities. Sequence analysis of the 16 S rRNA showed that the isolate has the highest similarity with the members of the gamma subclass of the Proteobacteria. The maximum similarity found was 86.1% with *Methylomonas methanica*. This indicates that the isolate belongs to a new genus. It has therefore provisionally been named *Methylophaga sulfidovorans*.

Cultivation of *M. sulfidovorans* at Various pH Values and O_2 Concentrations

The pH was found to significantly affect DMS oxidation by *M. sulfidovorans*. Different oxygen tensions had no significant effect on DMS oxidation in the chemostat (between 10 and 100% air saturation). Growth did not occur under anoxic conditions with nitrate as alternative electron acceptor.

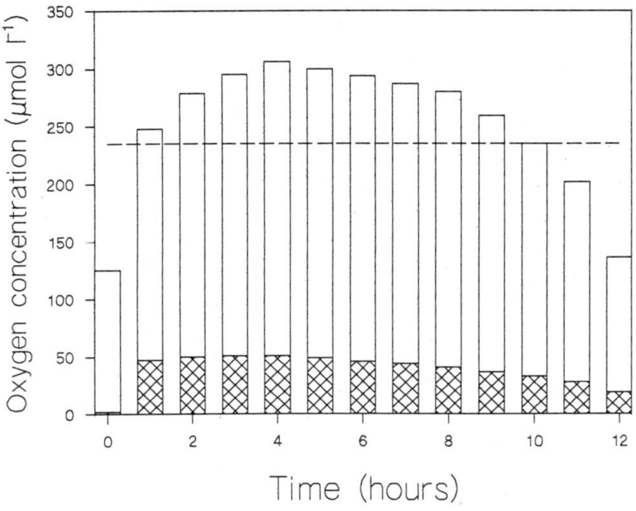

Figure 5. Simulated O_2 concentration in the phototrophic compartment (open bars) and in the deeper layers of the mat (closed bars) during a 12-hour light period.

The maximum oxidation rate of DMS rapidly decreased at pH-values higher than 8 (figure 4). For low pH values (4-7) the activity was relatively high. The pH value in marine microbial mats varies from 7 to 10 (24), and aerobic DMS metabolism would thus come to a halt at the high pH values, observed at the end of the light period. The affinity constant responded in a similar fashion, as shown in figure 4. The value steeply increased with higher pH values, and less steeply increased with low pH values.

In batch cultures in mineral medium with sulfide and DMS, diauxic use of sulfide and DMS was observed. Sulfide was first oxidized at a low rate (estimated at 100 (± 30) nmol H_2S mg^{-1} biomass min^{-1}), with no detectable biomass production, after which the DMS was oxidized. The v_{max} and k_s were not influenced by the concentration of sulfide (up to 6 mM H_2S) present in the batch cultures. The rates of conversion of DMS were independent of the amount of H_2S utilized before DMS consumption began.

In some of the batch and continuous culture experiments, a low level of a persistent satellite population, *Stenotrophomonas maltophilia* (identified by the Delft Culture Collection) was detected (< 5%). This infection could not use DMS (or the methanol, used as substrate in the continuous culture experiments) and did not affect the observed kinetic parameters for DMS oxidation.

Model Description of a Microbial Mat Ecosystem

The model was used to calculate the effects of microbial activity and transport processes on the concentration of compounds in the microbial compartments over a 12 hour light period. All parameters used in the model (e.g. the saturation constants, yields and maximum oxidation rate constants for the different organisms, buffer concentration, mass transfer coefficients, etc.) were tested for parameter sensitivity. From this analysis, it appeared that biological and physical constants significantly affected both reference variables. The most influential biological parameters were the k_s for DMS oxidation and the size of the DMS-oxidizing population. The most important physical constants were the mass

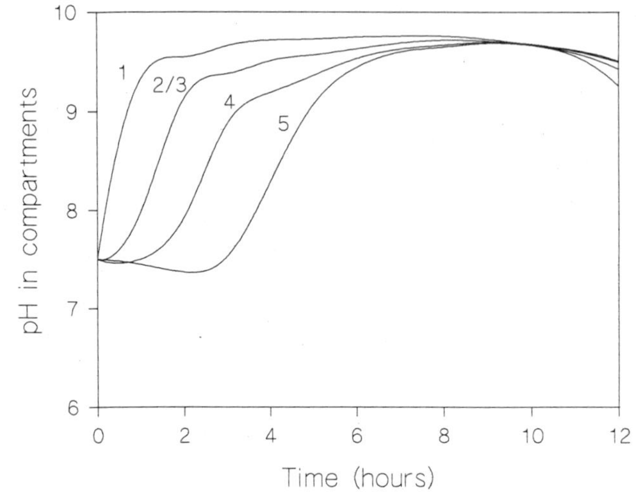

Figure 6. Simulation of pH changes in compartments 1 to 5 during a 12-hour light period. 1 to 5 represents compartments of the phototrophs, methylotrophs, chemolithoautotrophs, heterotrophs and the sulfate-reducing bacteria, respectively.

transfer coefficients of bicarbonate, H^+ and OH^- in the sediment. These values were obtained from the literature.

Figure 5 shows a simulation of the oxygen distribution through the mat for a light period of 12 hours. As can been seen, the oxygen concentration in the top layer of the mat rises above the air saturation of seawater (235 μmol O_2 l^{-1} at 25 °C). If the mass transfer of oxygen to the atmosphere was more effective (a higher mass transfer coefficient for oxygen), this would give a lower oxygen tension in the top layer of the mat.

The simulation of pH change in different layers of the mat (because of photosynthesis and settling, or evening out, of the pH) is shown in figure 6. It is assumed here that photosynthetic activity was maximum at neutral pH, with a linear decrease in activity for higher pH values (adjusted from (3)). The extent of the time delay before adjustment of the

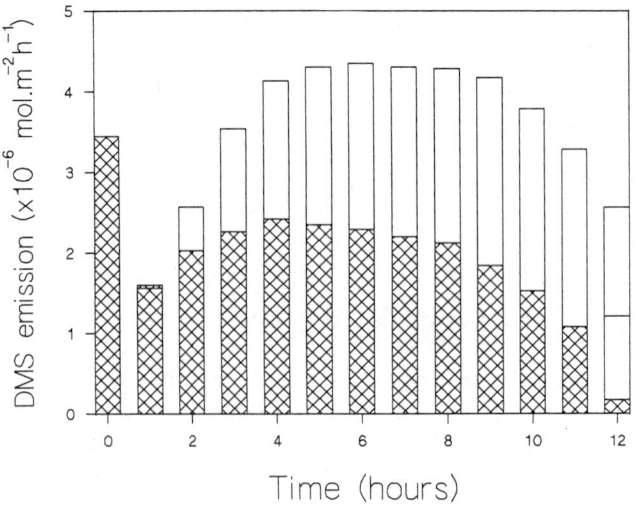

Figure 7. DMS emission from the phototrophic compartment during a 12-hour light period. The open bars represent the simulation results when the pH effect on DMS oxidation is incorporated. The closed bars represent the simulation results if this effect is neglected.

pH was mostly dependent on the value of the mass transfer coefficient, but also on the rate of photosynthesis, DMS oxidation (e.g. H_2SO_4 production), sulfate reduction, H_2S production etc. After 6 hours, the pH in the entire system reached the predicted value of 9 or higher.

The quantitative effect of a high pH value on the microbial activity of the DMS-oxidizing community was assumed to be similar to that of *M. sulfidovorans* (figure 4). A quantitative relation of the Michaelis–Menten parameters for DMS oxidation with pH was obtained and incorporated in the model.

Figure 7 shows the simulated DMS emission from the mat by diffusion of DMS from the photosynthetic compartment into the atmosphere. Simulations with and without the effect of pH on DMS-oxidation were made. Ignoring the increase in the pH gave relatively low DMS emission rates. The amount emitted was strongly related to the affinity (v_{max}/k_s) of the DMS oxidizing-community for DMS. If the effect of increasing pH on DMS oxidation was taken into account, the affinity for the substrate became very low. Emission was then higher, since at higher pH values (reached after 2-3 hours only (figure 6)) the microbial oxidation of DMS was negligible, leading to an increased DMS concentration in the compartments. Obviously, a higher concentration in the mat will give a higher emission rate.

DISCUSSION

DMS Oxidation in Sediments

Sediment samples from an intertidal region on the South-West coast of the Netherlands proved to be active in both DMS production and consumption. Figure 2 shows that aerobic and anaerobic consumption of DMS took place, and that a community of aerobic DMS oxidizing bacteria must therefore have been present. The aerobic DMS oxidation rate was estimated to be 1 μmol DMS g^{-1} d^{-1}. The value for anaerobic DMS removal agrees well with rates found by Kiene and Capone (14), who reported removal of about 100 μmoles DMS in 12 days by 25 ml sediment slurry. Recalculated for these experiments, this is about 0.33 μmol DMS ml^{-1} slurry d^{-1}.

The most probable number counts of aerobic DMS oxidizing bacteria in these sediment samples gave results that were similar to results obtained by Visscher et al. (24). They found that approximately 10^5 DMS-oxidizing cells cm^{-3} were present in a marine microbial mat on the Island of Texel (The Netherlands). Both of these counts were carried out with mineral medium, and clearly underestimate the aerobic DMS oxidizing bacteria in the sediment samples. If 5×10^4 bacteria $gram^{-1}$ sediment are responsible for the removal of 1 μmol DMS in 24 hours, assuming that 1 bacterial cell weighs 10^{-12} g, and that the yield of bacteria on DMS is ± 10 g biomass mol^{-1} DMS, the growth rate on DMS would have to be ± 8.5 h^{-1} (a division time of 5 min). Growth rates of known DMS-oxidizing bacteria, (5) have had maximum specific growth rates (μ_{max}) of around 0.05 h^{-1}. This indicates that enumeration underestimates the aerobic DMS oxidizing community present in the sediment.

Physiology and Taxonomy of Isolate

The new DMS-metabolizing marine isolate has a 16 S rRNA sequence that differs by at least 14% from known sequences of eubacteria (DSM analysis, sequence was compared with 4000 different sequences). In view of its methylotrophic and sulfide-oxidizing properties this proteobacterium has been provisionally named *Methylophaga sulfidovorans*. Thus far, *Hyphomicrobium spp.* (18, 20, 29) and *Thiobacillus spp.* (9, 11) have generally been found to be responsible for aerobic DMS degradation in (fresh water) waste water systems. From marine systems, *Thiobacillus spp.* capable of aerobic DMS degradation have been isolated(24, 27). *M. sulfidovorans* is an obligately methylotrophic bacterium that can degrade DMS aerobically. It represents a new group of bacteria capable of degrading DMS and methyl mercaptans. Kiene and Bates (15) have already concluded from inhibition experiments in sea water that mainly methylotrophic organisms (known to be inhibited by chloroform) are responsible for DMS degradation.

The DMS uptake characteristics of *M. sulfidovorans* have affinity constants and maximum growth rates comparable with other known DMS-utilizing bacteria. However, the biomass yield on DMS is comparatively low, about 10 g biomass mol^{-1} DMS. *Hyphomicrobium spp.* gave a yield of ± 18 g biomass mol^{-1} DMS, which is in the same range as the yield found for *Thiobacillus thioparus* T5.

Cultivation of *M. sulfidovorans* at Various pH Values, H_2S and O_2 Concentrations

The variation in the kinetics of DMS uptake as a function of pH indicated that at pH values above 8, the DMS-oxidizing capacity of *M. sulfidovorans* rapidly decreased. This has significance in microbial mats, since the pH during the day time may reach values above 10. This rapid decrease in DMS oxidation for pH values higher than 8 was also found in all natural samples tested.

The presence of relatively high concentrations of hydrogen sulfide and DMS in *M. sulfidovorans* cultures resulted in diauxic consumption of H_2S and DMS. Maximum hydrogen sulfide concentrations in mats vary from 0.1 to 0.5 mM H_2S (26), but are detected well below the oxic zone. At the interface of oxygen and hydrogen sulfide, the concentration of hydrogen sulfide is usually very low (< 50 μM) (26)), which makes simultaneous use of sulfide and DMS possible. *M. sulfidovorans* should therefore have its niche in a zone in the microbial mat, where during light periods, the H_2S concentration is sufficiently low.

Model Description of a Microbial Mat Ecosystem

The results obtained (figure 5) with the simulation experiment agreed well with reported observations of oxygen super-saturation during a light period in a microbial mat (4). Figure 5 indicates that a (stepwise) gradient from high oxygen tensions at the top to low oxygen concentrations at the bottom of the mat occur. The amount of oxygen in the different layers depends on the mass transfer of oxygen in the sediment, and the oxygen consumption rates of the colorless sulfur bacteria and the DMS-oxidizing bacteria.

The simulated rise in pH in the top layer (figure 6) was in accordance with experimental measurements of pH in the top layers of a mat (4). This rise can be explained by the depletion of bicarbonate by the phototrophs. The production of sulfuric acid by the colourless sulfur bacteria does not compensate for the rise in OH^- ions.

The simulated emission from the mat of 4 μmol DMS m^{-2} h^{-1}, fits with the observed maximum DMS emission rates of 335 μmol m^{-2} d^{-1} which were detected above a sediment from a brackish estuary (8). Assuming a constant emission of DMS during the day, this rate equals 14 μmol m^{-2} h^{-1} for the brackish estuary, which is in the same order of magnitude as predicted here. Furthermore, Jørgensen and Okholm–Hansen (8) found that DMS was emitted in the late afternoon. The model predicts this effect (figure 7), since pH rises in the mat during the day, which results in a decreased microbial activity for DMS removal.

The model presented here gives a simplistic view of a complex ecosystem. Relatively important sinks for hydrogen sulfide, including anoxygenic photosynthesis (23), were not taken into account. Furthermore, anoxygenic photosynthetic bacteria can use DMS as an electron donor which provides an additional sink for DMS under anaerobic conditions. As

DMS is produced in the top layer of the mat, the anaerobic conversions of DMS that would occur in the lower layers of such mats (14, 24) were not taken into account. It should be emphasized that this model design is of a very general nature, and can easily be extended with additional compartments without the need for new fundamental assumptions (6).

In spite of its simplicity, the model gives satisfactory predictions of, for example, oxygen distribution, pH-settling and DMS emissions. The model, together with the new isolate *M. sulfidovorans*, will help in quantitative understanding of processes in microbial mats, and will indicate the processes that must be examined in more detail in order to understand the regulation of the carbon and sulfur fluxes in a microbial mat.

REFERENCES

1. Charlson, C.D., Lovelock, J.E. and Andreae, M.O. 1987. Oceanic phytoplankton, atmospheric sulfur, cloud albedo and climate. Nature 326: 655-661.
2. Cohen, Y. 1989. Photosynthesis in Cyanobacterial mats and its relation to the sulfur cycle. A model for microbial sulfur interactions. In: 'Microbial mats; physiological ecology of benthic microbial communities.' Y. Cohen and E. Rosenberg (eds). Am. Soc. for Microbiol. Washington D.C. 1989, pp 22-36
3. Coleman, J.R. and B. Colman. 1981. Inorganic Carbon accumulation and photosynthesis in a blue green algae as a function of external pH. Plant Physiol. 67: 917-921.
4. D'Antoni D'Amelio, E., Y. Cohen, and D.J. Des Marais. 1989. Comparitive functional ultrastructure of two hypersaline submerged cyanobacterial mats. In: 'Microbial mats; physiological ecology of benthic microbial communities.' Y. Cohen and E. Rosenberg (eds). Am. Soc. for Microbiol., Washington D.C. 1989, pp 97-113
5. De Zwart, J.M.M., and J.G. Kuenen. 1992. C_1-cycle of sulfur compounds. Biodegradation 3: 37-59.
6. De Zwart, J.M.M., and J.G. Kuenen. 1995. Model for biological conversions of DMS, H_2S and organic carbon in a microbial mat. FEMS Microbiol. Ecol., in press.
7. De Zwart, J.M.M, and J.G. Kuenen, unpublished data.
8. Jørgensen, B.B., and B. Okholm-Hansen. 1985. Emissions of biogenic sulfur gases from a Danish estuary. Atmosph. Environ. 19: 1737-1749.
9. Kanagawa, T., and D.P. Kelly. 1986. Breakdown of DMS by mixed cultures and by *Thiobacillus thioparus*. FEMS Microbiol. Lett. 34: 13-19.
10. Keller, M.D., W.K. Bellows, and R.R.L. Guillard. 1989. Dimethyl sulfide production in marine phytoplankton. In: Biogenic sulfur in the environment. E.S. Saltzman and W.J. Cooper (eds). Am. Chem. Soc., Symposium Series 393, Washington DC, pp 167-181.
11. Kelly, D.P., and N.A. Smith. 1990. Organic sulfur compounds in the environment. Biogeochemistry, microbiology and ecological aspects. Adv. Microb. Ecol. 11: 345-385.
12. Kiene, R.P, R.S. Oremland, A. Catena, L.G. Miller, and D.G. Capone. 1986. Metabolism of reduced methylated sulfur compounds in anaerobic sediments and by a pure culture of an estuarine methanogen. Appl. Environ. Microbiol. 52: 1037-1045.
13. Kiene, R.P. 1988. Dimethylsulfide metabolism in salt marsh sediments. FEMS Microbiol. Ecol. 53: 71-78.
14. Kiene, R.P., and D.G. Capone. 1988. Microbial transformations of methylated sulfur compounds in anoxic salt marsh sediments. Microb. Ecol. 15: 275-291.
15. Kiene, R.P., and T.S. Bates. 1990. Biological removal of dimethyl sulphide from sea water. Nature 345: 702-705.
16. Ni, S., and D.R. Boone. 1991. Isolation and characterisation of a dimethylsulfide degrading methanogen *Methanolobus siciliae*. Int. J. Syst. Bacteriol. 41: 410-416.
17. Oremland, R.S., R.P. Kiene, I. Mathrani, M.J. Whiticar, D.R Boone. 1989. Desciption of an estuarine methylotrophic methanogen which grows on dimethyl sulfide. Appl. Environ. Microbiol. 55: 994-1002.
18. Pol, A., H. op den Camp, S.G.M. Mees, M.A.S.H. Kersten, and C. van der Drift. 1994. Isolations of a dimethylsulfide utilizing *Hyphomicrobium* species and its application in biofiltration of polluted air. Biodegradation 5: 105-112.
19. Przyjazny A.M, W. Janicki, W Chrzanowski, and R. Staszewski. 1983. Headspace gas chromatographic coefficients of selected organosulphur compounds and their dependence on some parameters. J. Chrom. 280: 249-260.
20. Suylen, G.M.H., and J.G. Kuenen. 1986. Chemostat enrichment of *Hyphomicrobium* EG. Ant. van Leeuwenhoek 52: 281-293.

21. Suylen, G.M.H., G.C. Stefess and J.G. Kuenen. 1986. Chemolithotrophic potential of a *Hyphomicrobium* species, capable of growth on methylated sulfur compounds. Arch. Microbiol. 146: 192-198.
22. Tanimoto, Y.,and F. Bak. 1994. Anaerobic Degradation of methylmercaptan and dimethylsulfide by newly isolated thermophilic sulfate-reducing bacteria. Appl. Environ. Microbiol. 60: 2450-2455.
23. Van Gemerden, H. 1993. Microbial mats; A joint venture. Marine Geology 113: 3-25.
24. Visscher, P.T., P. Quist, and H. van Gemerden. 1991. Methylated sulfur compounds in microbial mats: In situ concentrations and metabolism by a colourless sulfur bacterium. Appl. Environ. Microbiol. 57: 1758-1763.
25. Visscher, P.T., and H. van Gemerden. 1991. Production and consumption of dimethylsulfoniopropionate in marine microbial mats. Appl. Environ. Microbiol. 57: 3237-3242.
26. Visscher, P.T., and H. van Gemerden. 1993. Sulfur cycling in laminated marine ecosystems. In: Biogeochemistry of Global Change. R.S. Oremland (ed). Chapman and Hall, New York, NY, pp. 672-690.
27. Visscher, P.T., and B.F. Taylor. 1993. Aerobic and anaerobic degradation of a range of alkyl sulfides by a denitryfying bacterium. Appl. Environ. Microbiol. 59: 4083-4089.
28. Widdel, F., and N. Pfennig. 1981. Studies on dissimilatory sulfate reducing bacteria that decompose fatty acids. Arch. Microbiol. 129: 395-400.
29. Zhang, L., M. Hirai, and M. Shoda. 1992. Removal characteristics of dimethylsulfide by a mixture of *Hyphomicrobium* sp 155 and *Pseudomonas acidovorans* DMR-11. J. Ferm. Bioeng. 3: 174-178.
30. Zinder, S.H., and T.D. Brock. 1978. Production of methane and carbon dioxide from methane thiol and dimethyl sulfide by anaerobic lake sediments. Nature 273: 226-228.

INDEX